COMPUTATIONAL MECHANICS IN STRUCTURAL ENGINEERING

Recent Developments and Future Trends

Proceedings of Sino-US Joint Symposium/Workshop on Recent Developments and Future Trends of Computational Mechanics in Structural Engineering, Beijing, China, September 24-28, 1991.

COMPUTATIONAL MECHANICS IN STRUCTURAL ENGINEERING

Recent Developments and Future Trends

Edited by

FRANKLIN Y. CHENG
Curators' Professor of Civil Engineering
University of Missouri-Rolla,
Rolla, Missouri, USA

and

FU ZIZHI
Professor, Petroleum Design Institute,
Liupukang, Beijing, CHINA

CRC Press
Taylor & Francis Group
Boca Raton London New York

CRC Press is an imprint of the
Taylor & Francis Group, an **informa** business
A TAYLOR & FRANCIS BOOK

CRC Press
Taylor & Francis Group
6000 Broken Sound Parkway NW, Suite 300
Boca Raton, FL 33487-2742

First issued in paperback 2019

ISBN-13: 978-1-85166-867-0 (hbk)
ISBN-13: 978-0-367-86612-9 (pbk)

CIP Catalogue record for this book is available
is available from the British Libary

Library of Congress CIP data applied for

Visit the Taylor & Francis Web site at
http://www.taylorandfrancis.com

and the CRC Press Web site at
http://www.crcpress.com

PREFACE

It is generally acknowledged that the field of computational structural mechanics has enriched the rational application of mechanics and numerical methods in the practice of modern structural analysis, design and construction of building systems and other structures. In recent years, both the US and China have been actively engaged in research on computational structural mechanics for various engineering disciplines, and have sponsored cooperative research projects in areas of mutually beneficial interest. Researchers from both sides have shared a strong interest in exchanging state-of-the-art information in research and practice.

The Sino-US Symposium/Workshop on Recent Developments and Future Trends of Computational Mechanics in Structural Engineering was held during September 24–28, 1991, in Beijing, China. Its emphasis was to review the current state of knowledge and practice in computational structural engineering, to identify frontal research and development needs, and to stimulate cooperative research efforts among investigators in the US and China as well as other countries.

The technical program of the symposium included the following sessions: failure analysis and finite elements; shells and walls; structural control; computational strategies for buildings; nonlinear analysis; structural optimization; structural reliability; supercomputers, CAD and expert systems.

US delegates were:

Dr. Charles A. Anderson (in collaboration with Dr. Charles R. Farrar)—Los Alamos National Laboratory, New Mexico
Dr. A. H-S. Ang—University of California at Irvine
Dr. James E. Beavers—Center for Natural Phenomena Engineering, Martin Marietta Energy Systems, Inc.
Dr. Ted Belytschko—Northwestern University
Dr. Franklin Y. Cheng (Chairman)—University of Missouri-Rolla
Dr. K. P. Chong—National Science Foundation
Dr. S. C. Liu—National Science Foundation
Dr. Graham H. Powell—University of California-Berkeley
Dr. T. T. (Larry) Soong—State University of New York at Buffalo
Dr. Gerald A. Wempner—Georgia Institute of Technology

Chinese delegates were:

Professor Bao Shihua—Tsinghua University
Professor Ying-Jun Chen—Northern Jiao Tong University
Professor Deng Dahuan—Chinese Academy of Agricultural Mechanization Sciences
Professor Fu Zizhi (Chairman)—Petroleum Design Institute
Professor Huang Yih—Xian Institute of Metallurgy and Construction Engineering
Professor Li Guiqing—Institute of Aseismic Engineering Structures
Professor Li Mingrui—Beijing Agricultural Engineering University
Professor Li Xingsi—Dalian University of Technology
Professor Liang Qizhi—South China University of Technology
Professor Lin Shaopei—Shanghai Jiao Tong University
Professor Liu Guang-dong—Hunan University
Professor Lu Wenda—Tongji University
Professor Wang Guangyuan and Professor Ou Jinping—Harbin Architectural and Civil Engineering Institute
Assc Prof Wang Qizheng—Tsinghua University
Professor Xu Bohou—Zhejiang University
Professor Zhong Wanxie—Dalian University of Technology
Professor Zhou Fu-Lin—Guangzhou Urban Construction Institute

Guest observers and paper contributors for the symposium included: Professor Teizo Fujiwara and Professor Yoshikazu Yamada of Kyoto University, Dr. M. Shoji and Mr. Takashi Matsumoto of Kajima Corporation, Dr. S. C. Fan of Nanyang Technological University, Dr. D. S. Juang of National Central University, Dr. T. W. Lin of National Taiwan University, Dr. J. P. Pu of Feng Chia University, Dr. P. H. Wang of Chung-Yung Christian University, and Dr. Y. B. Yang of National Taiwan University.

The joint symposium/workshop was jointly sponsored by the US National Science Foundation, the Chinese Association of Computational Mechanics and the National Natural Science Foundation of China. As chairmen of the symposium/workshop and editors of the proceedings, we gratefully acknowledge their support.

FRANKLIN Y. CHENG
FU ZIZHI

CONTENTS

Computational Strategies for Buildings

Nonlinear Analysis

Structural Optimization

Structural Reliability

Supercomputer, CAD and Expert System

Group picture after opening ceremony

Visit to the National Science Foundation of China (NSFC)

Welcoming remarks at reception by Dr S. C. Liu,
Program Director, US NSF

Speech at banquet by Dr K.P. Chong, Program Director, US NSF

Welcoming remarks at banquet by Prof. Hu Zhaosen,
exec. vice-chairman, NSFC

Delegation friendship toast at banquet by Dr Franklin Y. Cheng
and Prof. Fu Zizhi

1

Opening Speech

Prof. Fu Zizhi, Vice–Chairman, CACM

Dear Colleagues and Friends,

The Sino–U.S. Joint Symposium / Workshop on Recent Development and Future Trends of Computational Mechanics in Structural Engineering is now declared to be opened. On behalf of the Chinese Association for Computational Mechanics (CACM) under CSTAM, let me express our warm welcome to all our Scholars present here today.

In this Symposium / Workshop, we have delegates from China, the United States of America, Japan and Singapore. They have been working in the foremost fronts of various domains of Computational Mechanics and are of the first class level. The purpose of this meeting is to promote mutual understanding through academic exchanges and discuss the future trends of Computational Mechanics as well as the contents and forms of cooperation. We sincerely hope that this meeting will serve as a starting point to promote a well planned long term exchange and promotion of Sino–U.S. cooperation in Computational Mechanics.

This Symposium / Workshop was first proposed by Professor K.P. Chong in 1989, sponsored by CACM and the National Science Foundation of the United States of America and also supported by CSTAM and National Natural Science Foundation of China. It is the result of two years of preparation. Professor K.P. Chong and Professor S.C. Liu, Program Directors, Engineering Directorate, NSF, Prof. Wu Jiaxiang, Director of the Bureau of International Cooperation of the NSFC, Prof. Jin Zhengmo, Head of the Mechanics Division of the same Foundation have lent us great support. We would like to mention especially Dr. Franklin Y. Cheng, Curators' Professor of the University of Missouri–Rolla who has spent great amount of energy and time in the preparation. Our cooperation in this respect has been very sincere and pleasant. To all those mentioned above, I hereby express our heartfelt thanks.

The meeting is short but it will produce deep effects in promoting the substantial cooperations in the future. I firmly believe that we can expect more cooperations between the Science Foundations of the two countries and the academic exchanges between our scholars.

I send my greeting to the success of this meeting and my best wishes to the happiness of all present here today.

Thanks for your kind attention.

OPENING REMARKS

Franklin Y. Cheng, Curators' Professor of Civil Engineering
University of Missouri-Rolla, USA

Good morning, distinguished participants, ladies and gentlemen.

On behalf of the US delegates, I am pleased that the Sino-U.S. Joint Symposium/Workshop on Recent Developments and Future Trends of Computational Mechanics in Structural Engineering has now formally opened. We welcome our guest observers wholeheartedly.

In recent years, researchers in the US and China have actively engaged in development and practice of computational structural mechanics. They share a strong interest in exchanging research information. The objectives of the symposium/workshop are to (1) review the current state of knowledge and practice in computational structural engineering; (2) identify the frontal joint research opportunities necessary to advance the current state of knowledge; and (3) stimulate future cooperative research and development in the areas of common needs in the US and China as well as other countries.

The symposium is jointly sponsored by the US National Science Foundation and the Chinese Association for Computational Mechanics with support of CSTAM and the National Natural Science Foundation of China. I gratefully acknowledge their sponsorship and support. We deem it a great honor that Dr. S.C. Liu and Dr. Ken P. Chong, program directors of the National Science Foundation, are here today. Dr. Henry J. Lagorio of NSF is unable to be with us, but he sends his best wishes for a successful meeting. We are thankful for the tireless efforts of Professor Fu Zizhi in arranging all local details.

Please permit me to remind you that the major task now is for us to diligently exchange research information and to successfully develop a long-term cooperative program. I think the resolutions to be established during the workshop this Friday will have a great impact on the success of this symposium. Let us work together closely this week to achieve our ultimate goal.

Thank you.

Welcome by Ken P. Chong, Director of Structural Systems and Construction Processes, National Science Foundation, USA

On behalf of the National Science Foundation I would like to welcome all of you to this prestigious workshop. As you know mechanics and structural engineering are the "backbones" of civil engineering, and computational mechanics play a key role in these areas. In the United States computational resources seem plentiful. However some of our super-computer centers are being strained. In some cases only 50 % of computer requests can be accommodated. On the other hand, researchers and engineers in China have always been very careful in minimizing computational labor and maximizing its efficiency. This is one of the areas where US - Chinese interaction would be beneficial.

I wish you all the success, and I would like to take this opportunity to thank the workshop co-chairmen, Professors Cheng and Fu, for their hard work over the last two years. Without their tireless effort we will not be here today.

Address to the China–US workshop on Structural Engineering and Computation

Prof. Hu Zhaosen
Exect. Vice Chairman, NSFC

Mr. Chairman, Ladies, Gentlemen and friends:

I feel great honoured to have this opportunity to meet so many distinguished experts from home and abroad working in the structural Engineering and Computation field.

On behalf of the National Natural Science Foundation of China to extend my sincere congratulations on the opening of the workshop on sturctural Engineering and Computation and express my warmest welcome to all the participants to come to Beijing.

The National Natural Science Foundation of China is an organization with both management and academic nature. In order to bring prosperity to science, economy and society of China, our foundation devotes all efforts to coordinate, support and grant basic research and part of applied research including research in the fields of Structural engineering and computation.

One of the tasks of our foundation is to support international scientific exchange and cooperation. These includs the joint research attended by Chinese and foreign scientists, supporting international conferences held in China and the attendance of Chinese scientists at international conferences held abroad, and other cooperative activities.

Therefore I am very pleased to see so many Chinese and foreign experts getting together to exchange experience and to discuss scientific problems of common intrest. I hope this will lead to the further cooperation between Chinese scientists and our foreign friends.

Finally, I sincerely wish the symposium a great success and all the delegates good health.

Thank you!

Ladies and Gentlemen,

The purpose of the "Sino–U.S. joint symposium / workshop on recent developments and future trends of computational mechanics in structural engineering" is to provide a forum for the scientists of both sides, to increase mutual understanding and to promote exchanges and coorporations. The contents of the symposium will be related to structural control, structural reliability, structural nonlinear properties, structural optimizations, CAD, expert systems and the application of supercomputers etc. All these reflect the latest developments of structural computational mechanics with high theoretical and practical values. We are confident that this symposium / workshop is a high level academic meeting.

The technical program consists of a symposium with three days presentations of research papers and a one day simultaneous workshop. We hope that this symposium / workshop will serve not only for academic exchange on the three days but also a starting point of coorporation between the colleagues of the two countries.

Now let us have a coffee break and then begin with the first general session.

A Talk by Prof. Zhong Wanxie, chairman, organizing committee

RESOLUTIONS OF SINO-US SYMPOSIUM WORKSHOP ON RECENT DEVELOPMENTS AND FUTURE TRENDS OF COMPUTATIONAL MECHANICS IN STRUCTURAL ENGINEERING

The Sino-U.S. Joint Symposium/Workshop on Recent Developments and Future Trends of Computational Mechanics in Structural Engineering was held during September 24-28, 1991, in Beijing, China. Participants conducted fruitful scientific exchanges in the areas of Buildings, Reliability, Structural Optimization and Control, Supercomputers, Shell Structures, Intelligent CAD and Expert Systems, and Nonlinear and Failure Analysis.

At the conclusion of the symposium/workshop, participants unanimously passed the following resolutions:

(A) Collaborative research between Chinese and U.S. researchers requires cross-flow of research results as published in journals and conference proceedings. While Chinese investigators have relatively easy access to international publications, Chinese research papers are often inaccessible to U.S. researchers because they are published either in Chinese or in journals not widely circulated in the West. Therefore: (1) Chinese researchers should be encouraged to publish in English and in international journals; (2) More Chinese-sponsored journals and conference proceedings should be published in the West; (3) Effective measures should be devised to disseminate research results to users; (4) Chinese researchers are encouraged to serve on editorial boards and as editors of international journals.

(B) Based on the success of this event as experienced by all participants, it is resolved that a joint symposium/workshop be continued at intervals of two or three years.

(C) Topics of mutual interest for further investigation, preferably as joint research activities, are listed in the following groups.

I. STRUCTURES, RELIABILITY, STRUCTURAL OPTIMIZATION AND CONTROL

The consensus of the group chaired by T.T. Soong and Li Guiqing was that reliability, optimization and control are closely related topics. A number of the papers presented at the workshop showed promise for collaborative research in this area. Specific topics discussed are noted below.

(a) Control - Although active control research is currently going on in China, passive control is considered more practical at this time. In the structural control area, the following topics are recommended for possible joint research.

1. More in-depth studies of merits as well as limitations of active control from the standpoint of reliability. Life safety enhancement and cost effectiveness through damage avoidance need to be carefully examined.

2. Computational issues in relation to the development of new control algorithms.

3. More emphasis on control of wind-induced vibrations. Structures that may benefit from such studies include radio towers in Shanghai and Nanjing and a number of cable-stayed or long-span suspension bridges.

4. Active control with other applications, such as construction processes.

(b) Reliability - The general feeling was that reliability research and applications have not been well developed in China. As an example, reliability-based codes for bridges are yet to be implemented there. The following were pointed out as potential joint research topics.

1. Implementational issues related to reliability-based codes, particularly those dealing with characterization of uncertainties.

2. Reliability as an integral part of the "full-range" analysis which includes design, optimization, analysis, identification and control.

(c) Optimization - Generally speaking, a strong base of optimization research exists in China. Joint research in the following directions could be particularly fruitful.

1. Basic work in nonlinear structural optimization and sensitivity analysis.

2. Engineering issues in optimization, e.g., reliability-based optimization and optimization with multiple objectives and multiple constraints.

II. SUPERCOMPUTERS, SHELL STRUCTURES, INTELLIGENT CAD AND EXPERT SYSTEMS

The group chaired by Graham Powell and Ou Jinping considered the above three topics. For each topic, key points from discussion are summarized below.

(a) Supercomputers - The use of supercomputers is driven by the need to perform extensive computations, most likely for large finite element models. Supercomputers are not currently available in China. As a result, Chinese researchers restrict themselves to problems which can be solved with modest computational effort. This has led to some interesting and innovative solution techniques for such structures as tall buildings. Parallel processing could be a viable alternative in China. US researchers in the group have access to supercomputers through their laboratories or NSF programs. A joint research possibility is in-depth studies of merits as well as limitations of analytical models used by Chinese and US researchers on aforementioned structures by comparing numerical results and computational efforts.

(b) Shell Structures - Chinese researchers in this area are interested mainly in reinforced concrete shells, and have attained a high level of expertise in modeling and analyzing shells. While US researchers and engineers tend to rely on finite element analyses, Chinese researchers pursue more traditional analytic methods. Their level of expertise in the analysis of cooling towers, for example, is very advanced. US researchers may not be aware of Chinese research. A possible joint effort is to develop analytical models and numerical algorithms with emphasis on computational efficiency.

(c) Intelligent CAD, Expert Systems - This is a broad area, ranging from conceptual design to routine draft. Several Chinese researchers are interested in different aspects of the broad problem, and are performing innovative research, such as intelligent draft, optimum design for seismic loading, structural design with hierarchical decomposition and multilevel optimization, design decision-making with application of the principles of fuzzy logic. US researchers are engaged in research on data models, data base design, and the structure of the design process. Chinese and US researchers share an interest in topics of intelligent CAD and optimum structural design. These could be fruitful areas for increased cooperative research.

III. NONLINEAR AND FAILURE ANALYSIS

This group chaired by T. Belytschko and Li Mingrui discussed areas of mutual interest in computational mechanics for nonlinear and failure analysis. In these discussions, an effort was made to identify areas for joint projects as follows.

(a) Nonlinear and Dynamic Stability Problems: Continuation Methods and Other Algorithms; Thin-Walled Structures - This topic seemed to be of most interest to Chinese participants. There was considerable interest in the development of more accurate and efficient algorithms for nonlinear problems, reflected in the subtopics under nonlinear dynamics of buildings and bridges.

(b) Adaptive Methods: Methods for Crack Growth and Localization; Error Criteria - US delegates felt that the Chinese research community, with its strong theoretical background, is in a position to make contributions to developing error criteria. Adaptive methods offer one of the few possibilities for solving important problems in failure such as crack propagation and strain localization.

(c) Nonlinear Dynamics of Buildings and Bridges: Efficient, Accurate Algorithms; Methodologies for Seismic Loading - There was considerable interest in this topic because of its practical impact. Chinese delegates also felt that significant improvements in algorithms are necessary to enable practical nonlinear transient analysis of buildings and bridges.

(d) Interaction and Coupled Problems: Fluid-Structure Interaction; Structure-Media Interaction - The focus of these problem areas is the safety of dams. Assessing the safety of dams involves many coupled field problems; for example, the seepage of fluid below a given dam foundation is a coupled problem of importance to safety.

(e) Boundary Elements for Nonlinear Problems: Software Development; Object-Oriented Programming - The boundary element method was felt to be useful in nonlinear problems such as crack propagation, where it can reduce the amount of remeshing. Contact problems were also mentioned as a viable application.

NSF PROGRAMS IN COMPUTATIONAL AND STRUCTURAL MECHANICS

K.P. CHONG, S.C. LIU and H.J. LAGORIO
Program Directors, Directorate of Engineering
National Science Foundation
Washington, DC 20550 USA

ABSTRACT

The U.S. National Science Foundation (NSF) spends over U.S. $40 million annually for basic research in civil engineering, covering areas in structures, construction processes, earthquake engineering, building systems, solid mechanics, geomechanics and environmental engineering. Significant amounts of basic knowledge and technology have been advanced through these research efforts. Computational and structural mechanics are important components of civil engineering research. In this paper, examples of current and recent research projects and findings are presented. Some significant developments, not necessarily funded by NSF, are also discussed.

INTRODUCTION

In a recent book by Boresi and Chong [1], it was indicated that finite element methods provide perhaps the most versatile tools available in numerical analysis of engineering problems in solid mechanics, fluid mechanics, flow in porous media, heat transfer, and so on. Finite element methods are constantly being developed and applied to ever more complex problems. New developments and applications of finite element methods include refined finite elements, advanced computing methods for stability and dynamics analyses, shell analysis, design sensitivity and optimization of nonlinear systems, viscoelasticity, and fracture. An extensive list of references is given.

Finite elements are widely used also in computer-aided engineering design [2] (CAE,CAD) and benchmark programs have been designed to evaluate finite element codes [3,4]. In finite element methods, the solution of simultaneous equations is very computationally intensive. The traditional Gaussian elimination method [5,6,7] is not very efficient compared with for example, iterative gradient methods in vectorized finite elements [8,9]. The availability of super computers and parallel computers [10,11,12] allows for vectorized processing and parallel processing, opening up powerful computational capabilities [13,14]. Substructuring or partitioning of matrices (for example, the stiffness matrix) are adaptable to parallel processing [15]. Efficient substructuring techniques have been developed for static, dynamic and stability analyses [9,13,14,16-24]. Unfortunately, nearly all finite element codes are written in sequential form which differ greatly from parallel procedures. Storaasli and Bergan [12] indicated that it is more effective to write new parallel algorithms than to modify existing algorithms. Corresponding developments in parallel processing hardware are also available. They include the NASA finite element machine [10], local memory multiprocessors [26] such as transputers which are widely used in Europe and in hypercube networks [9]. Local memory multiprocessors (as opposed to shared ones) possess individual memory for each processor. On the other hand, shared memory multiprocessors, which access a common memory, are more powerful and increasingly used in finite element analyses [27].

In most finite element formulations, the stiffness or displacement method is used in which the unknowns are nodal displacements. In the displacement method, stresses and strains are derived by differentiation which reduces accuracy. An alternative method is the force method in which member forces including moments are the primary variables (unknowns) and displacements are derived. Patnaik [28] developed an integrated force method (IFM) which uses all member forces including redundants as the unknowns. This method increases the accuracy of stress predictions with less computational labor [29,30]. A dual IFM is being developed to enable IFM elements to be introduced into existing displacement-based finite element codes such as NASTRAN, with minimal modification.

Based on Hamilton's principle, Marjadi [31] formulated an alternative force method for static, dynamic and buckling analyses, using isoparametric three-dimensional beam elements. Optimization and sensitivity derivatives have been investigated.

STABILITY, DYNAMICS, AND NONLINEAR ANALYSES

The following discussions in this section is taken from the book by Boresi and Chong [1]. Wempner [32] presented a comprehensive review on the mechanics and finite element analysis of shell structures, including developments in instability, inelasticity, nonlinearity, and excessive stiffness (locking) in modeling. About two hundred references were quoted. Shear-locking is also discussed by Ravichandran and Sridharan [33]. To investigate localized plastic collapse and post-collapse behavior of rotational shells, a mixture of general shell elements has been used (in localized regions with severe nonlinearities), with rotational shell elements elsewhere (Zienkiewicz and Taylor, [6]), and transitional shell elements in between [34,35].

In nonlinear and dynamic analyses, commercial programs such as ADINA (Automated Dynamic Incremental Nonlinear Analysis; Bathe [36]) and ABAQUS [37] have been used to model the impact responses and energy dissipation of metallic tubes [38,39], as well as the post-buckling of domes [40] and creep behavior. For thermoelastic-plastic and/or creep analyses, Snyder and Bathe [41] developed a finite element solution procedure, the so-called α-method which utilizes Euler forward and backward methods. The one-parameter integration method, which is stable, uses one time-step size for the computation of nodal displacements and another (smaller) time-step size for the integration of creep strains. Other effective time integration schemes have been presented by Zienkiewicz and Cormeau [42] and Bushnell [43].

Optimal structural design and design sensitivity of linear responses of structural systems, subject to static and dynamic loads, are well developed (see, for example, Haftka and Adelman [44], Haug and Arora [45], Lahey [46]). However, sensitivity analysis of nonlinear responses is still in its infancy [47]. Extensive computational efforts are needed for these sensitivity analyses.

The three-dimensional nonlinear finite element package for microcomputers, DIANA, (Borst et al., [48]) has been used to model crack propagation and fracture in concrete (Rots et al., [49]). Shear transfer and tension softening in the crack region (Chong et al. [50]) have been considered. Fracture parameters have also been investigated with boundary elements (Brebbia and Dominguez [51], Aliabadi [52], Blandford et al. [53]).

A finite element reliability method (Liu and Der Kiureghian [54]) has been developed, which combines the finite element method with the reliability method for reliability assessment of structural safety. Analyses of uncertainties and probabilities (Der Kiureghian and Ke, [55]) are included for geometrically nonlinear, elastic structures. Other investigators (Spanos and Ghanem [56]; Liu et al., [57]; Mahadevan and Haldar [58]) have contributed to stochastic finite element methods for structural dynamics, optimization and reliability.

One common denominator of numerical analyses is that elasticity and constitutive relationships always play a key role [59] in computation and modeling.

COMPUTATIONAL AND STRUCTURAL MECHANICS

The following is a list of examples of NSF-supported research projects in computational mechanics and structural mechanics with computational elements.

Finite Element Methods

Adaptive Finite Element Methods for Variational Inequalities with Application to Contact Problems and Plasticity
Oden, J.; University of Texas-Austin

To develop reliable a-posteriori error estimates of problems described by variational inequalities, development of adaptive strategies based on these error estimates and finally an algorithm development based on the developed strategies.

Mixed/Hybrid Plate and Shell Elements
Taylor, Robert; University of California-Berkeley

To develop new shell finite elements which can be used in a wider variety of problems than existing ones, to avoid artificial locking mechanisms, incorporate nonlinear elasticity, permit large rotations and some types of inelastic material behavior.

Computer Modeling and Examination of 3D Inelastic Beam-Column Joint Behavior in Steel Structures
Deierlein, Gregory; Cornell University

A finite element analysis procedure for including joint flexibility in the analysis and design of 3D frame structures will be developed. This procedure will be capable of geometric and/or material nonlinear analysis.

Large Deformation Analysis of Nonlinear Homogeneous and Heterogeneous Media Using an Adaptive Arbitrary Lagrangian Eulerian Finite Element Method with Applications in Industrial Forming
Ghosh, Somnath; University of Alabama, Tuscaloosa

To develop stable, accurate and efficient computational models for large deformations of metals and composites. The finite element method with arbitrary Lagrangian-Eulerian kinematic description will be implemented for industrial forming processes.

A Spectral Approach For Analyzing Structures With Random Parameters
Spanos, Pol D.; Rice University

Research involves structures with uncertain (random) properties and loads, approximate representation of random processes describing the material properties or the structural loads by a finite number of random variables, using the classical concepts of the Karhunen-Loeve and the Polynomials Chaos expansions, to be incorporated into spatial discretizations of structural elements such as beams and plates, in accordance with standard finite element procedures.

International Workshop on Finite Element Analysis of Reinforced Concrete
Meyer, Christian; Columbia University

An international workshop on finite element analysis and modeling of reinforced concrete structures, organized by the ACI/ASCE Committee 447 and its Japanese counter-part, the JCI Committee, were held in June, 1991. Objectives include critical review of various material models, assessing the state-of-the-art and identifying research needs.

Finite Element Reliability Methods
Der Kiureghian, Armen; University of California-Berkeley

To combine finite elements and structural reliability into a comprehensive methodology for safety and reliability assessment of non-linear structures with uncertain characteristics and subjected to random loads with spatial and/or temporal stochasticity.

Local-Global Analysis for Plastic Collapse of Shells
Gould, Phillip; Washington University, St. Louis

To develop an analytical model which can be used for the study of localized collapse of shells. General quadrilateral degenerate shell elements incorporating large deflection and plasticity effects are to be used to capture the localization of deformation.

Finite Element Analysis of Transverse Shear Failures in Reinforced Concrete Plates and Shells
Harmon, Thomas; Washington University, St. Louis

Study of the punching shear resistance of thick plates and shells made of reinforced concrete are being conducted.

Stochastic Finite Element Methods
Veneziano, Daniele; MIT

The objective was to develop new uncertainty-propagation, reliability-analysis, and variance-decomposition procedures for general-purpose FE codes. The methods were to be implemented on the NONSAP-C and NIKE2D programs and applied to a variety of linear and nonlinear, static and dynamic, time invariant and time-dependent problems. Project completed in 1987.

A Three-Dimensional Mixed Finite Element Method to Integrate the

Dynamic Crack Growth and Failure Mechanisms in a Composite Laminate
Vaicaitis, Rimas; Columbia University

To model the three-dimensional dynamic crack growth and failure mechanism in a composite laminate subjected to low velocity impact, incorporating variationally consistent fracture mechanics elements, multi-failure criteria and parallel processing algorithms.

Destructive Field Testing of a RC Bridge as a Benchmark To Validate FE Modeling and Analysis of RC Structures
Aktan, Ahmet; University of Cincinnati

To exploit tests for the advancement of analytical modeling and nonlinear analysis of reinforced concrete structures.

Dynamic Buckle Propagation in Submarine Pipelines
Carney, John; Vanderbilt University, Nashville

Pipeline buckle propagation and arrest are to be studied by means of a three-dimensional finite element model which will account for the energies dissipated by both circumferential deformations and longitudinal bending and shearing deformations.

Mixed Mode Fracture of Concrete
Hawkins, Neil; University of Washington

Shear transfer mechanism will be studied using beam specimens with short starter cracks and various loading histories which are crack-length dependent. Moire and loading data will be used interactively with a finite element code to deduce the crack closure and shear forces in the fracture process zone under mixed mode loading.

Development of Space-Time Finite Elements for Nonlinear Transient Response of Flexible Structures
Hodges, Dewey; Georgia Tech

Focuses on the computational difficulties associated with steep gradients in the velocity and displacement fields in structures subjected to impulsive loadings. The approach adopted herein, finite elements formulated in the space-time domain, offers the possibility of overcoming these computational problems.

Nonlinear and Probabilistic Theory for Concrete Creep
Bazant, Zdenek; Northwestern University

A numerical algorithm for step-by-step integration in time in finite element programs will be developed, in addition to creep tests.

Impact Damage in Curved Laminated Composites
Chang, Fu-Kuo; Stanford University

Stresses and strains in laminates will be calculated by a proposed finite element method which is based on transient dynamic analysis with the consideration of material and geometric nonlinearities.

Influence of Concrete Creep on the Load Capacity of Reinforced Concrete Shell Structures
Schnobrich, William C.; University of Illinois

With finite elements, it is possible to evaluate a failure load for concrete shell roofs, including creep, nonlinear geometry, and nonlinear material responses.

Computational Methods

Finite Rotations and Finite Elements of Thin Elastic Shells
Wempner, Gerald; Georgia Tech

The purpose of the proposal is to seek the answers about accuracy and convergence of numerical procedures by scrutinizing the theoretical bases, the alternative functionals stationary and/or external criteria. Complementary functionals provide means to that end. Computational programs for numerical and symbolic investigations are other means for exploring and for testing the practicality of the results.

Multi-Scale Methods for Structural Dynamics
Liu, Wing K.; Northwestern University

Multi-scale Galerkin methods will be developed for the study of structural/building systems in which the response is characterized by multiple-spatial and multiple-time scales, often dominated by the middle part of the frequency spectrum. New temporal and spatial discretization methods will be developed for the multi-time scale structural dynamic problems.

Post-Buckling Behavior And Imperfection Sensitivity of Space Trusses With Multiple Eigenmodes
Peek, Ralf; University of Michigan

To determine the worst shape of imperfection for any lattice type structure, by means of the Lyapunov-Schmidt-Koiter decomposition followed by asymptotic results, as well as by an iterative numerical technique.

Foundation and Special Reference of Inelasticity
Simo, Juan Carlos; Stanford University

Concerned with geometric and computational aspects of modern nonlinear continuum mechanics, with special emphasis on inelasticity, and suitable for large scale inelastic computation.

Hypersingular Boundary Integral Equations in Mechanics
Rudolphi, Thomas; Iowa State University

To evaluate the computational basis for evaluating hypersingular equations. The first application areas will be in calculating stresses near cracks in elastic bodies and the scattering of elastic waves from thin bodies.

An Automated Environment for Engineering Software Development
Chen, Wai-Fah; Purdue University

To develop prototype domain-specific software environment using structural engineering domain as a test case. This environment will be centered around the concept of software reuse and will be composed of Computer Aided Software Engineering (CASE) tools for managing and integrating reusable software components into applications.

Reliability and Uncertainty

Applications of ARMA and Related Random Process Models in Structural Engineering
Shinozuka, M.; Columbia University

Auto-regressive moving-average (ARMA) and related models were developed for stationary Gaussian multivariate random processes, constructed for multi-dimensional random processes with prescribed correlation functions, adapted to structural engineering applications.

Workshop on Reliability in Computational Mechanics
Oden, J.; University of Texas-Austin

This workshop sensitizes researchers to current methodologies for evaluating reliability as it applies to computational mechanics problems.

Performance-Optimized Reliability Analysis of Structural Systems
Corotis, Ross; Johns Hopkins University, Baltimore, MD

To develop practical procedures whereby characterization of limit state consequences can be incorporated into an analysis for reliability-constrained structural system optimization. Multiobjective mathematical programming will be used to identify the limit state surface for multiple, independently varying loads. Efficient methods of combining Monte Carlo directional simulation with mathematical programming, and especially linear programming, will be utilized to permit a comprehensive assessment of structural system reliability.

A New Approach - Convexity Modelling - to an Old Research Topic: Buckling of Viscoelastic and Nonlinear Elastic Structures with Uncertain Imperfections
Elishakoff, Isaac; Florida Atlantic University

To study the uncertainty in imperfections in viscoelastic plates and their effects on buckling. The main objective is to investigate the feasibility of convex modeling - a new concept of uncertainty in imperfection sensitivity analysis.

Sensitivity Analysis and Failure Cost Design of Random Structural Systems in Load Space
Soltani, Mehrdad; University of Miami

A nonlinear structural analysis program is utilized to derive ultimate and serviceability limit states in load space. System probability of failure is computed by integration of the joint density function of applied loads in the failure region in a three-dimensional load space.

Structural Engineering and Reliability Analysis
Tallin, Andrew; Polytechnic University, Brooklyn, NY

Research includes computer-based reliability analysis with special emphasis on systems with nonlinear dynamic response and hierarchial damage states, development of a hybrid reliability approach applied to seismic response and fatigue crack growth.

The First International Symposium on Uncertainty Modeling and Analysis: Fuzzy Reasoning, Probabilistic Methods, and Risk Management
Ayyub, Bilal; University of Maryland

Probabilistic methods, Bayesian approaches, statistical techniques as well as fuzzy logic and reasoning are currently being used by researchers to deal with uncertainty. In other countries, especially Japan, researchers and engineers are currently using all these methods more effectively in many engineering applications. The Symposium provides the opportunity for U.S. researchers and engineers to interact with counterparts from other countries.

Reliability of Existing Framed Structures Using Approximate Reasoning
Chou, Karen; Syracuse University

Research emphasizes the reliability and safety of existing and newly constructed plane frame structures. Classical reliability measure will be updated utilizing the fuzzy set logic to include the quality of connections and structural members through inspections.

Energy-Based Damage Model and Procedure for Safety Analysis of Structural Systems In Steel Buildings
Sohal, Iqbal Singh; Rutgers University

An energy-based damage model will be developed for computation of residual strength and stiffness of damaged steel structural members. For safety evaluation of structural systems, a damage analysis computer code is to be developed.

Integration of System Reliability in the Structural Design Process System Reliability-Based Design of Transmission Structures
Dagher, H. Joseph; University of Maine

To develop a computer program that designers and utilities can use to incorporate system reliability in the design of transmission structures.

Structural Optimization

Design Sensitivity Analysis and Optimization of Nonlinear Structural Systems
Arora, Jasbir; University of Iowa

The variables to be considered are: structural dynamics, nonlinear buckling loads, general nonlinearity, nonconservative aspects, elastic-plastic, viscoelastic, viscoplastic and related topics.

Design Sensitivities and Shape Optimization for Nonlinear Solid Mechanics Problems by Boundary Element Methods
Mukherjee, Subrata; Cornell University

To calculate design sensitivity coefficients for nonlinear solid mechanics problems and their use in design.

Optimal Design of Large-Scaled Structures with Multianalysis-Type Constraints
McGee, Oliver; Ohio State University

To develop efficient and reliable iterative procedures for minimum-weight design of large structures subject to multianalysis-type constraints (e.g., displacements, stresses, natural frequencies, and elastic stability).

A Decomposition Strategy for Optimization of Reinforced Concrete Systems
Balling, Richard; Brigham Young University

Decomposing the design into a system level and a component level. Continuous and discrete optimization techniques are applied at both level. Heuristic configuration search algorithms are used at the component level.

Chaos

Chaotic Dynamics of Elasto-plastic Structures: Experiments and Novel Computational Methods
Moon, Francis, C.; Cornell University

Attempt to develop radically new computational tools for analysis of nonlinear vibrations by inventing a way to put the computation on a chip using nonlinear cellular networks, as well as using both existing supercomputers.

Plastic Instability Phenomena with Aspects of Chaos of Fixed-Edge Structures Subjected to Short Pulse Loading
Symonds, Paul S.; Brown University

Anomalous phenomena have been studied in uniform beams by finite element analysis and indications of deterministic chaos appear. Their novelty and interest lie in their appearance with workless constraints and in the absence of harmonic or other sustained excitation. Research aims to confirm and amplify the preliminary results already obtained to gain deeper understanding by further analysis and extensions, and to provide methods for the estimation of parameter ranges in which they are likely to occur.

Chaos and Dimensionality in the Dynamics of Elastic Solids: Application of the Theory of Chaotic Vibrations to the Nonlinear

Modelling of Flexible Structures
Cusumano, Joseph; Pennsylvania State University

To develop better techniques for obtaining finite degree of freedom nonlinear models of flexible structures. Research includes testing, signal analysis and modeling.

Computational Modeling and Others

Intersection of Shells of Revolution
Steele, Charles; Stanford University

To develop more efficient methods of analysis of stress and deflections in intersecting shell structures. Finite element codes are intrinsically inefficient. The project is a combination of analytical and numerical methods.

Studies of Non-Linear Theory of Shear Deformable Elastic Shells
Reissner, Eric; University of California-San Diego

Based on the status of the existing nonlinear theory of shear-deformable elastic shells with independent descriptions for translational and rotational displacement states, it was proposed to undertake research on the form of two-dimensional constitutive relations for finite deformations of shells, and on possible new solutions of specific problems for shallow shells of revolutions and helicoidal shells. Completed in 1985.

Initiation and Growth of Tearing in Ductile Sheet Metal
Wierzbicki, Tomasz; MIT

To study plastic folding and tearing mechanisms of sheet metal and to develop models of fracture initiation and growth.

Application of Advanced Computer-Based Technologies in Civil Engineering
Rehak, Daniel; Carnegie Mellon University

Specific research includes finite element systems, database management systems, and integrated computer-aided engineering systems. To create a friendly, responsive, and powerful computer-based engineering environment.

A New Approach for the Analysis of Piece-Wise Periodic Structures of Finite Length
Lyrintzis, Constantinos; San Diego State University

A combination of finite elements, transfer matrices and wave propagation, the computations to proceed in the direction of wave propagation only, so that the results are always numerically stable.

Studies in Structural Mechanics
Miller, Gregory; University of Washington, Seattle

Emphasis on the application of advanced analytical and numerical techniques to the study of problems in materials characterization and structural response, including fracture mechanics, plasticity, fatigue analysis, micro-mechanics, and nonlinear dynamics.

SUMMARY AND CONCLUSION

An overview of the state of the art and NSF projects in computational and structural mechanics is presented. The authors hope that this paper will act as a catalyst, sparking interest and further research in these areas. This paper reflects the personal views of the authors, not necessarily those of the National Science Foundation.

REFERENCES

1. Boresi, A. P. and Chong, K. P., Approximate Solution Methods in Engineering Mechanics, Elsevier, London, 1991.

2. Chong, K. P., Sartori, H. J., Crouse, P. E., and Schmidt, R. J., Survey of CAD/CAM/CAE software, Proc. UPCAEDM, Fulton, R. E., and Craig, J. I. Eds., Georgia Tech, Atlanta, GA, 1988, pp. 94-8.

3. MacNeal, R. H., and Harder, R. L., A proposed set of problems to test finite element accuracy, Str., Dynamics, and Matl. Conf., Orlando, FL, 1984.

4. Fong, H. H., A commentary on commercial finite element systems, Finite Element Systems Handbook, 3rd edn, Brebbia, C. A. Ed., Springer-Verlag, New York, 1984.

5. Cook, R. D., Malkus, D. S., and Plesha, M. E., Concepts and Applications of Finite Element Analysis, 3rd edn, John Wiley, New York, 1989.

6. Zienkiewicz, O. C., and Taylor, R. L., The Finite Element Method, 4th edn, Vol. 1, Basic Formulation and Linear Problems, McGraw-Hill, New York, 1989.

7. Zienkiewicz, O. C., The finite element method: from intuition to generality, Appl. Mech. Rev., 1970, 23(3), 249-56.

8. Diekkamper, R., Vectorized finite element analysis of nonlinear problems in structural mechanics, Int. Conf. on Parallel Computing 83, Elsevier North-Holland, Amsterdam, 1984, pp. 293-8.

9. Farhat, C., and Wilson, E. L., Concurrent iterative solution of large finite element systems, Comm. Appl. Num. Meth., 1987, 3, 319-26.

10. Storaasli, O. O., Peebles, S. W., Crockett, T. W., Knott, J. D., and Adams L. The finite element maching: an experiment in parallel processing, NASA TM 84514, 1982.

11. Glouderman, J. F., Heinrich, C. W., and Hedge, J. C., The evolution of MSC/NASTRAN and the supercomputer for enhanced performance, NATO-ASI Series, High Speed Computation, Kowalik, J. S. Ed., Springer-Verlag, Berlin, 1984 F7, pp. 393-402.

12. Storaasli, O. O., and Bergan, P. Nonlinear substructuring method for concurrent processing computers, AIAA Journal, 1987, 25,6.

13. ParaSoft Corp., An Introduction to Parallel Programming and Transputers, MicroWay, Inc. (distributor), Kingston, MA., 1988.

14. Huttelmaier, H.P., and Epstein, M., A multilayered finite element formulation for vibration and stability analysis of plates, J. Eng. Mech., 1989, 115(2), 315-25.

15. Kung, H. T. The structure of parallel algorithms. Advances in Computers, Vol. 19, Academic Press, New York. 1980, pp. 65-112.

16. Bathe, K. J., and Graceneski, S., On nonlinear dynamic analysis using substructuring and mode superposition, Computers and Structures, 1981, 13, 699-707.

17. Noor, A. K., Kamel, H. A., and Fulton, R. E., Substructuring techniques-status and projections, Computers and Structures, 1978, 8, 1313-19.

18. Leung, A. Y. T., and Wong, W. K., Frame mode for thin-walled structures, J. Thin-Walled Structures, 1988, 6(2), 81-108.

19. Noor, A. K., and Peters, R. W., Recent advances in reduction methods for instability analysis of structures, Computers and Structures 1983, 16, 67-80.

20. Sabana, A. A., Substructure synthesis methods for dynamic

analysis of multi-body systems, Computers and Structures, 1985, 20(4), 737-44.

21. Guyan, R. J., Reduction of stiffness and mass matrices, AIAA Journal, 1965, 3,2.

22. Craig, R. R., and Chang, C. J., A review of substructure coupling for dynamic analysis, Adv. in Eng. Sci., NASA CP-2001, 1976, 2, 393-408.

23. Dodds, Jr., R. H., and Lopez, L. A., Substructuring in linear and nonlinear analysis, Int. J. Num. Meth. Eng., 1980, 15, 583-97.

24. Paz, M., Dynamic reduction methods, Structural Mechanics Software Series 5, Ed. Perrone, N. and Pilkey, W. University Press of Virginia, Charlottesville, VA, 1984.

25. Wilson, E. L., The static condensation algorithm, Int. J. Num. Meth. Eng. 1974, 8(1), 199-203.

26. Adeli, H., and Vishnubholta, P., Parallel processing, Microcomputers in Civil Engineering, 1987, 2(3), 257-69.

27. Farhat, C., and Crivelli, L., A general approach to nonlinear finite element computations shared memory multiprocessors, Computer Meth. Appl. Mech. and Eng, 1989, 72, 153-71.

28. Patnaik, S. N., The variational energy formulation for the integrated force method, AIAA Journal, 1986, 24(1), 129-37.

29. Patnaik, S. N., and Joseph, K. T., Generation of the compatibility matrix in the integrated force method, J. Comp. Appl. Mech. and Eng., 1986, 55, 239-57.

30. Vijayakumar, K., Murty, A. V., Krishna, P., and Patnaik, S.N., A basis for the analysis of solid continua using the integrated force method, AIAA Journal, 1988, 26, 628-9.

31. Marjadi, D., Force method of structural analyses for optimization, Ph.D. dissertation, University of Missouri, Columbia, MO., 1988.

32. Wempner, G., Mechanics and the finite elements of shells, Appl. Mech. Rev., 1989, 42(5), 129-42.

33. Ravichandran, R. V., and Sridharan, S., A comparative study of the performance of isoparametric shell elements in geometrically nonlinear analysis, Research Report No. 82, Department of Civil Engineering, Washington University, St. Louis, MO., 1989.

34. Lin, J. S., and Gould, P. L., Shells of revolution with local plasticity, Comp. Meth. Appl. Mech. Eng., 1987, 65, 127-45.

35. Gould, P. L., and Harintho, H., Nonlinear local-global analysis of shells of revolution, 3rd Joint ASCE-ASME Mechanics Conf., University of California, San Diego, Ca., 1989.

36. Bathe, K. J., ADINA finite element program for automatic dynamic incremental nonlinear analysis, Report 82448-1, Department of Mechanical Engineering, MIT, Cambridge, MA. 1975.

37. Habbitt, Karlson and Sorensen, Inc. Internal document, ABAQUS, Rhode Island, USA. 1984.

38. Carney III, J. F., The Influence of shock waves and strain rate effects on the energy dissipation characteristics of metallic tubes, ASCE Annual Convention, Boston, MA. 1986.

39. Carney III, J. F., Veillette II, J. R., and Reid, S. R., Hysteresis and energy dissipation in metallic tubes, Proc. 27th SDM Conf., San Antonio, Texas, 1986 Vol. 1, pp. 180-8.

40. Holzer, S. M., Watson, L. T., and Vu, P., On the structural behavior of reticulated domes, Proc. ASCE Annual Convention, 1981, pp. 179-209.

41. Snyder, M. D., and Bathe, K.-J., A solution procedure for thermoelastic-plastic and creep problems, Nuc. Eng. and Des., 1981, 64, 49-80.

42. Zienkiewicz, O.C., and Cormeau, I. C., Viscoplasticity-plasticity and creep in elastic solids-a unified numerical approach, Int. J. Num. Meth. Eng., 1974, 8, 821-45.

43. Bushnell, D., A strategy for the solution of problems involving large deflections, plasticity and creep, Int. J. Num. Meth. Eng., 1977, 11, 683-708.

44. Haftka, R. T., and Adelman, H. M., Recent advances in structural sensitivity analysis, 3rd Intl. Conf. on CAD/CAM, Robotics, and Factories of the Future, Southfield, MI, 1988.

45. Haug, E. J., and Arora, J. S., Applied Optimal Design, John Wiley, New York. 1979.

46. Lahey, R. S., Implementing design sensitivity analysis in MSC/NASTRAN, 18th Midwestern Mechanics Conf., University of Iowa, Iowa City, IA, 1984, pp. 349-51.

47. Arora, J. S., Computational design optimization: a review and future directions, Proceedings NSF Workshop on Research Needs in Structural Analysis, Design and Optimization, University of Colorado, Boulder, Colorado, 1988.

48. Borst, R. de, Kusters, G. M. A., Nauta, P., and de Witte, F. C., DIANA-a 3-D nonlinear finite element package on a microcomputer, Engineering Software for Microcomputers, Schrefler, B.A., et al., Eds., Pineridge Press, Swansea, UK., 1984.

49. Rots, J. G., Nauta, P., Kusters, G. M. A., and Blaauwendraad, J., Smeared crack approach and fracture localization in concrete, Heron (Delft University), 1985, 30(1), 3-48.

50. Chong, K. P., Li, V. C., and Einstein, H. H., Size effects, process zone, and tension softening behavior in fracture of geomaterials, J. Eng. Frac. Mech., 1989, 34(3), 669-78.

51. Brebbia, C. A., and Dominguez, J., Boundary elements, Comp. Mech. Pub., Computational Mechanics Institute, Ashurst, Southampton, United Kingdom, 1988.

52. Aliabadi, M. H., An enhanced boundary element method for determining fracture parameters, Proc. Intl. Conf. on Num. Meth in Frac. Mech., San Antonio, Texas, 1987, pp 27-39.

53. Blandford, G. E., Ingraffea, A. R., and Liggett, J. P., Two-dimensional stress intensity factor computations using the boundary element method, Int. J. Num. Meth. Eng., 1981, 17, 387-404.

54. Liu, P. L., and Der Kiureghian, A., Finite-element reliability methods for geometrically nonlinear stochastic structures, NSF supported research, Report No. UCB/SEMM-89/05, Department of Civil Engineering, University of California, Berkeley, CA, 1989.

55. Der Kiureghian, A., and Ke, B. J., The stochastic finite element method in structural reliability, Probabilistic Eng. Mech., 1988, 3(2), 83-91.

56. Spanos, P. D., and Ghanem, R., Stochastic finite element expansion for random media, J. Eng. Mech., 1989, 115(5), 1035-53.

57. Liu, W. K., Belytschko, T., and Mani, A., Probabilistic finite elements for nonlinear structural dynamics, Comp. Meth. Appl. Mech. Eng., 1989, 56, 61-86.

58. Mahadevan, S., and Haldar, A., Efficient algorithm for stochastic structural optimization, J. Struc. Eng., 1989, 115(7), 1579-98.

59. Boresi, A. P., and Chong, K. P., Elasticity in Engineering Mechanics, Elsevier, New York, 1987.

Funding Programs of the Computational Mechanics by National Natural Science Foundation of China (NSFC)

Jin Zhengmo Yang Lei

Department of Mathematical and Physical Sciences, NSFC.

I. General Survey of NSFC

• Duty: The basic duty of NSFC is to support basic research and parts of applied basic research. The areas of these researches are mathematics, physics, chemistry, astronomy, geology, biology, information science, engineering and material science, and so on.

The ratio of funds for basic and applied basic research is three to seven. This ratio is naturally formed.

• Organization: NSFC consists of one office, four bureaux and six science departments.

NSFC consists of about 200 staff , among whom there are about 80 persons who once worked in the field of science and technology.

• Sources of Funds: The funds of NSFC come from the governmental financial allocation. It has increased from eighty million yuan in 1986 to one hundred and seventy five million yuan in 1991. The annual increasing rate is seventy percent.

• Three Levels of Programs Supported:

The programs supported by NSFC are divided into three kinds: general programs, key programs and major programs. Their funds percentages are 75%, 15% and 10% respectively.

• Principles and steps of program evaluation: Applying programs are evaluated under the principle of "relying on experts, promoting democracy, supporting best programs,and being in a fair and reasonable way". After initial evaluation done by the academic committee of the unit where the applicant works, and peer-reviews, the subject evaluation panels give funding opinions to each program, and finally the Foundation ratifies the execution. The evaluation of the programs focuses on creativity, originality, frontiers and intersections of the disciplines.

Every year NSFC receives a great deal of applications of which about a quarter is usually supported. Table one is the survey of the general programs applied and supported in **1990**. As an example, it shows the outlook of the funds work.

• The international cooperation and academic exchange: Since the founding of NSFC, it has signed agreements or memorandum of understanding with fourteen countries and regions. In **1990**, NSFC has supported **505** programs which spent about **3.3** million yuan.

II. Brief Introduction of Mechanics Discipline

Mechanics is a basic as well as a technical discipline. It plays an important role in the development of science and technology in China.

• The main areas of research supported by Mechanics Section are solid mechanics, fluid mechanics, general mechanics, experimental mechanics, computational mechanics, rock and soil mechanics, biomechanics, explosion mechanics, physical mechanics, rational mechanics and so on.

• The areas of research encouraged are mechanical properties of materials including their constitutive relations; study of the nonlinear behavior in large deformation under various conditions of some typical materials with combined macroscopic, mesoscopic and microscopic methods; turbulence and its control; vortex-dominated flows and separated flows, boundary - layer theory and drag reduction, nonlinear waves and coupled fluid - solid problems; dynamic response and optimization theory of complex, nonlinear, large - scale structures; environmental mechanics and so on.

• General programs are evaluated once a year. Table two records the datum about application and support of mechanical programs in **1990**. Table three gives us the distribution of the applied and supported programs in the disciplines in **1990**.

• The key programs were not taken out from the general programs until **1990**.

During **1986 - 1990**, three major programs were supported:

• Mechanical Problems in Ocean Engineering;

• Complicated Flows Dominated by Vortex, Shock and Nonequilibrium Phenomenon;

• Fracture and Constitutive Relations of Metallic Materials.

III. Some Achievements in the Work of Funding for Computational Mechanics

In recent years, NSFC has supported computational mechanics in three directions. They are research in numerical methods, structure optimization design, and developing computer software and standardization. Chinese scholars have done a lot of work in these directions and got some achievements.

IV. Developing International Cooperation and Promoting the Development of Science

As basic research is generally characterized by international and opening features, NSFC pays a great deal of attention to international exchange and co-operation, encourages and supports academic exchanges. Of course, this co-operation should be sincere, equal, mutual beneficial and mutual respect. The results of the co-operative research should be shared by both sides. In recent years, we have six joint programs with England and Bulgaria which achieved satisfactory results.

With these results, we are confident that this symposium will expand and deepen international cooperation and exchanges, and we will make more contributions to the development of science and technology.

NSFC Organization Chart

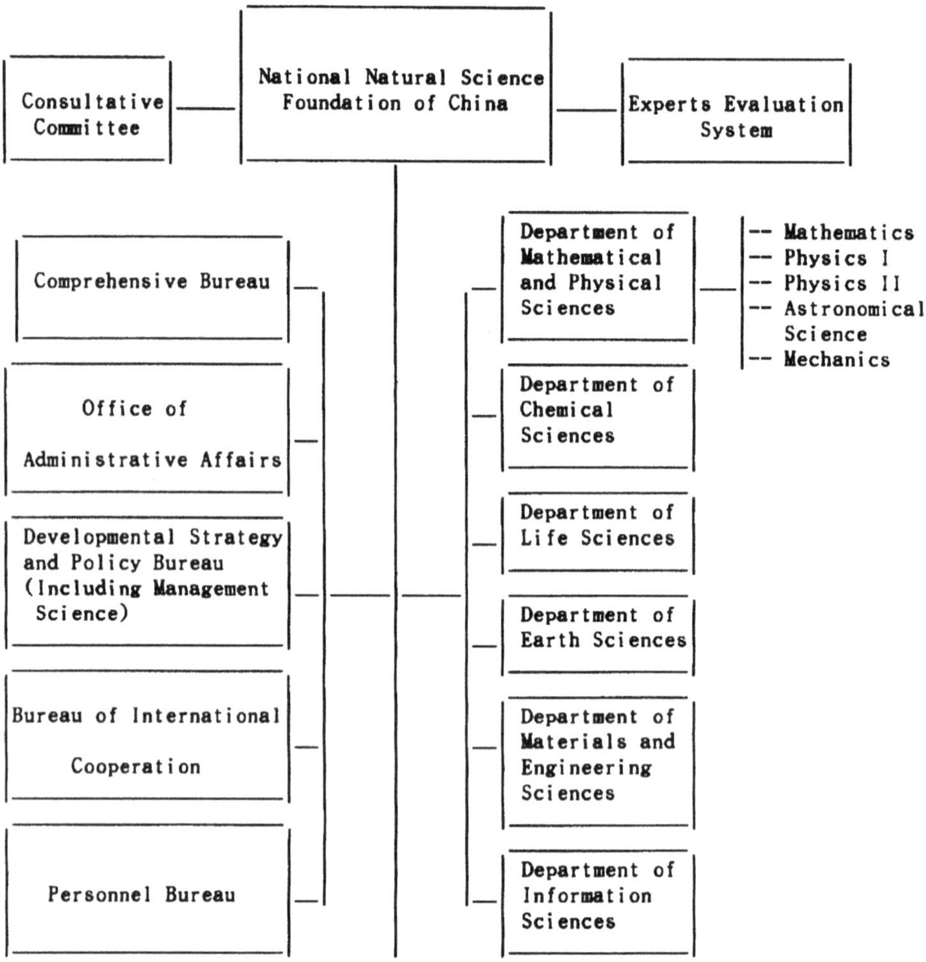

Table 1 **An Overview of the Amount of Support and Application for General Projects in 1990**

Unit : 1,000 yuan (RMB)

	Project	Amount	Intensity	Rate of Funding
Application	14438	759273		
Support	3478	119792	34.4	24.1%

Table 2 An Overview of the Amount of Support and Application for General Projects in Mechanics Discipline in 1990

Unit : 1,000 yuan (RMB)

	Project	Amount	Intensity	Rate of Funding
Application	422	24070		
Support	114	3650	32	27.0%

Table 3 The Distribution of the Applying and Support Disciplines (Percentage)

Branches of Mechanics	Applying	Supporting
Solid Mechanics	29.3	30.7
Fluid Mechanics	21.8	25.3
General Mechanics	10.7	12.0
Experimental Mechanics	10.0	9.3
Computational Mechanics	12.8	12.0
Others	15.4	10.7

Partial Differential Equations and Hamiltonian System

Zhong Wanxie
(Research Institute of Engineering Mechanics, Dalian Univ. of Tech., China)

Yang Zaishi
(Higher Education Press Beijing, china)

Abstract

In theory of mechanics and/or mathematical physics problems in prismatic domain, the method of separation of variables usually leads to the Sturm-Liouville type self-adjoint eigen-problems, and then the eigenfunction expansion method can be applied in equation solving. However, a number of important application problems cannot lead to self-adjoint operator for the transverse coordinate. From the minimum potential energy variational principle, by selecting the state and its dual variables, the generalized variational principle is derived, and then based on the analogy between the theory of structural mechanics and optimal control, the present paper derives the problem to Hamiltonian system. The finite dimensional theory for Hamiltonian system is extended to the corresponding theory of Hamiltonian operator matrix and adjoint simplectic spaces. The adjoint simplectic ortho-normality relation is proved for the whole state eigenfunction vectors, and then the expansion of an arbitrary whole state function vector by the eigenfunction vectors is established. Thus the classical method of separation of variables is greatly extended.

Keywords: *Separation of variables; Hamiltonian system; Simplectic; Eigenfunction Expansion; Plate bending; Whole state space.*

1. Introduction

The analogy theory between computational structural mechanics and optimal control had been given in some detail[1-5]. However, all the problems given in these papers are for finite dimensional state vector. Considered from the analogy between the semi-analytical method of elliptic partial differential equation (PDE) and the continuous-time optimal control, it is recognized that the dimension of the PDE is originally infinite, for the establishment of the analogy relation to finite dimensional control system, semi-analytical method is invoked. In treatises on semi-analytical method[6,7], the derivations are almost all based on the displacement method.

By using the finite dimensional subspace in the transverse coordinate, the semi-analytical method approximates the infinite dimensional space of the original PDE. Hence there is the problem of appropriate selection of subspace for better approximations. For the PDEs, for which the variables can be separated, it is natural to select the eigen-solutions of its Sturm-Liouville problem[8] as the basis functions of subspace, such that the ordinary differential equations derived are ultimately de-coupled. Unfortunately, the Sturm-Liouville problem can only be applied to self-adjoint operators after the separation of variables, which brings strict limitation to its extent of application. Thus semi-analytical method was proposed as the extension. However, the appropriate selection of the subspace remains a tremendous problem to be solved.

In the investigation of eigen-problems of large scale system theory, the various reduced state space methods[9] cannot ensure the eigenvalues of the original problem unchanged. For the invariance of eigen-solutions of the original system, it is necessary to select subspace from the whole state space, which corresponds to the theory of contact transformation of the Hamiltonian dynamic system, i.e. to select the adjoint simplectic subspace[10,11]. The main eigen- solutions dominate the long term property, so that in the numerical method for PDE, this factor must be considered.

The minimum potential energy variational principle corresponds to the displacement method[12]. From the analogy between structural mechanics and optimal control, the displacement corresponds to the state vector in control theory. Hence the minimum potential energy based semi-analytical method corresponds to the reduced state space method in large scale system theory. Unfortunately, this method corresponds only to the extended point transformation in Hamiltonian dynamic system theory, but not the contact transformation[13]. Generally speaking, the main eigen-solutions will be disturbed in the subspace for using only extended point transformations.

From Hamiltonian system theory, the general contact transformation should be executed in whole state space, which corresponds to the combination of spaces of displacement and internal force in structural mechanics. In PDE theory, the whole state space should be the combination of variables composed of the field function and its dual. From the viewpoint of variational methods in theory of elasticity, the generalized variational principle[12] should be used.

Applying the theory of contact transformation to the transformed PDE and introducing the eigen-problem of the Hamiltonian system, the eigenfunction expansion method can then be used to solve the PDE. The classical Sturm-Liouville problem and its eigen-function expansion for PDE is thus a special case of the present method.

2. The Hamiltonian operator matrix and variational principles

The principle described in the introduction is quite general, but the theory is better described with some typical problems, in which the principle given in the previous section should be reviewed frequently, so as to capture the essence of the present method.

For simplicity, the second order elliptic PDE is selected as the first problem, then the plate bending problem in a strip domain is solved by the eigen-function expansion method in the latter text. By using these two examples to describe the general method, it is conceivable that all the elastic analysis problem in a prismatic domain can be solved by the method of Hamiltonian operator matrix for whole state space, the generalized variational principle, the eigen-solutions in the adjoint simplectic space and the eigen-function expansion solution.

As the first problem, the second order elliptic PDE is considered

$$a\partial^2u/\partial x^2 + b\partial^2/\partial x\partial y + c\partial^2u/\partial y^2 - eu = 0 \qquad (2.1)$$

For simplicity, constant coefficients a, b, c, e are assumed, and $b^2-4ac<0$. The domain $\mathcal{D}$ is taken to be a strip of

$$\mathcal{D}: \qquad 0 < x < h\ , \qquad 0 < y < 1\ , \qquad h >> 1 \qquad (2.2)$$

x is the longitudinal direction. The boundary conditions are

$$u = 0\ , \qquad \text{when } y=0 \text{ or } y=1\ ; \qquad (2.3)$$

$$u = \text{given functions } u_{0*}(y) \text{ or } u_{h*}(y), \qquad \text{when } x=0 \text{ or } x=h \qquad (2.4)$$

The corresponding variational equation can be given as

$$\min_u \iint_{\mathcal{D}} \left[\frac{a}{2}\left(\frac{\partial u}{\partial x}\right)^2 + \frac{c}{2}\left(\frac{\partial u}{\partial y}\right)^2 + \frac{b}{2}\frac{\partial u}{\partial x}\frac{\partial u}{\partial y} + \frac{e}{2}u^2 \right] dx\, dy \qquad (2.5)$$

which is the minimum potential energy principle, where only the field function u(x,y) is in variation. The finite terms semi-analytical method based on the above variational equation[2,6] corresponds to the reduction from transverse infinite dimensional space of the PDE to finite dimensional subspace by the reduced state space method, for which the main eigen-solutions will be disturbed, which is not preferred. Hence, the dual function should be introduced.

By partially differentiating the integrand of (2.5) with respect to $(\partial u/\partial x)$ and taking negative, the 'internal force' dual function n(x,y) is introduced as

$$n(x,y) = -(a\partial u/\partial x + b/2\cdot\partial u/\partial y), \quad \text{or} \quad \partial u/\partial x = -(n - b/2\cdot\partial u/\partial y)/a \qquad (2.6)$$

Substituting the above equation into (2.1), it gives

$$\partial n/\partial x = -\left[e - (c-b^2/4a)\partial^2/\partial y^2\right]u - (b/2a)\cdot\partial n/\partial y \qquad (2.7)$$

Now the dual equations (2.6) and (2.7) can be written in operator form

$$\left.\begin{aligned} \partial u/\partial x &= -[(b/2a)\partial/\partial y]u \qquad - (1/a)n &&= \underset{\sim}{F}\, u - \underset{\sim}{G}\, n \\ \partial n/\partial x &= -[e-(c-b^2/4a)\partial^2/\partial y^2]u - [(b/2a)\partial/\partial y]n &&= -\underset{\sim}{Q}\, u - \underset{\sim}{F}^T n \end{aligned}\right\} \qquad (2.8)$$

where the operators $\underset{\sim}{F}$, $\underset{\sim}{G}$, $\underset{\sim}{Q}$ are

$$\underset{\sim}{F} = -(b/2a)\partial/\partial y, \qquad \underset{\sim}{F}^T = (b/2a)\partial/\partial y \qquad (2.9)$$

$$\underset{\sim}{G} = 1/a\ , \qquad \underset{\sim}{Q} = e-(c-b^2/4a)\partial^2/\partial y^2 \qquad (2.10)$$

The generalized variational principle of the above dual equations is necessary for clarifying the meaning of the operators. By integration by parts and taking account of the boundary conditions it can be verified that the variational equation of

$$\delta\int_0^h\int_0^1 \left[-n\frac{\partial u}{\partial x} - \frac{b}{2a}n\frac{\partial u}{\partial y} - \frac{n^2}{2a} + \frac{e}{2}u^2 - \frac{1}{2}(c-\frac{b^2}{4a})u\frac{\partial^2 u}{\partial y^2} \right] dy\, dx = 0\ ,$$

where u and n are considered as independent functions in the variational operation, can be used to derive the dual equation set (2.8). Based on the analogy of PDE with optimal control theory, the longitudinal coordinate x corresponds to the time coordinate in control. The above variational equation can be expressed in the operator form

$$\delta\int_0^h\int_0^1 [\ -n^T\dot{u} + n^T\underset{\sim}{F}u - \frac{1}{2}n^T\underset{\sim}{G}n + \frac{1}{2}u^T\underset{\sim}{Q}u\ ]\ dy\ dx = 0\ , \qquad \partial u/\partial x=\dot{u} \tag{2.11}$$

which can also be written in the form

$$\delta\int_0^h [\ -\langle n^T,\underset{\sim}{I},\dot{u}\rangle + \langle n^T,\underset{\sim}{F},u\rangle - \frac{1}{2}\langle n^T,\underset{\sim}{G},n\rangle + \frac{1}{2}\langle u^T,\underset{\sim}{Q},u\rangle\ ]\ dx = 0 \tag{2.12}$$

where $\underset{\sim}{I}$ is an identity operator and the operation $\langle\cdot,\cdot,\cdot\rangle$ is defined as

$$\langle n^T,\underset{\sim}{F},u\rangle = \int_0^1 n^T\ \underset{\sim}{F}\ u\ dy \tag{2.13}$$

etc., and ^{T} is a transpose operator, which gives no effect when acting on a scalar. By integration by parts and boundary conditions, it gives

$$\langle n^T,\underset{\sim}{F},u\rangle = \langle u^T,\underset{\sim}{F}^T,n\rangle \tag{2.14}$$

These operators $\underset{\sim}{F}$, $\underset{\sim}{G}$, $\underset{\sim}{Q}$ have specified sort of function to apply, such as the right to $\underset{\sim}{F}$ must be u, whereas left to $\underset{\sim}{F}$ must be n, etc.

Now express the dual equation with the Hamiltonian operator matrix $\underset{\sim}{H}$

$$\begin{Bmatrix} \dot{\underset{\sim}{u}} \\ \dot{\underset{\sim}{n}} \end{Bmatrix} = \underset{\sim}{H}\begin{Bmatrix} \underset{\sim}{u} \\ \underset{\sim}{n} \end{Bmatrix}, \qquad \text{where}\quad \underset{\sim}{H} = \begin{bmatrix} \underset{\sim}{F} & -\underset{\sim}{G} \\ -\underset{\sim}{Q} & -\underset{\sim}{F}^T \end{bmatrix} \tag{2.15}$$

The using of matrix formulation will be convenient for the comparison to the finite dimensional case. Thereafter, the field function $\underset{\sim}{u}$ (state) and the internal force function $\underset{\sim}{n}$ will be written in vector form. In fact, if the Riemann integration concept is used, the integration is considered as summation, and then the operation will be just the vectors and matrices etc. Thus,

$$\langle \underset{\sim}{n}^T,\underset{\sim}{F},\underset{\sim}{u}\rangle - \frac{1}{2}\langle \underset{\sim}{n}^T,\underset{\sim}{G},\underset{\sim}{n}\rangle + \frac{1}{2}\langle \underset{\sim}{u}^T,\underset{\sim}{Q},\underset{\sim}{u}\rangle = \frac{1}{2}\langle \begin{Bmatrix} \underset{\sim}{n} \\ -\underset{\sim}{u} \end{Bmatrix}^T,\ \underset{\sim}{H}\ ,\ \begin{Bmatrix} \underset{\sim}{u} \\ \underset{\sim}{n} \end{Bmatrix}\rangle \tag{2.16}$$

By introducing the rotation and exchanging operator matrix $\underset{\sim}{J}$ as

$$\underset{\sim}{J} = \begin{bmatrix} \underset{\sim}{0} & \underset{\sim}{I} \\ -\underset{\sim}{I} & \underset{\sim}{0} \end{bmatrix}, \qquad \underset{\sim}{J}^2 = \begin{bmatrix} -\underset{\sim}{I} & \underset{\sim}{0} \\ \underset{\sim}{0} & -\underset{\sim}{I} \end{bmatrix}, \qquad \underset{\sim}{J}^{-1} = -\underset{\sim}{J} = \underset{\sim}{J}^T \tag{2.17}$$

we have

$$\underset{\sim}{J}\underset{\sim}{H} = \begin{bmatrix} -\underset{\sim}{Q} & -\underset{\sim}{F}^T \\ -\underset{\sim}{F} & \underset{\sim}{G} \end{bmatrix} \tag{2.18}$$

$$\int_0^h \langle \underset{\sim}{n}^T,\ \underset{\sim}{I}\ ,\dot{\underset{\sim}{u}}\rangle\ dx = -\int_0^h \frac{1}{2}\langle \begin{Bmatrix} \underset{\sim}{u} \\ \underset{\sim}{n} \end{Bmatrix}^T,\ \underset{\sim}{J}\ ,\begin{Bmatrix} \dot{\underset{\sim}{u}} \\ \dot{\underset{\sim}{n}} \end{Bmatrix}\rangle\ dx + \frac{1}{2}[\underset{\sim}{n}^T\underset{\sim}{u}]_0^h \tag{2.19}$$

$$\frac{1}{2}\langle \begin{Bmatrix} \underset{\sim}{n} \\ -\underset{\sim}{u} \end{Bmatrix}^T,\underset{\sim}{H},\begin{Bmatrix} \underset{\sim}{u} \\ \underset{\sim}{n} \end{Bmatrix}\rangle = \frac{1}{2}\langle (\underset{\sim}{J}\begin{Bmatrix} \underset{\sim}{u} \\ \underset{\sim}{n} \end{Bmatrix})^T,\underset{\sim}{H},\begin{Bmatrix} \underset{\sim}{u} \\ \underset{\sim}{n} \end{Bmatrix}\rangle = -\frac{1}{2}\langle \begin{Bmatrix} \underset{\sim}{u} \\ \underset{\sim}{n} \end{Bmatrix}^T,\underset{\sim}{J}\underset{\sim}{H},\begin{Bmatrix} \underset{\sim}{u} \\ \underset{\sim}{n} \end{Bmatrix}\rangle \tag{2.20}$$

Now based on the concept of Hamiltonian system, the whole state vector

$$\underset{\sim}{v} = \begin{Bmatrix} \underset{\sim}{u} \\ \underset{\sim}{n} \end{Bmatrix} \tag{2.21}$$

can be introduced, where $\underset{\sim}{u}$ and $\underset{\sim}{n}$ are considered as independent function vectors. Thus the dual equation set can be concisely written as

$$\dot{\mathbf{v}} = \mathrm{H}\mathbf{v} \tag{2.15'}$$

and the variational equation (2.12) can be written as

$$\delta\int_0^h [\ \langle \mathbf{v}^T, \mathrm{J}, \dot{\mathbf{v}}\rangle - \langle \mathbf{v}^T, \mathrm{JH}, \mathbf{v}\rangle\ ]\ dx = 0 \tag{2.22}$$

To the operator matrices above, it is easily seen that the multiplication of two operator matrices is still an operator matrix. However, one must be aware to the operand of the operator matrix. The Hamiltonian operator matrix H is very important, for which

$$\mathrm{J}\ \mathrm{H}\ \mathrm{J} = \mathrm{H}^T \quad \left(= \begin{bmatrix} \mathrm{F}^T & -\mathrm{Q} \\ -\mathrm{G} & -\mathrm{F} \end{bmatrix}\right). \tag{2.23}$$

H transforms a whole state function vector to another whole state function vector. Its transpose operator matrix H^T transforms an adjoint whole state vector to an another adjoint whole state vector. J transforms a whole state function vector to an adjoint whole state function vector or transforms an adjoint whole state function vector back to a whole state function vector. It is defined that the operator matrix satisfying the equation (2.23) is called Hamiltonian. The operators Q and G must be symmetric, that the following relations are valid for arbitrary vectors satisfying the boundary conditions

$$\langle \mathbf{u}_2^T, \mathrm{Q}, \mathbf{u}_1\rangle = \langle \mathbf{u}_1^T, \mathrm{Q}, \mathbf{u}_2\rangle, \qquad \langle \mathbf{n}_2^T, \mathrm{G}, \mathbf{n}_1\rangle = \langle \mathbf{n}_1^T, \mathrm{G}, \mathbf{n}_2\rangle \tag{2.24}$$

By integration by parts and boundary conditions, the operators (2.10) obviously satisfy equation (2.24). For general PDE, the field function is not necessarily a scalar, it can be a vector function. The operator matrices H and H^T are written transpose to each other, which means an adjoint relation, so that the following equation for these two operators must be proved

$$\langle\ (\mathrm{J}\mathbf{v}_1)^T,\ \mathrm{H}\ ,\ \mathbf{v}_2\rangle = \langle\ \mathbf{v}_2^T,\ \mathrm{H}^T,\ (\mathrm{J}\mathbf{v}_1)\ \rangle \tag{2.25}$$

where the vectors $\mathbf{v}_1$ and $\mathbf{v}_2$ are arbitrary whole state vectors satisfying the boundary conditions. By integration by parts, the left side gives

$$\begin{aligned}
\langle(\mathrm{J}\mathbf{v}_1)^T, \mathrm{H}, \mathbf{v}_2\rangle &= \langle \mathbf{u}_1^T, \mathrm{Q}, \mathbf{u}_2\rangle - \langle \mathbf{n}_1^T, \mathrm{G}, \mathbf{n}_2\rangle + \langle \mathbf{n}_1^T, \mathrm{F}, \mathbf{u}_2\rangle + \langle \mathbf{u}_1^T, \mathrm{F}^T, \mathbf{n}_2\rangle \\
&= \langle \mathbf{u}_2^T, \mathrm{Q}, \mathbf{u}_1\rangle - \langle \mathbf{n}_2^T, \mathrm{G}, \mathbf{n}_1\rangle + \frac{b}{2a}\int_0^1 \left[-\mathbf{n}_1^T\frac{\partial \mathbf{u}_2}{\partial y} + \mathbf{u}_1^T\frac{\partial \mathbf{n}_2}{\partial y} \right] dy \\
&= \langle \mathbf{u}_2^T, \mathrm{Q}, \mathbf{u}_1\rangle - \langle \mathbf{n}_2^T, \mathrm{G}, \mathbf{n}_1\rangle + \frac{b}{2a}\int_0^1 \left[\mathbf{u}_2^T\frac{\partial \mathbf{n}_1}{\partial y} - \mathbf{n}_2^T\frac{\partial \mathbf{u}_1}{\partial y} \right] dy + \frac{b}{2a}\left[-\mathbf{n}_1^T\mathbf{u}_2 + \mathbf{u}_1^T\mathbf{n}_2 \right]_0^1
\end{aligned}$$

The boundary term equals zero for the boundary conditions, and the integration terms compose $\langle \mathbf{v}_2^T, \mathrm{H}^T, (\mathrm{J}\mathbf{v}_1)\rangle$, and (2.25) is verified.

3. Separation of variables and eigen-functions

To solve the equation (2.1), the usual method of separation of variables is valid only for the b=0 case. The traditional separation of variables gives an eigen-problem of self-adjoint operator, i.e. a Sturm-Liouville problem, all the eigenvalue will be real. The details can be found in reference [8]. However, the b term in (2.1) represents some gyroscopic effect, that the variational equation (2.5) still holds.

As mentioned in the previous sections, when b is not equal to zero, then based on the introduction of whole state vector and the generalized variational principle as well as the Hamiltonian system consideration, the method of separation of variables can still be derived. Thus the traditional method of separation of variables is extended. The separation of variables means to express the solution in the form as

$$\underset{\sim}{v} = \sum_{i=1}^{\infty} \underset{\sim}{Y}_i(y)\, X_i(x) \tag{3.1}$$

where any term of $(i=1,2,\ldots,\infty)$ is individually alone the solution of PDE (2.15'). Thus the functions should satisfy the equations

$$\dot{X}_i = \mu_i X_i \tag{3.2}$$

$$\underset{\sim}{H}\, \underset{\sim}{Y}_i(y) = \mu_i \underset{\sim}{Y}_i(y) \tag{3.3}$$

where μ_i is the eigenvalue to be determined. The equation (3.3) proposes the eigen-problem of the Hamiltonian operator matrix.

When the problem is reduced to be finite dimensional, the Hamiltonian operator matrix reduces to be a Hamiltonian matrix. In the high dimensional Hamiltonian matrix case, the eigen-problem should be solved by the adjoint simplectic subspace inverse substitution method, such that the subspace solutions can converge to the exact solutions of the high dimensional eigen-equation. The conclusions of the finite dimensional problem are quite instructive, see [10,11].

Based on the equation (2.23), from eigen-equation (3.3) it gives

$$-(\underset{\sim}{J}\underset{\sim}{H}\underset{\sim}{J})(\underset{\sim}{J}\underset{\sim}{Y}_i) = -\underset{\sim}{H}^T(\underset{\sim}{J}\underset{\sim}{Y}_i) = \mu_i(\underset{\sim}{J}\underset{\sim}{Y}_i) \tag{3.4}$$

hence the adjoint whole state vector $(\underset{\sim}{J}\underset{\sim}{Y}_i)$ is the eigenvector of the operator matrix $\underset{\sim}{H}^T$, and the eigenvalue is $-\mu_i$. So, if μ is an eigenvalue, then $-\mu$ is also an eigenvalue of $\underset{\sim}{H}$. Thus the eigenvalues of the operator matrix $\underset{\sim}{H}$ can be naturally subdivided into two groups

1. The a group $\quad \mu_{ai}, \quad Re(\mu_{ai}) \le 0 \quad$ (3.5)
2. The b group $\quad \mu_{bi}=-\mu_{ai}, \quad Re(\mu_{bi}) \ge 0 \quad$ (3.6)

The corresponding whole state eigen-function vectors can be nominated as $\underset{\sim}{Y}_{ai}$ and $\underset{\sim}{Y}_{bi}$ respectively, which are also called adjoint if the relation (3.6) holds for the corresponding eigenvalues.

4. Adjoint simplectic ortho-normalisation

Let the subscripts i and j represent two eigen-solutions of the Hamiltonian operator matrix respectively,

$$\underset{\sim}{H}\,\underset{\sim}{Y}_i = \mu_i \underset{\sim}{Y}_i, \qquad \underset{\sim}{H}\,\underset{\sim}{Y}_j = \mu_j \underset{\sim}{Y}_j \tag{4.1a,b}$$

Now based on the equation (2.23), it gives

$$-\underset{\sim}{H}^T(\underset{\sim}{J}\underset{\sim}{Y}_j) = \mu_j(\underset{\sim}{J}\underset{\sim}{Y}_j) \tag{4.2}$$

Multiplying the equation (4.1a) from left with $(\underset{\sim}{J}\underset{\sim}{Y}_j)^T$, then integrate from 0 to 1, by integration by parts it gives

$$\langle \underset{\sim}{Y}_j^T, \underset{\sim}{J}\underset{\sim}{H}, \underset{\sim}{Y}_i \rangle = \mu_i \langle \underset{\sim}{Y}_j^T, \underset{\sim}{J}, \underset{\sim}{Y}_i \rangle \tag{4.3}$$

Multiplying the equation (4.2) from left with $\underset{\sim}{Y}_i^T$ then integrate (0,1), by integration by parts, it gives

$$<\underset{\sim}{Y}_j^T \; , \; \underset{\sim}{JH} \; , \; \underset{\sim}{Y}_i> \; = -\mu_j \; <\underset{\sim}{Y}_j^T \; , \; \underset{\sim}{J} \; , \; \underset{\sim}{Y}_i> \tag{4.4}$$

thus

$$(\mu_i+\mu_j) \; <\underset{\sim}{Y}_j^T \; , \; \underset{\sim}{J} \; , \; \underset{\sim}{Y}_i> \; = 0 \tag{4.5}$$

The adjoint simplectic orthogonal relation is obtained as

$$\text{if } \mu_i \neq -\mu_j \; , \qquad <\underset{\sim}{Y}_j^T \; , \; \underset{\sim}{J} \; , \; \underset{\sim}{Y}_i> \; = 0 \; , \qquad <\underset{\sim}{Y}_i^T \; , \; \underset{\sim}{JH} \; , \; \underset{\sim}{Y}_j> \; = 0 \tag{4.6}$$

so the pair of $\mu_i=-\mu_j$ is called adjoint. From the above expression, it can be considered that $\underset{\sim}{J}$ is a metric operator matrix, but the $\underset{\sim}{J}$ metric is not positive definite. In finite dimensional case it is called as <u>simplectic</u>, and the term will also be used now for infinite dimensional case. Equation (4.6) means that the whole state eigen-function vectors are simplectically orthogonal each other include themself, except the adjoint one. Note that the eigen-solutions had been classified by (3.5) and (3.6) into two sets. To fix the constant multipliers for the eigen-solutions, the simplectic normalisation relation between adjoint eigen-solutions can be introduced. Because of

$$<\underset{\sim}{Y}_{ai}^T, \; \underset{\sim}{J} \; , \; \underset{\sim}{Y}_{bi}> \; \equiv - <\underset{\sim}{Y}_{bi}^T, \; \underset{\sim}{J} \; , \; \underset{\sim}{Y}_{ai}> \tag{4.7}$$

the simplectic normalisation relation can be defined as

$$<\underset{\sim}{Y}_{ai}^T, \; \underset{\sim}{J} \; , \; \underset{\sim}{Y}_{bi}> \; = 1 \tag{4.8}$$

that the adjoint eigen-solution of the set_a must be in set_b. Thus an appropriate arrangement of the eigen-solutions can make the i-th eigen-solution in set_a just adjoints the i-th eigen-solution in set_b, i.e. (4.8). Reversing the multiplication order, it gives

$$<\underset{\sim}{Y}_{bi}^T, \; \underset{\sim}{J} \; , \; \underset{\sim}{Y}_{ai}> \; = -1 \tag{4.8'}$$

Because both the adjoint eigen-function vectors have a constant factor to be fixed, one more equation can be supplied; it can be

$$\|\underset{\sim}{Y}_{ai}\| = \|\underset{\sim}{Y}_{bi}\| \tag{4.9}$$

where $\|\cdot\|$ is the Euclid norm, i.e.

$$\|\underset{\sim}{Y}_i\| = \| \begin{Bmatrix} \underset{\sim}{u}_i \\ \underset{\sim}{n}_i \end{Bmatrix} \| = <\bar{\underset{\sim}{Y}}_i^T \; , \; \underset{\sim}{I} \; , \; \underset{\sim}{Y}_i> \; = \int_0^1 (\underset{\sim}{u}_i^2+\underset{\sim}{n}_i^2) \; dy \tag{4.10}$$

where $\bar{Y}$ means the complex conjugate. The equations (4.9) and (4.8) will fix the two constant factors of real adjoint eigensolutions.

5. Expansion by the eigenfunction vectors

All the eigen-function vectors of the Hamiltonian operator matrix $\underset{\sim}{H}$ will construct a complete basis in the whole state space. The nature of the problem is similar to the case of self-adjoint Sturm-Liouville problem. Based on the Weierstrass theorem[8], any continuous function can be uniformly approximated by polynomials. In finite dimensional case, the n dimensional state space leads to 2n dimensional whole state space. Because of the adjoint simplectic ortho-normalisation, the 2n eigenvectors must be linearly independent, so that they must compose a complete set in the whole state space. Infinite dimensional problem can be approximated by finite dimensional problem based on the Weierstrass theorem; their natures are the same.

An arbitrary whole state function vector $\underset{\sim}{Y}$ can always be expanded with the eigen-solutions, that

$$\underset{\sim}{Y} = \sum_{i=1}^{\infty} (a_i \underset{\sim}{Y}_{ai} + b_i \underset{\sim}{Y}_{bi}) \tag{5.1}$$

where the coefficients a_i and b_i can be obtained with the adjoint simplectic ortho-normalisation relation

$$a_i = -<\underset{\sim}{Y}^T_{bi}, \underset{\sim}{J}, \underset{\sim}{Y}> , \qquad b_i = <\underset{\sim}{Y}^T_{ai}, \underset{\sim}{J}, \underset{\sim}{Y}> \tag{5.2}$$

The above equations give the eigen-function expansion.

It should be emphasized that $\underset{\sim}{Y}_{ai}$, $\underset{\sim}{Y}_{bi}$ can be used as the basis function vectors in the whole state function space; the coefficients a_i and b_i can be considered as the representation of a whole state. The eigen-solution expansion method is very effective, which is the extension of the traditional method of separation of variables. Now it is no longer limited only to self-adjoint operators.

6. Two end boundary conditions and its variational equation

Linear inhomogeneous equation can be solved by the superposition principle. Two approaches can be applied, namely directly expand the whole state $\underset{\sim}{v}$ with the transverse eigen-solutions or to represent the solution with the summation of a general homogeneous solution $\underset{\sim}{v}_H$ and a special inhomogeneous solution $\underset{\sim}{v}_0=\{\underset{\sim}{u}_0^T \ \underset{\sim}{n}_0^T\}^T$, i.e. $\underset{\sim}{v}=\underset{\sim}{v}_H+\underset{\sim}{v}_0$.

Now the homogeneous solution is considered, which can be expanded with the eigen-solutions; thus

$$\underset{\sim}{v}=\underset{\sim}{v}_0+\underset{\sim}{v}_H, \qquad \underset{\sim}{v}_H = \sum_{i=1}^{\infty} \left[X_{ai}(x)\times\underset{\sim}{Y}_{ai}(y) + X_{bi}(x)\times\underset{\sim}{Y}_{bi}(y) \right] \tag{6.1}$$

where X_{ai} and X_{bi} are functions to be determined. The equation for $\underset{\sim}{v}_H$ is (2.15') and the boundary condition at the two ends can be in various form, for example, the given displacement boundary conditions

$$\underset{\sim}{u} = \underset{\sim}{u}_{0*}(y) \text{ or } \underset{\sim}{u}_{h*}(y) \qquad \text{when } x=0 \text{ or } x=h \tag{6.2}$$

Substituting equation (6.2) into (2.15'), left multiplying with $(\underset{\sim}{J}\underset{\sim}{Y}_{bi})^T$ and then integrating for y, it gives

$$\dot{X}_{ai}=\mu_i X_{ai}, \qquad \text{thus} \quad X_{ai}= A_i exp(\mu_i x) \tag{6.3}$$

where A_i is an integration constant. Also

$$\dot{X}_{bi}=-\mu_i X_{bi}, \qquad \text{thus} \quad X_{bi}= B_i exp(-\mu_i x) \tag{6.4}$$

where B_i is also an integration constant. Now the problem turns to be how to determine these constants to satisfy (6.2). The integration constants are about the whole state, but the boundary condition (6.2) deals to only half of the whole state at both ends.

To derive the equations for the constants A_i and B_i, it is convenient to use the variational principle. For simplicity, using vector operation expressions, such as

$$\underset{\sim}{n}^T\underset{\sim}{u} \triangleq <\underset{\sim}{n}^T, \underset{\sim}{I}, \underset{\sim}{u}> \tag{6.5}$$

and dropping the subscript for $\underset{\sim}{v}_H$, the variational principle becomes

$$U = \int_0^h \left[\frac{1}{2}<\underset{\sim}{v}^T, \underset{\sim}{J}, \dot{\underset{\sim}{v}}> - \frac{1}{2}<\underset{\sim}{v}^T, \underset{\sim}{JH}, \underset{\sim}{v}>\right]dx + \left[\frac{1}{2}\underset{\sim}{n}^T\underset{\sim}{u} - \underset{\sim}{n}^T\underset{\sim}{u}_*\right]_0^h, \quad \delta U=0 \quad (6.6)$$

where $\underset{\sim}{u}_*$ is the given displacement vector. Executing the variational operation and integrating by parts, it gives

$$\delta U = \left[\ \delta n^T(u - u_*)\right]_0^h \quad (6.7)$$

which is the variational equation for the two end boundary conditions. It can be derived that

$$U = \sum_i (A_i\underset{\sim}{n}_{ai}^T + B_i\underset{\sim}{n}_{bi}^T)\underset{\sim}{u}_0 * [1-exp(\mu_i h)] - \frac{1}{2}\sum_i\sum_j A_iA_j\underset{\sim}{n}_{ai}\underset{\sim}{u}_{aj}[1-exp((\mu_i+\mu_j)*h]$$

$$- \frac{1}{2}\sum_i\sum_j B_iB_j\underset{\sim}{n}_{bi}^T\underset{\sim}{u}_{bj}[1-exp((\mu_i+\mu_j)*h] - \frac{1}{2}\sum_i\sum_j A_iB_j\underset{\sim}{n}_{ai}^T\underset{\sim}{u}_{bj}[1-exp((\mu_i-\mu_j)*h] \quad (6.8)$$

which is a quadratic form of A_i and B_i. By taking the partial differential with respect to A_i and B_i to zero, the resulted simultaneous equations can be used to solve these unknown constants. Note that the adjoint simplectic ortho-normalisation relation can also be expressed as

$$\underset{\sim}{n}_{ai}^T\underset{\sim}{u}_{aj} = \underset{\sim}{u}_{ai}^T\underset{\sim}{n}_{aj}, \quad \underset{\sim}{n}_{bi}^T\underset{\sim}{u}_{bj} = \underset{\sim}{u}_{bi}^T\underset{\sim}{n}_{bj}, \quad \underset{\sim}{u}_{ai}^T\underset{\sim}{n}_{bj} = \delta_{ij} + \underset{\sim}{n}_{aj}^T\underset{\sim}{u}_{bi} \quad (6.9)$$

so that the variational equation (6.7) will be symmetric.

However, the eigen-solutions are complex and the equations thus derived will be solved in complex arithmetic. However the original problem is in real expressions; it is hopefully to return in real arithmetic. Note that if (μ, Y) is a complex eigen-solution, then its complex conjugate ($\bar{\mu}$, $\bar{Y}$) will also be an eigen-solution. The complex conjugate solution is implicit in the sequences $\underset{\sim}{Y}_{ai}$ and $\underset{\sim}{Y}_{bi}$ (i=1,2,...). So if the subscripts i_1 and i_2 are complex conjugate eigen-solutions, then

$$\underset{\sim}{Y}_{ai1} = \bar{\underset{\sim}{Y}}_{ai2}, \quad \text{so} \quad X_{ai1} = \bar{X}_{ai2} \quad (6.10)$$

From equation (6.4) it gives

$$\mu_{i1} = \bar{\mu}_{i2}, \quad A_{i1} = \bar{A}_{i2} \quad \text{also} \quad B_{i1} = \bar{B}_{i2} \quad (6.11)$$

For returning real arithmetic, the case of complex conjugate eigenvalue is listed as follows, where the subscript r means real vectors both from the real or imaginary parts of the eigenvectors.

$\mu_1 = -\alpha_1 + j\beta_1$, $A_1 = a_1 + ja_2$ $\underset{\sim}{n}_{a1} = \underset{\sim}{n}_{ar1} + j\underset{\sim}{n}_{ar2}$, $\underset{\sim}{u}_{a1} = \underset{\sim}{u}_{ar1} + j\underset{\sim}{u}_{ar2}$	$\mu_2 = -\alpha_1 - j\beta_1$, $A_2 = a_1 - ja_2$ $\underset{\sim}{n}_{a2} = \underset{\sim}{n}_{ar1} - j\underset{\sim}{n}_{ar2}$, $\underset{\sim}{u}_{a2} = \underset{\sim}{u}_{ar1} - j\underset{\sim}{u}_{ar2}$
$-\mu_1 = \alpha_1 - j\beta_1$, $B_1 = b_1 + jb_2$ $\underset{\sim}{n}_{b1} = \underset{\sim}{n}_{br1} + j\underset{\sim}{n}_{br2}$, $\underset{\sim}{u}_{b1} = \underset{\sim}{u}_{br1} + j\underset{\sim}{u}_{br2}$	$-\mu_2 = \alpha_1 + j\beta_1$, $B_2 = b_1 - jb_2$ $\underset{\sim}{n}_{b2} = \underset{\sim}{n}_{br1} - j\underset{\sim}{n}_{br2}$, $\underset{\sim}{u}_{b2} = \underset{\sim}{u}_{br1} - j\underset{\sim}{u}_{br2}$
$\mu_3 = -\alpha_3 + j\beta_3$, $A_3 = a_3 + ja_4$ $\underset{\sim}{n}_{a3} = \underset{\sim}{n}_{ar3} + j\underset{\sim}{n}_{ar4}$, $\underset{\sim}{u}_{a3} = \underset{\sim}{u}_{ar3} + j\underset{\sim}{u}_{ar4}$	$\mu_4 = -\alpha_3 - j\beta_3$, $A_4 = a_3 - ja_4$ $\underset{\sim}{n}_{a4} = \underset{\sim}{n}_{ar3} - j\underset{\sim}{n}_{ar4}$, $\underset{\sim}{u}_{a4} = \underset{\sim}{u}_{ar3} - j\underset{\sim}{u}_{ar4}$
$-\mu_3 = \alpha_3 - j\beta_3$, $B_3 = b_3 + jb_4$ $\underset{\sim}{n}_{b4} = \underset{\sim}{n}_{br3} + j\underset{\sim}{n}_{br4}$, $\underset{\sim}{u}_{b3} = \underset{\sim}{u}_{br3} + j\underset{\sim}{u}_{br4}$	$-\mu_4 = \alpha_3 + j\beta_3$, $B_4 = b_3 - jb_4$ $\underset{\sim}{n}_{b4} = \underset{\sim}{n}_{br3} - j\underset{\sim}{n}_{br4}$, $\underset{\sim}{u}_{b4} = \underset{\sim}{u}_{br3} - j\underset{\sim}{u}_{br4}$
....	....

(6.12)

Substituting all the equations into the homogeneous solution (6.1), it gives the equation for internal force as

$$\underset{\sim H}{n}=2\sum_{i=1,3,..}\Big\{exp(-\alpha_i x)\left[a_i(\underset{\sim ari}{n}cos\beta_i x - \underset{\sim ar(i+1)}{n}sin\beta_i x) - a_{i+1}(\underset{\sim ari}{n}sin\beta_i x + \underset{\sim ar(i+1)}{n}cos\beta_i x)\right] + exp(\alpha_i x)\left[b_i(\underset{\sim bri}{n}cos\beta_i x - \underset{\sim br(i+1)}{n}sin\beta_i x) - b_{i+1}(\underset{\sim bri}{n}sin\beta_i x + \underset{\sim br(i+1)}{n}cos\beta_i x)\right]\Big\} \quad (6.13)$$

and the equation for displacement has the same form as above, except that instead of vector $\underset{\sim}{n}$, the vector $\underset{\sim}{u}$ should be used.

Now the variational equation of the two end boundary conditions (6.7) can be substituted with the representation (6.13), where the variational parameters are a_i and b_i. Thus the simultaneous linear equation can be established, which has been in real form.

Based on the adjoint simplectic ortho-normalisation relation, it can be proved that the simultaneous equations are symmetric. For the solution expressed as (6.13), it satisfies all the equations except at the two ends. The equation (6.7) represents two end boundary conditions, so that the equations derived from which are the canonical equations in the structural mechanics, which must be symmetric. As a matter of fact, it is the consequence of Betti reciprocal theorem. Obviously, (6.7) gives the force method canonical equation.

7. The solution of second order PDE (2.1)

The first step is to find the eigen-solutions. Expanding the equation (3.3), it gives

$$\begin{aligned} -(b/2a)du/dy \quad & - \; n/a = \mu\, u \\ \left[-e+(c-b^2/4a)d^2/dy^2\right]u \quad & - \; (b/2a)dn/dy = \mu\, n \end{aligned} \quad (7.1)$$

The boundary condition is (2.3). To solve the above differential equation set (7.1), let

$$u(y) = u_a exp(\lambda y)\ , \qquad n(y) = n_a exp(\lambda y) \quad (7.2)$$

Substituting into (7.1), the determinant of the resulted simultaneous equations must be zero, which gives the characteristic equation of λ, after solving it

$$\lambda_{1,2} = \left[\, -b \pm sqrt[\mu^2(b^2-4ac)+4ce]\, \right] / 2c \quad (7.3)$$

hence $$u(y) = A_1 exp(\lambda_1 y) + A_2 exp(\lambda_2 y) \quad (7.4)$$

Substituting into the boundary conditions (2.3), the transcendental characteristic equation for μ is obtained as $exp(\lambda_1) + exp(\lambda_2) = 0$,

thus $$\lambda_2 = \lambda_1 + \imath 2\rho\pi, \qquad \text{where } \rho \text{ is an integer} \quad (7.5)$$

From (7.3) it gives $sqrt(\mu_\rho^2(b^2-4ac)+4ce)/c = \imath 2\rho\pi$, or

$$\mu_\rho^2 = 4(c^2\rho^2\pi^2+ce)/(4ac-b^2) \quad (7.6)$$

For the present problem, all the eigenvalues are real numbers. It verifies that μ_ρ and $-\mu_\rho$ are both eigenvalues. After the eigenvalues have been found, the eigenvectors can be obtained as

$$\left.\begin{array}{l} u_\rho(y) = A\ exp[-b\mu_\rho y/(2c)]sin(\rho\pi y) \\ n_\rho(y) = -A\ exp[-b\mu_\rho y/(2c)]\left[(\frac{b^2}{4c} - a)\mu_\rho sin(\rho\pi y) - \frac{b}{2}\rho\pi cos(\rho\pi y)\right] \end{array}\right\} \quad (7.7)$$

which compose $\underset{\sim}{\mathbf{Y}}_\rho(y)=\{u_\rho^T\ n_\rho^T\}^T$, and the adjoint eigenvector $\mathbf{L}_\tau$ is

$$\underset{\sim}{\mathbf{L}}_\tau(y)=\begin{Bmatrix} n_\tau \\ -u_\tau \end{Bmatrix}=Aexp[-b\mu_\tau y/(2c)]\begin{Bmatrix} (\frac{b^2}{4c} - a)\mu_\tau sin(\tau\pi y) - \frac{b}{2}\tau\pi cos(\tau\pi y) \\ -sin(\tau\pi y) \end{Bmatrix} \quad (7.8)$$

where in (7.8), μ_τ and τ can take negative values of μ_ρ and ρ in (7.7). The simplectic orthogonality of these eigenvectors can be directly verified by integrating $\underset{\sim}{\mathbf{L}}_\tau^T\underset{\sim}{\mathbf{Y}}_\rho$, i.e. when $\tau+\rho\neq 0$; after some tedious derivation, the integration equals 0. When $\tau+\rho=0$, it gives

$$\underset{\sim}{\mathbf{L}}_{-\rho}^T\underset{\sim}{\mathbf{Y}}_\rho = A^2\ sqrt[(\pi^2\rho^2+e/c)*(4ac-b^2)]/2 \quad (7.9)$$

thus the normality condition gives

$$A = sqrt(2/sqrt[(\pi^2\rho^2+e/c)*(4ac-b^2)]) \quad (7.10)$$

Up to here the orthonormalised eigenvector basis has been obtained.

Now the two end boundary conditions can be expressed as

$$u = u_{0*}(y) \text{ or } u_{h*}(y) \qquad \text{when } x=0 \text{ or } x=h \quad (7.11)$$

The solution can be given as

$$\underset{\sim}{\mathbf{v}} = \sum_\rho [\ X_{a\rho}(x)*\underset{\sim}{\mathbf{Y}}_{a\rho}(y) + X_{b\rho}(x)*\underset{\sim}{\mathbf{Y}}_{b\rho}(y)\ ] \quad (7.12)$$

where $\underset{\sim}{\mathbf{Y}}_{a\rho}(y)$ is $\underset{\sim}{\mathbf{Y}}_\rho(y)$, and $\underset{\sim}{\mathbf{Y}}_{b\rho}(y)$ is $\underset{\sim}{\mathbf{Y}}_{-\rho}(y)$ in the equation (7.7).

To get some impression to the eigenfuncion-vectors, the curves of $u_\rho(y)$, $n_\rho(y)$ are plotted in fig.1 for the case of a=c=b=e=1.0.

8. Plate bending problem in strip domain

The plate bending problem in a strip domain is solved for illustration purpose. As being well-known, if both the support conditions for the opposite edges are simply supported, then the plate bending problem can be solved by the Levy single trigonometrical series expansion method[14], which means the variables have been separated. If either of the opposite edges is not simply supported such as both edges clamped, then the usual technique for separation of variables does not work. The bi-orthogonal expansion method can also work for plate bending problems[15]; however the Hamiltonian system method gives more general development. For simplicity, only the homogeneous equation is considered. The plate bending equation is

$$D\ [\partial^2/\partial x^2 + \partial^2/\partial y^2]^2\ w = 0 \quad (8.1)$$

The domain is $\mathcal{D}$: $0 < x < h, \quad 0 < y < 1$ (8.2)

The boundary conditions are

$$\partial w/\partial y=0, \quad \partial^3 w/\partial y^3 = 0\ , \qquad \text{when } y=0 \quad (8.3)$$

$$w=0, \quad \partial w/\partial y = 0\ , \qquad \text{when } y=1 \quad (8.4)$$

$$w,\ \partial w/\partial x = \text{given functions} \qquad \text{when } x=0 \text{ or } x=h \quad (8.5)$$

where, instead of the clamped boundary condition at y=-1, the symmetric condition is given at y=0, i.e. (8.3).

The variational principle can be written as

$$\int_0^h\int_0^1 \frac{D}{2}\left[\left(\frac{\partial^2}{\partial x^2}+\frac{\partial^2}{\partial y^2}\right)w\right]^2 dy\,dx = \min \qquad (8.6)$$

To reformulate the variational principle, let us introduce the state functions, i.e. w and the rotation

$$\theta = \partial w/\partial x \qquad (8.7)$$

Introducing the Lagrange parameter function q (the shearing force q_x), the variational equation becomes

$$\delta\int_0^h\int_0^1\left[\frac{D}{2}\left(\partial\theta/\partial x+\partial^2 w/\partial y^2\right)^2 - q\left(\partial w/\partial x-\theta\right)\right]dy\,dx = 0 \qquad (8.8)$$

where θ and w simulate the state function. Now introducing the moment further

$$m = -D\left(\partial\theta/\partial x+\partial^2 w/\partial y^2\right) \qquad (8.9)$$

the variational equation becomes further

$$\delta\int_0^h\int_0^1\left[-(q\dot{w}+m\dot{\theta}) + q\theta - m\frac{\partial^2 w}{\partial y^2} - m^2/(2D)\right]dy\,dx = 0 \qquad (8.10)$$

Instead of the boundary condition (8.3), it gives

$$\partial w/\partial y=0,\qquad \partial m/\partial y=0,\qquad \text{when}\quad y=0 \qquad (8.11)$$

Now the displacement function vector $\underset{\sim}{u}$ and the internal force function vector $\underset{\sim}{n}$ can be introduced as

$$\underset{\sim}{u} = \begin{Bmatrix} w \\ \theta \end{Bmatrix},\qquad \underset{\sim}{n} = \begin{Bmatrix} q \\ m \end{Bmatrix} \qquad (8.12)$$

The whole state function vector $\underset{\sim}{v}$ can be composed as (2.21). The variational equation gives

$$\begin{Bmatrix} \dot{w} \\ \dot{\theta} \\ \dot{q} \\ \dot{m} \end{Bmatrix} = \begin{bmatrix} 0 & 1 & 1 & 0 \\ -(\partial^2/\partial y^2) & 0 & 0 & -1/D \\ 0 & 0 & 0 & (\partial^2/\partial y^2) \\ 0 & 0 & -1 & 0 \end{bmatrix}\begin{Bmatrix} w \\ \theta \\ q \\ m \end{Bmatrix} \qquad (8.13)$$

Now introducing the operators

$$\underset{\sim}{Q} = \underset{\sim}{0},\qquad \underset{\sim}{F} = \begin{bmatrix} 0 & 1 \\ -(\partial^2/\partial y^2) & 0 \end{bmatrix},\qquad \underset{\sim}{G} = \begin{bmatrix} 0 & 0 \\ 0 & 1/D \end{bmatrix} \qquad (8.14)$$

the Hamiltonian operator matrix is thus composed according to equation (2.16), where

$$\langle \underset{\sim}{n}^T, \underset{\sim}{F}, \underset{\sim}{u}\rangle = \langle \underset{\sim}{u}^T, \underset{\sim}{F}^T, \underset{\sim}{n}\rangle,\qquad \text{thus}\quad \underset{\sim}{F}^T = \begin{bmatrix} 0 & -(\partial^2/\partial y^2) \\ 1 & 0 \end{bmatrix}$$

can be verified by integration by parts and boundary conditions. It is easily recognized that the variational equation (8.10) is in the form of (2.12).

The PDE set (8.13), i.e. (2.15), can be solved by using the separate variables equation (3.1); the corresponding eigen-equation is (3.3). It is best to solve the eigen-equation

$$\underset{\sim}{M}\ \underset{\sim}{Y} = \lambda\ \underset{\sim}{Y} \tag{8.15}$$

of the square Hamiltonian operator matrix[16,11]

$$\underset{\sim}{M} = \underset{\sim}{H}^2 = \begin{bmatrix} \underset{\sim}{F}^2 + \underset{\sim}{G}\underset{\sim}{Q} & -\underset{\sim}{F}\underset{\sim}{G} + \underset{\sim}{G}\underset{\sim}{F}^T \\ -\underset{\sim}{Q}\underset{\sim}{F} + \underset{\sim}{F}^T\underset{\sim}{Q} & \underset{\sim}{F}^{2T} + \underset{\sim}{Q}\underset{\sim}{G} \end{bmatrix} \tag{8.16}$$

first. To the present problem, the equation (8.16) is

$$\begin{bmatrix} -(\partial^2/\partial y^2) & 0 & 0 & -1/D \\ 0 & -(\partial^2/\partial y^2) & 1/D & 0 \\ 0 & 0 & -(\partial^2/\partial y^2) & 0 \\ 0 & 0 & 0 & -(\partial^2/\partial y^2) \end{bmatrix} \begin{Bmatrix} w \\ \theta \\ q \\ m \end{Bmatrix} = \lambda \begin{Bmatrix} w \\ \theta \\ q \\ m \end{Bmatrix} \tag{8.17}$$

The adjoint eigen-solutions of $\underset{\sim}{H}$ with eigenvalues μ and $-\mu$ must be the linear combinations of the two eigen-solutions of $\underset{\sim}{M}$ with eigenvalue $\lambda=\mu^2$. The two eigen-solutions of (8.17) can be expressed as

(1) $\{ w_1, \theta_1, 0, 0 \}^T$, where $d^2w_1/dy^2+\mu^2w_1=0$, $d^2\theta_1/dy^2+\mu^2\theta_1=0$,

(2) $\{ w_2, \theta_2, q_2, m_2\}^T$, where $d^2q_2/dy^2+\mu^2q_2=0$, $d^2m_2/dy^2+\mu^2m_2=0$,

$$d^2w_2/dy^2+\mu^2w_2 = -m_2/D, \qquad d^2\theta_2/dy^2+\mu^2\theta_2 = q_2/D, \tag{8.18}$$

Note that the solution (1) can be included in solution (2), of which the general solution can be expressed as

$$\left.\begin{aligned} &q_2 = A_q\cos\mu y + B_q\sin\mu y \qquad m_2 = A_m\cos\mu y + B_m\sin\mu y \\ &w_2 = A_w\cos\mu y + B_w\sin\mu y - (A_m/2\mu D)y\sin\mu y + (B_m/2\mu D)\cos\mu y \\ &\theta_2 = A_s\cos\mu y + B_s\sin\mu y + (A_q/2\mu D)y\sin\mu y - (B_q/2\mu D)\cos\mu y \end{aligned}\right\} \tag{8.19}$$

Now from the boundary condition (8.11), $B_w=B_s=0$, $B_m=B_q=0$. By returning the eigenequation of $\underset{\sim}{H}$, it gives

$$A_s=\mu A_w, \qquad A_q=-\mu A_m \tag{8.20}$$

Substituting w_2 into boundary condition (8.4), it gives

$$\left.\begin{aligned} \cos\mu\ A_w - \sin\mu\ (A_w/2\mu D) &= 0 \\ -\mu\sin\mu\ A_w - (\sin\mu-\mu\cos\mu)\ (A_m/2\mu D) &= 0 \end{aligned}\right\} \tag{8.21}$$

The determinant of the above equation must be zero; the transcendental equation for the eigenvalues is obtained as

$$\mu = -\sin\mu\cos\mu\ , \qquad \text{or} \qquad -\sin(2\mu) = (2\mu) \tag{8.21'}$$

The above equation has only complex conjugate roots except the useless root 0. Obviously, μ and $-\mu$ are roots simultaneously, as it must be for the Hamiltonian operator matrix $\underset{\sim}{H}$. The roots can be expressed as

$$\mu = \pm\ \alpha \pm \beta\iota \tag{8.22}$$

where it can be solved

α = $\pi/2$+0.5345	$3\pi/2$+0.6439	$5\pi/2$+0.6872	$7\pi/2$+0.7036	$9\pi/2$+0.7169
β = 1.1254	1.5516	1.7755	1.9294	2.04685

The corresponding eigen-function vector is

$$\mathbf{Y} = \begin{Bmatrix} w \\ \theta \\ q \\ m \end{Bmatrix} = \begin{Bmatrix} (\sin\mu\cos\mu y - y\cos\mu\sin\mu y)/\mu \\ (\sin\mu\cos\mu y - y\cos\mu\sin\mu y) \\ -2\mu D\cos\mu\cos\mu y \\ 2D\cos\mu\cos\mu y \end{Bmatrix} \tag{8.23}$$

Because μ is complex so that the eigen-function vector is also complex, but from which the real vectors $\tilde{n}_{ar}$, $\tilde{u}_{ar}$, $\tilde{n}_{br}$, $\tilde{u}_{br}$ can be extracted easily, such as

$$\tilde{n}_{ari} = Re\begin{Bmatrix} -2D\mu_i\cos\mu_i\cos\mu_i y \\ 2D\cos\mu_i\cos\mu_i y \end{Bmatrix}, \qquad \tilde{n}_{ar(i+1)} = Im\begin{Bmatrix} -2D\mu_i\cos\mu_i\cos\mu_i y \\ 2D\cos\mu_i\cos\mu_i y \end{Bmatrix} \tag{8.24}$$

and the same procedure for $\tilde{u}$. The subscript a means the eigen-solution decaying when $x\to\infty$, and the μ_i selects the negative α_i and positive β_i; the subscript b is for α_i positive and β_i negative.

The isotropic plate bending is symmetric for reflection transformation, when the adjoint eigenvalue $\mu_{bi}=-\mu_i$ substitutes μ_i, it gives

$$w_{ai}=w_{bi}, \qquad \theta_{ai}=-\theta_{bi}, \qquad q_{ai}=-q_{bi}, \qquad m_{ai}=m_{bi} \tag{8.25}$$

for the adjoint eigenfunctions. To such kinds of problems the semi-analytical method can keep the main eigen-solutions in the subspace unchanged. The general anisotropic plate bending has no such characteristics.

So far the homogeneous solution has been considered. For four edges clamped strip plate under uniform lateral pressure g, the solution can be the summation of an inhomogeneous solution $\tilde{v}_0$ and a homogeneous solution $\tilde{v}_H$. The inhomogeneous special solution is easily obtained as $\tilde{v}_0=\{w_0, \theta_0, q_0, m_0\}^T$, where

$$w_0=(1-y^2)^2\times g/(24\times D), \qquad \theta_0=0, \qquad q_0=0, \qquad m_0=4(3y^2-1)\times g \tag{8.26}$$

The end boundary conditions for the homogeneous solution will be

$$\theta_H=0, \qquad w_H=-w_0 \qquad \text{when } x=0 \text{ or } x=h \tag{8.27}$$

Now the variational equation (6.7) can be used to establish the canonical simultaneous linear equations to solve the unknown constants a_i and b_i. The flowchart can be described as

1. **From equation (8.21'), solve a number of eigenvalues.**
2. **Find the corresponding real eigenfunction vectors by (8.24).**
3. **Based on the variational equation (6.7), establish the canonical simultaneous equations, and then solve a_i and b_i.**
4. **Based on (6.13), compute the internal forces and displacements, and then superpose the inhomogeneous solution.**

As an example, the semi-infinitely long clamped strip plate under uniform lateral load g is solved. The end conditions are

$$\left.\begin{aligned} &w_H=-(1-y^2)^2g/(24D), \qquad \theta_H=0 \quad \text{when } x=0 \\ &w_H\to 0, \qquad \theta_H\to 0 \qquad \text{when } x\to\infty \end{aligned}\right\} \tag{8.28}$$

Thus the decaying condition dictates that $b_i=0$ (i=1,2,...). To solve a_i, let us select μ_i (i=1,...,5). As the eigenvalues have been listed before, the corresponding eigen-functions can be found from (8.23) and (8.24). Now the canonical simultaneous equations are

$$\underset{\sim}{C}\,\underset{\sim}{a} = \underset{\sim}{p} \tag{8.29}$$

where the matrix $\underset{\sim}{C}$ is

$$\begin{bmatrix}
0.9629 & & & & & & & & & \\
0.0 & 0.9629 & & & & & \text{SYMMETRY} & & & \\
0.4873 & 0.8529 & 1.7724 & & & & & & & \\
-.1143 & 0.6323 & 0.0 & 1.7724 & & & & & & \\
0.2704 & 0.7001 & 1.0603 & 1.3982 & 2.4460 & & & & & \\
-.0873 & 0.3203 & -.4848 & 1.1766 & 0.0 & 2.4460 & & & & \\
0.1182 & 0.5653 & 0.5470 & 1.2778 & 1.5656 & 1.7582 & 3.0691 & & & \\
-.0545 & 0.1935 & -.4340 & 0.6197 & -.8020 & 1.6537 & 0.0 & 3.0691 & & \\
0.1467 & 0.4697 & 0.3431 & 1.0860 & 0.8085 & 1.6847 & 2.0426 & 2.0607 & 3.6619 & \\
-.0355 & 0.1302 & -.3155 & 0.3843 & -.7703 & 0.8740 & -1.0872 & 2.1126 & 0.0 & 3.6619
\end{bmatrix}$$

and the righthand vector is

$$\{-.0599\ 0.1103\ 0.0692\ 0.0746\ .0607\ 0.0371\ 0.0493\ .0218\ .0408\ .0143\}^T$$

It solves the a_i as

$$\{-.06212\ .11842\ -.0016\ -.0032\ -.0005\ -.0023\ .0013\ -.0001\ .0000\ .0002\}^T$$

Now the displacement and moment distribution can be computed. The end moment distribution is plotted in fig.2, and the midspan moment is 0.2277gD. According to [14], it is 0.2284gD, which is within the range of precision.

9. Concluding remarks

The above plate bending problem is solved analytically, for the equation is simple enough. However in various applications, numerical method is unavoidable. Hence the Hamiltonian operator matrix must be reduced to 2n dimensional Hamiltonian matrix to carry out the numerical procedure, which requires discretization. It is very important to select appropriately the discretization basis functions.

Based on [10,11], the selection of the subspace should be in the whole state space, but not in state space only such as in the traditional semi-analytical method[6], which is usually derived from the minimum potential energy variational principle. Because selecting subspaces from the whole state space can include the dominant eigen-solutions of the original problem in the subspace, even if the subspace selected is of limited dimension, for which the potential energy variational principle based traditional semi- analytical method cannot attain for general case.

The analogy theory brings the state space method, optimal control and Hamiltonian system theory into the theory of elasticity and partial differential equations, from which the problems in a prismatic domain can be solved by the separation of variables method. The present paper introduces a new methodology, which far extends the range of traditional method of separation of variables. The Hamiltonian operator matrix eigen-solution expansion breaks the limitation of self-adjoint Sturm-Liouville problem, and leads the problem to simplectic geometric system instead of Euclidean.

Numerically, it is almost impossible to achieve very precise solution for one try of transverse discretization. The adjoint simplectic subspace inverse substitution algorithm[10,11] had proposed iterative method for dominant eigen-solutions, but it is for fixed 2n dimensional problems. For PDEs, the recently developed finite element adaptive mesh[7] method can be used, which can help to select better mesh and more precise basis functions with iterations. One must be aware of the numerical integration by finite difference scheme instead of differential equations; the simplectic property of the operator must be preserved carefully, otherwise the error will be increasing very rapidly[17,18].

The present paper is only a beginning exploration on this way, more research is anticipated.

References

[1] *Zhong Wanxie* and *Zhong Xiangxiang*: Computational structural mechanics optimal control and semi-analytical method for PDE. Computers and structures, 1990, vol.37, No.6, pp.993-1004.

[2] *Zhong Wanxie* and *Zhong Xiangxiang*: Elliptic Partial Differential equation and optimal control. Accepted to publish on Numerical Method for Partial Differential Equations.

[3] *Zhong Wan-xie, Lin Jia-hao, Qiu Chun-hang*: The Eigenvalue Problem of the Chain of Identical Substructures and the Expansion Method Solution Base on the Eigenvectors. Acta Mechanica Sinica, vol.23, pp.72-81, 1991, (in Chinese)

[4] *Zhong Wanxie, Cheng Gengdong*: Regularization of singular control and stiffness shifting method. To publish on Information and Control. (in Chinese).

[5] *Zhong Wanxie, Lin Jiahao, Qiu Chunhang*: Computational structural mechanics and optimal control--The simulation of structural chain theory to linear quadratic optimal control problems--. To be published on Intern. J. Num. Meth. Eng.

[6] *L.V. Kantorowich and V.E. Kryloff*: Approximate method of higher analysis. 1962, (in Russian), English translation, Groningen, Noordhoff, 1958.

[7] *O. C. Zienkiewicz*: The finite element method. McGraw-Hill, NY, 1977.

[8] *R. Courant, D. Hilbert*: Methods of mathematical physics. vol.I, Interscience. New York, 1953.

[9] *P. N. Paraskevopoulos*: Techniques in model reduction for large scale systems. Control and dynamic systems, (Ed. C.T. Leondes) vol.23, pp.165-193, Academic Press, New York, 1986.

[10] *Zhong wanxie, Zhong Xiangxiang*: On the adjoint simplectic subspace iteration method for the dominant eigen-solutions of the Hamiltonian matrix. To publish on Science progress (in Chinese)

[11] *Zhong Wanxie, Zhong Xiangxiang*: On the adjoint simplectic inverse substitution method for main eigen-solutions of large Hamiltonian matrix. Accepted to publish on Journ. System Eng.

[12] *Hu Haichang*: Variational Principles of theory of elasticity with applications. Science Press, Beijing, 1984.

[13] *E. T. Whittaker*: A Treatise on the analytical dynamics. Cambridge University Press, 4th ed., 1960.

[14] *S. P. Timoshenko, S. Woinowsky-Krieger*: Theory of plates and shells. McGraw-Hill, NY, 1959.

[15] *B.S.R. Rao, B.G. Prakash*: Generalized orthogonality relations for the flexure of rectangular plates and a cantilever problem for semi-infinite plates. J. Struct. Mech. vol.9, pp.51-70, 1981.

[16] *C. F. Van Loan*: A simplectic method for approximating all the eigenvalues of a Hamiltonian matrix. Linear Algeb. and Appl., 1984, vol.61, pp.233-251.

[17] *Feng Kang*: Difference schemes for Hamiltonian formalism and simplectic geometry. J. Comp. Math. vol.4, pp.279-289, 1986.

[18] *Feng Kang, Wu Hua-mo, Qin Meng-zhao, Wang Dao-liu*: Construction of canonical difference schemes for Hamiltonian formalism via generating functions. J. Comp. Math. vol.7, no.1, 1989.

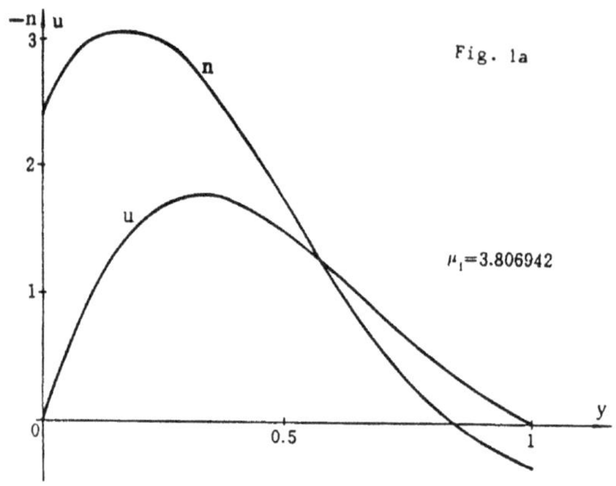

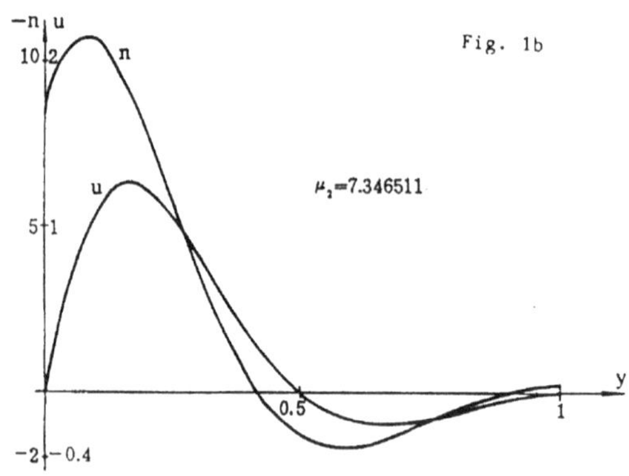

Fig. 1 The eigenfunction-vectors

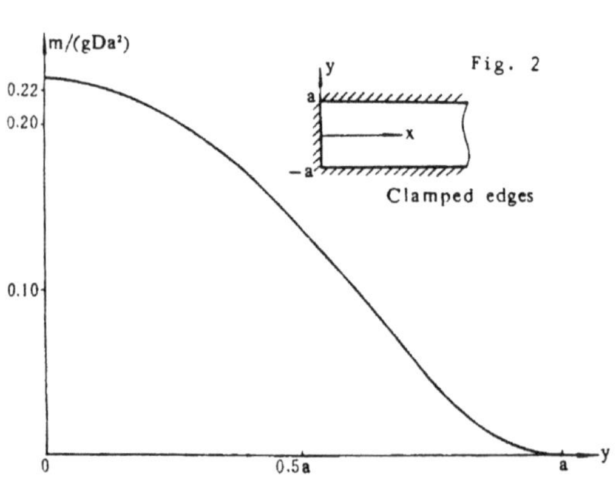

Fig. 2. The end moment distribution

GENERALIZED OPTIMAL ACTIVE CONTROL ALGORITHM OF SEISMIC STRUCTURES AND RELATED SOIL-STRUCTURE FORMULATION

Franklin Y. Cheng
Curators' Professor
Peter Tian and Sukit Suthiwong
Ph.D. Candidates
Civil Engineering Department
University of Missouri-Rolla
Rolla, MO 65401 USA

ABSTRACT

The paper presents an optimal active control algorithm for seismic structures. By using an intelligent control device in the feedback gain matrix, the algorithm can be automatically adjusted, taking account of actual earthquakes on site, to provide effective controlling results. Numerical examples are provided to show the influences of the gain matrix parameters on structural response and to show the advantages of using this technique over other approaches currently in vogue. The presentation also includes a brief review of structural optimization related to active control as well as active control with soil-structure interaction.

INTRODUCTION

Optimal control as applied to civil engineering structures affords protection from the damaging effects of destructive seismic force or from human discomfort over structural motion induced by strong wind and other types of vibrations [1,2,3,4]. Optimal structural control can be achieved by using passive or active control devices or a combination of both. Passive devices utilize the fact that an energy dissipating mechanism can be activated by the motion of the structure itself.

Active control devices require external energy for their operation. In addition to the power they need, their maintenance and per-unit cost determine the effectiveness and economics of a protection system. In this light, studies on optimal number and optimal placement of active control devices are significant and have been reported elsewhere by Cheng and his associate, among others [5].

Recent studies also show the importance of combining structural optimization with active control [6,7]. Structural optimization allows one to design and build a structure to satisfy an economical and serviceable objective, which may include cost, performance, safety, service life, maintenance, or future adaptability [8]. Active control structures, however, are not designed on the basis of an optimization algorithm to satisfy design constraints and objective functions. Combining structural optimization with optimal control gives access to the strong points of both techniques [4,9,10,11]. A review of seismic structural optimization may be found in articles by Cheng [10] and of active controls by Soong [12,13].

For active control seismic structures, an algorithm currently in vogue is the instantaneous optimal active control (IOAC) [3,14] for which excitation can be measured real-time on-line by acceleration sensors on basement floors. Current studies indicate that the algorithm is time-dependent on whatever incremental time intervals are used in the response analysis. Using different time intervals yields various control forces and structural responses of a given structure subjected to the same earthquake. Thus, control effectiveness cannot be ensured for a structure subjected to different earthqukes during its lifetime.

This paper presents an algorithm called generalized intelligent optimal active control (GOAC) which is able to adjust the feedback gain matrix to achieve better controllability. It is suitable for real-time on-line sensors but not time-dependent on incremental intervals. Weighting elements in the gain matrix can be chosen by taking account of actual earthquakes on site, which then provides effective controlling results by adjusting stiffness or damping or a combination of both. The presentation includes mathematical formulations and numerical results as well as further consideration of soil-structure interaction.

STATE-VECTOR FOR ACTIVE TENDON CONTROL

The motion equation of the active tendon control shown in Fig. 1 may be expressed as

$$[M]\{\ddot{x}(t)\}+[C]\{\dot{x}(t)\}+[K]\{x(t)\}=[\gamma]\{u(t)\}+\{\delta\}\ddot{x}_g(t) \tag{1}$$

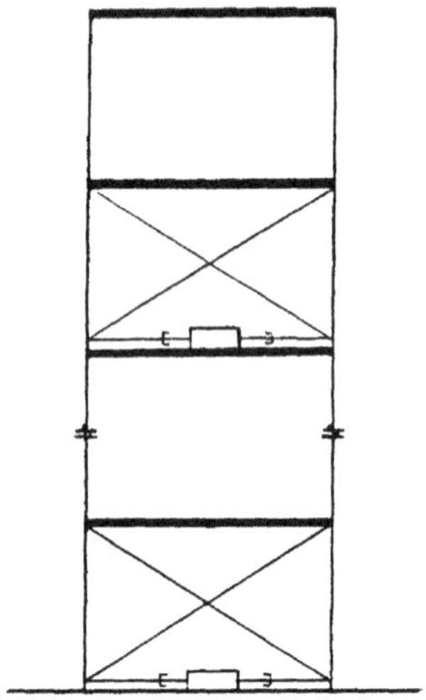

Figure 1. Active tendon control model

where [M], [C], and [K] of (NxN) are mass, damping, and stiffness matrices, respectively; {u(t)} of (rx1) is the control force vector, where r = number of controllers; $[\gamma]$ of (Nxr) is the matrix relating to the location of the controllers on the structure; and $\{\delta\}$ of (Nx1) is a coefficient vector for the earthquake excitation, $\ddot{x}_g(t)$. By defining the state-vector

$$\{Z(t)\}=\left\{\frac{\{x(t)\}}{\{\dot{x}(t)\}}\right\} \tag{2}$$

then Eqs. (1) and (2) can be combined as

$$\left\{\frac{\{\dot{x}(t)\}}{\{\ddot{x}(t)\}}\right\}=\left[\frac{[0]}{-[M]^{-1}[K]}\middle|\frac{[I]}{-[M]^{-1}[C]}\right]\left\{\frac{\{x(t)\}}{\{\dot{x}(t)\}}\right\}+\left[\frac{[0]}{[M]^{-1}[\gamma]}\right]\{u(t)\}+\left\{\frac{\{0\}}{[M]^{-1}\{\gamma\}}\right\}\ddot{x}_g(t) \tag{3}$$

In compact form, Eq. (3) becomes

$$\{\dot{Z}(t)\}=[A]\{Z(t)\}+[B]\{u(t)\}+\{C\}\ddot{x}_g(t) \tag{4}$$

TRANSVERSALITY CONDITION

Consider a system governed by Eq. (1) with end conditions given by

$$\left.\begin{array}{l} t_{i-1}=t_o+(i-1)\Delta t \\ \{Z(t_{i-1})\}=\{Z_{i-1}\} \\ t_i=t_o+(i)\Delta t \end{array}\right\} \tag{5}$$

where

t_o = initial time-instant;

$\Delta t = \frac{1}{N}(t_f - t_o)$, i = 1,2,...,N;

t_f = final time-instant; and

$\{Z_{i-1}\}$ = the value of $\{Z(t)\}$ at t_{i-1}.

Equation (5) can be written in the following vector form,

$$\{\Omega\}=\left\{\begin{array}{c}\Omega_1\\ \{\Omega_2\}\\ \Omega_3\end{array}\right\}=\left\{\begin{array}{l}(t_{i-1}-t_o)-(i-1)\Delta t\\ \{Z(t_{i-1})\}-\{Z_{i-1}\}\\ (t_i-t_o)-(i)\Delta t\end{array}\right\}=\{0\} \tag{6}$$

Let the performance index be expressed as

$$\begin{aligned} J_i&=\frac{1}{2}\{Z(t_i)\}^T[S]\{Z(t_i)\}+\frac{1}{2}\int_{t_{i-1}}^{t_i}(\{Z(t)\}^T[Q]\{Z(t)\}+\{u(t)\}^T[R]\{u(t)\})dt\\ &=g+\int_{t_{i-1}}^{t_i}\bar{f}(t)\,dt \end{aligned} \tag{7}$$

By introducing multipliers $\{\mu\}$ and $\{\lambda(t)\}$ and forming

$$G = g + \{\mu\}^T\{\Omega\} \tag{8}$$

$$\begin{aligned} F &= \bar{f} + \{\lambda(t)\}^T([A]\{Z(t)\}+[B]\{u(t)\}+\{C\}\ddot{x}_g(t)-\{\dot{Z}(t)\})\\ &= \bar{f} + \{\lambda(t)\}^T\{f(t)\} \end{aligned} \tag{9}$$

then the transversality condition can be expressed as

$$dG-\left(\left\{\frac{\partial F}{\partial\{\dot{Z}(t)\}}\right\}^T\{\dot{Z}(t)\}-F\right)dt\Bigg|_{t_{i-1}}^{t_i}+\left\{\frac{\partial F}{\partial\{\dot{Z}(t)\}}\right\}^T d\{Z(t)\}\Bigg|_{t_{i-1}}^{t_i}=0 \tag{10}$$

Applying Eq. (10) to Eqs. (8) and (9) yields

$$[S]\{Z(t_i)\}-\{\lambda(t_i)\}=\{0\} \tag{11}$$

SOLUTION TECHNIQUE

For a closed-loop control system, the relation between the state vector {Z(t)} and the control force vector can be given by

$$\{u(t)\}=[G]\{Z(t)\} \tag{12}$$

where [G] is called feedback gain matrix. Then, the motion equation of Eq. (4) can be changed as follows

$$\{\dot{Z}(t)\}=[D]\{Z(t)\}+\{C\}\ddot{x}_g(t) \tag{13}$$

where

$$[D]=[A]+[B][G] \tag{14}$$

Let {Z(t)} be expressed in terms of the modal transformation matrix, [T], of the plant matrix, [D], i.e.,

$$\{Z(t)\}=[T]\{\psi(t)\} \tag{15}$$

and [T] has the following property

$$[T]^{-1}[D][T]=[\phi] \tag{16}$$

Substituting Eq. (15) into Eq. (13) and premultiplying by $[T]^{-1}$ yields

$$\{\dot{\psi}(t)\}=[\phi]\{\psi(t)\}+\{\Gamma(t)\} \tag{17}$$

where

$$\{\Gamma(t)\}=[T]^{-1}\{C\}\ddot{x}_g(t) \tag{18}$$

Assuming the initial conditions are

$$\{\psi(0)\}=\{0\},\quad \{u(0)\}=\{0\},\ and\ \ddot{x}_g(0)=0$$

the solution of Eq. (18) can be given by

$$\{\psi(t)\}=\int_0^t \exp([\phi](t-\tau)\{\Gamma(\tau)\}d\tau \tag{19}$$

which can be performed by numerical integration. After {Ψ(t)} is determined, {Z(t)} and {u(t)} can be obtained by using Eqs. (15) and (12), respectively.

COMMENTS ON GOAC ALGORITHM

By employing Euler equation to Eq. (9), the following characteristic equations can be obtained

$$[Q]\{Z(t)\}+[A]^T\{\lambda(t)\}+\{\dot{\lambda}(t)\}=\{0\} \tag{20}$$

$$[R]\{u(t)\}+[B]^T\{\lambda(t)\}=\{0\} \tag{21}$$

Eq. (21) can be rewritten as

$$\{u(t)\}=-[R]^{-1}[B]^T\{\lambda(t)\} \tag{22}$$

At each end point $t = t_i$ ($i = 1,2,\ldots N-1$), Eqs. (22) and (12) can be written as

$$\{u(t_i)\}=-[R]^{-1}[B]^T\{\lambda(t_i)\} \tag{23}$$

$$\{u(t_i)\}=[G(t_i)]\{Z(t_i)\} \tag{24}$$

Substituting transversality condition Eq. (11) into Eq. (23) yields

$$\{u(t_i)\}=-[R]^{-1}[B]^T[S]\{Z(t_i)\} \tag{25}$$

By comparing Eq. (25) with Eq. (24), the feedback gain matrix at each end point t_i is obtained as

$$[G(t_i)]=-[R]^{-1}[B]^T[S] \tag{26}$$

It is noted from Eq. (26) that $[G](t_i)]$ is actually a constant matrix. Hence, Eq. (21) can be written as

$$[G]=-[R]^{-1}[B]^T[S] \tag{27}$$

which is valid at every end point t_i. Note that it is neither a function of time t_i nor a function of time increment Δt. Therefore, during the computation process, Δt can be arbitrarily changed within the range of precision. It can also be noted that if [S] is chosen to be the Algebraic Ricatti Matrix [P], i.e., [S] = [P], the feedback gain matrix can be written as

$$[G]=-[R]^{-1}[B]^T[P] \tag{28}$$

that is the same as the gain matrix of the Ricatti closed-loop control algorithm. Therefore, the Ricatti closed-loop control algorithm is actually included in this generalized optimal control algorithm. Expanding Eq. (27)

$$\begin{aligned}[G]_{rx2N}&=-[R]^{-1}_{rxr}\begin{bmatrix}[0]_{Nxr}\\ [M]^{-1}_{NxN}[\gamma]_{Nxr}\end{bmatrix}^T\begin{bmatrix}[S_{11}]_{NxN} & [S_{12}]_{NxN}\\ [S_{12}]_{NxN} & [S_{22}]_{NxN}\end{bmatrix}\\ &=-\left[[R]^{-1}[\gamma]^T[M]^{-1}[S_{12}] \quad [R]^{-1}[\gamma]^T[M]^{-1}[S_{22}]\right]\end{aligned} \tag{29a}$$

the following expression for the control force {u(t)} can be obtained as

$$\{u(t)\}=-[R]^{-1}[\gamma]^{T}[M]^{-1}([S_{12}]\{x(t)\}+[S_{22}]\{\dot{x}(t)\}) \tag{29b}$$

Studies of [G] lead to two conclusions:

(1) One can construct matrix [S] to fit the requirement for the gain matrix [G]. For example, if the period of the earthquake is small, the velocity of the structure will be larger than its displacement. In this case, a diagonal [S] matrix could be chosen to produce a large control force. On the other hand, if the excitation period is large, the velocity of the structure will be smaller. In this case, a nondiagnal [S] matrix should be employed.

(2) Since the excitation period cannot be predicted before a real earthquake, an intelligent program could be used to change the elements in [S] during the control procedure so as to obtain the required control force.

SOIL-STRUCTURE INTERACTION

A multi-story building equipped with an active tendon system of Fig. 2 sits on a flexible mat foundation which is embedded in an unbounded soil. The soil consists of horizontal layers resting on a half-space, both consisting of isotropic viscoelastic material with hysteretic damping. The soil properties vary with depth but remain constant within the individual layers. If the soil around the surface of the foundation is irregular, the analysis can still be performed by introducing artificial interface.

In substructure method of Figure 3, the total system is divided into two subsystems by the artificial interface. The first, called structure system, consists of the structure and foundation, including the irregular soil region adjacent to it which will be analyzed on the basis of time-domain procedure. The second subsystem, called ground system, is the unbounded soil with excavation. Since the properties of this unbounded domain are frequency-dependent, the boundary-element method in frequency domain is appropriately applied [15].

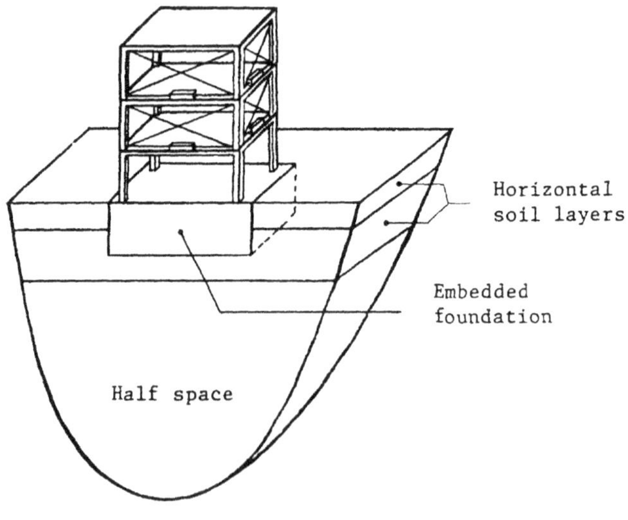

Figure 2. Soil-structure interaction with active control

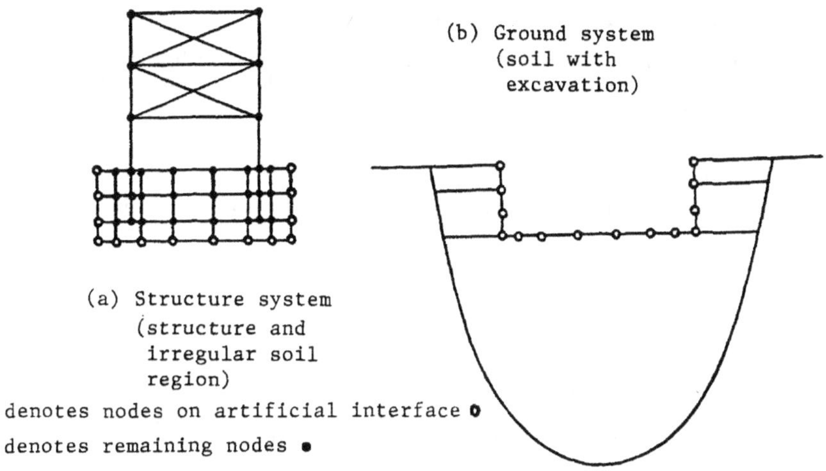

Figure 3. Structure and irregular soil region, soil with excavation

The unbounded soil, which has infinite length in x-direction, can be analyzed by two uncouple models, the out-of-plane model and in-plane model sketched in Fig. 4. The former is analogous to a pure shear problem in y-direction. The latter is similar to a plane strain problem on xz-plane.

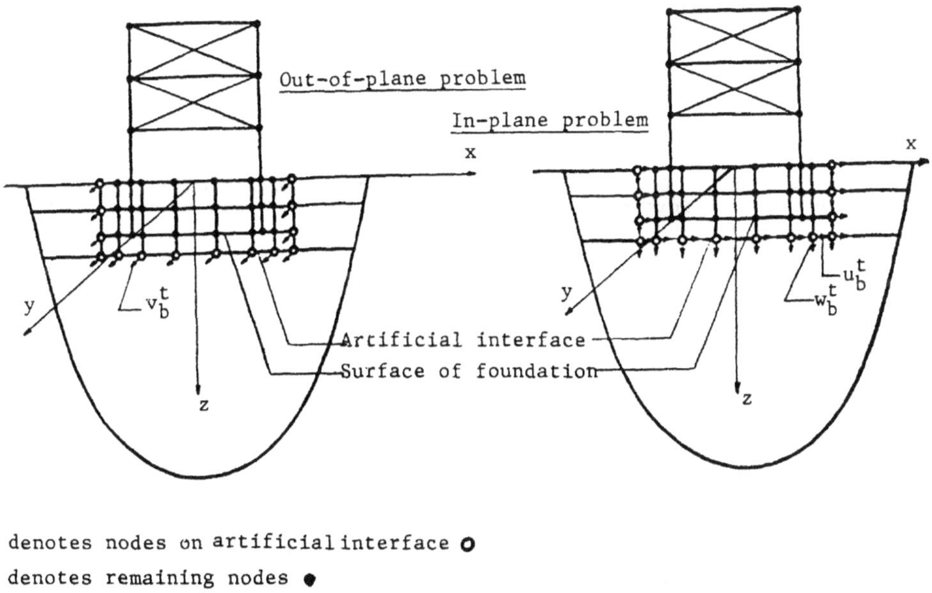

Figure 4. In-plane and out-of-plane model

The Equation of Motion for Structure System

The structure system in Fig. 3a is formulated in discrete time domain. The irregular finite soil domain is also included in this system. At the nth time step, the equation of motion can be expressed as

$$\begin{bmatrix} [M_{ss}] & [M_{sb}] \\ [M_{bs}] & [M_{bb}] \end{bmatrix} \begin{Bmatrix} \{\ddot{x}_s^t(n)\} \\ \{\ddot{x}_b^t(n)\} \end{Bmatrix} + \begin{bmatrix} [C_{ss}] & [C_{sb}] \\ [C_{bs}] & [C_{bb}] \end{bmatrix} \begin{Bmatrix} \{\dot{x}_s^t(n)\} \\ \{\dot{x}_b^t(n)\} \end{Bmatrix} + \begin{bmatrix} [K_{ss}] & [K_{sb}] \\ [K_{bs}] & [K_{bb}] \end{bmatrix} \begin{Bmatrix} \{x_s^t(n)\} \\ \{x_b^t(n)\} \end{Bmatrix}$$

$$= \begin{Bmatrix} \{P_s(n)\} \\ -\{R_b(n)\} \end{Bmatrix} + \begin{Bmatrix} [\gamma]\{u(n)\} \\ [0] \end{Bmatrix} \tag{30}$$

where [M], [C] and [K] are the mass, damping and stiffness matrix, respectively; and {x} is the displacement vector. {P} and {R} are the external force vector and the interaction force vector, respectively. $[\gamma]$ is the matrix relating to location of the controllers on the structure; and {u} is the control force vector. Subscript b stands for the degree-of-freedom along the artificial interface and s for those in the rest of the system. Superscript t stands for the total system.

$\{x_b^t(t)\}$ and $\{R_b(t)\}$ are related to the flexibility matrix of ground system by the convolution integral as

$$\{x(t)\} = \int_0^{\infty} [F(\tau)]\{R(t-\tau)\}d\tau \tag{31}$$

where $\{x(t)\} = \{x_b^t(t)\} - \{x_b^g(t)\}$ and $\{x_b^g(t)\}$ is the displacement vector of ground system on the nodes along the artificial interface due to seismic excitation. $[F(\tau)]$ is the dynamic-flexibility matrix of ground system. In a discrete time system with time interval equal to Δt, the interaction forces can be formulated as

$$\{R_b(n)\} = [KR]\{x_b^t(n)\} - \{RR\} \tag{32}$$

where [KR]is constant and {RR} depends on the displacement of ground system and the interaction forces in the previous steps. Both can be expressed as

$$[KR] = \frac{1}{\Delta t}[F(0)]^{-1} \tag{32.a}$$

$$\{RR\} = \frac{1}{\Delta t}[F(0)]^{-1} \left(\{x_b^g(n)\} + \sum_{k=1}^{N} \Delta t \, [F(k)]\{R(n-k)\} \right) \tag{32.b}$$

By using the interaction forces in Eq. (32), the acceleration vector in Eq. (30) can be written as

$$\begin{Bmatrix} \{\ddot{x}_s^t(n)\} \\ \{\ddot{x}_b^t(n)\} \end{Bmatrix} = -\begin{bmatrix} [M_{ss}] & [M_{sb}] \\ [M_{bs}] & [M_{bb}] \end{bmatrix}^{-1} \begin{bmatrix} [C_{ss}] & [C_{sb}] \\ [C_{bs}] & [C_{bb}] \end{bmatrix} \begin{Bmatrix} \{\dot{x}_s^t(n)\} \\ \{\dot{x}_b^t(n)\} \end{Bmatrix} - \begin{bmatrix} [M_{ss}] & [M_{sb}] \\ [M_{bs}] & [M_{bb}] \end{bmatrix}^{-1} \begin{bmatrix} [K_{ss}] & [K_{sb}] \\ [K_{bs}] & [K_{bb}] + [KR] \end{bmatrix} \begin{Bmatrix} \{x_s^t(n)\} \\ \{x_b^t(n)\} \end{Bmatrix}$$

$$+ \begin{bmatrix} [M_{ss}] & [M_{sb}] \\ [M_{bs}] & [M_{bb}] \end{bmatrix}^{-1} \begin{Bmatrix} \{P_s(n)\} \\ \{RR(n)\} \end{Bmatrix} + \begin{bmatrix} [M_{ss}] & [M_{sb}] \\ [M_{bs}] & [M_{bb}] \end{bmatrix}^{-1} \begin{Bmatrix} [\gamma]\{u(n)\} \\ \{0\} \end{Bmatrix} \tag{33}$$

To express Eq. (33) in compact form, one introduces the state vector as

$$\{Z(n)\}=\begin{Bmatrix}\{x_s^t(n)\}\\ \{x_b^t(n)\}\\ \{\dot{x}_s^t(n)\}\\ \{\dot{x}_b^t(n)\}\end{Bmatrix} \tag{34}$$

And noting that

$$\begin{Bmatrix}\{\dot{x}_s^t(n)\}\\ \{\dot{x}_b^t(n)\}\end{Bmatrix}=[I]\begin{Bmatrix}\{\dot{x}_s^t(n)\}\\ \{\dot{x}_b^t(n)\}\end{Bmatrix} \tag{35}$$

Eq. (33) can be written as

$$\{\dot{Z}(n)\}=[A]\{Z(n)\}+[B]+[C] \tag{36}$$

for which the aforementioned solution procedures can be similarly applied to Eq. (36).

NUMERICAL ILLUSTRATIONS

Comparison of response by IOAC with GOAC

A single DOF structural model is used to illustrate the fundamental behavior of the control algorithm. The structural properties are: floor mass M=345.6 tons; translational stiffness K=3.404 x 10^5 kN/m; and linear viscous damping coefficient C=734.3 kN sec/m. The undamped frequency and damping ratio are 5HZ and 0.034, respectively. Two earthquake acceleration records are used as ground motion input: N-S component of El-Centrol earthquake of May 18, 1940, and Mexico earthquake (SCT1) of September 19, 1985. The structural model is equipped with an active tendon, and the weighting matrices are chosen to be: $Q_{11}=Q_{22}=10^6$, $Q_{12}=Q_{21}=0$, and $Q_{22}/R=10^9$. Both IOAC and GOAC of closed-loop algorithms are employed in the study.

Figures 5(a) and (b) show the influence of time-increment Δt on floor displacement and control force of the structure. It can be seen that for IOAC algorithm the influence of Δt is remarkable. When two different time increments of 0.0025 sec. and 0.01 sec. are used, the responses of displacements and control forces associated with both Δt's are significantly different. However, Δt does not influence the response of the new GOAC algorithm when two different Δt's are employed. Apparently, IOAC is time-dependent and sensitive to time-increment while GOAC is independent of varying time-increments. In engineering practice, a structure may be subjected to earthquakes with different excitation periods. Reponse and control force of the structure should both be free from the variable Δt.

Influence of feedback gain matrix of GOAC

The influence of gain matrix, G, on the response of a system is mainly due to elements S_{12} and S_{22} of [S]. Figures 6(a) and (b) show how these elements with various coefficients affect the displacement of the structural model. Similarly, Figures 7(a),(b) and 8(a),(b) show the influence on control force and velocity, respectively. Comparisons of displacement and velocity are summarized in Table 1 where D and V signify displacement and velocity, respectively.

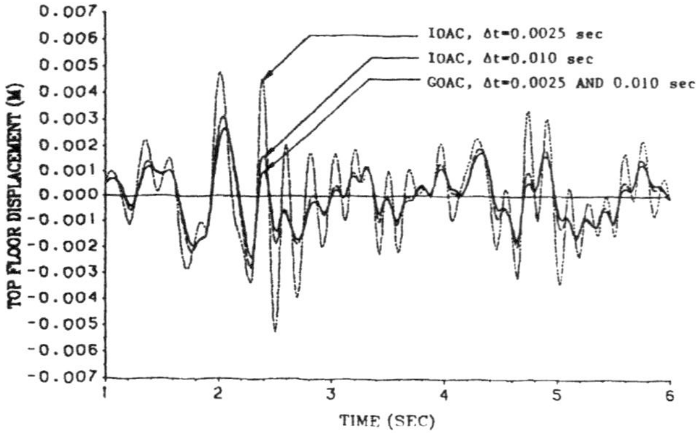

Figure 5a. Time-increment effect on displacement

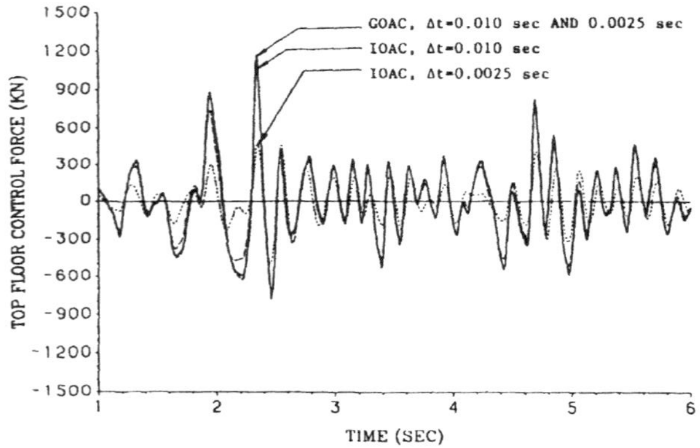

Figure 5b. Time-increment effect on control force

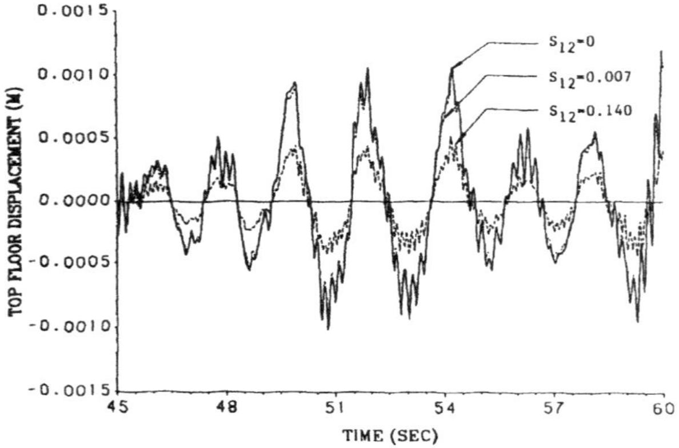

Figure 6a. S_{12} vs displacement with S_{22}=0

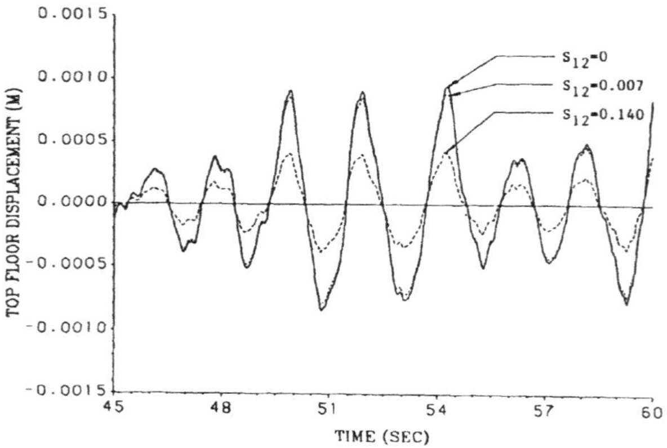

Figure 6b. S_{12} vs displacement with S_{22}=0.007

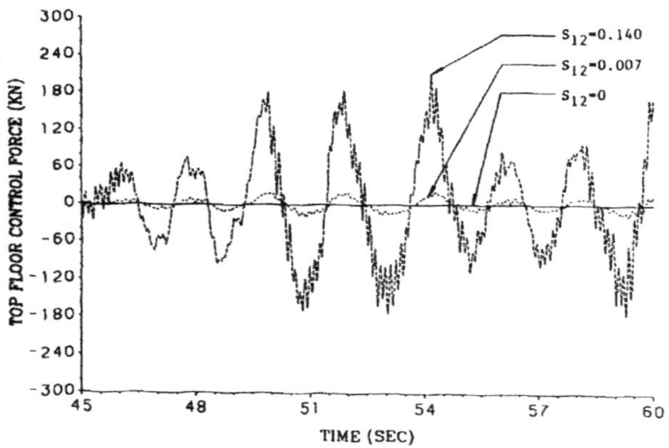

Figure 7a. S_{12} vs control force with $S_{22}=0$

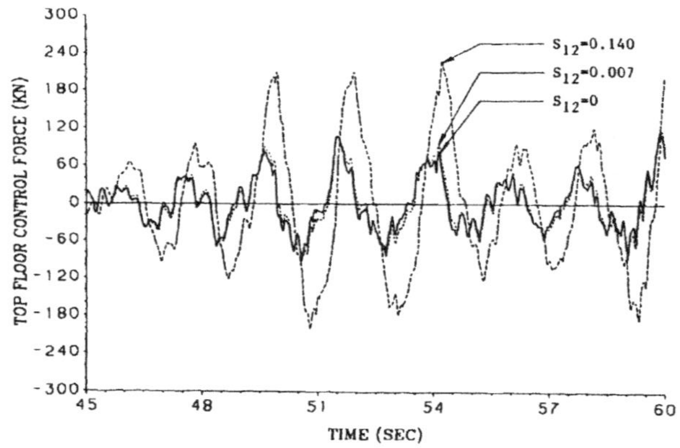

Figure 7b. S_{12} vs control force with $S_{22}=0.007$

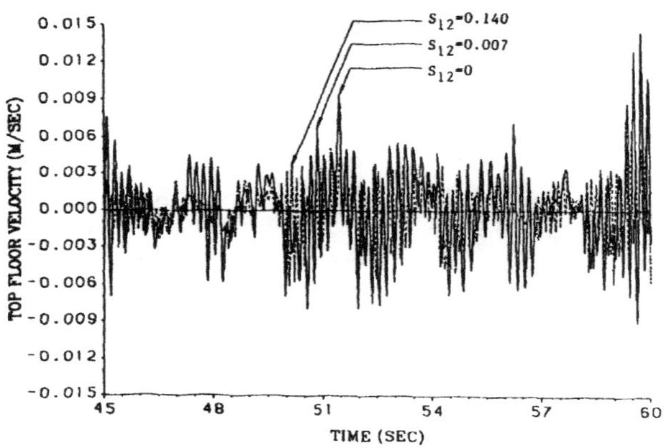

Figure 8a. S_{12} vs velocity with $S_{22}=0$

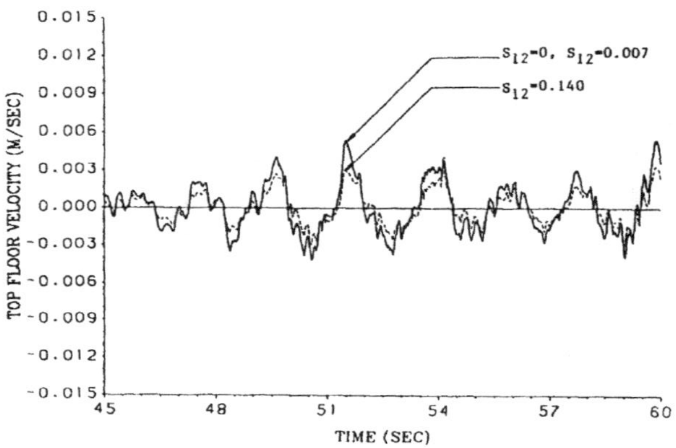

Figure 8b. S_{12} vs velocity with $S_{22}=0.007$

TABLE 1
Comparison of response VS S_{12} and S_{22}

RESPONSE (S_{12} / S_{22})	(I) 0		(II) 0.140×10^6		(II)/(I)	
S_{22}	D_I(cm)	V_I(cm/s)	D_{II}(cm)	V_{II}(cm/s)	D_{II}/D_I	V_{II}/V_I
(a), o	0.119	1.426	0.052	0.665	0.437	0.466
(b), 0.007×10^6	0.094	0.591	0.043	0.394	0.457	0.667
(b)/(a)	0.793	0.414	0.827	0.592	---	---

The above table shows that:

(1) For a given S_{12} value, when S_{22} changes from 0 to 0.007, the displacement shows little change. For S_{12}=0 and 0.140×10^6, displacement is reduced to 20.7% and 17.3%, respectively. Meanwhile, velocity changes greatly. For S_{12}=0 and 0.140×10^6, velocity is reduced to 58.6% and 33.3%, respectively.

(2) When the value of S_{12} increases, both displacement and velocity are reduced. When $S_{22}=0.007 \times 10^6$, displacement and velocity are reduced to 54.3% and 33.3%, respectively. Note that an increase in the value of S_{12} can reduce the maximum value of velocity but not its severe fluctuation, as shown in Figure 8(a). Only a reduction in S_{22} can achieve this, as demonstrated by Figure 8(b).

Comparing Figure 6(a) with (b) shows that the contribution of S_{22} is mainly to reduce the high-frequency vibration in the displacement response but not its magnitude. S_{12}, however, contributes significantly to reduction of the magnitude of displacement. The reason is that as S_{12} and S_{22} increase, so does stiffness and damping of the system.

Comparing Figures 6, 7 and 8 where $S_{12}=S_{22}=0$ (i.e., no active control) and $S_{12}=0$, $S_{22}=0.007Q_{22}$, one may find that the reduction of displacement is small, but reduction of velocity and control force is significant. It is because S_{22} in the feedback gain matrix can only affect system velocity, but reduction of velocity does not necessarily reduce displacement.

In the Mexico earthquake, the excitation period is long. Although displacement response is large, velocity response is not. Therefore, S_{22} cannot significantly reduce displacement but S_{12} can. The latter is very effective in displacement reduction and becomes an important factor in the GOAC algorithm.

CONCLUSIONS

The paper presents a generalized optimal active control algorithm for seismic structure. By means of an intelligent feedback gain matrix, the algorithm can be automatically adjusted, taking account of actual earthquakes on site for effective control. Unlike the instantaneous algorithm, this approach is not dependent on incremental time intervals. Its ability to generalize is also enhanced by results that match the algebraic Ricatti matrix method. Given a certain element in gain matrix, the two methods yield the same result. Thus, the proposed algorithm has advantages over other methods currently in vogue. As the numerical illustrations show, selecting the elements S_{12} and S_{22} in gain matrix yield desirable results in control forces and structural responses.

The presentation also reviews state-of-the-art structural optimization of seismic structures. Future directions could be a combination of structural optimization and control as well as further application of the proposed method for development of other innovative algorithms to achieve an intelligent control system covering nonlinear behavior. Other studies needed on instrumentation, reliability of energy supply, and experimental work may be found in NSF documents [16,17].

ACKNOWLEDGMENTS

The work reported herein is partially supported by the National Center for Earthquake Engineering Research and the National Science Foundation. Their financial support is gratefully acknowledged.

REFERENCES

1. Cheng, F.Y. and Pantelides, C.P., Optimum seismic structural design with tendon and mass damper controls and random process. ASCE Structures Congress, 1986, 5, 40-53.

2. Cheng, F.Y. and Pantelides, C.P., Optimization of structures and controls under seismic loading. Proceedings of the International Conference on Computational Mechanics, Tokyo, 1986, 10, 135-140.

3. Cheng, F.Y. and Pantelides, C.P., Optimal active control of wind structures using instantaneous algorithm. In Vibration Control and Active Vibration Suppression, eds. D.J. Inman and J.C. Simons, American Society of Mechanical Engineers, 1987, 21-28.

4. Cheng, F.Y. and Pantelides, C.P., Algorithm Development for Using Optimal Control in Structural Optimization Subjected to Seismic and Wind Forces. NSF Report, U.S. Department of Commerce, National Technical Information Service, NTIS No. PB90-133471/AS, 1988, 283 pages.

5. Pantelides, C.P. and Cheng, F.Y., Optimal placement of controllers for seismic structures. International Journal of Engineering Structures, 1990, 12, 254-262.

6. Cheng, F.Y. and Pantelides, C.P., Combining Structural Optimization and Structural Control, Technical Report, SUNY-Buffalo, National Center for Earthquake Engineering Research, NCEER-88-0006, 1988, 82 pages.

7. Cheng, F.Y., Response control based on structural optimization and its combination with active protection. Proceedings of Ninth World Conference on Earthquake Engineering, Tokyo-Kyoto, 1988, 8, 471-476.

8. Cheng, F.Y. and Chang, C.C., Safety-Based Optimum Design of Nondeterministic Structures Subjected to Various Types of Seismic Loads, NSF Report, U.S. Department of Commerce, National Technical Information Service, NTIS No. PB90-133489/AS, 1988, 326 pages.

9. Cheng, F.Y., Choi, C.K., and Juang, D.S., Developments of optimum design and control of seismic structures. International Workshop on Intelligent Structures, Elsevier Science Publishers, 1990, 264-281.

10. Cheng, F.Y., Optimum design of seismic structures: methodologies and results. Proceedings of Fourth U.S. National Conference on Earthquake Engineering, 1990, 2, 917-927.

11. Truman, K.Z. and Cheng, F.Y., Optimum assessment of irregular 3-D seismic buildings. Journal of Structural Engineering, ASCE, 1990, 116, 3324-3337.

12. Soong, T.T., Masri, S.F. and Housner, G.W., An overview of active structural control under seismic loads. Earthquake Spectra, 1991, 7, 483-505.

13. Soong, T.T., Active Structural Control: Theory and Practice, Longman, London, and Wiley, New York, 1990.

14. Yang, J.N., Akbarpour, A., and Ghaemmaghami, P., Instantaneous Optimal Control Laws for Building Structures under Earthquake Excitations, Technical Report, SUNY-Buffalo, National Center for Earthquake Engineering Research, NCEER-TR-87-0007, 1987.

15. Wolf, John P., Soil-Structure Interaction Analysis in Time Domain, Prentice-Hall, New York, 1987.

16. Masri, S., ed., Proceedings of the U.S. National Workshop on Structural Control Research. Department of Civil Engineering, University of Southern California, Los Angeles, 1990.

17. National Science Foundation, Structural Control Research for Performance, Safety and Hazard Mitigation: Announcement of Program Initiative. Washington, D.C., 1991.

NONLINEAR FINITE ELEMENT ALGORITHMS FOR MASSIVELY PARALLEL SIMD COMPUTERS

T. BELYTSCHKO[1], L. P. BINDEMAN[1] and E. PLASKACZ[2]

1. Department of Civil and Mechanical Engineering, Robert R. McCormick School of Engineering and Applied Science, Northwestern University, Evanston, IL 60208-3109

2. Argonne National Laboratory, Argonne, IL 60439-4845

ABSTRACT

The implementation of nonlinear finite element algorithms based on vector methods of solution for massively parallel SIMD computers is described; these include explicit methods for transient problems and conjugate gradient methods for static and implicit problems. For both classes of problems, direct methods which do not require the evaluation or storage of a stiffness matrix are used. A feature of these algorithms is an exchange process for the nodal forces so that communication is minimized. The resulting algorithm is capable of processing approximately 1.6×10^6 element-cycles per second of clock time on the 64K processor parallel machine. Element design plays an important role in making such calculations feasible, particularly one point quadrature and effective means of stabilization. Stabilization techniques based on mixed variational principles are described and their performance is examined.

INTRODUCTION

Nonlinear finite element procedures are becoming of increasing importance in industry and research. For example, one of the most rapidly growing fields today is the simulation of manufacturing processes, such as sheet metal forming, extrusion and forging; an entire issue of International Journal for Numerical Methods in Engineering was recently devoted to this topic. Another important area is prototype testing, such as in the crashworthiness analysis of automobiles and helicopters. These solutions generally involve 20,000 to 50,000 elements and require 20 or more hours on the latest supercomputers, even though the meshes are in fact quite coarse compared to the resolution which would be needed to accurately capture the deformation modes. It is readily apparent from studies of such problems that the general solution of nonlinear problems poses severe difficulties, particularly when one is concerned with failure.

The major source of difficulties in nonlinear problems can be appreciated by considering the phenomena which occur in failure. The following is a list of the most common phenomena found in failure:

1. buckling, which for many materials, such as elastic plastic, results in hingelines;
2. fracture, particularly with growing cracks;
3. shear banding, which arises from the loss of material stability.

All of the above phenomena involve localization of deformation, which is accompanied by the development of severe gradients in the displacement field.

A finite element mesh must be extremely fine to resolve these types of phenomena. For example, shear bands are typically from 10 to 300 microns in width. The profile of the strain in a shear band is similar in character to a sinusoidal wave, so to achieve adequate resolution, if the same guidelines which are used for waves are applied, then on the order of 5 to 10 constant strain elements would be needed across the width of the band. This corresponds to an element size of 1 micron (10^{-6} meters) for the worst case. Hence, for a specimen of dimension 10^{-1} meters, a mesh of $10^5 \times 10^5$ elements would be needed to adequately resolve such a problem. Considerable savings can be achieved by adaptive methods in which the small elements are confined to the shear band, but even then for a problem such as this on the order of 10^6 to 10^7 elements would be required in a two dimensional problem! Similar situations exist for hingelines and the modelling of arbitrary crack growth. It may be argued that many nonlinear problems are not concerned with failure, and hence avoid these phenomena. However, when nonlinear situations are used to guide engineering design, they must be used to avoid failures, so the engineer must know how close he is to the failure of a process. In some types of simulation, such as crashworthiness, the entire process revolves about the controlled failure of a structure so that accelerations of the occupant are minimized.

Thus, in nonlinear analysis, it is anticipated that the proper solution of many problems will require 10^6 to 10^8 elements even in two dimensions. Current supercomputers cannot solve such problems in a reasonable amount of time; for example, in the best circumstances, in an explicit calculation, a supercomputer with a real speed of 100 mflops (roughly the speed of an explicit finite element calculation on a supercomputer with 1000 mflops peak performance) can handle an element cycle in an explicit calculation in 10^{-6} sec. A mesh of 10^8 elements then requires 100 sec per time step, and since a typical explicit calculation for a mesh of this size requires 10^5 time steps, the CPU time for the calculation would be 10^7 sec (more than 100 days).

Because single processor supercomputers are asymptotically approaching their limits of performance, it is widely acknowledged that the supercomputers of the next decade will be massively parallel machines, either single-instruction multiple-data stream (SIMD) machines with about 10^5 or more processors or multiple-instruction multiple-data stream (MIMD) machines with on order of 10^3 processors. It is anticipated that the memory in these machines will be local and with each local memory accessible only to a single processor. A recent New York Times article (November 25,1990) indicated that many supercomputer manufacturers are now developing massively parallel machines. These machines will probably use 100 to 100,000 processors with a MIMD or SIMD architecture. Many of the manufacturers appear to be leaning to MIMD architectures because they can deal more effectively with a large variety of tasks and require less reprogramming of single processor software, but for large-scale finite element or finite difference computations, SIMD architectures have many advantages.

At the present time, among the available massively parallel machines are the SIMD Connection Machine (CM) manufactured by Thinking Machines and the MIMD

Touchstone manufactured by Intel Corporation. The Connection Machine (CM) consists of a front-end computer, such as a SUN workstation, and a parallel processing unit of 16K to 64K data processors, each with 64K bits of local memory, and (optionally) an I/O system that supports mass storage and graphics display devices. Communication between processors is accomplished over a hypercube network, and as usual with these machines, it is the weak link because it is very slow! (regardless of what the manufacturers claim).

This paper describes some of our experiences and findings in the development of vector finite element algorithms for nonlinear analysis on the CM. The difficulties encountered in implementing these algorithms on SIMD machines are somewhat different from those associated with MIMD machines, where the number of processors tends to be smaller and the memories of each processor are much larger. In MIMD machines load balancing is the major concern, but minimizing communication is of importance in both types of architectures. Moreover, the structure of explicit finite element programs for SIMD machines possesses such innate simplicity and elegance that for very large problems at least one level of SIMD architecture is irresistible.

In a SIMD machine, each processor must perform exactly the same operation at the same time. This means that the effectiveness of the machine hinges on the homogeneity of various aspects of the problem. If the mesh consists of a single type of element and single type of material and if the number of elements is equal to or slightly less than an integer multiple m of the number of processors), the machine is very effective because all elements are processed in m cycles. On the other hand, if the mesh has different types of elements or different types of materials, the efficiency decreases, for an element cycle is required for each type of element or material. For example, if a 64K mesh consisting of 5 types of elements is run on a 64K machine, only 20% of the speed on a uniform mesh can be achieved, since the five types of elements have to be processed sequentially.

The difficulties are akin to those found in vectorization except they are more severe. In vectorized machines, inefficiencies arise if homogeneous groups of elements (same type of element and material) are smaller than the optimal vector length, whereas in SIMD machines, difficulties arise when the size of homogeneous element groups is smaller than the number of processors. Since optimal vector length in modern supercomputers is on the order of 10^2 whereas the number of processors in massively parallel SIMD machines is of the order 10^4, the difficulties are amplified in the latter.

An even more egregious obstacle in the effective implementation of massively parallel machines is the locality of data storage. By locality we mean that each processor possesses a local memory, and a processor can readily access only the data which resides in its own local memory; these architectures are often said to have *distributed* memories. Access to local data of other processors must be made through the slow communication network. Furthermore, in a problem with little structure, it is difficult to locate the processor in which the necessary data resides. This makes the treatment of contact-impact very difficult, as has been discussed in a companion paper, Belytschko, Plaskacz, and Chiang [9].

The outline of this paper is as follows. In the next section, the finite element algorithm which has been developed for explicit methods on massively parallel computers with local memory will be described. Section 3 describes some recent improvements in one-point quadrature elements, which incidentally are of tremendous advantage in current parallel computers. Some results and timings on the CM are given in Section 4.

EXPLICIT FINITE ELEMENT ALGORITHM

In Belytschko, Plaskacz, Kennedy and Greenwell [10], the implementation of the direct vector method for explicit time integration on a SIMD was described for two-dimensional elastic continua (Belytschko, Chiapetta, and Bartell [5], Flanagan and Belytschko [12]. Substantial redesign of the von Neumann algorithm and underlying data structure was needed to adapt the explicit finite element algorithm for the CM. A key step was the development of an *exchange* procedure which minimizes interprocessor communication in explicit schemes at the expense of additional computations and storage. An understanding of this procedure and its data structures is needed to explain the extension to contact-impact problems.

The governing equations for a finite element semidiscretization by a Lagrangian mesh are:

momentum equation

$$\mathbf{M}\dot{\mathbf{v}}(t) = \mathbf{f}(t) \tag{2.1}$$

$$\mathbf{f}(t) = \mathbf{f}_{ext}(t) - \mathbf{f}_{int}(t) \tag{2.2}$$

$$\mathbf{f}_{ext}(t) = \sum_{e} \mathbf{L}_e^T \mathbf{f}_{ext}^e(t) \qquad \mathbf{f}_{int}(t) = \sum_{e} \mathbf{L}_e^T \mathbf{f}_{int}^e(t) \tag{2.3}$$

$$\mathbf{f}_{int}^e(t) = \int_{\Omega_e} \mathbf{B}^T(\mathbf{x},t)\, \boldsymbol{\sigma}(\mathbf{x},t)\, d\Omega \tag{2.4}$$

measure of deformation

$$\dot{\boldsymbol{\varepsilon}}(\mathbf{x},t) = \mathbf{B}(\mathbf{x},t)\mathbf{v}_e(t) \tag{2.5}$$

$$\mathbf{v}_e(t) = \mathbf{L}_e\mathbf{v}(t) \tag{2.6}$$

constitutive equation

$$\overset{\nabla}{\boldsymbol{\sigma}}(\mathbf{x},t) = \text{function } (\dot{\boldsymbol{\varepsilon}}(t), \boldsymbol{\sigma}\ ) \tag{2.7}$$

In the above

t = time;
$\mathbf{x}$ = Eulerian coordinates;
$\mathbf{M}$ = mass matrix;
$\mathbf{v}(t)$ = nodal velocity matrix;
$\mathbf{f}(t)$, $\mathbf{f}_{ext}(t)$, $\mathbf{f}_{int}(t)$ = resultant, external and internal nodal forces, respectively
$\mathbf{B}(\mathbf{x},t)$ = semidiscrete form of the symmetric part of the gradient operator;
$\dot{\boldsymbol{\varepsilon}}$ = rate-of-deformation (velocity strain) tensor;
$\boldsymbol{\sigma}$ = Cauchy stress tensor;
$\mathbf{L}_e$ = Boolean connectivity matrix of element e.

A natural allocation of operations between processors is to assign one or more elements to each processor. Thus, if one element is assigned per processor, processor e performs the calculations associated with equations (2.4), (2.5) and (2.7) for element e.

The operations indicated by Eqs. (2.3) and (2.6) are often called the assembly and gather operations, respectively. These operations require an element to access arbitrary nodal values. This is very inefficient with a SIMD computer if the node-related and element-related data are stored in separate processors. In that case, the nodal displacements and forces cannot be accessed by an element-processor without considerable communication, and as stressed previously, communication is the slowest part of distributed memory parallel computers today.

In the SIMD algorithm described in Belytschko, Plaskacz, Kennedy and Greenwell [10], these two operations are replaced by a single operation: nodal forces are exchanged between elements with common nodes and the accelerations of all nodes of an element are integrated by the processor associated with that element. This procedure is shown in Table 1; note that this flow is simultaneously executed on all processors of the SIMD computer.

This algorithm of course entails certain redundancies in calculations and storage: the nodal variables of any node are stored as often as the number of elements connected to the node. However, in massively parallel computers such as the CM, the most important step in maximizing speed is to reduce communication. The exchange algorithm described in Table 1 minimizes communication between processors for it exchanges only the resultant forces between processors and this exchange occurs only once per time step (the nodal coordinates have been added in this version for purposes of the contact-impact algorithm).

TABLE 1
Flowchart for Explicit Integration in SIMD Computer

1. Initial conditions: $t = 0$; $\mathbf{v}(\mathbf{x},0) = \mathbf{v}_0(\mathbf{x})$, $\sigma(\mathbf{x},0) = \sigma_0(\mathbf{x})$
2. FOR all elements e:
    a. Evaluate velocity strains
    b. Evaluate stresses by constitutive law
    c. Compute element force, $\hat{\mathbf{f}}_e$

    EXCHANGE $\mathbf{f}_e$ between elements

    d. Perform contact-impact operations
    e. Obtain accelerations from equations of motion and integrate accelerations for all nodes of element e

    END of element computations
3. $t \rightarrow t + \Delta t$; $j \rightarrow j + 1$; go to 2

ONE-POINT QUADRILATERAL ELEMENT

For low-order elements such as 4 node quadrilaterals and 8 node hexahedrals, one-point quadrature is essential in nonlinear analysis on the CM, and undoubtedly on future computers of this type. For example, in the hexahedron, 2x2x2 quadrature can achieve only about 13% of the speed of one point quadrature. Furthermore, the element with 2x2x2 quadrature does not even fit on the memory of a CM processor if the constitutive equation is moderately complex.

It has been shown in Belytschko and Bachrach [3] that one-point quadrature elements have the same rate of convergence as full quadrature elements; in some cases such as incompressible materials, they are in fact far better! However one-point quadrature elements possess spurious singular modes, which result in physically unrealistic response called hourglassing.

Flanagan and Belytschko [12] have developed a consistent stabilization which satisfies the patch test. However, the Flanagan-Belytschko stabilization entails an arbitrary user-set parameter, so it is often called perturbation stabilization. Belytschko and Bindeman [4] have recently, based on Belytschko and Bachrach [3], developed an hourglass control which is based on mixed variational principles and requires no user-set parameters.

The development of this hourglass control is based on the nonlinear counterpart of the Simo-Hughes [15] principle, which has been given by Fish and Belytschko [11] as the following weak form:

$$0 = \delta\Pi = \int_\Omega \delta\dot{\varepsilon}^T \sigma(\dot{\varepsilon},\sigma,\ldots)d\Omega + \delta\int_\Omega \tau^T(\nabla_s v - \dot{\varepsilon})d\Omega - \delta\dot{W}_{ext} \tag{3.1}$$

$$\delta\dot{W}_{ext} = \delta \mathbf{v}^T \mathbf{f}_{ext} \tag{3.2}$$

where $\dot{\varepsilon}$ is the interpolated velocity strain (rate of deformation), σ the Cauchy stress which is computed from the velocity strain and other state variables by the constitutive equation, τ the interpolated Cauchy stress, and $\nabla_s \mathbf{v}$ is the symmetric part of the velocity gradient; the latter would be equivalent to the velocity strain in a standard displacement method, but in mixed methods, the velocity gradient is projected on a smaller space to avoid locking. The essential feature of the Simo-Hughes approach is that the stress field τ is chosen to be orthogonal to the difference between $\dot{\varepsilon}$ and $\nabla_s v$ so that the second integral of (3.1) vanishes.

The velocity is interpolated by

$$\mathbf{v}(\mathbf{x}) = \sum_I^m N_I(\mathbf{x})\mathbf{v}_I \equiv \mathbf{N}(\mathbf{x})\mathbf{v} \tag{3.3}$$

where $\mathbf{N}$ are the standard isoparametric shape functions and m is the number of nodes per element. We define the standard $\mathbf{B}$ matrix by

$$\nabla_s \mathbf{v} = \sum_I^m \nabla_s N_I \mathbf{v}_I \equiv \mathbf{B}\mathbf{v} \tag{3.4}$$

We also define the $\overline{\mathbf{B}}$ matrix by

$$\dot{\varepsilon}(\mathbf{x}) = \sum_I^m \overline{\mathbf{B}}_I \mathbf{v}_I \equiv \overline{\mathbf{B}}\mathbf{v} \tag{3.5}$$

which differs from $\mathbf{B}$ since $\dot{\varepsilon}(\mathbf{x})$ differs from $\nabla_s \mathbf{v}$. Substituting (3.3–3.5) into (3.1) and using the orthogonality condition for τ gives

$$0 = \delta\Pi = \delta\mathbf{v}^T \int_\Omega \overline{\mathbf{B}}^T \sigma(\dot{\varepsilon})d\Omega - \delta\mathbf{v}^T \mathbf{f}_{ext} \tag{3.6}$$

Exploiting the arbitrariness of $\delta\mathbf{v}$ we obtain the discrete equilibrium equations

$$\mathbf{f}_{int} - \mathbf{f}_{ext} = 0 \tag{3.7}$$

$$\mathbf{f}_{int} = \int_\Omega \overline{\mathbf{B}}^T \sigma d\Omega \tag{3.8}$$

where the stress is given by some nonlinear constitutive equation

$$\sigma = \mathcal{S}(\dot{\varepsilon},\sigma,\ldots) = \mathcal{S}(\overline{\mathbf{B}}\mathbf{v},\sigma,\ldots) \tag{3.9}$$

The above formulation is applicable to problems with both material and geometric nonlinearities. In applying the assumed strain stabilization procedure, it is desirable to use a corotational formulation where the Cauchy stresses and velocity strains are expressed in terms of a coordinate system $(\hat{x}, \hat{y})$ which rotates with the element. The corotational coordinate system can be embedded by various techniques, and for anisotropic materials, it is advantageous to embed the coordinate system so that it coincides with axes of orthotropy or other directional features of the material. All subsequent equations are in the corotational coordinate system, but the hats have been omitted to simplify the notation.

The rate form of the constitutive equation can be written

$$\dot{\sigma} = \mathbf{C}\dot{\varepsilon} \tag{3.10}$$

where C is a matrix which depends on the stress and other state variables. In developing the hourglass resistance based on physical parameters, two assumptions must be made:

1. the spin is constant within the element
2. the material response tensor $\hat{C}$ is constant within the element.

For the 4 node quadrilateral, the velocity is given in a form similar to that developed in Belytschko and Bachrach [3]

$$v_i = (\Delta^T + x\mathbf{b}_x^T + y\mathbf{b}_y^T + h\gamma^T)\mathbf{v}_i \tag{3.11}$$

where $h=\xi\eta$, and (ξ,η) are the referential coordinates, $\xi\in[-1, +1]$, $\eta\in[-1, +1]$. Δ, $\mathbf{b}_x$, $\mathbf{b}_y$, and γ are all given in [3]. The symmetric gradient of the velocity field is obtained by taking derivatives of (3.11) which gives

$$\nabla_s \mathbf{v} = \begin{Bmatrix} v_{x,x} \\ v_{y,y} \\ v_{x,y}+v_{y,x} \end{Bmatrix} = \begin{bmatrix} \mathbf{b}_x^T+h_{,x}\gamma^T & \mathbf{0} \\ \mathbf{0} & \mathbf{b}_y^T+h_{,y}\gamma^T \\ \mathbf{b}_y^T+h_{,y}\gamma^T & \mathbf{b}_x^T+h_{,x}\gamma^T \end{bmatrix} \begin{Bmatrix} \mathbf{v}_x \\ \mathbf{v}_y \end{Bmatrix} \equiv \mathbf{B}\mathbf{v} \tag{3.12}$$

where commas denote derivatives with respect to the variables which follow.

An element based on the symmetric displacement gradient will exhibit volumetric locking. This can be explained by considering a mesh of rectangular elements in plane strain; the left and bottom boundaries are fixed as shown in Fig. 1a. Consider the element in the lower left-hand corner, element 1. The nodal displacements of the element for an incompressible material must preserve the total volume of the element (or to be specific, the area in plane strain, since constant volume implies that the area be constant). The total volume of the element must be unchanged if each point of the element is incompressible. The only displacements of node 3 which maintain constant area are

$$u_{x3} = -\alpha a \qquad u_{y3} = +\alpha b \tag{3.13}$$

where α is a constant; all other nodal displacements of element 1 must vanish because of the boundary conditions.

If we consider these nodal displacements of Eq. (3.13), it can be shown that the dilatation is given by

$$\Delta = u_{x,x} + u_{y,y} = \frac{1}{4}\alpha(bh_{,y} - ah_{,x}) \tag{3.14}$$

which for a rectangle is given by

$$\Delta = \frac{\alpha}{ab}(b\bar{x} - a\bar{y}) \tag{3.15}$$

where $(\bar{x}, \bar{y})$ is a local coordinate system in the element shown in Fig. 1a. Thus the dilatation only vanishes at the origin of the element. If the element is fully integrated, the volumetric energy is nonzero everywhere except at the origin and along the line $\bar{y} = (b/a)\bar{x}$; therefore, for an incompressible material the volumetric energy will be infinite. Thus node 3 will not be able to move; nodes 2 and 3 then provide a rigid boundary for the left hand side of element 2, and it can similarly be shown that node 6 cannot move. This argument can then be repeated for all nodes of the mesh to show that deformation of the mesh is impossible. It can be shown similarly that if the first element is a quadrilateral such as shown in Fig. 1b, Eqs. (3.13) are still the only constant area displacements and the dilatation is given by

$$\Delta = \frac{1}{4}\alpha(bh_{,y} - ah_{,x})[1 - b(x_3 - x_2) - a(y_3 - y_4)] \tag{3.16}$$

which in general will not vanish throughout the element.

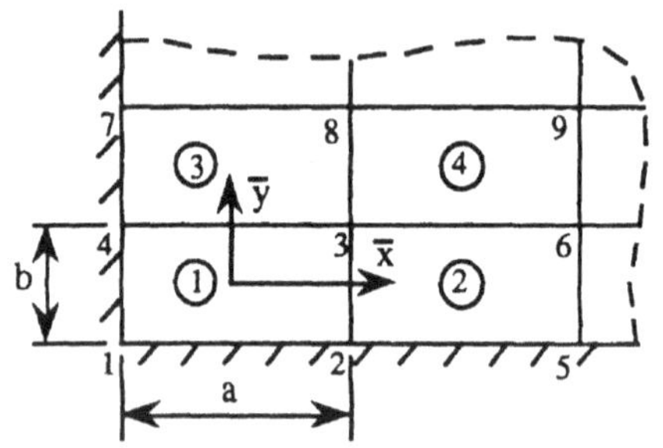

Figure 1a

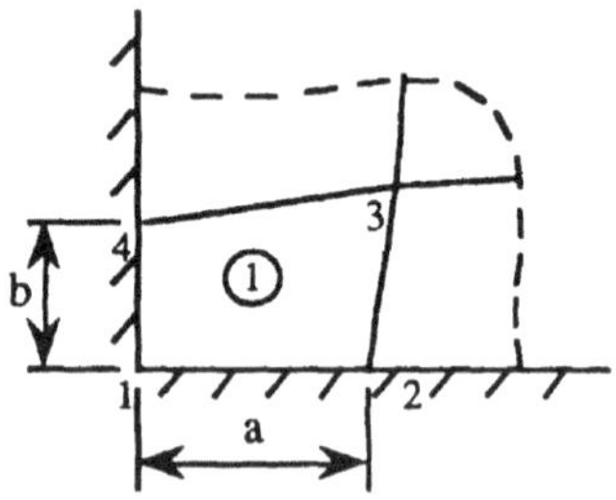

Figure 1b

A second type of locking, more accurately called excess shear stiffness, is caused by nonzero shear strain in elements that deform in the hourglass mode. The hourglass mode represents pure bending, so nonzero shear strains cause the solution to be too stiff. To prevent excess shear stiffness, the γ terms are eliminated from the third row of **B**.

To eliminate locking, an assumed strain field which is defined by a $\bar{\mathbf{B}}$ matrix is used to formulate the element. The following $\bar{\mathbf{B}}$ eliminates both types of locking when the arbitrary constants e_1 and e_2 satisfy the relationship $e_2 = -e_1$.

$$\bar{\mathbf{B}} = \begin{bmatrix} \mathbf{b}_x^T + e_1 h_{,x}\gamma^T & e_2 h_{,y}\gamma^T \\ e_2 h_{,x}\gamma^T & \mathbf{b}_y^T + e_1 h_{,y}\gamma^T \\ \mathbf{b}_y^T & \mathbf{b}_x^T \end{bmatrix} \tag{3.17}$$

$\bar{\mathbf{B}}$ can be broken into its constant and nonconstant parts.

$$\bar{\mathbf{B}} = \bar{\mathbf{B}}_o + \bar{\mathbf{B}}_1 \tag{3.18a}$$

$$\overline{\mathbf{B}}_o = \begin{bmatrix} \mathbf{b}_x^T & \mathbf{0} \\ \mathbf{0} & \mathbf{b}_y^T \\ \mathbf{b}_y^T & \mathbf{b}_x^T \end{bmatrix} \qquad \overline{\mathbf{B}}_1 = \begin{bmatrix} e_1 h_{,x} \boldsymbol{\gamma}^T & e_2 h_{,y} \boldsymbol{\gamma}^T \\ e_2 h_{,x} \boldsymbol{\gamma}^T & e_1 h_{,y} \boldsymbol{\gamma}^T \\ \mathbf{0} & \mathbf{0} \end{bmatrix} \tag{3.18b}$$

The stress rate can also be split into constant and nonconstant parts.

$$\dot{\sigma} = \dot{\sigma}_o + \dot{\sigma}_1 \tag{3.19a}$$

$$= \mathbf{C}\overline{\mathbf{B}}_o\mathbf{v} + \mathbf{C}\overline{\mathbf{B}}_1\mathbf{v} \tag{3.19b}$$

Taking advantage of this form of the stress field, and inserting (3.18) and (3.19) into (3.8) and taking advantage of the orthogonality properties of $h_{,x}$ and $h_{,y}$

$$\int_{\Omega e} h_{,x} d\Omega = \int_{\Omega e} h_{,y} d\Omega = 0 \tag{3.17}$$

and the fact that **C** is constant in the element, gives

$$\mathbf{f}_{int} = A\mathbf{B}_o^T \sigma_o + \mathbf{f}_H \tag{3.20}$$

where $\mathbf{f}_H$ are the hourglass (stabilization) nodal forces, which are given by

$$\mathbf{f}_H = \int_{\Omega e} \overline{\mathbf{B}}_1^T \sigma_1 d\Omega \tag{3.21}$$

and A is the area of the element.

ASQBI stabilization of Belytschko and Bindeman [4] is obtained by setting $e_1 = 1$ and $e_2 = -\bar{\nu}$ in (3.17), where $\bar{\nu} = \nu$ for plane stress and $\bar{\nu} = \nu/(1-\nu)$ for plane strain. ASQBI stabilization is very accurate for linear materials. ASOI(1/2) stabilization, also in [4], is obtained by setting $e_1 = 0.5$ and $e_2 = -0.5$ in (3.17). It is not accurate in bending for coarse meshes, but it converges at the same rate as ASQBI, and the stabilization terms are independent of the Poisson's ratio of the material. Test results for these two assumed strain elements follow. Also shown for comparison are the fully integrated element by the displacement method (Quad4), and the mean dilatation element of Nagtegaal et al. [13], and the Pian-Sumihara [14] element.

Several test problems were run in [4] and three are summarized here. The first test problem, a cantilever shown in Fig. 2, demonstrates the accuracy of the elements in a problem where bending dominates the solution. Compressible and nearly incompressible elastic materials were run. The results in Table 2 are for the four element mesh. The L_2 norm of the displacement error for more refined meshes is plotted in Fig. 3 for the nearly incompressible material.

As expected, ASQBI stabilization is very accurate while ASOI(1/2) stabilization is less accurate for a coarse mesh. ASOI(1/2) converges at the same rate as ASQBI. The Quad4 element is less accurate and converges slowly for incompressible material. The mean dilatation element of Nagtegaal et al. [13] is also less accurate in bending.

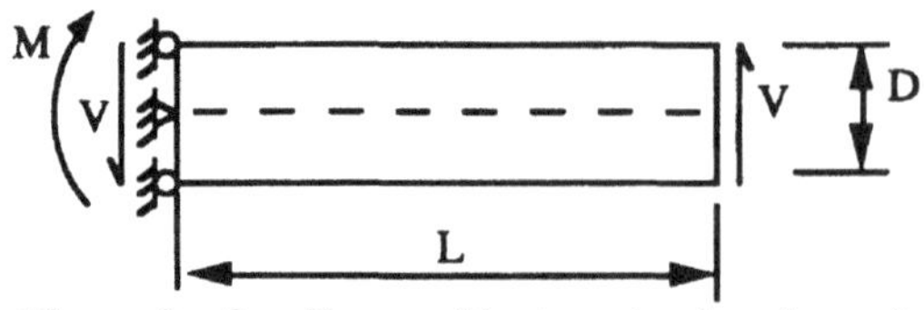

Figure 2. Cantilever with shear load at the end

TABLE 2
Normalized displacement at point A in 4 element mesh of Fig. 2b

	Quad4	Mean dilatation	ASQBI or Pian-Sumihara	ASOI(1/2)
plane stress; ν= 0.25	0.71	0.80	0.99	1.16
plane strain; ν = 0.4999	0.06	0.94	0.98	1.21

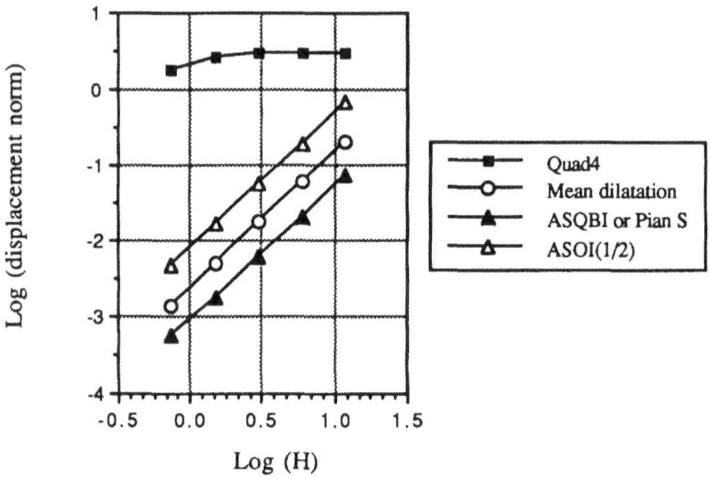

Figure 3. Convergence plot of displacements; plane strain with ν = 0.4999

The third example is a two-dimensional wave problem in an elastic-plastic material, one quarter of which is shown in Fig. 4. An axisymmetric pressure wave is applied to the interior of the hole which rises linearly with time from zero to a maximum that exceeds the elastic limit of the material. The load then decreases linearly to zero.

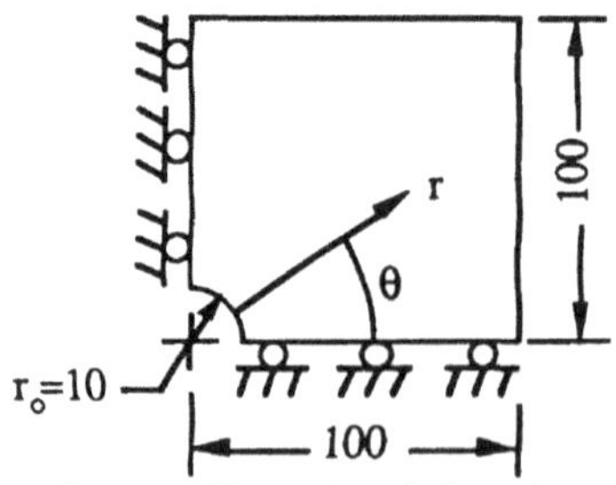

Figure 4. One quarter of a two-dimensional domain with a hole at its center

This problem was modeled by quadrilateral elements and also by a very fine mesh of axisymmetric elements. Before the wave reflected off the mesh boundary, the displacements along radial lines of $\theta=0°$ and $\theta=45°$ were noted, and an error norm was calculated by comparing these displacements to the solution by axisymmetric elements. Table 3 shows that all of the elements perform similarly.

TABLE 3
Normalized L_2 norms of error in displacements

θ	Quad4	Mean dilatation	ASQBI	ASOI(1/2)
0°	.0063	.0061	.0061.	.0063
45°	.0069	.0086	.0088	.0088

RESULTS ON MASSIVELY PARALLEL MACHINES

The finite element procedure was first programmed for the CM in C* for the quadrilateral two-dimensional element and the four-node quadrilateral shell described by Belytschko, Lin and Tsay [6]. The problems reported here were run on a 16K processor CM. Subsequently, the plane quadrilateral has been rewritten in the CM FORTRAN.

The first problem reported here concerns two structured meshes, one with 8192 elements, the second with 16,384 elements. In the structured mesh, the exchange process is accomplished with NEWS (NORTH, EAST, WEST, SOUTH), a communication procedure which exploits the ability of a hypercube communication network to store adjacent elements within one communication link. NEWS communication is significantly faster than Router, which is applicable to arbitrary unstructured meshes.

The running times for these meshes are reported in Table 4 for both C* and FORTRAN versions. It can be seen that the FORTRAN version is about twice as fast. Extrapolating these results to a 64 K machine, for which the running time for a 64K element mesh should be identical to the results for the 16K mesh, gives an element cycle time of 0.56×10^{-6} sec. An element cycle time is the time required to update one element for a single time step, so it is obtained by dividing total wall-clock time by the product of the number of elements and number of time steps. This element cycle time for the CM is approximately 16 times faster than that observed on a vectorized version of the program on the CRAY X-MP.

TABLE 4
2D elastic wave propagation problem with 4 node quadrilaterals; 7500 time steps, wall clock times in seconds

Number of elements	C version	FORTRAN version
8192	563	286
16384	652	321

To illustrate the relative performance of the NEWS and Router communication schemes, timings for the elastic-plastic large displacement response of an impulsively-loaded, cylindrical panel are given. Details of the problem can be found in Belytschko, Lin, and Tsay [6]. The running times are summarized in Table 5. Also shown are timings for the case where the cylindrical panel was subdivided into two structured subdomains and NEWS was used within each subdomain and the router between subdomains. The motivation for this hybrid approach to the exchange process is that large meshes often consist of structured subdomains because many finite element mesh generators construct structured meshes. It can be seen that with two subdomains, the hybrid process is only slightly slower than NEWS. A more detailed description of the hybrid NEWS-Routes Interprocessor Communication scheme may be found in Belytschko and Plaskacz [8].

TABLE 5
Execution Time of Impulsively Loaded Cylindrical Panel; 16384 elements; 5000 time steps; Wall-Clock Time In Seconds

Process	Router	News	Hybrid
Initialize Processors	707.7	617.4	625.3
Compute Internal Forces	1235.2	1172.0	1156.9
Exchange Internal Forces	2604.4	113.8	339.6
Integrate Equation of Motion	69.6	98.8	70.9
Enforce Boundary Conditions	61.5	58.0	60.1
Update Nodal Triads	97.1	90.0	90.0
Total	5029.4	2401.4	2625.0

TABLE 6
Cylindrical Panel Results

Mesh for half-panel	Maximum Displacement (inch) at $z = -6.28$	$z = -9.42$
6x16	1.001	0.494
8x16	1.106	0.526
10x20	1.126	0.527
12x24	1.146	0.531
16x32	1.194	0.585
64x128	1.264	0.636
128x256	1.264	0.636
Balmer and Witmer [1]	1.28	0.68

One interesting application of the CM is the development of benchmark solutions. Elastic-plastic shell problems are notoriously difficult because they exhibit hinge-line formation, so that a uniform mesh cannot achieve a converged solution unless it is extremely fine. While the ultimate test of a physical theory are experimental results, it is difficult to distinguish errors arising from deficiencies in the material model and boundary conditions from insufficient resolution of the finite element mesh. The cylindrical panel problem had previously been solved with uniform meshes of 6x16 elements to 16x32 elements for the half-panel (twofold symmetry was used). On the CM, 64x128 and 128x256 meshes were used to solve the problem; results are summarized in Table 6. From the results, it is evident that the CM solutions are converged solutions. Note the substantial change from a uniform 16x32 mesh, which most analysts would consider an adequate mesh for this problem.

DISCUSSION AND CONCLUSIONS

The CM, particularly with the new FORTRAN compiler, is a powerful machine for the solution of homogeneous finite element problems by explicit methods. The element cycle time on a 4 node quadrilateral in a homogeneous, structured problem is an order of magnitude less than on a CRAY X-MP. In assessing the performance of the CM, it should be noted that the computations reported here are highly homogeneous: the entire mesh consists of the same element and material. Thus the tenfold increase in speed is deceptive. For inhomogeneous problems with many different types of elements and materials, the CM is less effective. It is particularly ineffective for problems with a few elements of a different type. In many models, often a few elements of a different type are present. The effectiveness of the CM could then be enhanced immensely if it were possible to do some calculations in parallel on the front-end machine or on a machine attached to the front-end. Other alternative options, such as being able to drive blocks of 8K processors independently, would also be useful.

An alternative way to treat different types of elements on the CM is to mimic some elements by subsets of instructions for other elements, as proposed in Belytschko, Plaskacz et al. [10]. For example, the instruction for a two-dimensional two node C^0 beam can be developed from a subset of the instructions for the quadrilateral if some additional instructions are added.

In implementing a program on the CM, only computationally intensive portions need to be recoded. Functions such as input are performed on the front-end (in our case a SUN workstation) and need not be recoded. However, the design of the computationally intensive portions needs to be rethought extensively. Because of the locality of memory and the high cost of communication, standard finite element data structures and algorithms are not effective. Thus, we found that in an explicit scheme the data had to be reorganized with all data for one element, including all of its nodal variables, stored in each processor. An exchange of nodal forces between the adjacent elements then provides the required communication between processors. The structure of an iterative solver, such as a conjugate gradient method, would be similar.

The biggest obstacle in the use of the CM for problems such as contact-impact is the absence of any global memory. If certain addresses of the processors were automatically routed to a global memory, the difficulties we dealt with in the contact problem would vanish. Such a partial global memory would also enable the CM to deal with tabulated constitutive equations. The current CM can not deal with large tabulated constitutive equations, such as steam tables, since the table would have to be replicated in each processor and it would not fit in the memory of a processor. The development of a contact-impact algorithm for vectorized and SIMD machines has been described in Belytschko and Neal [7].

Many of the major supercomputer manufacturers, such as Fujitsu, Convex and Cray, are now designing massively parallel machines, and Thinking Machines is developing a new generation of machines. Among the first three, rumors suggest that a MIMD character may predominate, whereas the latter may retain a strong SIMD personality. Although many of the manufacturers claim to be developing compilers which will run existing von Neumann FORTRAN, this is hard to believe, unless they are content with exploiting 1% of the machine's efficiency. Thus the development of effective nonlinear finite element tools for these programs will remain a challenge.

ACKNOWLEDGMENT

The support of the U.S. Army Research Office and the National Science Foundation to Northwestern University is gratefully acknowledged.

REFERENCES

1. Balmer, H. A. and Witmer, E. A., Theoretical-experimental correlation of large dynamic and permanent deformation of impulsively loaded simple structures. In *Air Force Flight Dynamics Laboratory, Report FDP-TDR-64-108.*, (1964).

2. Belytschko, T., An overview of semidiscretization and time integration procedures. In *Computational Methods for Transient Analysis*, eds. T. Belytschko and T. J. R. Hughes, North Holland, Amsterdam (1983) 1-65.

3. Belytschko, T. and Bachrach, W. E., Efficient implementation of quadrilaterals with high coarse mesh accuracy, *Comput. Meths. Appl. Mech. Engrg.* (1986) **54** 279-301.

4. Belytschko, T. and Bindeman, L. P., Assumed strain stabilization of the 4-node quadrilateral with 1-point quadrature for nonlinear problems, *Comput. Meths. Appl. Mech. Engrg.* (1991) **88** 311-340.

5. Belytschko, T., Chiapetta, R. L., and Bartell, H., Efficient large scale nonlinear transient analysis by finite elements, *Internat. J. Numer. Meths. Engrg.* (1976) **10** 579-596.

6. Belytschko, T., Lin, J. I., and Tsay, C. S., Explicit algorithms for the nonlinear dynamics of shells, *Comp. Meths. Appl. Mech. Engrg.* (1984) **42** 225-251.

7. Belytschko, T. and Neal, M. O., Contact-Impact by the pinball algorithm with penalty and Lagrangian methods, *Internat. J. Numer. Meths. Engrg.* (1991) **31** 547-572.

8. Belytschko, T., Plaskacz, E. J., SIMD implementation of a nonlinear transient shell program with partially structured meshes, To appear in *Internat. J. Numer. Meths. Engrg.* (1992).

9. Belytschko, T., Plaskacz, E. J., and Chiang, H.-Y., Explicit finite element methods with contact-impact on SIMD computers, *Computing Systems in Engineering* (1991) vol 2/3 269-276.

10. Belytschko, T., Plaskacz, E. J., Kennedy, J. M., and Greenwell, D. L., Finite element analysis on the CONNECTION machine, *Comp. Meths. Appl. Mech. Engrg.* (1990) **81**(2) 229-254.

11. Fish, J. and Belytschko, T., Elements with embedded localization zones for large deformation problems, *Computers and Structures* (1988) **30** (1/2) 247-256.

12. Flanagan, D. P. and Belytschko, T. A uniform strain hexahedron and quadrilateral with orthogonal hourglass control, *Internat. J. Numer. Meths. Engrg.* (1981) **17** 679-706.

13. Nagtegaal, J. C., Parks, D. M., and Rice, J. R., On numerically accurate finite element solutions in the fully plastic range, *Comput. Meths. Appl. Mech. Engrg.* (1974) **4** 153-178.

14. Pian, T. H. H. and Sumihara K., Rational approach for assumed stress finite elements, *Internat. J. Numer. Meths. Engrg.* (1984) **20** 1685-1695.

15. Simo, J. C. and Hughes, T. J. R., On the variational foundations of assumed strain methods, *J. Appl. Mech.* (1986) **53** no. 1, 51-54.

SPLINE FINITE ELEMENT METHOD - RECENT DEVELOPMENTS

S. C. FAN and M.H. LUAH

School of Civil & Structural Engineering, Nanyang Technological University, Nanyang Avenue, Singapore 2263

ABSTRACT

The application of spline functions in engineering analysis is increasingly popular among many researchers in recent years. Of the various approaches utilizing spline functions, the spline finite element method has the striking advantages in that it inherits not only the higher accuracy and efficiency of spline interpolation, but the generality and flexibility of the finite element method as well. The versatility of spline finite elements for plate and shell problems has been reported in the literature. The approach is also found to be particularly suitable for elements requiring C^1 continuity of displacement. In this paper, a brief review on the application of spline functions in engineering mechanics is given. The numerical procedure for deriving B-spline shape functions is discussed and two spline finite elements, recently developed by the authors for the analysis of plates and shells are outlined. Sample numerical examples to show the performance of the elements are also presented.

INTRODUCTION

This paper reviews briefly the application of spline functions in engineering mechanics and introduces a new interpolating scheme for the finite element method using B_3-splines. Two elements, which have been called spline finite elements, developed for the analysis of plates and shells are presented. The original research on these elements has been reported in references [1-6].

The modern mathematical theory of spline approximation was first introduced by Schoenberg [7,8]. Initially, B-splines were used in approximation problems such as surface fitting, curve design, and interpolating. Recently, they have been extended to solve differential equations and applied to plate and shell problems. The interest in using spline functions for engineering analysis may be attributed to their mathematical properties which are advantageous in numerical work. Firstly, B-splines have small local support, that is, each B-spline function is only non-zero over a few sections of the domain. As a consequence, B-spline approximation is a local approximation scheme. At any point the approximation takes into account only the local behaviour of the approximated function. This enables, for instance, B-spline approximation to reproduce abrupt variations in any local portions of the approximated function. Secondly, the differentiability of B-splines is striking compared to other piecewise interpolating functions. For instance, the cubic B-spline has C^2 continuity, whereas cubic Lagrange and cubic Hermite have only C^0 and C^1 continuity respectively. Thirdly, B-spline approximation enjoys the variation diminishing property. That is, the curve approximated by B-splines is like a beam which minimizes its internal flexural strain energy. This property is simply the observation that of all functions interpolating a given set of data, the B-spline function minimizes the integral of the square of the curvature. In this sense the cubic B-spline function is the smoothest function of all C^2 interpolating functions (such as Lagrangian or Hermitian functions) interpolating a given set of data. B-splines are well documented by deBoor [9] and Schumaker [10].

Many spline functions are available, but in engineering application, cubic B-spline is the most popular. Use of fundamental bi-cubic splines and the Ritz method to analyse bending of rectangular plates was reported as early as 1974 [11]. Later, Mizusawa et al. [12,13] used B-spline functions of various orders as coordinate functions in the Rayleigh-Ritz method to solve the problems of vibration and buckling of skew plate structures. Fujii [14] and Fujii and Hoshino [15] applied the B-spline functions in the discrete and non- discrete mixed methods for bending as well as eigenvalue problems of plates. Shen and Wang [16,17] investigated the static and vibration analysis of flat shells using B_3 and B_5 spline functions and beam functions. A spline strip method, in which the displacement function is expressed as the product of basic function series in the longitudinal direction and B-spline functions in the other direction was also developed [18,19]. The advantages of splines interpolation such as higher accuracy and efficiency in the solution have been reported.

A more significant development in the application of spline functions for engineering analysis was perhaps the development of spline finite strip method by Cheung et al.[20]. In this method, the trigonometric series employed in the conventional finite strip method were replaced by B_3-spline functions and the many problems resulted from mixed boundary conditions, concentrated loads, and continuous spans previously encountered in the finite strip method were thus overcome. Cheung and his collaborators as well as other authors have then developed the method continuously and it is now capable of solving a broad range of structural problems [21-38].

More recently, use of other types of spline functions to enhance certain interpolating properties of conventional spline functions has also been attempted. Clenshaw and Negus [39] described the formulation of the cubic X-spline and demonstrated its application to interpolation. Two advantages were described. One of them is that the X-spline is a generalization of the conventional cubic spline in which the second derivative of the X-spline is allowed to possess minor discontinuities at internal knots. This makes the cubic X-spline very useful in classical interpolation problems compared to conventional cubic splines. The advantage derives from the fact that the computational labour is being minimized with only a slight loss of C^2 continuity. Another is that the cubic X-spline can be made more accurate than the conventional cubic spline with no corresponding increase in computational effort. Therefore, the cubic X-spline combines the theoretical advantage with the computational advantages of a piecewise cubic function. Yang and Chong [40] introduced the use of X-spline functions in the finite strip method for plate bending problems. Yang and Zhang [41] also developed an X-spline finite strip based on an assumed stress hybrid method.

In order to introduce greater flexibility in the spline finite strip method, some authors proposed the use of unequally spaced splines instead of conventional spline functions for the method [42-44]. Compared with an equally spaced B-spline, unequally spaced spline functions is ideal for localized discretization and is therefore more efficient in dealing with concentrated loads and reactions or support members that are either isolated or in irregular locations. With an equally spaced spline, it is not possible to independently modify knot spaces in the region of high stress gradients or locations of abrupt change in the order of continuity condition. Thus, excessive mesh refinement may have to be utilized throughout the field to obtain the required accuracy at critical locations. However, even with the use of unequally spaced splines, the spline finite strip method is still lacking in generality and flexibility compared to the finite element method.

The use of spline functions in the finite element procedure has been introduced only lately. A number of researchers have employed spline functions in various ways in finite element analysis. Moore et al.[45] used B-spline functions and rational B-spline functions for geometric representation of shell element and they had shown that with this technique, complicated shell surfaces can be modelled accurately or even exactly. The present authors first proposed the use of spline functions as displacement shape functions in finite element formulation[1]. This approach inherits both the merits of spline interpolation and the generality and flexibility of conventional finite element method. Two spline finite elements have since been developed, namely a curved axisymmetric thin shell element and a 9-node quadrilateral thin plate element of curved geometry. Both static and free vibration analysis using the elements have been reported and the numerical results obtained clearly showed the accuracy and efficiency of the method [1-6]. A few other authors have also attempted to develop spline finite element in various ways [46-48] but their approaches do not conform exactly to the standard finite element procedure.

In this paper, the spline finite elements proposed by the authors are briefly discussed and some numerical examples are presented to illustrate the advantages of the elements.

DEVELOPMENT OF B-SPLINE SHAPE FUNCTIONS

In general, a cubic spline function along any one-dimensional curvilinear axis S may be expressed as[49],

$$f(S) = \sum_{i=-1}^{m+1} \alpha_i \varphi_i(S) \tag{1}$$

where m is the number of equal sections into which the domain is discretized, φ_i is a local B_3-spline spanning over four consecutive sections with $S = S_i$ as the centre and α_i is the coefficient of φ_i. A local B_3-spline is defined by

$$\varphi_i(S) = \frac{1}{6h^3} \begin{cases} 0 & ; S < S_{i-2} \\ (S-S_{i-2})^3 & ; S_{i-2} \le S \le S_{i-1} \\ h^3 + 3h^2(S-S_{i-1}) + 3h(S-S_{i-1})^2 - 3(S-S_{i-1})^3 & ; S_{i-1} \le S \le S_i \\ h^3 + 3h^2(S_{i+1}-S) + 3h(S_{i+1}-S)^2 - 3(S_{i+1}-S)^3 & ; S_i \le S \le S_{i+1} \\ (S_{i+2}-S)^3 & ; S_{i+1} \le S \le S_{i+2} \\ 0 & ; S_{i+2} < S \end{cases} \tag{2}$$

where h denotes the length of a section. Fig.1 shows a series of local B_3-spline constituting a spline function over a domain.

For the purpose of illustration, consider a spline function with only two consecutive sections (i.e. m = 2) of the form

$$f(S) = \sum_{i=-1}^{3} \alpha_i \varphi_i(S) \tag{3}$$

with the locations at S_0, S_1 and S_2 taken as nodal points and are denoted by 0, 1 and 2 respectively. A transformation procedure is now applied to express Eq.3 in terms of a set of one-dimensional shape functions and a corresponding set of nodal parameters. By choosing f_0, f_0', f_1, f_2 and f_2' as the nodal parameters in which f_i and f_i' are the functional value and the value of the derivative with respect to S respectively at node i, the following set of equations can be written

$$\begin{Bmatrix} f_0 \\ f_0' \\ f_1 \\ f_2 \\ f_2' \end{Bmatrix} = \begin{bmatrix} \varphi_{-1}(S_0) & \varphi_0(S_0) & \varphi_1(S_0) & \varphi_2(S_0) & \varphi_3(S_0) \\ \varphi'_{-1}(S_0) & \varphi'_0(S_0) & \varphi'_1(S_0) & \varphi'_2(S_0) & \varphi'_3(S_0) \\ \varphi_{-1}(S_1) & \varphi_0(S_1) & \varphi_1(S_1) & \varphi_2(S_1) & \varphi_3(S_1) \\ \varphi_{-1}(S_2) & \varphi_0(S_2) & \varphi_1(S_2) & \varphi_2(S_2) & \varphi_3(S_2) \\ \varphi'_{-1}(S_2) & \varphi'_0(S_2) & \varphi'_1(S_2) & \varphi'_2(S_2) & \varphi'_3(S_2) \end{bmatrix} \begin{Bmatrix} \alpha_{-1} \\ \alpha_0 \\ \alpha_1 \\ \alpha_2 \\ \alpha_3 \end{Bmatrix} = [C] \begin{Bmatrix} \alpha_{-1} \\ \alpha_0 \\ \alpha_1 \\ \alpha_2 \\ \alpha_3 \end{Bmatrix} \tag{4}$$

Inverting the matrix [C] gives

$$[\alpha_{-1} \ \alpha_0 \ \alpha_1 \ \alpha_2 \ \alpha_3]^T = [C]^{-1} [f_0 \ f_0' \ f_1 \ f_2 \ f_2']^T \tag{5}$$

By substituting Eq.5 into Eq.3, the spline function becomes

$$\begin{aligned} f(S) &= [\varphi_{-1} \ \varphi_0 \ \varphi_1 \ \varphi_2 \ \varphi_3] [C]^{-1} [f_0 \ f_0' \ f_1 \ f_2 \ f_2']^T \\ &= [H_{01} \ H_{02} \ H_{11} \ H_{21} \ H_{22}] [f_0 \ f_0' \ f_1 \ f_2 \ f_2']^T \end{aligned} \tag{6}$$

in which H_{01}, H_{02}, H_{11}, H_{21} and H_{22} are each linear combination of φ_i and they represent the one-dimensional shape functions corresponding to the nodal parameters.

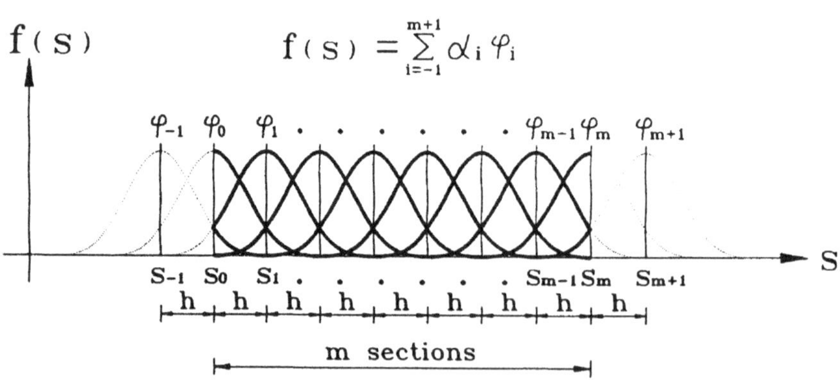

Figure 1. A cubic spline function of m equal sections formed by linear combination of m+3 local B_3-splines.

It can be shown that

$$
\begin{aligned}
H_{01} &= -\frac{1}{2}\varphi_{-1} + \frac{7}{4}\varphi_0 - \frac{1}{2}\varphi_1 + \frac{1}{4}\varphi_2 - \frac{1}{2}\varphi_3 \\
H_{02} &= h\left[-\frac{13}{6}\varphi_{-1} + \frac{7}{12}\varphi_0 - \frac{1}{6}\varphi_1 + \frac{1}{12}\varphi_2 - \frac{1}{6}\varphi_3\right] \\
H_{11} &= 2\varphi_{-1} - \varphi_0 + 2\varphi_1 - \varphi_2 + 2\varphi_3 \\
H_{21} &= -\frac{1}{2}\varphi_{-1} + \frac{1}{4}\varphi_0 - \frac{1}{2}\varphi_1 + \frac{7}{4}\varphi_2 - \frac{1}{2}\varphi_3 \\
H_{22} &= h\left[\frac{1}{6}\varphi_{-1} - \frac{1}{12}\varphi_0 + \frac{1}{6}\varphi_1 - \frac{7}{12}\varphi_2 + \frac{13}{6}\varphi_3\right]
\end{aligned}
\tag{7}
$$

The extension of the one-dimensional shape functions into two dimensions can be achieved in the usual manner.

If by considering a spline function with three consecutive sections (i.e. $m = 3$) with the locations at S_0 and S_3 taken as nodal points and f_0, f_0', f_0'', f_3, f_3' and f_3'' as nodal parameters, the spline function is

$$
f(S) = [H_{01}\ H_{02}\ H_{03}\ H_{31}\ H_{32}\ H_{33}]\ [f_0\ f_0'\ f_0''\ f_3\ f_3'\ f_3'']^T \tag{8}
$$

where

$$
\begin{aligned}
H_{01} &= \varphi_{-1} + \varphi_0 + \varphi_1 \\
H_{02} &= h[-\varphi_{-1} + \varphi_1] \\
H_{03} &= \frac{h^2}{3}[\varphi_{-1} - \frac{1}{2}\varphi_0 + \varphi_1] \\
H_{31} &= \varphi_2 + \varphi_3 + \varphi_4 \\
H_{32} &= h[-\varphi_2 + \varphi_4] \\
H_{33} &= \frac{h^2}{3}[\varphi_2 - \frac{1}{2}\varphi_3 + \varphi_4]
\end{aligned}
\tag{9}
$$

The one-dimensional shape functions given in Eq.9 and the two-dimensional shape functions extended from Eq.7 are used in the formulations of axisymmetric thin shell element and quadrilateral thin plate element respectively.

A 9-NODE QUADRILATERAL THIN PLATE ELEMENT

The spline finite element discussed in this section is a 9-node quadrilateral thin plate element of arbitrary geometry (see Fig.2). The element is sub-parametric, with biquadratic Lagrangian shape functions for co-ordinate transformations and B-spline shape functions for the interpolation of displacements. In the present analysis, only small deformations are considered. The classical Kirchhoff's plate theory is used and the material property is assumed to be linearly elastic, isotropic and homogeneous. Four co-ordinate systems as shown in Fig.2 are used in the formulation and these are the global Cartesian co-ordinate system (x, y, z), element local curvilinear co-ordinate system (ξ, η), nodal local co-ordinate system (ξ', n or n, η'), and local Cartesian co-ordinate system (x', y').

In the sub-parametric formulation, the global co-ordinates of a point in the arbitrarily shaped quadrilateral element are expressed in terms of the curvilinear co-ordinates as

$$\begin{Bmatrix} x \\ y \end{Bmatrix} = \sum_{i=1}^{9} L_i(\xi,\eta) \begin{Bmatrix} x_i \\ y_i \end{Bmatrix} \tag{10}$$

where (x_i , y_i) are the global co-ordinates of node i and $L_i(\xi,\eta)$ is the corresponding biquadratic Lagrangian shape function.

The displacement field of the element is expressed in terms of B-spline shape functions as

$$\begin{aligned} w &= \sum_{i=1}^{4} [N_{i1} w_i + N_{i2} (\frac{\partial w}{\partial \xi})_i + N_{i3} (\frac{\partial w}{\partial \eta})_i + N_{i4} (\frac{\partial^2 w}{\partial \xi \partial \eta})_i] \\ &+ \sum_{i=5,7}^{7} [N_{i1} w_i + N_{i2} (\frac{\partial w}{\partial \eta})_i] + \sum_{i=6,8}^{8} [N_{i1} w_i + N_{i2} (\frac{\partial w}{\partial \xi})_i] + N_{91} w_9 \\ &= [N] \{\delta'\} \end{aligned} \tag{11}$$

in which w_i , $(\partial w/\partial \xi)_i$, $(\partial w/\partial \eta)_i$ and $(\partial^2 w/\partial \xi \partial \eta)_i$ are the local degrees-of-freedom associated with node i and N_{ij}'s are the corresponding B-spline shape functions. The matrix [N] is the usual matrix of shape functions and { δ' } is the local displacement vector. The shape functions N_{ij}'s are extended from the one-dimensional component functions of Eq.7 (see ref.[4] for details).

In the present formulation, the thickness of the element is assumed to be varying according to the function

$$t = \sum_{i=1}^{9} F_i(\xi,\eta) t_i \tag{12}$$

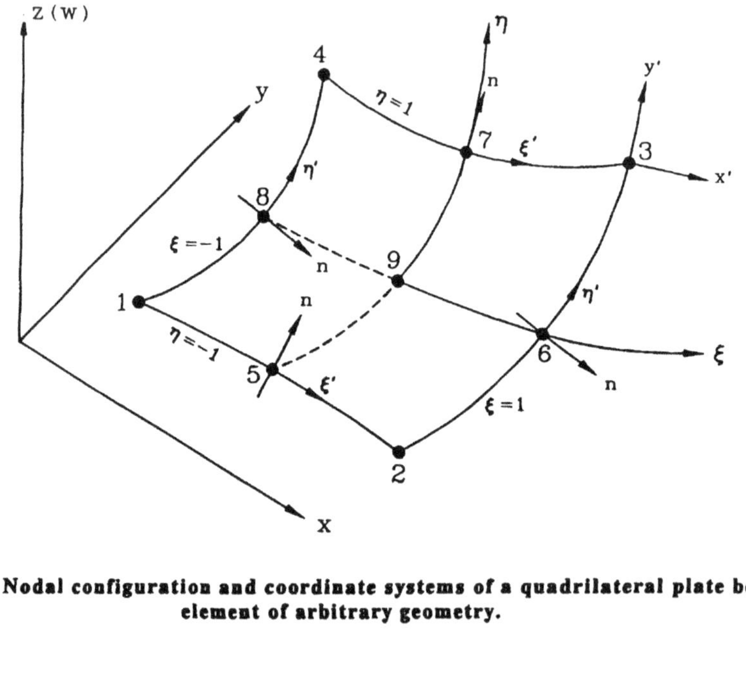

Figure 2. Nodal configuration and coordinate systems of a quadrilateral plate bending element of arbitrary geometry.

where t_i is the thickness of the element at node i and $F_i(\xi, \eta)$ is the corresponding shape function consisting of bilinear splines. Similarly, the load intensity of the element is

$$q = \sum_{i=1}^{9} F_i(\xi, \eta) q_i \tag{13}$$

where q_i is the load intensity of the element at node i .

For the purpose of allowing irregular element meshes to be used in the analysis, the element displacement parameters in the curvilinear co-ordinate system as given in Eq.11 must be expressed in some other co-ordinate systems for compatibility requirements. This necessitates the parameters $(\partial^2 w/\partial\xi\partial\eta)_i$ at the four corner nodes to be eliminated because transformation of second derivatives between two co-ordinate systems is possible only if all the components of second derivative in the two systems are involved. The eliminations can be achieved by applying some approximate displacement constraints [4] and the original 25 degrees-of-freedom of the element will be reduced to 21. Table 1 gives a summary of the original and final degrees-of-freedom associated with the various nodes of the element.

As the present formulation follows the standard displacement approach, all the nodal parameters of the element are displacement degrees-of- freedom. The boundary conditions for a simply supported edge are

$$w=0 \quad , \quad \frac{\partial w}{\partial S} = 0 \tag{14}$$

where S denotes the axis along the boundary and for a clamped edge

$$w=0 \quad , \quad \frac{\partial w}{\partial x} = 0 \quad , \quad \frac{\partial w}{\partial y} = 0 \tag{15}$$

For curved or skew edges, additional transformations must be applied to the element matrices in order to impose the boundary conditions of Eq.14.

A fuller description of the element formulation is available in reference [4].

TABLE 1

Original and final degrees-of-freedom for plate element (see Fig.2)

Node No.	Original D.O.F.	Final D.O.F.
1, 2, 3, 4	$w, \frac{\partial w}{\partial \xi}, \frac{\partial w}{\partial \eta}, \frac{\partial^2 w}{\partial \xi \partial \eta}$	$w, \frac{\partial w}{\partial x}, \frac{\partial w}{\partial y}$
5, 7	$w, \frac{\partial w}{\partial \eta}$	$w, \frac{\partial w}{\partial n}$
6, 8	$w, \frac{\partial w}{\partial \xi}$	$w, \frac{\partial w}{\partial n}$
9	w	w
Total No. of D.O.F.	25	21

AN AXISYMMETRIC THIN SHELL ELEMENT

In this section, a curved axisymmetric thin shell element is presented. The analysis is restricted to linearly elastic, isotropic and homogeneous materials with small deformations only. The classical thin shell theory due to Novozhilov [50] is assumed. A fuller description of the element formulation is available in reference[1].

For axisymmetric shells, the geometry is fully defined by the meridional curve. By using the shape functions given in Eq.9, the meridional curve of a typical element (see Fig.3) may be described parametrically as

$$\begin{Bmatrix} r \\ z \end{Bmatrix} = \begin{bmatrix} N_1 & 0 & N_2 & 0 & N_3 & 0 & N_4 & 0 & N_5 & 0 & N_6 & 0 \\ 0 & N_1 & 0 & N_2 & 0 & N_3 & 0 & N_4 & 0 & N_5 & 0 & N_6 \end{bmatrix} \{\zeta_i\} \tag{16}$$

where

$$N_1 = H_{01}\ ;\ N_2 = H_{02}\ ;\ N_3 = H_{03}\ ; N_4 = H_{31}\ ; N_5 = H_{32}\ ; N_6 = H_{33} \tag{17}$$

and

$$\{\zeta_i\} = \{r_1\ \ z_1\ \ r_1'\ \ z_1'\ \ r_1''\ \ z_1''\ \ r_2\ \ z_2\ \ r_2'\ \ z_2'\ \ r_2''\ \ z_2''\}^T \tag{18}$$

is a set of nodal parameters in which r and z are respectively the radial and vertical coordinates. The primes denote derivatives with respect to the meridional length S.

The displacement function for the axisymmetric shell element (Fig.3) is expressed as a product of B-spline shape functions (Eq.9) in the meridional direction and harmonic functions in the circumferential direction. In concise matrix form,

$$\{ f_j \} = [C_j][N]\{\delta_{ij}\} \tag{19}$$

where

$$\{ f_j \} = [\ u_j\ \ w_j\ \ v_j\ ]^T \tag{20}$$

$$[C_j] = \begin{bmatrix} \cos j\theta & 0 & 0 \\ 0 & \cos j\theta & 0 \\ 0 & 0 & \sin j\theta \end{bmatrix} \tag{21}$$

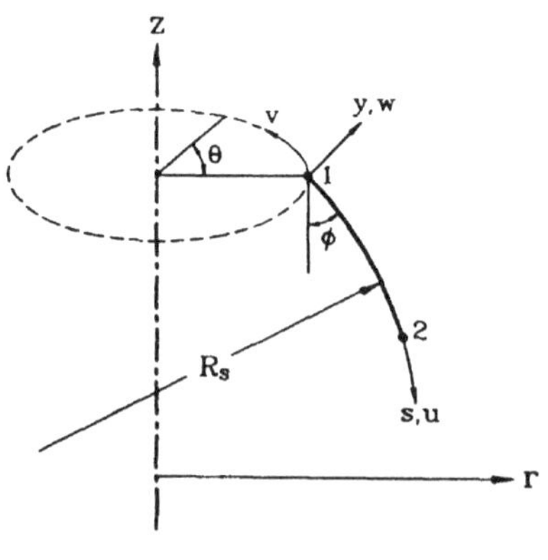

Figure 3. Curved axisymmetric shell element.

$$[N] = \begin{bmatrix} N_1\ 0\ 0 & N_2\ 0\ 0 & N_3\ 0\ 0 & N_4\ 0\ 0 & N_5\ 0\ 0 & N_6\ 0\ 0 \\ 0\ N_1\ 0 & 0\ N_2\ 0 & 0\ N_3\ 0 & 0\ N_4\ 0 & 0\ N_5\ 0 & 0\ N_6\ 0 \\ 0\ 0\ N_1 & 0\ 0\ N_2 & 0\ 0\ N_3 & 0\ 0\ N_4 & 0\ 0\ N_5 & 0\ 0\ N_6 \end{bmatrix} \tag{22}$$

and

$$\{\delta_{ij}\} = [u_{1j}\ w_{1j}\ v_{1j}\ u_{1j}'\ w_{1j}'\ v_{1j}'\ u_{1j}''\ w_{1j}''\ v_{1j}''\ u_{2j}\ w_{2j}\ v_{2j}\ u_{2j}'\ w_{2j}'\ v_{2j}'\ u_{2j}''\ w_{2j}''\ v_{2j}'']^T \tag{23}$$

in which u_j, w_j and v_j are respectively the displacement components in the meridional, normal and circumferential directions of the element corresponding to j circumferential waves and θ is the circumferential angle. The displacement components with subscripts ij represent the values of the displacements at node i with j circumferential waves along the reference meridians where the associated harmonic functions take the value of unity.

NUMERICAL EXAMPLES

Two numerical examples, one for each of the elements discussed are presented to show the performance of the element. The common notations used in the examples for structural properties are: E for Young's modulus, ν for Poisson's ratio, ρ for mass density, ω for frequency in radians per second, and t for thickness of the element.

Free Vibration and Static Analysis of Circular Plate

The capability of a general plate element to analyse circular plates using irregular meshes is of great importance as it implies that any plate of arbitrary geometry can likewise be analysed. In this example, a quadrant of the plate is modelled using the 9-node quadrilateral element and the mesh pattern employed is shown in Fig.4. Table 2 presents the frequency parameters obtained for the lowest 8 modes of vibration of circular plates with simply supported edge. The theoretical results of Leissa and Narita [51] are quoted for comparison purposes. The accuracy of the present element is clearly exhibited as even with the 3 elements scheme; the errors are rather negligible. In the present approach, since only one quadrant of the plate is used for analysis, the boundary conditions along the lines of symmetry of the plate have to be adjusted accordingly to obtain the various modes of vibration.

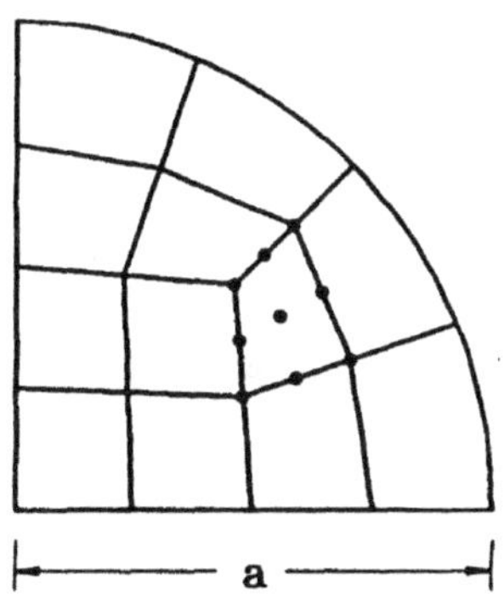

Figure 4. A typical 12-element mesh for a quadrant of a circular plate.

Table 2

Natural frequency parameters λ for simply supported circular plate [n = no. of nodal diameters, s = no. of interior nodal circles, $\lambda^2 = 12(1-\nu^2)\rho\omega^2 a^4/(Et^2)$, $\nu = 0.3$]

Mode Sequence (s , n)	3-Element λ	3-Element % error	12-Element λ	12-Element % error	27-Element λ	27-Element % error	Leissa & Narita[51]
(0,0)	4.92684	-0.17	4.93378	-0.03	4.93461	-0.01	4.93515
(0,1)	13.8631	-0.25	13.8893	-0.06	13.8943	-0.03	13.8982
(0,2)	25.5133	-0.39	25.5951	-0.07	25.6055	-0.03	25.6133
(1,0)	29.6706	-0.17	29.6938	-0.09	29.7084	-0.04	29.7200
(0,3)	39.7436	-0.53	39.9288	-0.07	39.9448	-0.03	39.9573
(1,1)	48.4204	-0.12	48.4551	-0.05	48.4619	-0.04	48.4789
(0,4)	55.7973	-1.84	56.8063	-0.06	56.8237	-0.03	56.8416
(1,2)	70.1728	0.08	70.1822	0.09	70.1108	-0.01	70.1170

For the static analysis, both simply supported and clamped edges are considered, and the loading is either a uniformly distributed load or a central point load. The central deflection normalized by exact result from thin plate theory[52] for each of the cases analysed is given in Table 3. It can be seen that very accurate results are obtained even if only one element is used.

Table 3

Normalized central deflection (W/W_{exact}) of circular plate under various loading and boundary conditions

No. of Elements	Uniformly Distributed Load		Central Point Load	
	Simply Supported	Clamped	Simply Supported	Clamped
1	0.9979	0.9964	0.9837	0.9618
3	1.0028	1.0059	0.9988	0.9945
12	1.0004	1.0006	0.9994	0.9980
27	1.0002	1.0003	0.9998	0.9992

Free Vibration Analysis of Fixed-Free Hyperboloidal Shell

The natural frequencies of a sample hyperboloidal shell given by Carter et al.[53] were investigated. The geometry of the shell is shown in Fig.5 and the following geometric and material properties were used in the analysis: a = 25.603m; b = 63.906m; d = 82.194m; L = 100.787m; t = 127mm; E = 2.069×10^4 MN/m^2; $\nu = 0.15$ and $\rho = 2405$kg/m^3. The results of the present study using 3 and 8 axisymmetric elements are given in Table 4 together with those obtained by Carter et al.[53] using a numerical integration technique. Remarkable agreement in the results is observed. However, the torsional mode obtained by the present method was not reported in the previous studies[53-55].

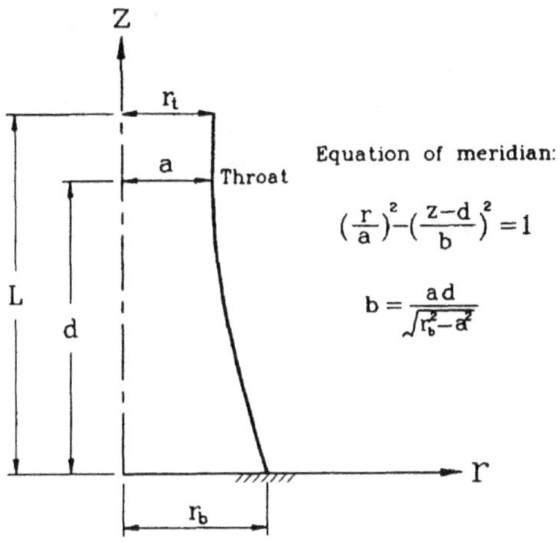

Figure 5. Geometry of hyperboloidal shell.

Table 4
Natural Frequency in Hertz of a fixed-free hyperboloidal shell

j	Mode number	Carter et al.[53]	3 elements	8 elements
0	Torsional mode	-	6.2297	6.2305
	1	7.7494	7.7502	7.7490
	2	11.4167	11.7729	11.4162
	3	11.9023	12.8201	11.9010
1	1	3.2884	3.2894	3.2885
	2	6.7905	6.7926	6.7906
	3	10.5207	10.5475	10.5182
2	1	1.7654	1.7672	1.7657
	2	3.6931	3.6898	3.6880
	3	6.9562	6.9512	6.9529
3	1	1.3749	1.3786	1.3755
	2	1.9904	1.9904	1.9928
	3	4.3254	4.3346	4.3353
4	1	1.1808	1.1776	1.1809
	2	1.4475	1.4531	1.4474
	3	2.7777	2.7772	2.7768
5	1	1.0348	1.0282	1.0331
	2	1.4293	1.4442	1.4290
	3	2.0559	2.0522	2.0562
6	1	1.1467	1.1477	1.1478
	2	1.3231	1.3335	1.3250
	3	2.0141	2.0089	2.0148
7	1	1.3015	1.3079	1.3027
	2	1.5134	1.5097	1.5140
	3	1.9217	1.9111	1.9221

CONCLUDING REMARKS

This paper has given a brief review on the application of spline functions in engineering analysis and introduced a new interpolating scheme for finite element formulation using B_3-spline functions. Two spline finite elements for the analysis of plates and shells were outlined and sample numerical results were given. From the theoretical considerations and numerical performance of the elements, the following remarks can be made.

1. The spline finite element method incorporates both the merits of spline interpolation and the generality and flexibility of conventional finite element method. The elements developed are therefore highly accurate and efficient without any loss of general applicability and simplicity.

2. The present approach is particularly suitable for thin plate and shell elements requiring C^1 continuity of the displacements.

In recent years, most of the research endeavour has been directed towards the development of Mindlin-type elements with the capability of thin plate and shell analysis. However, it is believed that for thin plate and shell problems, Kirchhoff-type elements are more efficient and reliable. As thin plate and shell structures are very commonly used in engineering practice, the development of highly efficient C^1 elements exclusively for their analysis is worthwhile. Spline functions have the advantages of higher accuracy and continuity with lower order compared to conventional polynomials and these are the characteristics particularly desirable for elements with C^1 requirement. The use of spline functions to improve the potential of C^1 elements is promising and the research carried out by the authors represents an effort in this direction.

REFERENCES

1. Fan,S.C. and Luah,M.H., New spline finite element for analysis of shells of revolution. J. Engng. Mech., ASCE,1990, **116**, 709-726.
2. Fan,S.C. and Luah,M.H.,Spline finite element for axisymmetric free vibrations of shells of revolution. J.Sound Vibr.,1989, **132**, 61-72.
3. Luah,M.H. and Fan,S.C.,General free vibration analysis of shells of revolution using the spline finite element method. Comput. Struct.,1989, **33**, 1153-1162.
4. Fan,S.C. and Luah,M.H., A new spline finite element for plate bending. J. Engng. Mech., ASCE, **(to appear)**.
5. Fan,S.C. and Luah,M.H., A 9-node spline element for free vibration analysis of general plates. J.Sound Vibr., **(to appear)**.
6. Luah,M.H. and Fan,S.C., Analysis of complicated plates by a 9-node spline plate element. Comput. Struct., **(to appear)**.
7. Schoenberg,I.J.,Contributions to the problem of approximation of equidistant data by analytic functions. Part A : on the problems of smoothing of graduation. A first class of analytical approximation formulae. Quarterly J. Appl. Math., 1946, **4**, 45-99.
8. Schoenberg,I.J.,Contributions to the problem of approximation of equidistant data by analytic functions. Part B : on the problem of osculatory interpolation. A second class of analytical approximation formulae. Quarterly J. Appl. Math., 1946, **4**, 112-141.
9. DeBoor,C., A practical guide to the splines (Appl. Math. Sci.,Vol.27). Springer, New York, 1978.

10. Schumaker,L.L., Spline Functions : Basic Theory, John Wiley, New York,1981.

11. Antes,H., Bicubic fundamental splines in plate bending. Int. J. Numer. Meth. Engng., 1974, **8**, 503-511.

12. Mizusawa,T., Kajita,T. and Naruoka,M., Vibration of stiffened skew plates by using B-spline functions. Comput. Struct., 1979, **10**, 821-826.

13. Mizusawa,T., Kajita,T. and Naruoka,M., Buckling of skew plate structures using B-spline functions. Int. J. Numer. Meth. Engng., 1980, **15**, 87-96.

14. Fujii,F., Discrete and non-discrete mixed methods for plate bending analysis. Int. J. Numer. Meth. Engng., 1981, **17**, 1843-1859.

15. Fujii,F. and Hoshino,T., Discrete and non-discrete mixed methods applied to eigenvalue problems of plates. J. Sound Vibr., 1983, **87**, 525-534.

16. Shen,P.C. and Wang,J.G., Vibration analysis of flat shells by using B-spline functions. Comput. Struct., 1987, **25**, 1-10.

17. Shen,P.C. and Wang,J.G., Static analysis of cylindrical shells by using B-spline functions. Comput. Struct., 1987, **25**, 809-816.

18. Mizusawa,T. and Kajita,T., Vibration and buckling of skew plates with edges elastically restrained against rotation. Comput. Struct., 1986, **22**, 987-994 .

19. Mizusawa,T., Application of spline strip method to analyse vibration of open cylindrical shells. Int. J. Numer. Meth. Engng., 1988, **26**, 663-676.

20. Cheung,Y.K., Fan,S.C. and Wu,C.Q., Spline finite strip in structural analysis. Proc. of Int. Conf. on Finite Element Methods, Shanghai, China, 1982,704-709.

21. Cheung,Y.K. and Fan,S.C., Static analysis of right box girder bridges by spline finite strip method. Proc. Instn. Civ. Engrs., Part 2, 1983, **75**, 311-323.

22. Fan,S.C. and Cheung,Y.K., Analysis of shallow shells by spline finite strip method. Engng. Struct., 1983, **5**, 255-263.

23. Fan,S.C. and Cheung,Y.K., Flexural free vibration of rectangular plates with complex support conditions. J.Sound Vibr., 1984, **93**, 81-94.

24. Tham,L.G., Li,W.Y. and Cheung,Y.K., Bending of skew plates by spline-finite-strip method. Comput. Struct., 1986, **22**, 31-38.

25. Li,W.Y., Cheung,Y.K. and Tham,L.G., Spline finite strip analysis of general plates. J. Engng. Mech., ASCE, 1986, **112**, 43-54.

26. Cheung,Y.K., Tham,L.G. and Li,W.Y., Application of spline-finite-strip method in the analysis of curved slab bridge. Proc. Instn. Civ. Engrs., Part 2, 1986, **81**, 111-124.

27. Lau,S.C.W. and Hancock,G.J., Buckling of thin flat-walled structures by a spline finite strip method. Thin-walled Struct., 1986, **4**, 269-294.

28. Cheung,Y.K. and Zhu,D.S., Large deflection analysis of arbitrary shaped thin plates. Comput. Struct., 1987, **26**, 811-814.

29. Cheung,Y.K., Tham,L.G. and Li,W.Y., Free vibration and static analysis of general plate by spline finite strip. Computational Mech., 1988, **3**, 187-197.

30. Cheung,Y.K., Li,W.Y. and Tham,L.G., Free vibration analysis of single curved shell by spline finite strip method. J. Sound Vibr., 1989, **115**, 411-422.

31. Cheung,Y.K. and Zhu,D.S., Postbuckling analysis of shells by spline finite strip method. Comput. Struct., 1989, **31**, 357-364.

32. Lau,S.C.W. and Hancock,G.J., Inelastic buckling analyses of beams, columns and plates using the spline finite strip method. Thin-walled Struct., 1989, **7**, 213-238.

33. Tham,L.G. and Szeto,H.Y., Buckling analysis of arbitrarily shaped plates by spline finite strip method. Comput. Struct., 1990, **36**, 729-735.

34. Vanerp,G.M. and Menken,C.M.,The spline finite-strip method in the buckling analysis of thin-walled structures. Communications in Appl. Numer. Meth., 1990, **6**, 477-484.

35. Li,W.Y., Tham,L.G., Cheung,Y.K. and Fan,S.C., Free vibration analysis of doubly curved shells by spline finite strip method. J. Sound Vibr., 1990, **140**, 39-53.

36. Geannakakes,G.N. and Wang,P.C., Moving load analysis of arbitrarily shaped plates using the B_3-spline finite strip method. J. Sound Vibr., 1990, **141**, 127-142.

37. Geannakakes,G.N. and Wang,P.C., Vibration analysis of arbitrarily-shaped shell panels using B_3-spline finite strips. Comput. Struct., 1991, **39**, 489-492.

38. Kwon,Y.B. and Hancock,G.J., A nonlinear elastic spline finite strip analysis for thin-walled sections. Thin-walled Struct., 1991, **12**, 295-319.

39. Clenshaw,C.W. and Negus,B., The cubic X-spline and its application to interpolation. J. Inst. Math. Appl., 1978, **22**, 109-119.

40. Yang,H.Y. and Chong,K.P., Finite strip method with X-spline functions. Comput. Struct., 1984, **18**, 127-132.

41. Yang,H.Y. and Zhang,Y.T., Assumed stress hybrid model in X-spline finite strip method. Comput. Struct., 1987, **25**, 83-93.

42. Chen,C.J., Gutkowski,R.M. and Puckett,J.A., B-spline compound strip analysis of stiffened plates under transverse loading. Comput. Struct., 1990, **34**, 337-347.

43. Gutkowski,R.M., Chen,C.J. and Puckett,J.A., Plate bending analysis by unequally spaced splines. Thin-walled Struct., 1991, **11**, 409-430.

44. Chen,C.J., Gutkowski,R.M. and Puckett,J.A., Spline compound strip analysis of folded plate structures with intermediate supports. Comput. Struct., 1991, **39**, 369-379.

45. Moore,C.J., Yang,T.Y. and Anderson,D.C.,A new 48 D.O.F quadrilateral shell element with variable-order polynomial and rational B-spline geometries with rigid body modes. Int. J. Numer. Meth. Engng., 1984, **20**, 2121-2141.

46. Kim,J., B-spline Finite Elements Analysis of Arbitrarily Loaded Shells of Revolution. PhD Thesis, Cornell University, 1983.

47. Leung,A.Y.T. and Au,F.T.K., Spline finite elements for beam and plate. Comput. Struct., 1990, **37**, 717-729.

48. Gupta,A., Kiusalaas,J. and Saraph,M., Cubic B-spline for finite element analysis of axisymmetric shells. Comput. Struct., 1991, **38**, 463-468.

49. Prenter,P.M., Splines and Variational Methods. John Wiley, New York, 1975.

50. Novozhilov,V.V., Thin Shell Theory. Translated from the second Russian edition by P.G. Lowe, P. Noordhoff Ltd, Groningen, The Netherlands, 1964.

51. Leissa,A.W. and Narita,Y., Natural frequencies of simply supported circular plates. J. Sound Vibr., 1980, **70**, 221-229.

52. Timoshenko,S.P. and Woinowsky-Krieger,S., Theory of Plates and Shells. 2nd Ed., McGraw-Hill, New York, 1959.

53. Carter,R.L., Robinson,A.R. and Schnobrich,W.C., Free vibrations of hyperboloidal shells of revolution. J. Engng. Mech., ASCE, 1969, **95**, 1033-1052.

54. Hashish,M.G. and Abu-Sitta,S.H., Free vibration of hyperbolic cooling towers. J. Engng. Mech., ASCE, 1971, **97**, 253-269.

55. Sen,S.K. and Gould,P.L., Free vibration of shells of revolution using FEM. J. Engng. Mech., ASCE, 1974, **100**, 283-303.

[1]THE RITZ-LANCZOS ALGORITHM IN STRUCTURAL DYNAMIC ANALYSIS

LI MINGRUI FENG YUNTIAN

BEIJING AGRICULTURAL ENGINEERING UNIVERSITY,
BEIJING, 100083, CHINA

ABSTRACT

A self-adaptive Ritz-Lanczos algorithm for finding a specified number of eigenpairs is proposed. The error estimation of dynamic response is thoroughly discussed. A shift method for high frequency excitation is suggested.

INTRODUCTION

In structural dynamic analysis people are usually concerned about two basic problems: (a) How to solve the eigenvalue problem of a structure effectively? (b) What is the dynamic response of a structure when the excitation is known?

The commonly used algorithms for dynamic response can be classified into two types: one is the mode superposition method and the other is the step by step integration method. The first one is also related to the eigenvalue problem. Since Wilson et al.[1] suggested that the Ritz vectors can be used instead of the eigen vectors in the mode superposition method, and the first several Ritz vectors are good approximate eigenvectors, one

[1]This project is supported by the National Natural Science Foundation of China.

now knows that the eigenvectors are a kind of Ritz vectors, and the mode superposition method is also a kind of Ritz analysis. As the ordinary mode superposition method considers only the dynamic characteristic of a structure, i.e. stiffness and mass, while the Ritz vector superposition method considers also the spatial distribution of the load vector, so we may use fewer vectors to obtain higher accuracy. People now concentrate on these fundamental problems. (1) Which kind of Ritz vector is supposed to be optimal? (2) Which algorithm will be the fastest for finding a specified number of eigenpairs ? (3) What is the accuracy of the dynamic response when the structure is reduced by a given number of Ritz vectors? We shall answer these questions in this paper but will not talk about other methods such as mode synthesis.

THE SELF-ADAPTIVE RITZ-LANCZOS ALGORITHM FOR EIGENVALUE PROBLEMS

The eigenvalue problem of very large structures is always an important and basic problem in the field of dynamic analysis. In 1950 Lanczos[2] proposed an algorithm of three terms recurrence formula to produce a set of orthogonal vectors, and then to transform the original matrix to a tridiagonal matrix. As there are many efficient solvers for the eigenvalue problem of a tridiagonal matrix such as the QL method, the Lanczos algorithm seems to be the fastest. Unfortunately it has a fatal defect, i.e. numerical instability caused by the loss of orthogonality, which has been thoroughly studied by Paige[3] in his thesis, and solved by other researchers[4,5] with various reorthogonalization methods. The following problem now remains: the Lanczos method produced a set of orthogonalized vectors, in which the first several vectors are good approximations to the eigenvectors. When the tolerance of eigenvalue error is prescribed and the number l of Lanczos vectors is required, the number of Lanczos vectors j is uncertain. It has been suggested[6] that $j = 2l$. Apparently this value does not relate to the prescribed tolerance and in most cases it is not sufficient. We propose the following Ritz-Lanczos (R-L) algorithm, slightly different from [1] but more

efficient.
The generalized eigenvalue problem is :

$$K\,\phi_i = M\,\phi_i\,\Omega_i^2 \quad (i=1,2,\ldots,n) \tag{2.1}$$

where K and M are stiffness and mass matrices respectively. (Ω_i^2, ϕ_i) is the ith eigen pair. The R-L algorithm is:

a)Take a vector q_1 as the initial vector, where q_1 is a M-orthogonalized vector defined by

$$||q_1||_M = (q_1^T\,M\,q_1)^{1/2} = 1$$

For the eigenvalue problem, q_1 should be a random vector including the possible components of all mode shapes.

b)Form an iteration process (i=1,2,...,j)

$$\beta_{i+1}\,q_{i+1} = K^{-1}\,M\,q_i - \alpha_i q_i - \beta_i q_{i-1} \tag{2.2}$$

where $\alpha_i = q_i^T\,M\,K^{-1}\,M\,q_i$; β_{i+1} is the M normalized factor of q_{i+1}, and $\beta_1 = 0$.

The above iteration process can be written in Matrix form:

$$K^{-1}\,M\,Q_j = Q_j\,T_j + \beta_{j+1}\,q_{j+1}\,e_j^T \tag{2.3}$$

where $Q_j = \{\, q_1,\ q_2,\ \ldots,\ q_j \,\}$

$$Q_j^T\,M\,Q_j = I_j; \qquad Q_j^T\,M\,q_{j+1} = 0 \tag{2.4}$$

I_j is an identity matrix of order j,

$$T_j = \begin{bmatrix} \alpha_1 & & & & \\ \beta_2 & \alpha_2 & & \text{sym} & \\ & \cdot & \cdot & & \\ & & \cdot & \cdot & \\ & & & \cdot & \cdot \\ & & & \beta_j & \alpha_j \end{bmatrix} \tag{2.5}$$

and e_j is the jth column vector of I_j. We will call q_i (i= 1,2, ...j) the Lanczos vectors in the following. (2.2) may be replaced by a reorthogonalized step if necessary.

c) Solve the eigenvalue problem of the tridiagonal matrix T_j

$$T_j \; S_j = S_j \; \Sigma_j \qquad (2.6)$$
$$S_j^T \; S_j = I_j$$
$$\Sigma_j = \text{diag} \{ \sigma_1, \sigma_2, \dots, \sigma_j \}$$

d) Compute vectors $X_j = \{ x_1, x_2, \dots, x_j \}$

$$X_j = Q_j \; S_j \qquad (2.7)$$

We will call x_i $(i=1,2,\dots,j)$ the Ritz vectors in the following.
e) Check the accuracy. Define the relative error

$$\mu_i^{(j)} = | (1/\sigma_i^{(j)} - \Omega_i^2) \sigma_i^{(j)} | = | 1 - \Omega_i^2 \sigma_i^{(j)} | \qquad (2.8)$$

where Ω_i^2 is the ith unknown eigenvalue of (2.1). If $\mu_i^{(j)} < \epsilon$, where ϵ is a prescribed tolerance for eigenvalue error, then $(1/\sigma_i^{(j)}, x_i)$ is the ith approximate eigenpair of the original problem (2.1). There are a number of papers suggesting various methods for computing $\mu_i^{(j)}$, but we recommend [6] as a more preferable method. The algorithm can be summed up as follows: if $\sigma_i^{(j)}$ is the ith eigenvalue of T_j, we could find an eigenvalue $\sigma_i^{(k)}$ of T_k satisfying

$$| \sigma_i^{(j)} - \sigma_i^{(k)} | \le | \beta_{j+1} s_{ji} | \qquad (2.9)$$

where $k > j$ and s_{ji} is the ith component of the last row of S_j. We know[8] that when Lanczos algorithm converges accurately at the mth step, i.e. $\beta_{m+1} = 0$, the reciprocals of all eigenvalues of T_m are the exact eigenvalues of(2.1). Obviously we will have $m \ge j$. From (2.9) we may deduce

$$| \sigma_i^{(j)} - 1/\Omega_i^2 | \le | \beta_{j+1} s_{ji} | = c_{ji} .$$

Then the relative error may be expressed as

$$\mu_i^{(j)} = c_{ji} \Omega_i^2 = c_{ji} / \sigma_i^{(j)} \qquad (2.10)$$

Calculate the number of $\sigma_i^{(j)}$ $(i=1,2,\dots, j)$ which satisfy $\mu_i^{(j)} < \epsilon$, and is denoted by $m(j,\epsilon)$. We may thus present the self-adaptive Ritz-Lanczos (R-L) algorithm:

1) For a prescribed tolerance ϵ and a number l of eigen pairs to

be found, if the iteration number $j < l$, the convergence test is not required.

2) As $j \geq l$, compute the last row of S_j, while the eigenvectors do not need to be found. The amount of work for finding the last row of S_j is very little . This can be done by made a small modification of the standard QL algorithm.

3) When $m(j,\varepsilon) = l$, compute all the eigenpairs. The first l eigenpairs are wanted.

In practice, this self-adaptive R-L algorithm is about 5 to 10 times faster than any of the modified or improved subspace iteration algorithms. The larger the problem is, the more efficient the algorithm.

DYNAMIC RESPONSE AND ERROR ESTIMATION

Now we will study the following dynamic governing equation.

$$M\ u'' + C\ u' + K\ u = F(x,t) \tag{3.1}$$

Assume that the excitation takes the form $F(x,t) = f(x)g(t)$. We will call $f(x)$ the spatial distribution vector of load. Also assume that the damping matrix C is a Rayleigh damping, i.e.

$$C = \alpha\ M + \beta\ K \tag{3.2}$$

From (2.3) we will have

$$Q_j^T\ M\ K^{-1}\ M\ Q_j = T_j \tag{3.3}$$

where j is the number of Lanczos vectors we have taken to reduce the structure. This time we will take the initial vector q_1 of R-L algorithm to satisfy :

$$K^{-1}\ f = \beta_1\ q_1 = \beta_1\ Q_j\ e_1 \tag{3.4}$$

that means q_1 is the static deformation generated by $f(x)$. Substitute the approximate response u to $Q_j\ y$, then multiply

(3.1) by $Q_j^T M K^{-1}$ from the left side, and make use of (3.2), (3.3), (3.4); thus we may transform (3.1) to

$$T_j y'' + (\alpha T_j + \beta) y' + y = \beta_1 e_1 g(t) \tag{3.5}$$

Since we use only j Lanczos vectors to approximate the actual response u^*, the approximate response u will not satisfy (3.1), and we will have

$$M u'' + C u' + K u = F(x,t) - dF \tag{3.6}$$

Here dF is the unbalanced force, and the error vector du will be $du = u^* - u$; then we will have

$$M du'' + C du' + K du = dF \tag{3.7}$$

Starting from the R-L algorithm (2.2), and making use of (3.2)-(3.6), we have

$$M u'' + C u' + K u = M Q_j y'' + (\alpha M + \beta K) Q_j y' + K Q_j y$$

$$= K[K^{-1} M Q_j y'' + (\alpha K^{-1} M + \beta) Q_j y'' + Q_j y']$$

$$= K[Q_j T_j y'' + Q_j(\alpha T_j + \beta) y' + Q_j y + \beta_{j+1} q_{j+1} e_j^T (y'' + \alpha y')]$$

$$= K Q_j \beta_1 e_1 g(t) + \beta_{j+1} K q_{j+1} e_j^T (y'' + \alpha y')$$

$$= f(x) g(t) + \beta_{j+1} K q_{j+1} e_j^T (y'' + \alpha y')$$

So

$$dF = f(x) g(t) - (Mu'' + Cu' + Ku)$$

$$= -\beta_{j+1} K q_{j+1} \delta_j \tag{3.8}$$

where

$$\delta_j = e_j^T (y'' + \alpha y') \tag{3.9}$$

We may see that the unbalanced force dF consists of two parts: the first part $\beta_{j+1} K q_{j+1}$ indicates the error of how the Lanczos vectors approximate the static loading vector f(x), and the second part δ_j represents the influence of the dynamic response of the reduced structure. From (3.8) we may obtain the first error estimation formula:

$$\mu_1 = || K^{-1} dF ||_M / || K^{-1} F ||_M$$

$$= (\beta_{j+1} / \beta_1)*|\delta_j / g(t)| \qquad (3.10)$$

In the following two special cases we might obtain an accurate response, i.e. $\mu_1 = 0$.

1) R-L algorithm converges accurately, i.e. $\beta_{j+1} = 0$.

2) The external load is only static, i.e. $\delta_j = 0$.

Generally speaking, μ_1 represents the relative error of static deformation caused by the unbalanced force. Because the dynamic magnification has not yet been taken into consideration, the estimation (3.10) should be improved furthermore. Let us study a structural response, free of damping, excited by a sinusoidal loading

$$M u'' + K u = g(t) = \sin(pt).$$

The steady response could be solved precisely,

$$u^* = (K - p^2 M)^{-1} g(t).$$

If we use the R-L algorithm and truncate at j steps, the unbalanced force will be $dF = -\beta_{j+1} (K - p^2 M)^{-1} K q_{j+1} \delta_j$, where $\delta_j = -p^2 e_j^T(I - p^2 T_j)^{-1} e_1 g(t)$.

The displacement excited by dF will be

$$du = -\beta_{j+1} (K - p^2 M)^{-1} K q_{j+1} \delta_j.$$

Thus we may acquire an accurate error estimation

$$\mu = || du ||_M / || u ||_M .$$

Assuming that the R-L algorithm converges at the mth step, we may have

$$|| du ||_M = || (I - p^2 T_m)^{-1} e_{j+1} || * \beta_{j+1} * | \delta_j |$$

$$|| u ||_M = || (I - p^2 T_m)^{-1} e_1 || * \beta_1 * | g(t) |$$

So

$$\mu = c(p,T_m) * (\beta_{j+1} / \beta_1) * | \delta_j / g(t) | \qquad (3.11)$$

where $c(p,T_m) = || (I-p^2 T_m)^{-1} e_{j+1} || / || (I-p^2 T_m)^{-1} e_1 ||$ may be called the dynamic magnification factor.

Comparing (3.11) with (3.10) we can see that the only modification made is to add a dynamic magnification factor to (3.10). In practical computation, the T_m is not known and may be replaced by T_j. Moreover, the loading may consist of several frequencies; we have to use p^* to replace p, where p^* is defined as

$$p^* = || y'' || / || y || = || u'' || / || u || \quad (3.12)$$

Here, p^* is called the current equivalent frequency. In case g(t) may take zero value at a certain time, g(t) may also be replaced by $g(t)^*$, where $g(t)^*$ may be defined as the mean or maximum value in the time interval $t_i < t \leq t_i+dt$ to be evaluated. Finally the improved error estimation can be written in the form:

$$\mu = c(p^*, T_j) * (\beta_{j+1} / \beta_1) * | \delta_j / g(t)^* | \quad (3.13)$$

By means of (3.13) we may estimate the error of the dynamic response at each time step. Numerical results [9] show that this is a very good error estimation formula, especially in the low frequency range.

THE SHIFT R-L ALGORITHM

In the above R-L algorithm, the Lanczos vectors are generated by considering the structural dynamic characteristic (K,M) and the load spatial distribution (f(x)), but the frequency characteristic of the load has not yet been taken into consideration. Because of the power property of the R-L algorithm, the vectors thus generated could simulate the dynamic property of the original structure in the low frequency range satisfactorily. When the load frequency rises from low to high, the difference between the reduced structure and the original structure may be apparent. One way out is to increase the number of Lanczos vectors , but this will greatly increase the amount of work. Especially in the case of narrow-banded high frequency external excitation, the above method is obviously not cost-worthy. Instead, we suggest the following shift R-L algorithm. The basic idea is to make a shift of p^2 to the stiffness matrix in advance,

$$K^* = K - p^2 M \tag{4.1}$$

Here p is a frequency at which the excitation mainly works. The only change made in the R-L algorithm is to replace K by K^*. The equation (3.5) then becomes

$$T_j y'' + [\alpha T_j + \beta(I_j + p^2 T_j)]y' + (I_j + p^2 T_j)y$$

$$= \beta_1 e_1 g(t) \tag{4.2}$$

Equation (2.9) will be replaced by

$$\delta_j = e_j^T [y'' + (\alpha + \beta p^2)y' + p^2 y] \tag{4.3}$$

Numerical results[9] show that when the excitation frequency approaches the shift frequency, very good accuracy could be obtained. When the deviation of the two frequencies increases, the accuracy goes down. When the two frequencies deviate far apart, there will be no difference between the shift R-L and the ordinary R-L algorithm. This is exactly what we expected. So the shift R-L algorithm can handle the structural dynamic response excited by narrow-banded high frequency loading very well. If the excitation undergoes a broad frequency range but varies slowly, we may use the shift R-L algorithm in different ranges.

CONCLUSIONS AND PROSPECTS

In this paper we have proposed the R-L algorithm which solves the following problems satisfactorily:

1) Compute the number of Lanczos vectors automatically to generate a set of eigenvectors satisfying a prescribed error tolerance.

2) Estimate the error of dynamic response at each time step by using a specified number of Lanczos vectors to reduce the struc-

ture.

3) Handle the narrow-banded high frequency excitation.

By means of the above presentation we may treat any kind of Ritz vectors superposition method as well. In other words, the theoretical foundation of the Ritz vector superposition method is thus formed. The error estimation method we have suggested here is a kind of post-estimation. That means: firstly, obtain j Lanczos vectors; secondly, find out the dynamic response; thirdly, perform the error estimation. A better scheme may be called the pre-estimation is to do the error estimation beforehand or perform the eigenvalue solving process and the error estimation alternately. This will be a hard task but very important. Hope it will be solved soon.

REFERENCES

1. E.L.Wilson, M.Yuan, J.M.Dickens, Dynamic analysis by direct superposition of Ritz vectors, Earthquake Eng. Struct. Dyn. 10,-813-821, 1983.

2. C.Lanczos, An iteration method for the solution of the eigenvalue problem of linear differential and integral operators, J. Res. Natl. Bur. Standards, 45,255-282, 1950.

3. C.C.Paige, The computation of eigenvalues and eigenvectors of very large sparse matrices, Ph.D. Thesis, Univ. of London, 1971.

4. H.D.Simon, The Lanczos algorithm for solving symmetric linear system, Ph.D. Thesis, Univ. of California, Berkeley, 1982.

5. B.N.Parlette, The Lanczos algorithm with selective orthogonalization, Math. Comp. 33,217-228, 1979.

6. M.Yuan, The Manual of SAP-84, Beijing Univ., 1989.

7. Erxiong Jiang, The computation of symmetric matrix. Shanghai Science and Technology Press. 1984.

8. Feng Yuntian, Li Mingrui, A study on the direct superposition of Ritz vectors, J. of Beijing Agricultural Eng. Univ. (English edition) Vol.1, 1987.

9. Feng Yuntian, The study on the direct superposition of Ritz vectors and its implementation in structural dynamic analysis, Ph.D. Thesis, Beijing Agricultural Eng. Univ., 1988.

RIGID BODY CRITERIA IN NONLINEAR ELEMENT FORMULATION

YEONG-BIN YANG
Department of Civil Engineering
National Taiwan University
Taipei, Taiwan 10764, China

ABSTRACT

According to the rule of rigid body motion, a body initially stressed by a set of forces in equilibrium will remain both in shape and in equilibrium when subjected to a rigid body motion. The result is that the magnitude of the forces acting on the body will remain unchanged, while their direction will follow that of the rigid body rotation. Such a rule represents the minimum requirement that must be satisfied by a nonlinear finite element for it to be qualified in an incremental-iterative nonlinear analysis. Whenever such a rule is violated, artificial forces may be induced in the finite element during the rigid body motion. Imbalances due to such artificial forces may then be accumulated and carried over to the iteration process, resulting in various computational problems such as incorrect load-deflection curves, and slow convergence or numerical overflow in the iteration. The purpose of the paper is not to derive any new theory or present any numerical model, but to highlight the importance of rigid body criteria in the formulation of nonlinear finite elements, through application of the rigid body test and general eigenvalue test. For this purpose, the two-dimensional beam element derived in a previous study is selected as the vehicle of illustration. Emphasis of the paper is placed on the physical interpretation of the test methods and their potential applications to other nonlinear problems.

INTRODUCTION

The rigid body concept has played a very important role in the 1960s and 1970s in the development of new finite elements to meet various engineering needs [1-7]. In most of the applications, the concept has been interpreted as the condition that a finite element must exhibit zero strains and zero forces when subjected to rigid body motions. Such a condition was implied in the famous patch test originated by Irons [1], which requires a patch of finite elements to display the same state of constant stress as the one consistently applied at the boundary nodes. It is easy to see that the patch test includes the rigid body condition as the special case of zero stress. The rigid body condition also constitutes the underlying principle of the conventional eigenvalue test, which requires the correct number of zero eigenvalues or rigid body modes to be solved from the stiffness matrix of an unrestrained element [8].

One feature with the previous applications of rigid body concept is that they are restricted to linear problems, in which a finite element is assumed to be free of any initial stresses or nodal forces. In a geometrically nonlinear analysis, however, a finite element will in general be deformed and acted upon by a set of nodal forces in equilibrium at the beginning of the incremental step. For this reason, the patch test and eigenvalue test, which have been originally devised for linear elements, cannot be directly applied to the nonlinear cases.

When the geometric nonlinearity of a structure is concerned, it is necessary to introduce a geometric stiffness matrix in addition to the elastic stiffness matrix. The elastic stiffness matrix, being derived from the strain energy, will generate zero element forces when subjected to rigid body motion, if it is properly derived or if it passes the patch test. The geometric stiffness matrix, which represents the effect of initial forces on the finite element due to geometry change, will in general produce nonzero forces when subjected to rigid body motion. Thus, the geometric stiffness matrix contains the most important message for describing the behavior of a preloaded element during a rigid body motion. However, to examine the quality of a nonlinear finite element, one cannot rely solely on the geometric stiffness matrix, but must consider the incremental stiffness equation as a whole to see if it can really cope with the rigid body motion [9].

To the writer's knowledge, the problem of element quality for nonlinear problems has not received adequate attention from researchers in the past. In reference [9], the rigid body test has been proposed for testing a two-dimensional beam element using the incremental stiffness equation. The curved beam elements derived in [10] have been demonstrated to pass such a test [11]. Of interest is the study on the stretching and rigid body behaviors of a higher-order truss element presented in [12], which indicates that whether a higher-order term should be included in the stiffness formulation of a finite element depends not only on the order of the term itself, but also on how it affects the rigid body behavior. In Reference [13], the concept of rigid body rotation was used to derive an external stiffness matrix for an initially stressed element. This matrix, together with the elastic and geometric stiffness matrices, was then used in a general eigenvalue test for checking the number of rigid body modes contained by the initially stressed finite element.

In this paper, the concept of rigid body motion will first be reviewed for an initially stressed element, along with introduction of the rigid body test. The general eigenvalue test proposed in [13] considering the effects of initial forces will also be summarized. For the purpose of illustration, an updated Lagrangian formulation will be adopted for deriving the incremental stiffness equation for nonlinear finite elements, based on the principle of virtual displacements. With this formulation, all physical relations are defined for the current deformed configuration, but referred to the last calculated configuration. Though the present discussion is restricted to elements of the beam type, the procedures are so general that they can be applied to elements of other types. In the concluding remarks, highlights are given on a number of research items that remain to be conducted in the future concerning the rigid body aspects of nonlinear finite element analysis.

Before ending this section, it should be noted that the rigid body concept serves not only as a useful tool for detecting the theoretical flaws of a finite element in the formulation stage, but as a debugging aid for finding the coding errors in the implementation of a nonlinear analysis program.

PRINCIPLE OF VIRTUAL DISPLACEMENTS

As shown in Fig. 1, the history of motion of a body can be described by three typical configurations, i.e., the initial undeformed state (C_0), last (known) equilibrium state (C_1), and current (desired) equilibrium state (C_2). In the present study, a left superscript of a quantity indicates the configuration in which the quantity occurs, and a subscript the configuration in which the quantity is measured. The absence of a superscript indicates that the associated quantity is an increment between C_1 and C_2.

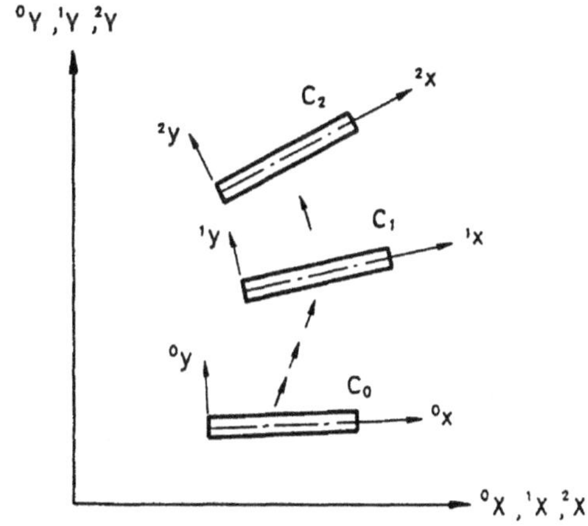

Figure 1. Motion of a beam element.

Neglecting the body forces, the equilibrium of a body at the C_2 state can be expressed with reference to the known C_1 state as follows [14]:

$$\int_{^1V} {}^2_1S_{ij}\,\delta_1\varepsilon_{ij}\,d^1V = {}^2_1R \qquad (1)$$

in which the external virtual work ${}^2_1R$ is

$${}^2_1R = \int_{^1S} {}^2_1t_i\,\delta u_i\,d^1S \qquad (2)$$

In Eqs. 1 and 2, δ denotes the variational operator; ${}^2_1S_{ij}$ the 2nd-Piola-Kirchhoff stresses; ${}_1\epsilon_{ij}$ the Green-Lagrange strains; ${}^2_1t_i$ the surface tractions; u_i the displacement components; and ${}^1V$ and ${}^1S$ the volume and surface area of the body at C_1, respectively. In an incremental formulation, the stresses ${}^2_1S_{ij}$ at each step can be related to the initial (Cauchy) stresses ${}^1\tau_{ij}$ and the strain increments ${}_1\epsilon_{kl}$ by

$$ {}^2_1S_{ij} = {}^1\tau_{ij} + {}_1C_{ijkl}\; {}_1\epsilon_{kl} \tag{3} $$

where ${}_1C_{ijkl}$ denotes the components of the constitutive tensor. The strains ${}_1\epsilon_{ij}$ can be further divided into the linear and nonlinear parts,

$$ {}_1\epsilon_{ij} = {}_1e_{ij} + {}_1\eta_{ij} \tag{4} $$

Substituting Eq. 3 into Eq. 1, and making use of Eq. 4, one obtains

$$ \begin{aligned} &\int_{{}^1V} {}_1C_{ijkl}\; {}_1e_{kl}\, \delta_1 e_{ij}\, d^1V + \int_{{}^1V} {}_1C_{ijkl}\; {}_1\eta_{kl}\, \delta_1 e_{ij}\, d^1V + \int_{{}^1V} {}_1C_{ijkl}\; {}_1e_{kl}\, \delta_1 \eta_{ij}\, d^1V \\ &+ \int_{{}^1V} {}^1C_{ijkl}\; {}_1\eta_{kl}\, \delta_1 \eta_{ij}\, d^1V + \int_{{}^1V} {}^1\tau_{ij}\, \delta_1 \eta_{ij}\, d^1V + \int_{{}^1V} {}^1\tau_{ij}\, \delta_1 e_{ij}\, d^1V = {}^2_1R \end{aligned} \tag{5} $$

It should be noted that Eq. 5, though expressed in an incremental form, remains an exact statement of the equilibrium of the body at the C_2 state. How the terms on the left-hand side of Eq. 5 interact with each other to represent the effects of stretching and rigid body motions has been analyzed in [12] for a higher-order truss element. For the present purpose, it is sufficient to say that the first through fourth integrals in Eq. 5 account for the change in the strain energy of the body under consideration during the increment from C_1 to C_2. The fifth integral relates to the change in the potential of the initially stressed body during the same increment. And the sixth integral represents the virtual work done by the surface tractions ${}^1_1t_i$, which are already existing on the body at C_1, during the increment from C_1 to C_2, as can be appreciated from the following equality for a body in equilibrium at C_1:

$$ {}^1_1R = \int_{{}^1V} {}^1\tau_{ij}\, \delta_1 e_{ij}\, d^1V = \int_{{}^1S} {}^1_1t_i\, \delta u_i\, d^1S \tag{6} $$

By dropping the second through the fourth integrals on the left-hand side of Eq. 5, one obtains the following linearized version,

$$ \int_{{}^1V} {}_1C_{ijkl}\; {}_1e_{kl}\, \delta_1 e_{ij}\, d^1V + \int_{{}^1V} {}^1\tau_{ij}\, \delta_1 \eta_{ij}\, d^1V + \int_{{}^1V} {}^1\tau_{ij}\, \delta_1 e_{ij}\, d^1V = {}^2_1R \tag{7} $$

This equation serves as a valid basis for deriving the incremental stiffness equation for most nonlinear problems.

INCREMENTAL STIFFNESS EQUATION FOR FINITE ELEMENTS

By relating the strains to displacements at a generic point of the finite element under consideration, and then to the nodal displacements through interpolation functions, the terms in Eq. 7 can be represented by products of matrices and vectors as follows [9,10,12,13]:

$$ \begin{aligned} &\int_{{}^1V} {}_1C_{ijkl}\; {}_1e_{kl}\, \delta_1 e_{ij}\, d^1V = \{\delta u\}^T [k_e]\, \{u\} \\ &\int_{{}^1V} {}^1\tau_{ij}\, \delta_1 \eta_{ij}\, d^1V = \{\delta u\}^T [k_g]\, \{u\} \\ &\int_{{}^1V} {}^1\tau_{ij}\, \delta_1 e_{ij}\, d^1V = \{\delta u\}^T \{{}^1f\} \end{aligned} \tag{8} $$

$$ {}^{2}_{1}R = \{\delta u\}^{T}\{{}^{2}f\} $$

where $[k_e]$ is the elastic stiffness matrix, $[k_g]$ the geometric stiffness matrix, $\{u\}$ the element displacement increments, $\{{}^1f\}$ the forces already existing on the element at C_1, and $\{{}^2f\}$ the total forces acting on the element at C_2. Substituting the preceding expressions into Eq. 7 and taking the arbitrary nature of virtual displacements $\{\delta u\}$, one obtains the following incremental equation of equilibrium:

$$ [k_e]\{u\} + [k_g]\{u\} + \{{}^1f\} = \{{}^2f\} \tag{9} $$

A physical interpretation of Eq. 9 can be given as follows: By increasing the forces acting on the element from $\{{}^1f\}$ to $\{{}^2f\}$, further deformations $\{u\}$ occur in the element such that it will move from the C_1 to the C_2 state. During this deformation process, the applied force increments $\{{}^2f\}-\{{}^1f\}$ will be resisted not only by the elastic actions generated by the $[k_e]$ matrix, but also by the forces resulting from the change in geometry as represented by the $[k_g]$ matrix.

The only assumption with the incremental stiffness equation, i.e., Eq. 9, is that the strains occurring within each incremental step should be small enough so that the second through fourth integrals in Eq. 5 can be neglected. Beyond this, no restriction has been placed on the type of motion for the element to move from C_1 to C_2. As a result, the incremental stiffness equation should be able to cope with any type of motions for finite elements that are initially stressed, which should include, in particular, the rigid body motion.

With regard to the rigid body motion, the following comments can be made. First, a finite element satisfying the patch test or its equivalent in the linear sense does not necessarily imply its satisfaction of the rigid body criteria in the nonlinear sense, if one realizes that the elastic stiffness matrix $[k_e]$ and geometric stiffness matrix $[k_g]$ are derived from terms of different nature in the virtual work expression. Second, just as a linear element is not born to pass the quality test, a nonlinear finite element may likewise fail to meet the rigid body requirements. Third, just as the quality of a linear element can be assured only if it passes certain tests, the quality of a nonlinear element cannot be considered reliable whenever the rigid body criteria have been violated. Finally, although the above discussions have been restricted primarily to problems with small strains, the same rigid body criteria should be satisfied by large-strain elements as the condition of small strains is approached.

RIGID BODY LAW AND ASSOCIATED TEST

Consider a bar sitting on the surface of the earth. Suppose the bar is subjected a gravitational force P at the top. For equilibrium, a reaction of the same magnitude P will appear at the bottom (Fig. 2a). This is the situation with the bar in the C_1 state. Now, suppose the earth moves from the C_1 state to the C_2 state by a rigid body rotation θ_r. The result will be the one shown in Fig. 2b. An observation of this figure shows that the line of action of force P rotates with the movement, but its magnitude remains unchanged. An overall result is the maintenance of equilibrium of the bar in the C_2 state.

The essential point here is that if the bar is represented by a finite element and loaded by an arbitrary set of forces in equilibrium, which is the typical situation to be encountered in an incremental analysis, the finite element must be able to duly account for such a rigid body effect. In other words, when a finite element is subjected to a rigid body motion, the initial forces acting on the element in the C_1 state must rotate or translate with the rigid body motion, while their magnitudes should remain unchanged, so as to preserve the equilibrium of the element in the C_2 state, i.e., after the rigid body motion. This is exactly the philosophy behind the rigid body test proposed in Reference [9] for nonlinear elements.

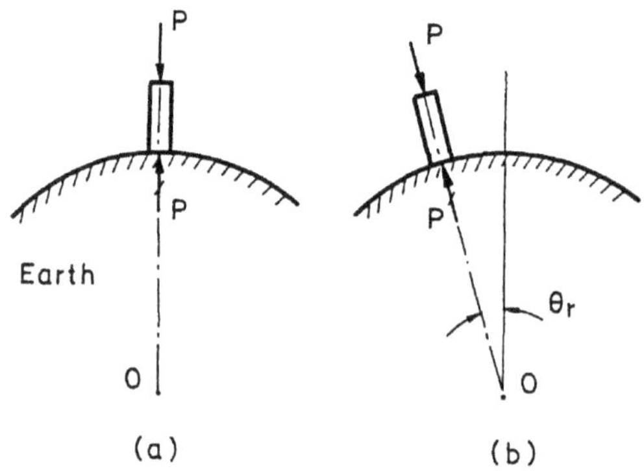

Figure 2. Initially loaded bar before and after rigid body rotation.

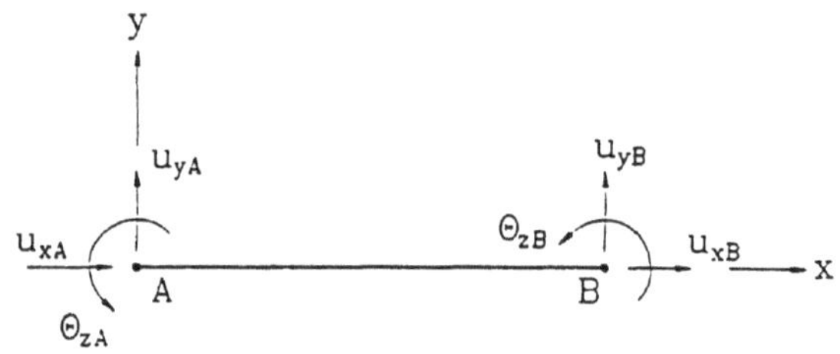

Figure 3. Two-dimensional beam.

For the present purposes, the two-dimensional beam element shown in Fig. 3 serves as an effective vehicle for illustrating the rigid body test and the general eigenvalue test to follow. The displacement vector {u} for the element is

$$\{u\}^T = \{u_{xA}\ u_{yA}\ \theta_{zA}\ u_{xB}\ u_{yB}\ \theta_{zB}\} \tag{10}$$

For convenience, the elastic stiffness matrix $[k_e]$ and geometric stiffness matrix $[k_g]$ reproduced from [9] are given in the Appendix. It should be noted that the forces appearing in the $[k_g]$ matrix are components of the initial force vector $\{^1f\}$ acting on the element in the C_1 state. One important characteristic with such components is that they are in equilibrium, as shown in Fig. 4a. Based on the condition of equilibrium, the initial forces $\{^1f\}$ acting on the element can be expressed as

$$\{^1f\}^T = \left\{ -F_{xB} \quad \frac{M_{zA}+M_{zB}}{L} \quad M_{zA} \quad F_{xB} \quad -\frac{M_{zA}+M_{zB}}{L} \quad M_{zB} \right\} \tag{11}$$

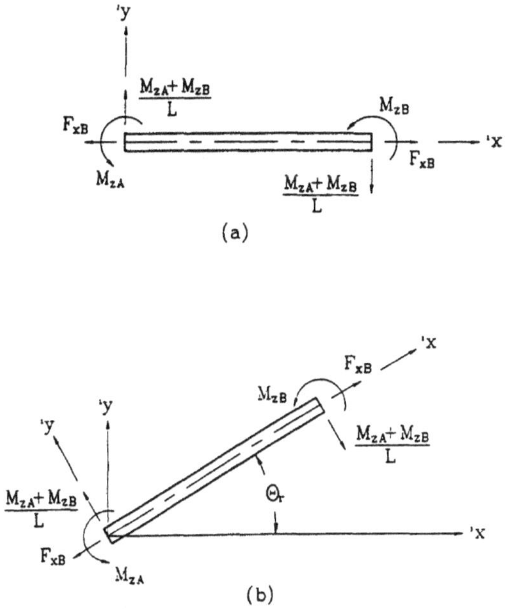

Figure 4. Initially stressed finite element before and after rigid body rotation.

To conduct the rigid body test, one first assume that the element is subjected to a rigid body rotation θ_r, which, assumed to be small in magnitude, can be described by the following displacement vector $\{u\}_r$:

$$\{u\}_r^T = \{0 \quad 0 \quad \theta_r \quad 0 \quad L\theta_r \quad \theta_r\} \tag{12}$$

For the present element, it is easy to show that the elastic forces generated by the $[k_e]$ matrix during the rigid body rotation are zero,

$$[k_e]\{u\}_r = \{0\} \tag{13}$$

Such a result is not surprising, if one realizes that the elastic stiffness matrix $[k_e]$ given in the Appendix is exact for beam elements with no distributed loads. In contrast, the forces generated by the geometric stiffness matrix $[k_g]$ during the rigid body rotation are neither zero nor in equilibrium, as can be seen below,

$$[k_g]\{u\}_r = \left\{-\frac{M_{zA}+M_{zB}}{L}\theta_r \quad -F_{xB}\theta_r \quad 0 \quad \frac{M_{zA}+M_{zB}}{L}\theta_r \quad F_{xB}\theta_r \quad 0\right\}^T \tag{14}$$

The final step in the test is see if the resulting forces $\{^2f\}$ calculated according to Eq. 9 for the element in the C_2 state, i.e., after the rigid body rotation θ_r, can really conform to the rigid body law. By substituting Eqs. 11, 13, and 14 into the left-hand side of Eq. 9, the forces $\{^2f\}$ can be calculated accordingly,

$$\{^2f\}^T = \left\{-F_{xB}-\frac{M_{zA}+M_{zB}}{L}\theta_r \quad -F_{xB}\theta_r+\frac{M_{zA}+M_{zB}}{L} \quad M_{zA} \right.$$

$$\left. F_{xB}+\frac{M_{zA}+M_{zB}}{L}\theta_r \quad F_{xB}\theta_r-\frac{M_{zA}+M_{zB}}{L} \quad M_{zB}\right\} \tag{15}$$

These are exactly the forces shown in Fig. 4b. As the forces acting on the finite element have rotated with the rigid body rotation, while their magnitudes are kept unchanged, the result predicted by the finite element appears to be consistent with the rigid body law. Thus, the ability of the finite element to deal with rigid body motions is confirmed.

The following remarks can be made regarding the rigid body test for nonlinear finite elements:

(1) The rigid body law is not restricted to *small* rigid rotations. By making the rigid rotation θ_r *small*, we are talking about the *minimum* requirements that must be satisfied by a finite element. Nevertheless, it is realized that the small-rotation assumption does represent a quite wide range of nonlinear problems in practice.

(2) The above test has been performed for the condition of rigid rotation, simply because most elements can easily cope with the condition of rigid translation. Theoretically, the test should be conducted for all types of rigid motion for a finite element.

(3) The two-dimensional beam element derived in [9] represents merely one of very few elements that have been proved to pass the rigid body test. It should be warned that most of the nonlinear elements existing in the literature remain non-tested.

(4) For elements whose stiffness matrices can be written in an algebraic form, such as those given in the Appendix, the rigid body test can be carried out by hand calculation as described above. However, for elements whose stiffness matrices can only be constructed by numerical integration, the test has to be performed instead by a computer program. In this regard, the selection of the initial forces $\{^1f\}$, which must be in equilibrium by themselves before rigid rotation, to cover all possible load combinations becomes very important.

(5) For elements involving rotational degrees of freedom in the three-dimensional sense, such as three-dimensional beams, the moments induced by the initial moments undergoing rotations have to be considered in the expression of the force vector $\{^1f\}$ in the C_2 state [15].

(6) Formulation of finite elements based on the incremental virtual work equation (Eq. 7) is not a foolproof procedure. Even starting from the

same basis, many factors, such as the selection of nodal degrees of freedom, shape function and even the kinematic displacement fields, may affect the quality of the derived element. It is not unusual in the finite element derivation that considerations made from the computational aspects may not meet the mechanics requirements, including those concerning the rigid body motion.

GENERAL EIGENVALUE TEST

In the rigid body test, the finite element is assumed to be acted upon by a set of initial forces $\{^1f\}$ in equilibrium. Since the initial forces $\{^1f\}$ are known in advance, the geometric stiffness matrix $[k_g]$ can therefore be determined. The displacement vector $\{u\}_r$ can also be determined for a given rigid rotation θ_r. Since all the terms on the left-hand side of Eq. 9, including the $[k_e]$ matrix, are known, one can readily obtain the force vector $\{^2f\}$ as the right-hand side of the equation, which represents the forces acting on the element after the rigid rotation. According to the rigid body law, the forces $\{^2f\}$ have to be identical in magnitudes to the initial forces $\{^1f\}$, but different in the acting directions due to rigid body rotation. This serves as the criteria for judging the legitimacy of the derived element.

The general eigenvalue test takes a different view from that of the rigid body test. First, for an element acted upon by a set of initial forces $\{^1f\}$ and subjected to a rigid rotation θ_r, the resulting forces $\{^2f\}$ can be determined directly from the rigid body law. Since all the terms on both sides of Eq. 9 are known, and since the equation has to be satisfied, it is possible to rewrite the equation in an eigenvalue form. The legitimacy of the derived stiffness matrices can then be judged from the nature of the eigenroots, in particular, the zero roots, solved from this equation. The following is a summary of the procedure given in [13], again with the two-dimensional beam element serving as the vehicle of illustration.

For a two-dimensional beam element, the initial forces $\{^1f\}$ acting on the element that meet the condition of equilibrium are those given in Eq. 11. According to the rigid body law, the forces $\{^2f\}$ acting on the element after the rigid body rotation θ_r are those given in Eq. 15. For small rigid body rotation $\theta_r = (u_{yB}-u_{yA})/L$, the difference between the two force vectors $\{^2f\}$ and $\{^1f\}$ can be related to the displacement vector $\{u\}$ as

$$\{^2f\} - \{^1f\} = [k_g]_e\{u\} \tag{16}$$

which represents the force components generated by the rigid body rotation. The external stiffness matrix $[k_g]_e$ in Eq. 16, which has been reproduced from [13] in the Appendix, stands for the effects of the force components generated by the initial forces $\{^1f\}$ upon the rigid body motion from C_1 to C_2 [17]. One feature with the $[k_g]_e$ matrix given above is that it can be derived directly from the force vectors $\{^1f\}$ and $\{^2f\}$, having nothing to do with the stiffness matrices $[k_e]$ and $[k_g]$ or the shape functions used in deriving these matrices.

By substituting Eq. 16 into Eq. 9, one shows that the following equation *must* be satisfied by the finite element for each rigid body mode:

$$([k_e] + [k_g] - [k_g]_e)\{u\} = \{0\} \tag{17}$$

To detect the rigid body modes implied by the element, the following eigenvalue equation can be solved,

$$([k_e] + [k_g] - [k_g]_e - \lambda[I])\{u\} = \{0\} \qquad (18)$$

where λ is the eigenvalue and [I] a unit matrix. The preceding equation has been referred to as the general eigenvalue test in [13]. It differs from the following one in that the effect of the initial forces $\{^1f\}$ has been duly taken into account:

$$([k_e] + [k_g] - \lambda[I])\{u\} = \{0\} \qquad (19)$$

The latter should be avoided from practice due to its lack of consistency. For the special case where the element is free of any initial stresses, $\{^2f\} = \{^1f\} = \{0\}$, $[k_g]_e = [k_g] = [0]$, the general eigenvalue test reduces to the conventional one concerning the quality of the $[k_e]$ matrix,

$$([k_e] - \lambda[I])\{u\} = \{0\} \qquad (20)$$

Since a rigid body motion causes no energy input to the initially stressed element, it is required that a zero eigenvalue λ be obtained for each of the rigid body modes. As a consequence, three zero roots corresponding to the three rigid body modes should be obtained for a two-dimensional element. In fact, an eigenvalue analysis based on Eq. 18 has specifically shown that the present beam element passes such a test. This has the indication that all the assumptions employed in deriving the $[k_e]$ and $[k_g]$ matrices are consistent with rigid body requirements, at least in the limit of small rigid body rotations.

CONCLUDING REMARKS

The concept of rigid body motion has not received adequate attention in the past from researchers working on nonlinear problems. Just as a linear finite element has to be thoroughly tested before it can be applied to practical problems, a nonlinear finite element has to be subjected at least to the same level of testing, if not stricter, before it can be considered acceptable. A dangerous approach in nonlinear analysis is to regard the formulation of nonlinear finite elements as a foolproof process, based on the virtual work expression of Eq. 7 or any equivalent form, but say nothing about the quality of the derived element and the associated analysis procedure.

The rigid body concept described in this paper serves not only as a useful tool in the formulation stage for detecting the theoretical flaws of a new finite element, but also as a debugging aid in the programming stage for finding the coding errors. Although only the quality of nonlinear elements is concerned in this paper, many other factors may affect the reliability of a step-by-step nonlinear analysis. One such factor is the procedure for recovering and updating the element forces in each incremental step. Numerical experiments have shown that bad results may appear if the rigid body effect is not properly treated in the force recovery procedure [9]. Another factor relates to the step-by-step updating of the nodal orientations of each element involving three-dimensional rotations, in which the non-commutative nature of rigid rotations is of concern [16].

Other problems that remain to be investigated include, for instance, the ability of higher-order elements to deal with large as well as small rigid body rotations, and the application of the rigid body concepts to problems involving material nonlinearities.

ACKNOWLEDGEMENTS

The results presented herein were generated from a series of research projects conducted by the writer in recent years on the stability and nonlinear behaviors of space frames. These research projects have been sponsored in a consecutive fashion by the National Science Council particularly through the following grant numbers: NSC78-0410-E002-30, NSC79-0410-E002-07, NSC79-0410-E002-56, NSC80-0410-E002-18. The writer would like to thank many of his former graduate students, including S. R. Kuo, H. T. Chiou, L. J. Leu, Y. D. Cherng, and J. H. Chou, among others, for their contribution to these projects. Any opinions expressed herein are those of the writer and not of the sponsor.

REFERENCES

1. Bazeley, G. P., Cheung, Y. K., Irons, B. M., and Zienkiewicz, O. C., Triangular elements in plate bending—conforming and non-conforming solutions, *Proc. Conf. Matrix Meth. in Struct. Mech.*, Wright-Patterson Air Force Base, Ohio, 1965, 547-576.

2. Haisler, W. E., and Stricklin, J. A., Rigid body displacements of curved elements in the analysis of shells by the matrix-displacement method, *AIAA J.*, 1967, **5**(8), 1525-1527.

3. Cantin, G., and Clough, R. W., A curved, cylindrical-shell, finite element, *AIAA J.*, 1968, **6**(6), 1057-1062.

4. Cantin, G., Rigid body motions in curved finite elements, *AIAA J.*, 1970, **8**(7), 1252-1255.

5. Mebane, P. M., and Stricklin, J. A., Implicit rigid body motion in curved finite elements, *AIAA J.*, 1971, **9**(2), 344-345.

6. Rigby, G. L., and McNiece, G. M., A strain energy basis for studies of element stiffness matrices, *AIAA J.*, 1972, **10**(11), 1490-1493.

7. Fonder, G. A., and Clough, R. W., Explicit addition of rigid-body motions in curved finite elements, *AIAA J.*, 1973, **11**(3), 305-312.

8. Cook, R. D., Malkus, D. S., and Plesha, M. E., *Concepts and Applications of Finite Element Analysis*, 3rd Ed., John Wiley & Sons, New York, 1989.

9. Yang, Y. B., and Chiou, H. T., Rigid body motion test for nonlinear analysis with beam elements, *J. Eng. Mech.*, ASCE, 1987, **113**(9), 1404-1419.

10. Yang, Y. B., Kuo, S. R., and Cherng, Y. D., Curved beam elements for nonlinear analysis, *J. Eng. mech.*, ASCE, 1989, **115**(4), 840-855.

11. Cherng, Y. D., Stability analysis and rigid body test of thin-walled

curved beams, thesis presented to Depart. of Civil Eng., National Taiwan Univ., in Taipei, Taiwan, in partial fulfillment of the requirements for the degree of Master of Science, June 1987 (In Chinese).

12. Leu, L. J., and Yang, Y. B., Effects of rigid body and stretching on nonlinear analysis of trusses, *J. Struct. Eng.*, ASCE, 1990, **116**(10), 2582-2598.

13. Yang, Y. B., Chou, J. H., and Leu, L. J., Rigid Body Considerations for Nonlinear Finite Element Analysis, accepted for publication in *Int. J. Num. Meth. Eng.* on January 28, 1991.

14. Malvern, L. E., *Introduction to the mechanics of a continuum medium*, Prentice-Hall, Englewood Cliffs, N.J., 1969.

15. Yang, Y. B., Rationality of Nonlinear Analysis, *Proc. Third Int. Conf. Civil and Struct. Eng. Computing*, Editor: Topping, B. H. V., Civil-Comp Press, Edinburgh, U.K., September 1987, pp. 169-175.

16. Kuo, S. R., Theory of static and dynamic stability of space frames, thesis presented to Depart. of Civil Eng., National Taiwan Univ., in Taipei, Taiwan, in partial fulfillment of the requirements for the degree of Doctor of Philosophy, June 1991 (In Chinese).

17. Yang, Y. B., and McGuire, W., Joint Rotation and Geometric Nonlinear Analysis, *J. Struct. Eng.*, ASCE, 1986, **112**(4), 879-905.

APPENDIX

The following stiffness matrices have been reproduced from References [9] and [13]:

- Elastic stiffness matrix $[k_e]$:

$$\begin{bmatrix} \frac{EA}{L} & & & & \textit{Symmetric} & \\ 0 & \frac{12EI_z}{L^3} & & & & \\ 0 & \frac{6EI_z}{L^2} & \frac{4EI_z}{L} & & & \\ -\frac{EA}{L} & 0 & 0 & \frac{EA}{L} & & \\ 0 & -\frac{12EI_z}{L^3} & -\frac{6EI_z}{L^2} & 0 & \frac{12EI_z}{L^3} & \\ 0 & \frac{6EI_z}{L^2} & \frac{2EI_z}{L} & 0 & -\frac{6EI_z}{L^2} & \frac{4EI_z}{L} \end{bmatrix}$$

- Geometric stiffness matrix $[k_g]$:

$$\begin{bmatrix} \frac{F_{xB}}{L} & & & & & \text{Symmetric} \\ 0 & \frac{6F_{xB}}{5L}+\frac{12I_zF_{xB}}{AL^3} & & & & \\ -\frac{M_{zA}}{L} & \frac{F_{xB}}{10}+\frac{6I_zF_{xB}}{AL^2} & \frac{2LF_{xB}}{15}+\frac{4I_zF_{xB}}{AL} & & & \\ -\frac{F_{xB}}{L} & 0 & \frac{M_{zA}}{L} & \frac{F_{xB}}{L} & & \\ 0 & -\frac{6F_{xB}}{5L}-\frac{12I_zF_{xB}}{AL^3} & -\frac{F_{xB}}{10}-\frac{6I_zF_{xB}}{AL^2} & 0 & \frac{6F_{xB}}{5L}+\frac{12I_zF_{xB}}{AL^3} & \\ -\frac{M_{zB}}{L} & \frac{F_{xB}}{10}+\frac{6I_zF_{xB}}{AL^2} & -\frac{LF_{xB}}{30}+\frac{2I_zF_{xB}}{AL} & \frac{M_{zB}}{L} & -\frac{F_{xB}}{10}-\frac{6I_zF_{xB}}{AL^2} & \frac{2LF_{xB}}{15}+\frac{4I_zF_{xB}}{AL} \end{bmatrix}$$

- External stiffness matrix $[k_g]_e$

$$\begin{bmatrix} 0 & & & & & \text{Symmetric} \\ \frac{M_{zA}+M_{zB}}{L^2} & \frac{F_{xB}}{L} & & & & \\ 0 & 0 & 0 & & & \\ 0 & -\frac{M_{zA}+M_{zB}}{L^2} & 0 & 0 & & \\ -\frac{M_{zA}+M_{zB}}{L^2} & -\frac{F_{xB}}{L} & 0 & \frac{M_{zA}+M_{zB}}{L^2} & \frac{F_{xB}}{L} & \\ 0 & 0 & 0 & 0 & 0 & 0 \end{bmatrix}$$

SHELLS: MECHANICS AND APPROXIMATION - LINEAR AND NONLINEAR ASPECTS

GERALD A. WEMPNER
Professor
School of Civil Engineering
Georgia Institute of Technology
Atlanta, GA 30332

ABSTRACT

The following paper focuses on the special attributes of shells, various mechanical peculiarities and related problems. The presentation addresses theories and approximations which are applicable to thin and thick shells, geometrical nonlinearities and inelastic deformations. Attention is given to the origins of particular phenomena, to viewpoints and methods which appeal to basic concepts of mechanics.

INTRODUCTION

Although the theories and approximations of shells are governed by the underlying mechanics of all solids and structures, the geometry of such thin curved bodies introduces peculiar attributes which, in turn, require special methods of approximation and computation. The thinness of the body admits flexural strains, changes of curvature and large rotations; the curvature of the body provides stiffness to support transverse loads, but also admits the possibilities of instabilities at critical loads and imperfection sensitivity. The flexure implies large gradients of strain, and stress, in the transverse direction; such gradients pose special difficulties in approximation, particularly when inelastic deformations occur. Here, our aim is to focus upon those aspects of the mechanics, approximation and computation which are peculiar to shells.

Certain peculiarities of the shell (plate or beam) can be attributed to thinness which necessitates special attention to the smaller dimension when formulating the mathematical model, the shell theory and any discrete finite elements. These are generally present in all circumstances, elastic or inelastic behavior, small or large deformations, linear or nonlinear. They stem from the severe strain gradients in the transverse direction, and the important role of such flexural strains compared to the minor role of transverse shear strains. The continuum theory of thin shells accommodates this peculiarity via the hypothesis of Kirchhoff-Love [1,2]. Counterparts are needed in the formulation of discrete finite element; such counterparts, and alternatives, constitute an initial part of our presentation.

Another peculiarity of shells (plates or beams) is also a consequence of thinness. Because of the flexure of such bodies, large changes of curvature and rotations can occur, though strains may remain small. Indeed, in practical circumstances such large changes can occur though the behavior remains hookean. Consequently, the theory and any approximations are governed by nonlinear equations; the nonlinearities are strictly geometrical in origin. Methods to accommodate such nonlinearities constitute the second part of our presentation.

Though geometrical nonlinearities pose computational difficulties, they are primarily time-consuming. Errors in rendering numerical approximation can be minimized when the response is elastic. The existence of a potential, and the nonlinear equations of state, enable the assessment and correction of error. The difficulties of approximating inelastic response pose far greater difficulties. In general, the non-conservative system requires that one trace the history but, unfortunately, the path-dependence of the state leaves no sure means to assess the error of a numerical approximation. Of course, that is a problem common to all inelastic bodies.

The literature is replete with the mechanics, theory and approximation of elastic shells. A multitude of theories describe elastic shells, thin and thick, linear and nonlinear. By comparison few have attempted comparable theories of inelastic shells, theories which describe the inelasticity in the two-dimensional space of a surface. Such theory is elusive: It must describe the incremental strains and stresses which evolve during flexure and extension. Again, the flexural strain (and stress) causes initial yielding at a surface of the thin body and a progression through the wall. The inelastic theory must accommodate that evolution of inelastic deformation. Questions of plasticity and some alternative approaches are cited in our presentation.

FIRST APPROXIMATION

Any discussion of shells must recount the hypothesis of Kirchhoff-Love [1,2] which effectively reduces the mechanics of the thin three-dimensional body to two dimensions: Accordingly, lines initially normal to an intermediate surface remain normal to the surface during deformation. Additionally, the transverse normal stress is neglected. Then the essential kinematics concerns the deformation of the surface. Specifically, the differential geometry of the surface, and indirectly the shell also, is defined by two symmetrical tensors, a metric and a curvature tensor. Changes of the former imply extensional strain and changes of the latter imply flexure of the surface. Let $a_{\alpha\beta}$ and $A_{\alpha\beta}$ denote the components of the metric tensor in the initial and current states, $b_{\alpha\beta}$ and $B_{\alpha\beta}$ the respective components of the curvature tensor. These serve to describe the strain throughout the shell (extension of the normal plays no role). Simple definitions of the extensional and flexural strains are components of two symmetric tensors:

$$\gamma_{\alpha\beta} = \frac{1}{2} (A_{\alpha\beta} - a_{\alpha\beta})$$

$$\kappa_{\alpha\beta} = (B_{\alpha\beta} - b_{\alpha\beta})$$

The virtual work of stresses upon incremental (virtual) strains ($\gamma_{\alpha\beta}$ and $\kappa_{\alpha\beta}$) has the form:

$$\dot{w} = m^{\alpha\beta}\dot{\kappa}_{\alpha\beta} + n^{\alpha\beta}\dot{\gamma}_{\alpha\beta} \quad . \tag{1}$$

As always, the statement (1) serves to identify precisely [3,4] two stresses: Components $m^{\alpha\beta}$ are essentially the bending couples, $n^{\alpha\beta}$ the membrane forces. These too are components of symmetrical tensors. Note that the foregoing commentary presumes no specific properties, but only the continuity of the surface and cohesion of the material. From our present position, this "first approximation" remains applicable to elastic and inelastic shells. In the former, conservation implies that the virtual work is the differential of the potential $\phi(\kappa_{\alpha\beta},\gamma_{\alpha\beta})$:

$$m^{\alpha\beta}\dot{\kappa}_{\alpha\beta} + n^{\alpha\beta}\dot{\gamma}_{\alpha\beta} = \frac{\partial\phi}{\partial\kappa_{\alpha\beta}}\dot{\kappa}_{\alpha\beta} + \frac{\partial\phi}{\partial\gamma_{\alpha\beta}}\dot{\gamma}_{\alpha\beta}$$

Since the strains are independent, the stress-strain relations follow:

$$m^{\alpha\beta} = \frac{\partial\phi}{\partial\kappa_{\alpha\beta}} , \qquad n^{\alpha\beta} = \frac{\partial\phi}{\partial\gamma_{\alpha\beta}} \tag{2}$$

The relations of elasticity and, particularly, hookean behavior have been studied extensively. The interested reader can consult the definitive work of Koiter [5] and references cited therein.

The classical theory of plasticity is founded upon concepts of work and, specifically, work dissipated upon inelastic strain increments $(\dot{\kappa}_{\alpha\beta},\dot{\gamma}_{\alpha\beta})$. There is only the assumption that a part of any incremental strain is recovered (elastic) and another is not recovered (plastic) during the application and removal of incremental stresses $(m^{\alpha\beta},n^{\alpha\beta})$. One might debate the validity of the decomposition and the assumption, yet the classical theory has proven effective in practical problems of time-independent responses. Then the plastic increments are related to a yield, or loading, condition $\psi(m^{\alpha\beta},n^{\alpha\beta})$:

$$\dot{\kappa}^{P}_{\alpha\beta} = \dot{\lambda}\,\frac{\partial\psi}{\partial m^{\alpha\beta}} , \qquad \dot{\gamma}^{P}_{\alpha\beta} = \dot{\lambda}\,\frac{\partial\psi}{\partial n^{\alpha\beta}} \tag{3}$$

where the scalar λ is dependent upon the hardening, i.e., the evolution of the function ψ. Theories which follow such concepts of plasticity are rare. Indeed, the author is only familiar with one, in which Bieniek [6] postulated a quadratic form for the loading function ψ, together with a form for the evolution of the function as yielding progresses. Because yielding of the three-dimensional shell is initiated at an outer surface and progresses inward, the evolution of the function is indeed complicated. Such direct approaches require much insight, keen observations, and correlations with numerical results obtained by incremental procedures of computation which trace the progression of yielding through the thickness.

The long history of the first approximation has established the validity of the hypothesis. The literature is replete with studies, refinements and specific solutions for hookean shells. Theories which accommodate inelastic behavior, even classical plasticity, are sadly lacking. Some attempts were made to derive approximations by adopting the first approximation (the kinematical assumption of Kirchhoff-Love), but admitting alternative approximations for the distribution of stresses through the thickness. The interested reader can consult an earlier work [7].

HIGHER APPROXIMATIONS

Though transverse shear stresses are present in most shells (thin membranes excepted), they are workless under the first approximation. In certain practical circumstances (e.g., shells of moderate thickness or layered composition), relaxation of the Kirchhoff-Love hypothesis admits transverse shear strain and leads to "higher approximations". Early theories were advanced by Reissner [8], Mindlin [9] and others. More recently other theories were formulated; some expressly accommodate layers which might possess varying degrees of stiffness [10]. An account of higher approximations, various theories and approaches, is beyond the scope of the present account. Nonetheless, it is relevant to note that some advantages accrue to approximations which might also accommodate thicker shells, layered shells and also a transition to the three-dimensional theory. The question is particularly relevant to subsequent approximation via finite elements.

TWO-DIMENSIONAL THEORIES IN PRACTICE

The two-dimensional theory according to Love [2], or the various refinements, apply to hookean shells. Such theories call for the solution of a system of partial differential equations and boundary conditions. If formulated in terms of the twelve components of stress and strain, the governing equations are a "complete eight-order set of twelve equations and twelve unknowns" (six stress-strain equations, three equilibrium equations and three compatibility equations) [11]. These are linear or geometrically nonlinear, if changes of curvature and rotation are appreciable. Clearly, complete solutions of actual problems are possible in only the simplest circumstances, shells of regular form (e.g., spheres, cylinders, cones), simple loadings and supports (e.g. uniform or symmetrical loadings upon axi-symmetrical shells). Still, valuable results have been obtained and much has been learned about the effects of localized loads [12] and edge conditions. Comparisons with results of three-dimensional solutions provide information about the limitations of the theories. Indeed, the practitioner must always remember that every approximation has limitations. Figure 1 illustrates the point. At intermediate positions of a long thin cylindrical tube, the simple membrane stress ($S = pr/h$) is quite adequate. Conditions at an edge or a juncture usually invalidate the simple ("interior") solutions. Indeed, stresses at the re-entrant corners of intersections are only predictable by a theory of three dimensions. Solutions involving nonlinearities, buckling or post-buckling, pose near insurmountable difficulties in all but the simplest cases.

As the engineer turns to the simulation, analysis and, ultimately, to the design of actual shells, the hulls of ships and aircraft, or the dome-like structures of modern architecture, further approximations are necessary. The subsequent (or simultaneous) approximation of the two-(or three) dimensional shell into finite elements has proven most effective. The effectiveness can be attributed in part to the intimate correlations between the discrete mathematical models and the mechanical attributes of the elements. More important, the advent and rapid evolution of the electronic computer has paved the way for our rapid handling of huge systems of numerical, algebraic and symbolic equations and data. In deference to these factors our attention turns to finite elements and certain problems which are

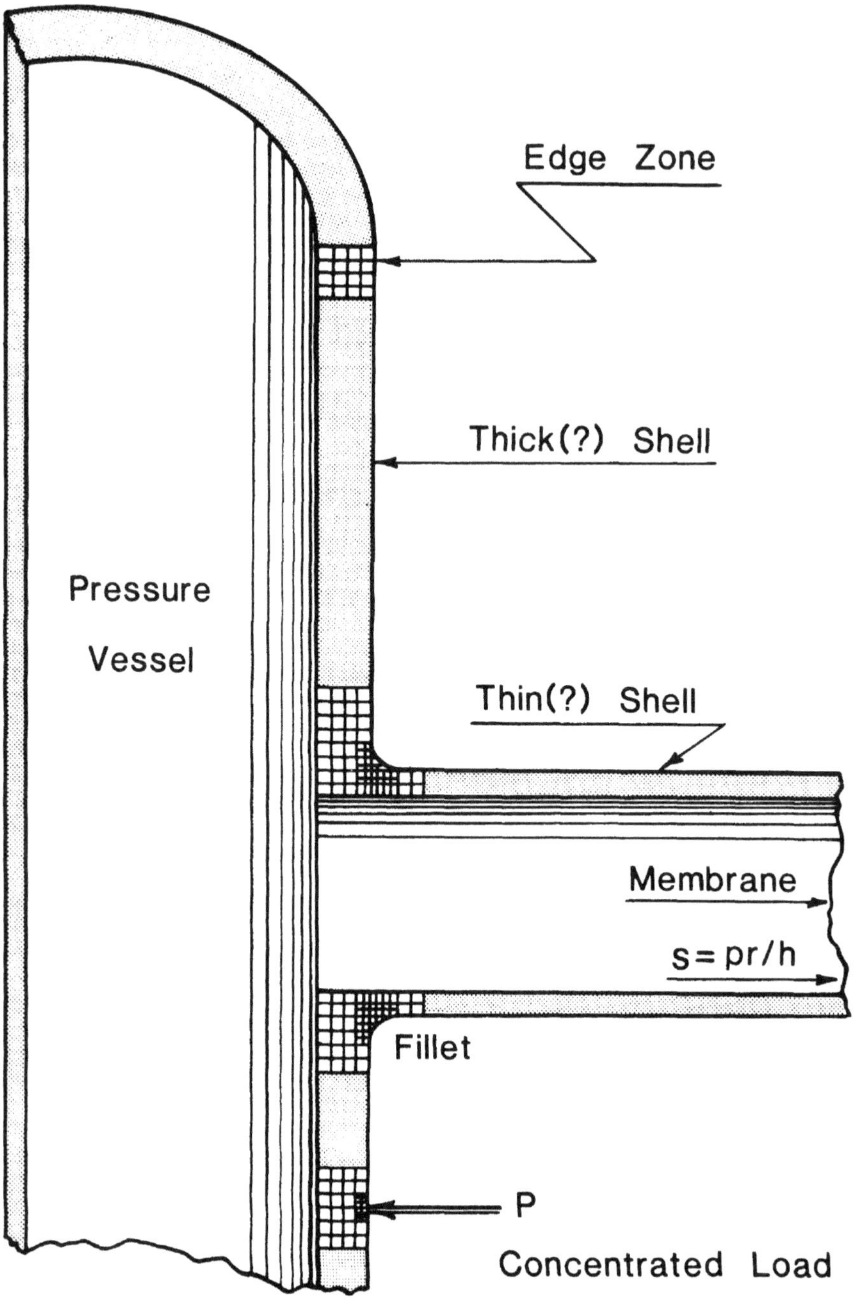
Edge Zone
Thick(?) Shell
Pressure
Vessel
Thin(?) Shell
Membrane
s = pr/h
Fillet
P
Concentrated Load

Figure 1

peculiar to shells (plates or beams).

FINITE ELEMENTS

Finite elements, concept and application, require no introduction. Still, certain attributes of such discrete approximations warrant reiteration. Advantages and disadvantages accrue when the discrete model of an elastic body is derived from the principle of minimum potential. The approach assures the convergence to the solution for a continuous body in stable equilibrium [13]. On the other hand, such discrete models are excessively stiff. In applications to shells (plates or beams) straightforward application of the principle often produces models which do converge but are so stiff and so slowly convergent that the model is practically worthless. This is most simply illustrated by the beam of Fig. 2, subjected to simple bending. A bilinear approximation of the displacement in each rectangular element provides a valid and convergent model; each element deforms to a similar trapezoid. By the method of stationary potential the discrete model is impractical, because of the excessive stiffness and very slow convergence. One can understand the problem by examining the strain energies associated with the extensional strain ($\epsilon_{11} = x_2\Delta\phi/\ell$) and shear strain ($\epsilon_{12} = x_1\Delta\phi/2\ell$), where the local coordinates (x_1,x_2) originate at the middle of the element. Both converge to the correct values ($\ell\rightarrow 0$), i.e., $\epsilon_{11} \rightarrow x_2\kappa$ (κ = curvature), $\epsilon_{12} \rightarrow 0$; however, in any useful finite element ($\ell > h$, length > thickness), shear energy U_s dominates bending energy U_B; $U_S/U_B \approx (\ell/h)^2$. Stiffness is so excessive that some invoke the term "locking", implying that little, or no, deformation is admitted. Many schemes have been introduced to overcome the inherent problem. An effective means to avoid excessive energy is the discrete counterpart of the Kirchhoff assumption. Such constraints ensure that the discrete model converges to the continuous theory of Kirchhoff-Love; then, associated energy is deleted and also the excessive stiffness. The notion and a simple method of enforcement were introduced by the author [14]. Though effective, the method is specifically intended for thin shells; such elements do not admit transverse shear, nor transition to a three-dimensional element, e.g. the refinement needed near the re-entrant corner of Fig. 1.

Various schemes have been introduced to circumvent excessive stiffness, e.g. "shear locking". A prevalent device is "reduced integration" wherein the internal energy, e.g. U_s, is evaluated by approximating the strains. For example, one might approximate the shear energy U_s for an element in Fig. 2 by using only the value $\epsilon_{12}(0) = 0$, at the mid-point; that is the mean value, the limiting value ($\ell \rightarrow 0$), and the correct value, in this case. Such schemes admit the shear deformation and are also applicable to three-dimensional elements. They are founded upon mathematical arguments. The interested reader can consult the article by Zienkiewicz, Taylor, and Too [15]. Here, we recount an alternative, which is founded upon a modified potential, mechanical and kinematical arguments.

The functional of Hu-Washizu [16,17] is best defined as the potential augmented by all kinematical constraints, imposed via Lagrange multipliers. The latter are the stresses (S^{ij} in the interior, t on the boundary). If U denotes the potential, then the functional of Hu-Washizu follows

$$H(V, S^{ij}, \epsilon_{ij}) = U(V, S^{ij}, \epsilon_{ij}) \quad \ldots\ldots \tag{4}$$

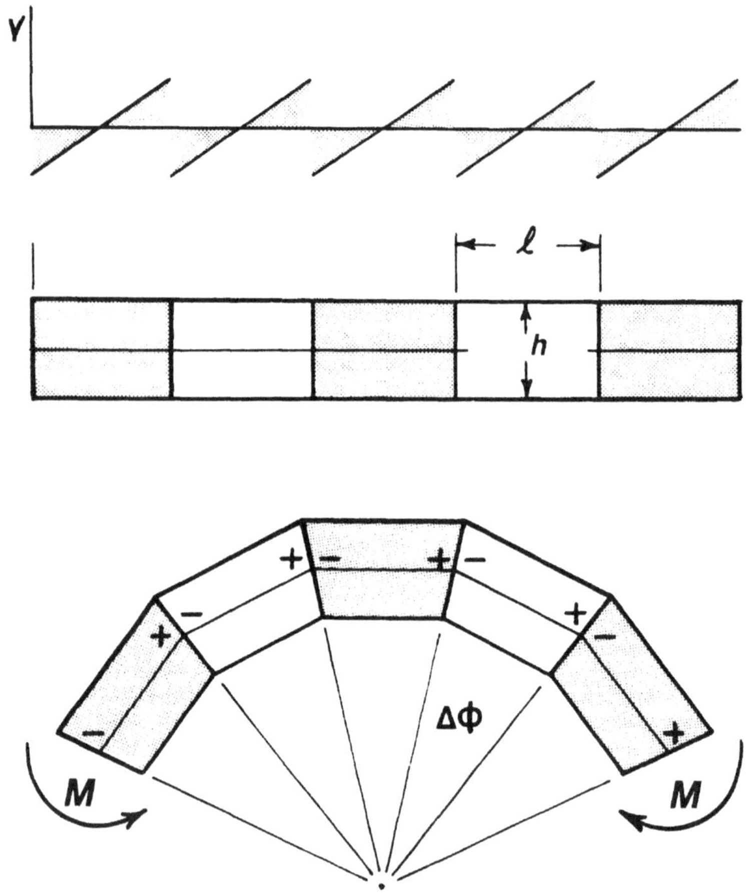

Figure 2

$$- \iiint_{v} S^{ij} \, [\epsilon_{ij} - O_{ij}(\mathbf{V})] \, dv$$

$$- \iiint_{S_v} \mathbf{t} \bullet (\mathbf{V} - \tilde{\mathbf{V}}) \, ds$$

Here, $\mathbf{V}$ denotes displacement, $\tilde{\mathbf{V}}$ prescribed displacement on portion s_v of the bounding surface and strain-displacement equations (linear or nonlinear) are (symbolically)

$$\epsilon_{ij} = O_{ij}(\mathbf{V})$$

where O_{ij} denotes the appropriate differential operator. The explicit expression (4) gives the internal potential U_I in terms of the strain components, i.e. $U_I = U_I\ (\epsilon_{ij})$. The theorem of Hu-Washizu provides the equations of equilibrium, kinematical equations and constitutive equations, as the stationary conditions for the functional H. An important attribute of the functional is noteworthy. The functional approaches the potential as approximations of strains and displacements approach compatibility. Indeed, _if_ the fields are fully compatible, then H = U.

From the viewpoint of discrete approximation via finite elements, the functional H and the associated theorem admit independent approximations of the three fields: stress (S^{ij}), strain (ϵ_{ij}) and displacement ($\mathbf{V} = \mathbf{R} - \mathbf{r}$). To illustrate, suppose that the body is modeled via rectangular elements. The simplest polynomial which preserves interelement continuity is the "trilinear" (cubic). If x^i denotes the local rectangular Cartesian coordinate originating at the center, then the initial and subsequent positions within the element are given by vectors $\mathbf{r}$ and $\mathbf{R}$ (the approximation):

$$\mathbf{r} = \mathbf{r}_o + \mathbf{I}_i \, x^i$$

$$\mathbf{R} = \mathbf{R}_o + \mathbf{G}_i \, x^i + \frac{1}{2} \mathbf{R}_{ij} x^i \, x^j + \mathbf{R}_{123} x^1 x^2 x^3$$

where $\mathbf{R}_{ij} = \mathbf{R}_{ji}$. ($= 0$ if $i = j$) The element has 24 degrees-of-freedom. The motion can be decomposed into a rigid motion and a deformation. The displacement of the center $\mathbf{V}_o = \mathbf{R}_o - \mathbf{r}_o$ represents a translation. A rigid rotation carries the orthogonal triad $\mathbf{I}_i$ to the similar triad $\mathbf{J}_i$ such that

$$\mathbf{J}_j \bullet \mathbf{G}_i = \mathbf{J}_i \bullet \mathbf{G}_j$$

This is consistent with the usual decomposition; the rotation is that of the principal lines at the origin. The element has six constant components of strain, viz.

$$\bar{\epsilon}_{ij} = \frac{1}{2} (\mathbf{J}_i \bullet \mathbf{G}_j + \mathbf{J}_j \bullet \mathbf{G}_i)$$

Additionally, the element possesses higher-order deformational modes; the approximation for each component of strain contains corresponding higher-order terms ("HOT" terms):

$$\epsilon_{ij} = \bar{\epsilon}_{ij} + \frac{1}{2}(J_i \bullet R_{jP}\, x^P + J_j \bullet R_{iP}\, x^P) + \cdots$$

$$+ \frac{1}{2}\left[J_i \bullet R_{123}\,(x^1x^2x^3)_{,j} + J_j \bullet R_{123}(x^1x^2x^3)_{,i} \right]$$

The element possesses twelve HOT modes; these are determined by the constants:

$$J_i \bullet R_{jk} \quad (R_{jk} = R_{kj}, = 0 \text{ if } i = j)$$

$$J_i \bullet R_{123}$$

Each of the HOT modes appears as a coefficient in two (or more) components of strain. For example, the components which play a role in the beam of Fig. 2 are

$$\epsilon_{11} = \bar{\epsilon}_{11} + \underline{J_1 \bullet R_{12}\, x^2} + J_1 \bullet R_{13}\, x^3 + \underline{J_1 \bullet R_{123}\, x^2x^3}$$

$$\epsilon_{12} = \bar{\epsilon}_{12} + \frac{1}{2}(J_1 \bullet R_{21}\, x^1 + J_2 \bullet R_{12} x^2)$$

$$+ \frac{1}{2}(J_1 \bullet R_{23}\, x^3 + J_2 \bullet R_{13}\, x^3)$$

$$+ \frac{1}{2}(J_1 \bullet R_{123} x^1x^3 + J_2 \bullet R_{123}\, x^2x^3)$$

The first underlined terms of ϵ_{11} and ϵ_{12} correspond to the bending mode of Fig. 2.

Two properties of any such approximation must be observed. Firstly, the constants ϵ_{ij} are dominant; they provide mean values and account for the energy in the limit, as the element shrinks. Secondly, each higher mode must be included in one, but <u>only</u> one, component in the approximation of strain. Accordingly, it is sufficient that our approximation for the strains in the beam of Fig. 2 take the form:

$$\epsilon_{11} = \bar{\epsilon}_{11} + \kappa \quad x^2$$

$$\epsilon_{12} = \bar{\epsilon}_{12}$$

NOTE: The uniform flexure requires that $J_2 \bullet R_{12} = 0$. The essential requisites are fulfilled. The one higher-order mode is inhibited by the energy in flexure (κ). The excessive and redundant energy in shear is suppressed; no "locking" occurs. In general, such elimination of unnecessary potential improves convergence.

The application to shells has been demonstrated in previous works

[18,19]. The Hu-Washizu functional is not restricted to small strains; versions applicable to finite strains of shells were presented in a previous article [20].

GEOMETRICAL NONLINEARITIES

Geometrical nonlinearities in the theory of thin shells have been studied exhaustively. The interested reader can consult the works of Koiter [21], Libai and Simmonds [22] and Pietraszkiewicz [23]. Theories and experiments have revealed that the most relevant nonlinearities can be attributed to the occurrence of moderately large rotations and changes of curvature. The latter are usually coupled with membrane forces which play the dominant role in providing the desired stiffness of shells and shell-like structures. The desirable properties, curvature and membrane action which resists transverse loads, are also the attributes which account for the instability at critical loads and sensitivity to imperfections [24,25].

From a practical view toward formulation of discrete models (finite elements) and subsequent numerical computation, the origins of the nonlinearities are relevant. In particular, most practical problems (buckling, post buckling and small inelastic deformations) occur with small strain. This implies that relative rotations within small finite elements are also small. Accordingly, the modeling of the element requires only geometrically linear equations. Then, if the element is hookean, the model is governed entirely by linear equations. Finite or moderate rotations can be accommodated in the assembly. This can be accomplished by introducing the relative, but small, rotations of adjoining elements. These questions have been addressed [26,27], but require closer scrutiny.

From the practical view toward implementing computations, one must resort to incremental and corrective procedures. In the computations for shells, subsequent steps require special attention to the changes of rotation and curvature. Additionally, procedures must also provide for the possibility of limit points. Such unstable states of equilibrium are characterized by vanishing, or negative, increments of load. A method of "generalized arc length" was introduced by the author to accommodate such situations [28]. Briefly, one defines an increment of "arc length" dS^2 as follows. If $X_i(i = 1,...N)$ denotes the totality of state variables (stresses and/or strains and/or displacements) and λ the loading parameter, then

$$dS^2 = \Sigma\, dX_i\, dX_i + d\lambda^2 \tag{5}$$

Now, the change of "arc length" (S) is the control parameter; the change of load (λ) is one additional unknown, while a version, usually a linearization of (5), is the additional equation. Any change of state provides a change of "arc length" and $d\lambda \le 0$ is always an acceptable result, e.g. $d\lambda = 0$ at a limit point. The interested reader can consult the author's original work [28] or many subsequent articles which have effectively utilized the method.

PLASTICITY OF SHELLS

The shell is a three-dimensional body, albeit, thin in one dimension.

Accordingly, one can treat the shell entirely in accordance with the general (3-D) theories of inelastic bodies; however, that approach is impractical and seldom necessary. Indeed, the Kirchhoff-Love hypothsis is probably adequate for many circumstances of inelastic shells. Then equation (1) clearly identifies conjugate stresses ($m^{\alpha\beta}$, $n^{\alpha\beta}$) and strains ($\kappa_{\alpha\beta}, \gamma_{\alpha\beta}$), and admits the development of two-dimensional theories. This latter avenue remains largely unexplored. Between the extremes (3-D and 2-D) lie numerous alternatives. Here, we sketch a few which have been employed with varying degrees of success.

The most widely used method is based upon the Kirchhoff-Love assumption which serves to describe the strain distribution through the thickness. Transverse shear deformation is neglected and transverse normal stress also. At any point the state is one of "plane" stress. The stress distribution is monitored through the thickness and the appropriate yield and flow equations are enforced during the deformation. This requires incremental and numerical procedures which are costly in terms of computational time and storage of data. Such procedures are tantamount to subdividing the shell into numerous layers, interrelated via the underlying (K-L) assumption, but responding differently according to the state of that layer.

An extreme view of the shell as a layered body is the ideal sandwich wherein the shell consists simply of two discrete layers. Then, a shell subjected only to bending yields simultaneously at both surfaces and the initial yielding of the ideally-plastic shell coincides with the limit state. An attempt to reach a better, yet practical, compromise was given by the author [29]: The "club" sandwich replaces the two layers of the ideal sandwich by four. This provides for a piecewise transition from the initial yielding at a surface, to the subsequent yielding of intermediate surfaces, and to the limit state if the material is ideally plastic. Of course, approximation by four layers applies to other materials and hardening.

Two very different approaches to two-dimensional theories were taken by the author [7] and by Bieniek [6]. The former derives a two-dimensional theory from a three-dimensional theory; the latter postulates a form of the two-dimensional yield function $\psi(m^{\alpha\beta}, n^{\alpha\beta})$ and equations governing the evolution of hardening parameters. Both are founded on the K-L hypothesis and were implemented for ideally plastic behavior.

Briefly, the author's derived theory expresses the distribution of normal stresses in terms of Legendre polynomials, viz

$$S^{\alpha\beta} = \Sigma\, m_i^{\alpha\beta}\, P_i(z)$$

wherein $P_i(z)$ is the polynomial of degree 'i' in the thickness coordinate 'z'. The usual difficulties of monitoring stress, initial yielding, etc., were circumvented by adopting the so-called "endochronic" theory [30], a strain-dependent form of plasticity. It is interesting to note that only the stresses $m_0^{\alpha\beta}$, $m_1^{\alpha\beta}$ (force and couple) do work in accordance with equation (1). Higher-order stresses are workless. Numerical results were obtained using a Ramberg-Osgood form of stress-strain equation. Comparisons were given with the results obtained by a many-layered model (22 layers); moment-curvature histories and also stress distributions were compared for various paths of loading, unloading and reloading.

Bieniek's direct approach postulates a quadratic form of the yield condition:

$$\psi = Q_{00} + \frac{2}{3} \mid Q_{01} \mid + 3\, Q^{*}_{11} = 1$$

where

$$Q_{ij} = \frac{3}{2}\, m_{i\alpha\beta}\, m^{\alpha\beta}_{i} - \frac{1}{2}\, m^{\alpha}_{i\alpha}\, m^{\beta}_{i\beta}$$

$$Q^{*}_{11} = Q_{11}\, (m^{\alpha\beta}_{1} - \bar{m}^{\alpha\beta}_{1})$$

The parameters $\bar{m}^{\alpha\beta}_{1}$ evolve with the occurrence of yielding and account for a translation of the yield surface which accompanies the bending; again, $m^{\alpha\beta}_{1}$ are essentially the bending couples. The parameters $\bar{m}^{\alpha\beta}_{1}$ evolve incrementally with the plastic increments of curvature. Since the theory was developed for ideal plasticity, the equation of evolution incorporates a limit condition. The theory, though largely empirical, possesses the essential attributes and offers an effective means for numerical treatment of ideally elastic-plastic shells.

SUMMARY

The foregoing account of shells, mechanics and approximation is intended to provide a perspective of the topic, to reveal the various facets, and to give some insights into the origins and mechanics of those features which are peculiar to these thin bodies. The viewpoints taken are admittedly those of the author, since a lucid account can only draw upon intimate personal experiences. Details are necessarily lacking; the works cited are more specific and provide additional references. Other investigators have studied various aspects of the subject, often from very different viewpoints. The author's earlier review [31] includes references to alternative theories and methods.

REFERENCES

1. Kirchhoff, G., Über das Gleichgewicht und die Bewegung einer elastichen Scheibe. J. Math. (Crelle) 40, 51-58, 1850.

2. Love, A. E. H., On the small free vibrations and deformations of thin elastic shells. Philo. Trans. Roy. Soc., Ser. A, 179, 491-546, 1888.

3. Leonard, R. W., Nonlinear first-approximation thin shell and membrane theory. Thesis, Virginia Polytechnic Institute, 1961.

4. Sanders, J. L., Jr., Nonlinear theories for thin shells. Quart. J. Appl. Mech., 21, 21-36, 1963.

5. Koiter, W. T., A consistent first approximation in the general theory of thin elastic shells. Proc. IUTAM Symp., North-Holland, Delft, Netherlands, 12-33, 1960.

6. Bieniek, M. P. and Funaro, J. R., Elasto-plastic behavior of plates and shells. Tech. Rep. DNA 3954T, Defense Nuclear Agency, Washington, D.C., 1976.

7. Wempner, G. and Hwang, C-M., A derived theory of elastic-plastic shells. Int. J. Solids Structures, 13, 1123-1132, 1977.

8. Reissner, E., The effect of transverse shear deformations on the bending of elastic plates. J. Appl. Mech., 12, A69-A77, 1945.

9. Mindlin, R. D., Influence of rotary inertia and shear on flexural motions of isotropic, elastic plates. J. Appl. Mech., 18, 31-38, 1951.

10. Librescu, L., Elastostatics and Kinetics of anisotropic and hetrogeneous shell-type structures. Nordhoff, Leyden, Netherlands, 1975.

11. Koiter, W. T. and Simmonds, J. G., Foundations of shell theory. Proc. IUTAM symp., Springer, N. Y., 150-176, 1973.

12. Lukasiewicz, S., Local loads in plates and shells. Noordhoff, Leyden, Netherlands, 1979.

13. Johnson, M. W., Jr. and McLay, R. W., Convergence of the finite element method in the theory of elasticity. J. Appl. Mechanics, 35, 274-278, 1968.

14. Wempner, G., New concepts for finite elements of shells. Z. Angew, Math. Mech., 48, T174-T176, 1968.

15. Zienkiewicz, O. C., Taylor, R. L. and Too, J. M., Reduced integraton methods in the general analysis of plates and shells. Int. J. Num. Methods Eng., 3, 275-290, 1971.

16. Hu, H. C., On some variational principles in the theory of elasticity and plasticity. Scientia Sinica, 4, 1955.

17. Washizu, K., On the variational principles of elasticity and plasticity. Tech. Rep. No. 25-18, Massachusetts Institute Technology, 1955.

18. Wempner, G., Talaslidis, D. and Hwang, C-M., A simple and efficient approximation of shells via finite quatrilateral elements. J. Appl. Mech., 49, 115-120, 1982.

19. Wempner, G., Finite elements via the theorem of Hu-Washizu. Comp. Meth. Mech. Eng., 2, 1982.

20. Wempner, G., A general theory of shells and the complementary potentials. J. Appl. Mech., 53, 881-885, 1986.

21. Koiter, W. T., On the nonlinear theory of thin elastic shells. Proc. Kon. Ned. Ak. Wet. 69B, 1-54, 1956.

22. Libai, A. and Simmonds, J. G., The nonlinear theory of shells. Academic, N. Y., 1988.

23. Pietraszkiewicz, W., Finite rotations in shells. in Theory of Shells, E. Becker and G. K. Mikhailov, Eds., North-Holland, Amsterdam, 445-571, 1980.

24. Koiter, W. T., On the stability of equilibrium (in Dutch). Thesis, Tech. Univ. Delft, 1945 [NASA rep. TTF-10 (English Transl.), 1967].

25. Budiansky, B. and Hutchinson, J. W., Buckling, progress and challenge, Trends in solid mechanics, Delft Univ., Delft, Netherlands, 1979.

26. Wempner, G. and Kross, D., Finite deflections, buckling and postbuckling of an arch. Developments in mechanics, 5, Iowa State Univ., Ames, 439-450, 1969.

27. Wempner, G., Finite elements finite rotations and small strains of flexible shells. Int. J. Solids Structures, 5, 117-153, 1969.

28. Wempner, G., Discrete approximations related to nonlinear theories of solids. Int. J. Solids Structures, 7, 1581-1599, 1971.

29. Wempner, G. and Hwang, C-M., Elasto-plasticity of the club sandwich. Int. J. Solids Structures, 16, 161-165, 1980.

30. Valanis, K., C., A theory of viscoplasticity without a yield surface. Arch. Mech. Stos., 23, 1971.

31. Wempner, G., Mechanics and finite elements of shells. Appl. Mech. Rev., 42, 129-142, 1989.

STATICS, DYNAMICS AND STABILITY OF SHELL STRUCTURES

Weng Zhiyuan Lu Wenda
(Institute of Engineering Mechanics, Tongji University
1239 Siping Road, Shanghai 200092, China)

ABSTRACT

Plate and shell structures are widely used in civil engineering. Their static, dynamic behavior and the stability are the most interested topics since long. Advances in several aspects are reviewed in this paper.

INTERACTION OF STRUCTURES AND CONTINUOUS MEDIA

This subject has a broad background in the underground protection engineering, offshore engineering, nuclear power engineering and municipal life-line engineering. For a long time, we have focused our attention on the dynamic interaction of soil/fluid-structures and structural responses under shock waves or seismic waves, and a great achievement has been made.

Buried thin-wall structures with various shapes, such as thin and thick cylindrical shells, double-decker cylinders connected by annulus and shallow shells, under the action of explosive waves have been investigated theoretically and experimentally. In the theoretical study, the combining use of analytical method with numerical method or comprehensive application of different kind of numerical method were the distinguishing features [1-5].

The dynamic response of the vertical well, which can be treated as a buried thick cylinder, under the impact loads has been successfully solved by combining use of analytic and boundary element method. First, the dynamic response of a thick cylinder with time-dependent boundary was studied by using the general analytic method combining with the transformation method of spatial variable. Then the three-dimensional wave propagation in a semi-infinite medium with a cavity was solved by the semi-analytic boundary element method, so that a set of coupled equation was established with the consideration of compatible condition between wall and cavity, that is a set of algebraic equation for shell

$$\{u_s\} = [B]\{t_s\} + \{C\} \tag{1}$$

The boundary integral equation for the medium at the time $t = n\Delta t$ will be

$$\begin{bmatrix} H_{11} & H_{12} \\ H_{21} & H_{22} \end{bmatrix} \begin{Bmatrix} \{u^s_m\}_2 \\ \{u^s_m\}_2 \end{Bmatrix} = \begin{bmatrix} G_{11} & G_{12} \\ G_{21} & G_{22} \end{bmatrix} \begin{Bmatrix} \{t^s_m\}_2 \\ \{t^s_m\}_2 \end{Bmatrix} + \begin{Bmatrix} \{F_1\} \\ \{F_2\} \end{Bmatrix} \tag{2}$$

and the compatible conditions for stresses and displacements on the boundary are

$$t^s_k = - t_{mk} = - (t^I_{mk} + t^s_{mk}) \tag{3a}$$

$$u^s_k = - u_{mk} = - (u^I_{mk} + u^s_{mk}) \tag{3b}$$

Numerical results were verified by the dynamic model test. The main figure of this work is that it can be carried out with a microcomputer for such a complex problem.

Free vibration of offshore towers with consideration of the fluid-soil-pile interaction has been studied, and the effects of surface waves and compressibility of the water were included. Some contributions to the vibration of elliptical column partially submerged in water were also made [6-7].

The seismic response and structure-soil interaction of the nuclear reactor containment under strong earthquake was another topic we have concerned with. [8-9]. In this aspect, a curved shell element was used and discretization in the generatrix direction was made. The impedance of buried base in an elastic half-space was adopted. It is solved in time domain. Great reduction in error can be expected in contrast to the lumped-mass system or beam model which were generally used in engineering practice. Details of the dynamic characteristics as well as response of the shell structure under seismic loads were obtained.

The equation of motion for the analysis of containment-foundation interaction under seismic action is written as

$$\begin{bmatrix} [M] & \{M_u\} & \{M_\varphi\} \\ \{M_u\}^T & M_{ss}+M_b/\pi & M_{sg} \\ \{M_\varphi\}^T & M_{sg} & M_{ss}+(I^b_\varphi+I_\varphi)/\pi \end{bmatrix} \begin{Bmatrix} \ddot{\delta} \\ \ddot{u}_0 \\ \ddot{\varphi}_0 \end{Bmatrix} + \begin{bmatrix} [C] & \{0\} & \{0\} \\ \{0\}^T & C_H/\pi & C_{H\varphi}/\pi \\ \{0\}^T & C_{H\varphi}/\pi & C_H/\pi \end{bmatrix} \begin{Bmatrix} \dot{\delta} \\ \dot{u}_0 \\ \dot{\varphi}_0 \end{Bmatrix} +$$

$$\begin{bmatrix} [K] & \{0\} & \{0\} \\ \{0\}^T & K_H/\pi & K_{H\varphi}/\pi \\ \{0\}^T & K_{H\varphi}/\pi & K_H/\pi \end{bmatrix} \begin{Bmatrix} \delta \\ u_0 \\ \varphi_0 \end{Bmatrix} = - \begin{Bmatrix} \{M_u\} \\ M_{ss}+M_b/\pi \\ M_{sg} \end{Bmatrix} \ddot{u}_g \tag{4}$$

It be will reduced to the dynamic equation of shell on a rigid foundation when the effect of soil-structure interaction is ignored.

$$[M]\{\ddot{\delta}\} + [C]\{\dot{\delta}\} + [K]\{\delta\} = -\{M_u\}\ddot{u}_g \qquad (5)$$

Results of the buried pipeline and tunnel under seismic wave were obtained by combining application of wave function expansion and numerical calculation [10].

DYNAMIC ANALYSIS OF STRUCTURES UNDER THE ACTION OF SHOCK WAVES, MISSILES OR AIRCRAFT IMPACTS

In this aspect, several investigations have been made on the dynamic response of hyperbolic shallow shells under shock waves, dynamic response of buried thick plates, nonlinear dynamic response of elasto-plastic thin plates and dynamic response of nuclear reactor containment [11-12].

Elastic or elasto-plastic dynamic response of shells with various shapes, such as shallow shell, semi-spherical shell and combined shells of semi-spherical and cylinder, and the rupture effects of reinforced concrete beams and plates are the other topics that we are interested in [13-14].

The dynamic response of a shallow spherical dome to arbitrary impact was studied under the classical shallow shell theory. By using the plane wave decomposing method with the small parameter related to the scale of the shallow shell, dynamic response was decomposed into the two parts composed of quasi-static solution and mode solution. After the generalized coordinates $Q(\tau)$ of the system were obtained, the dynamic response of displacements and interior forces can then be expressed as

$$\begin{aligned} u_r(r,\varphi,\tau) &= u_r^{(s)}(r,\varphi)\cdot G(\tau) + \sum_{i=1}^{\infty} u_r^{(i)}(r,\varphi)\cdot Q_i(\tau) \\ N_r(r,\varphi,\tau) &= N_r^{(s)}(r,\varphi)\cdot G(\tau) + \sum_{i=1}^{\infty} N_r^{(i)}(r,\varphi)\cdot Q_i(\tau) \end{aligned} \qquad (6)$$

where the first and second term on the right-hand side represent the quasi-static solution and mode solution respectively. Numerical results show that peak value of displacement takes place when impact load acts at the pole of the shell, and it occurs after unloading.

In the study of dynamic response of nuclear reactor containment to aircraft impact, the model of aircraft impacting on rigid target was employed and its elasto-plastic response was obtained. Furthermore, the verified loads have been established with the effects of local rupture in impact region to be taken into consideration. Besides, some efforts on the determination of plasto-dynamic factor and the reasonable shape of the containment have also been made [15,16].

Moreover, some numerous work on the basic theory of thick plate and shell and its practical application has been contributed. Those include the vibration of elastic thick plate, dynamic response analysis of elasto-plastic thick plate, vibration of orthogonal anisotropic thick plate, the general equation of motion for elastic thick shells of arbitrary shape and the vibration of elastic thick cylinders [17-19].

The mentionable contribution was that, by using the displacement and rotation as the independent variables, the equation of motion and relevant natural boundary conditions for elastic thick shells of arbitrary shape were deduced. The Hamilton's principle was used, and not only the common terms of membrane forces and moments but also the effects of transverse shear deformation, inertia of rotation and transverse extrusion deformation were included. It is easy to come to the equation of motion for thick revolution shell, spherical shell cylindric shell and thick plate as the special cases.

RANDOM RESPONSES OF COOLING TOWERS TO TURBULENT GUST WIND

Under the sponsor of Science Committee of China, ever since 1976, the theoretical researches and the full-scale tests on the wind pressure distributions and gust responses of hyperbolic cooling towers have been developed. A mean wind pressure distribution curve which is valid for smooth surface cooling towers has been suggested for the first time [20,21].

$$C_p(Z,\theta) = 1 - (1 - \min C_p)\sin(\frac{90}{75}\theta)^{\gamma} \tag{7}$$

where $\min C_p = -0.32(\frac{Z}{90})^{-2\alpha} - 1.22$, $\alpha = 0.2 \sim 0.25$. A simplified physical model for power spectral and cross-spectral density functions of fluctuating wind pressure $S_{p_A p_B}(r_A, r_B; f)$ has been put forward [21].

During the years of 1986 to 1990, a further study of the dynamic soil-hyperbolic cooling tower interaction was made theoretically as well as experimentally by full-scale test and laser holographic model test [22-24]. Numerical study of such a complex soil-structure system is based on the subregion variational principle. It is verified that the dynamic variational principle is valid in subregions as long as the continuity of the displacement on the boundary is satisfied at arbitrary time t. Because of the axisymmetricity of the structure, the displacement u_i, the stresses σ_{ij} and the loads $\bar{P}_i$ can be expanded into Fourier series

$$X(r,Z,\theta) = \sum_{k=0}^{\infty} [X_k(r,Z)\cos k\theta + X'_k(r,Z)\sin k\theta] \tag{8}$$

The curved rotational shell element is used and the infinite soil medium is expressed as an equivalent dynamic boundary system [23]. At an arbitrary point (s,θ), the spectral and rms response of the field variable under the action of random gust wind should be

$$S_{\varphi}(s,\theta;f) = \int_0^{2\pi}\int_0^{2\pi}\int_0^{s_0}\int_0^{s_0} H^*_A(jf)H_B(jf)S_{p_A p_B}(r_A, r_B; f) \cdot r(s_A)r(s_B)ds_A ds_B d\theta_A d\theta_B \tag{9}$$

and

$$\sigma_{\varphi}(s,\theta) = [2\int_0^{\infty} S_{\varphi}(s,\theta;f)df]^{1/2} \tag{10}$$

The gust response factor is defined as

$$G_{\varphi} = 1 + g \frac{\sigma_{\varphi}(s,\theta)}{\bar{\varphi}(s,\theta)} \tag{11}$$

where $\bar{\varphi}(s,\theta)$ is the mean response and g is the peak factor,

$$g = \sqrt{2\ln\gamma T} + \frac{0.577}{\sqrt{2\ln\gamma T}}$$

Numerical results show that the gust response factor will decrease with the softening of the foundation. For a considerably hard foundation, it can be treated as clamped. The feasibility of this proposed model for dynamic tower-foundation interaction is verified not only by a model test of free vibration [24] but also by the full-scale measuring [22,25].

For the general rotational shell structures with sparsely distributed stiffening ribs, the mass and stiffness matrices of the ribs can be considered as an increment to the initial matrix of towers with smooth surface. The asymptotic incremental expansion of natural frequencies and normal modes for the stiffened shell can be obtained by asymptotic perturbation analysis [26].

$$\lambda = \lambda_0 + \varepsilon\lambda_1 + \varepsilon^2\lambda_2 + \cdots, \quad U = U_0 + \varepsilon U_1 + \varepsilon^2 U_2 + \cdots \tag{12}$$

where $\lambda_1 = \dfrac{U_0^T(K_1 - \lambda_0 M_1)U_0}{U_0^T M_0 U_0}$, $\cdots$; and the corresponding increment of the normal mode will be

$$U_1^i = \sum_{\substack{j=1 \\ j\neq i}}^{n} \frac{1}{\lambda_0^i - \lambda_0^j} \left\{ \lambda_0^j \frac{U_0^{iT} K_1 U_0^j}{U_0^{jT} K_0 U_0^j} - \lambda_0^i \frac{U_0^{iT} M_1 U_0^j}{U_0^{jT} M_0 U_0^j} \right\} U_0^j \quad (\, i = 1,2,\cdots,n) \tag{13}$$

in which the superscript expresses the order of normal mode. λ_0 and U_0 are the frequency factor and the normal mode of the corresponding smooth shell. For accuracy, the total normal mode (U_0+U_1) should be calculated by using the known natural frequency $(\lambda_0+\lambda_0)$. Similarly, the gust response of the stiffened shell should be

$$\hat{G}_W = \hat{G}_W^f + \hat{G}_W^u, \qquad \hat{G}_W^f = g\cdot\frac{\hat{\sigma}_W^f}{\bar{W}} - \frac{\hat{\sigma}_W}{\bar{W}}\cdot\frac{\hat{W}_W^f}{\bar{W}}\cdot g \tag{14a}$$

$$\hat{G}_W^u = g\cdot\frac{\hat{\sigma}_W^u}{\bar{W}} - \frac{\hat{\sigma}_W}{\bar{W}}\cdot\frac{\hat{W}_W^u}{\bar{W}}\cdot g, \qquad \hat{\sigma}_W = \hat{\sigma}_W^f + \hat{\sigma}_W^u \tag{14b}$$

This is the so-called perturbation finite element solution for the dynamic characteristics and gust response of stiffened shells.

DYNAMIC RESPONSE ANALYSIS OF HYPERBOLIC COOLING TOWERS WITH LOCAL GEOMETRICAL IMPERFECTION

Since the geometric imperfection is insignificant in contrast to the overall dimension of tower, the dynamic equation of the tower with geometrical imperfection can then be established by using the initial displacement method or imperfection element.

$$[M_{ii}]\{\ddot{\alpha}_i\} + [K_{ii}]\{\alpha_i\} + \varepsilon \sum_{j=0}^{M} [K'_{ij}]\{\alpha\} = \{f_i\} \quad (i=0,1,2,\cdots,N) \tag{15}$$

in which $\varepsilon[K'_{ij}]$ is a weakly coupled additional stiffness matrix, ε is a small parameter. The corresponding eigenvalue equation will be

$$-\omega_r^2[M]\{\phi_r\} + [K]\{\phi_r\} + \varepsilon[K']\{\phi_r\} = 0 \quad \{r=1,2,\cdots,(M+1)N\} \tag{16}$$

Expanding ω_r and $\{\phi_r\}$ in accordance with the small parameter, a set of perturbation equation is then resulted. The first order approximation of ω_r and $\{\phi_r\}$ can be easily obtained [27-31]

$$\omega'_{ik} = K'^{i}_{k}\omega^{0}_{ik}/(2K^{i}_{k}) \quad (i=0,1,2,\cdots,M;\ k=1,2,\cdots,N) \tag{17}$$

$$\{\phi'_{ik}\} = \sum_{l=1}^{R} C^{i}_{kl}\{\phi^{0}_{il}\}; \qquad \{\phi'_{ijk}\} = \sum_{l=1}^{R} C^{ij}_{kl}\{\phi^{0}_{il}\} \tag{18}$$

and the first order approximation of response can be written as

$$\{\alpha\} = \sum_{j=0}^{M} \sum_{k=1}^{N} \sum_{l=1}^{N} \beta_{jk}\bar{C}^{ij}_{kl}\{\phi^{0}_{il}\} \quad (i=0,1,2,\cdots,M) \tag{19}$$

STABILITY ANALYSIS OF SHELL STRUCTURES

Linear Stability Analysis of Cooling Tower with Ring Ribs

For the linear stability analysis of hyperbolic cooling towers which has some weak nonlinear characteristics, we have proposed a feasible method so called the theory of linear pre-buckling consistent stability analysis. It has the advantages of high accuracy and less computation work. In case of the combined action of dead weight and wind pressure, the basic equation will be simplified into a standard eigenvalue problem

$$(K + K_q)^{-1}K_p\alpha + \alpha/\lambda_p = 0 \tag{20}$$

Numerical analysis shows that [32,33] our results are in good agreement with those of nonlinear analysis by Chan-Firmin and the difference is within 0.7%. The ring ribs will not only cause an increase in its critical wind pressure but also reduce the influence of the dead weight to its stability. It is found that the location of the sparsely distributed stiffening ribs plays a more important role than the number of ribs.

Nonlinear Analysis of RC Shells

In this study, the nonlinear construction relationship, presented by Liu, and the smeared cracking model are adopted. The cracked concrete element is taken to be an orthogonal anisotropic material and the effect of cracking is simulated by changing in stiffness of the layered shell element. The tension stiffening of concrete after cracking is estimated by additional stiffness of the reinforcement. The change of cracking direction is permitted by using a material geometrical matrix. The bilinear hardening relationship is used for the reinforced steel bars and the modified Lagrangian formula is employed to calculate the geometric nonlinearity.

$$\int \Delta_t S_{ij} \delta \Delta_t \varepsilon_{ij} dV_t + \int \tau_{ij} \delta \Delta_t \eta_{ij} dV_t = \delta W - \int \tau_{ij} \delta \Delta_t \varepsilon_{ij} dV_t \tag{21}$$

Combining use of load and displacement increment method is adopted to ensure the consistent convergence near the buckling points [34]. The ultimate loading capacity of cooling towers, undergone wind and dead weight, and the containment shell, undergone inner pressure, are analyzed. The results show that the main factor to consider the safety of cooling tower undergone wing should be the ultimate load, not the buckling load [35]. The work by Mang, Milford and Gupta [36] verified this conclusion.

Asymptotic Solution of Nonlinear Stability Critical Load for Shells

For nonlinear stability analysis of structures, a set of equilibrium equation and nonlinear eigen equation will be resulted.

$$\begin{aligned} &[K_e + N_1(V) + N_2(V^2)]V = \mu P_{ref} \\ &\det[K_e + K_1(V) + K_2(V^2)] = 0 \end{aligned} \tag{22}$$

Define $P = \mu P_{ref}$, $V = \lambda V_{ref}$. We have $K_1(V_{ref}) >> K_2(V^2_{ref})$, for the general shell structures; then it it can be transformed into a nonlinear quadratic eigenvalue problem with a small parameter ε [37].

$$[A + \lambda B + \lambda^2 \varepsilon C]X = 0 \tag{23}$$

An asymptotic solution can be obtained by perturbation method

$$\lambda = \lambda_0 + \varepsilon\lambda_1 + \varepsilon^2\lambda_2 + \cdots, \quad X = X_0 + \varepsilon X_1 + \varepsilon^2 X_2 + \cdots \tag{24}$$

where $\lambda_1 = -\lambda_0^2 \dfrac{X_0^T C X_0}{X_0^T B X_0}$, $X_k = \sum_{\substack{j=1 \\ j \neq m}}^{m} \dfrac{X_0^j}{\lambda_0^m - \lambda_0^j} \dfrac{(X_0^j)^T (P_k^m)}{(X_0^j)^T B (X_0^j)}$. The resulted proportional factor for displacement will not satisfy the nonlinear equilibrium equation in general. This is the so-called incomplete asymptotic analysis method.

However, λ can be transformed into a proportional loading factor with the help of nonlinear equilibrium equations, that is $\mu = (1+\gamma)\lambda$. γ is the ratio of the incremental strain energy corresponding to the nonlinear deformation and linear deformation. It has the order as same as ($\lambda/\lambda_0 - 1$). So the hypothesis of $\mu=\lambda$ is unreasonable. If we solve these two sets of nonlinear equation simultaneously by perturbation method, then we have a new quadratic eigenvalue problem which is similar to the previous ones.

$$\det\{K_e + \mu\varepsilon K_1(V_1^*) + \mu^2\varepsilon^2[K_2(V_1^{2*}) + K_1(V_2^*)]\} = 0 \tag{25}$$

This is the so-called complete asymptotic analysis method. Only the linear calculation is needed for these two asymptotic approach mentioned above , and less storage and small amount of computation work are needed in contrast to the methods usually used.

PLASTIC POST-BUCKLING AND IMPERFECTION SENSITIVITY ANALYSIS OF STRUCTURES

The main difficulty for plastic post-buckling analysis is due to the moving boundary caused by unloading. Since the unloading area extends with the increase in loading, taking the characteristic size within unloading area to be the perturbational parameter [39], the higher order asymptotic solution can be expected and the imperfection sensitivity analysis can then be carried out. For Hutchinson's model, the size of the unloading area corresponding to the maximum load is a constant when the bilinear hardening construction relationship is adopted. If we choose the relative quantity, which is the change in size of the unloading area over this constant value mentioned above, as a parameter, make a secondary perturbation near the maximum load and connect these two perturbation solution, then an asymptotic solution which is valid in a considerable large range can be expected [39].

$$P = P_c[\ 1+0.5625\zeta^2-0.4843\zeta^3+\cdots\ ] \qquad (\zeta\leq\zeta_{max}/2)$$

$$P = P_c[\ 1.0504-0.8417(\zeta-\zeta_{max})^2-1.73(\zeta-\zeta_{max})^3+\cdots\ ], \qquad (\zeta\geq\zeta_{max}/2)$$

The imperfection sensitivity can be expressed as

$$\bar{P}_m = P_m - 2K_1E^tL\frac{2\bar{S}}{\bar{U}}e(\bar{\theta})^{1/2} \tag{26}$$

It shows that the ultimate loading capacity is decrease in power of $(\bar{\theta})^{1/2}$ to the imperfection. For the post-buckling analysis of plate and shell structures, the unloading area will not only extend in thickness but also vary in the direction of length and width. These unloading behaviors cause the post-buckling solution to have a boundary layer effect similar to that occurred in visco- elastic fluid, so an amplification in scale should be made. The continuity of the field variable along the moving boundary should always be contended with. For simplifying the procedure, a transformation is introduced to the no-unloading area which maintains the form of the solution even when the boundary undergoes a changing. The high-order approximate solution and imperfection sensitivity analysis of a clamped circular plate in the post-buckling range have been investigated by the proposed systematic perturbation method [40]

$$\sigma/\sigma_c = 1 + \lambda_1\eta + \lambda_2\eta^{4/3} + \lambda_3\eta^{5/3} + \cdots \tag{27}$$

where $\eta = -\ [1-J_0(K)][a_6\zeta^6/6 + a_8\zeta^8/8 + \cdots]$ is the amplitude of the normal mode,

$$\sigma = \hat{\sigma}\ \frac{1}{12}\ \frac{\hat{E}}{1-\nu}\ \frac{t^2}{R^2}K^2\{a_6^1\zeta^6/2+3(a_8^1-K^2a_6^1/4)\zeta^8/8+3a_2^1\varepsilon^{1/2}\zeta^2/2+\cdots\} \tag{28}$$

in which $\hat{\sigma}$ is the value of the load when unloading takes place, and ε

is a parameter of imperfection. It can be found that, in difference from the elastic post-buckling, the higher order terms of non-integral power appear in this post-buckling asymptotic solution and it plays a key role in the plastic post-bucking analysis.

FUTURE DIRECTIONS OF RESEARCH

As an extension of the previous work, the following topics would be the focus of our future attention.

(1) Structural Vibration Control and Seismic Base Isolation of the Foundation
   (a) seismic base isolation of the foundation for nuclear reactor containment (including the fast breeder reactor) and vibration control of the facilities in nuclear power plant.
   (b) vibration control and isolation analysis for high rise structures, such as cooling towers and funnels.
   (c) The measures of seismic isolation and vibration control for the supports of bridges.

(2) Enhancement of the Structural Capacity to Impact and Investigation on the Rupture Mechanism of Relevant Materials.
   (a) Studies on the constructional measures and reasonable shape of the structure to increase its anti-impact capacity.
   (b) Determination of the load curve for flexible missiles impact to reinforced concrete target.
   (c) Studies on the rupture mechanism of reinforced concrete and fiber reinforced composite beams, plates and shells to impact loads.
   (d) Dynamic response of beams, plates and shells under verified loads.

(3) Plastic Post-buckling and Imperfection Sensitivity Analysis of Structures
   (a) development of systematic perturbation method for plastic post-buckling and imperfection sensitivity analysis of two-dimensional structures with the characteristic size of unloading area to be the perturbation parameter.
   (b) development of the perturbation finite element method corresponding to the systematic perturbation method.

(4) Reliability Analysis of Large Reinforced Concrete Structures
   (a) The probabilistic ultimate strength and reliability analysis of reactor containments and hyperbolic cooling towers with concrete cracking.
   (b) The dynamic reliability of reactor containments and hyperbolic cooling towers.

REFERENCES

[1] Weng Zhiyuan and Zeng Deshun, Unsteady interaction of plates and shells with continuous medium under explosive waves, Transactions of Int. Conf. on SMiRT-9, Lausanne, Swiss. 1987.

[2] Ni Haiying and Weng Zhiyuan, Semi-analytic boundary element method for analysis of 3-D wave motion on semi-infinite media, Proc. of 1st Japan-China Symposium on BEM, 1987, Japan.

[3] Le Rong and Weng Zhiyuan, Dynamic response of an axially symmetric moderate thick cylindrical shell buried in medium, Transactions of the Int. Conf. on Appl. Mech., Beijing China, 1989.

[4] Fang Yizhe and Weng Zhiyuan, Dynamic response of double-decker cylindrial shell buried in elastic medium by action of shock wave, Comp. Struct. Mech. and Applications, 6(1989), 1-11.

[5] Ni Haiying and Weng Zhiyuan, AS-BE coupling method for 3-D dynamic response of underground vertical well, Transactions of Int. Conf. on SMiRT-10, Vol. K, 1989 Los Angeles, USA.

[6] Zhu Yongyi, Weng Zhiyuan and Wu Jialong, Water-soil-cylinder interaction in offshore tower structures - free vibration, J. Civil Eng., 22(1989), 55-64.

[7] Zhu Yongyi, Weng Zhiyuan and Wu Jialong, The coupling vibration between column and water considering the effects of surface wave and compressibility of water, Acta Mech. Sinica, 21(1989), 658-667.

[8] Weng Zhiyuan and Xie Ning, Seismic response analysis for a containment of nuclear power plant in consideration of soil-structure interaction, Transactions of Int. Conf. on SMiRT-11, Vol. K, 1991, Tokyo, Japan.

[9] Wang Mingyang, Weng Zhiyuan And Qian Jiang, Dynamic structure-foundation interaction analysis for nuclear containment by FE-BE coupling method, Proc. of the 4th China-Japan Symposium on BEM, Beijing, 1991.

[10] Zhou Deliang and Weng Zhiyuan, The dynamic response of two parallel structures in elastic half space under SH waves, Trans. of Int. Conf. on SMiRT-11, Vol.K, 1991, Tokyo, Japan.

[11] Weng Zhiyuan and Huang Xiaogang, The dynamic response of doubly-curved shallow shell subjected to shock waves, Transactions of the Int. Symposium on Strong Shock and its Effects, Beijing, China, 1986.

[12] Wang Xintian, Hong Shantao and Weng Zhiyuan, Nonlinear elasto-plastic dynamic response of thin plates to impact loads, J. of Solid Mech., 3(1985).

[13] Shen Keling, Weng Zhiyuan and Hong Shantao, Dynamic response of

combined shell structures to axi-symmetric pulse. Vibration and Shocks, 3(1988).

[14] Pan Huaping, Weng Zhiyuan and Wang Yuangong, Dynamic response of shallow spherical dome to the arbitrary pulse, Vibration and Shocks, 4(1989).

[15] Wang Yuangong and Yu Aiping, Nuclear Power Engineering, special volume, 1988.

[16] Hong Shantao and Wei Lin, Nuclear Power Engineering, special volume, 1988.

[17] Weng Zhiyuan, Chen Xuechao and Zeng Deshun, Dynamic response of elasto-plastic thick plates, Sino-American Symposium on Bridge and Structural Engineering, Beijing, China, 1982.

[18] Weng Zhiyuan, Wang Yuangong and Zeng Deshun, A general equation of motion for thick elastic shell with arbitrary shape, Transactions of Int. Conf. on SMiRT-8, Vol. B, Brussels, Belqium, 1985.

[19] Weng Zhiyuan, Wang Yuangong and Zeng Deshun, Vibration of elastic thick cylindrical shells, Transactions of Int. Conf. on SMiRT-9, Lausanne, Swiss, 1987.

[20] Lu Wenda et al., Response of the hyperbolic cooling tower to turbulent wind, J of Electrical Eng., CSEE, 1(1982), 45-61.

[21] Lu Wenda et al., Proc. of the 2nd Int. Symp. of Natural Draught Cooling Towers, IASS-RUB, Springer-Verleg Berlin Heidelberg, Germany (1984), 136-154.

[22] Wang Shu and Lu Wenda, Gust factors for hyperbolic cooling towers on soils, Eng. Struct., 13(1991), 21-26.

[23] Lu Wenda and Wang Shu, Analysis of dynamic interaction between hyperbolic cooling towers and soils, Comp. Struct. Mech. & Applications, 7(1990), 43-49.

[24] Wang Shu and Lu Wenda, Theoretical and experimental solution of cooling-tower-soil system, J. Engng. Mech., ASCE, 116(1990), 862-869.

[25] Lu Wenda, Wang Shu and H.J. Niemann, Sonderforschungsbereich Tragwerksdynamik, SFB-151, Nr. 17.(1990), Ruhr-Universtat Bochum, Germany, 37-72.

[26] Li Longyuan and Lu Wenda, Asymptotic analysis of dynamic response of hyperbolic cooling tower shells with ring-stiffeners -- perturbation finite element solution, Applied. Math. and Mech., 8(1987), 603-610.

[27] Lu Wenda and Lin Baoging, The effect of axisymmetric geometric imperfections on dynamic response to turbulent wind in a cooling

tower, Progress in Applied Mechanics, Martinus Nijhoff, Netherlands, 1987, 163-180,

[28] Gao Shigiao and Lu Wenda, Dynamic solution of hyperbolic cooling towers with nonaxisymmetric imperfection by finite element method, Applied Math. and Mech., 9(1988), 515-519.

[29] Gao Shigiao and Lu Wenda, Analysis of random wind response for hyperbolic cooling tower shell with meridional random geometric imperfections, J. of Beijing Inst. Tech. (in Chinese), 9(1989), 61-66.

[30] Lu Wenda and Gao Shigiao, The effect of nonaxisymmetric geometric imperfection of hyperbolic cooling tower on its turbulent wind response, Proc. the 3rd Int. Sym. on Natural Draught Cooling Towers, IASS-Paris, France, (1989), 217-224.

[31] Gao Shigiao and Lu Wenda, Perturbation method to solve the couple problem among harmonic wave, Applied Math. and Mech.,9(1988), 975-980.

[32] J. F. Abel, et al., Comparison of complete and simplified elastic buckling analysis for cooling tower shells, Eng.Struct., 8(1986), 25-28.

[33] Lu Wenda and Gu Haozhong, Buckling of cooling tower shells with ring-stiffeners, Applied Math. and Mech., 10(1989),559-567.

[34] Lu Wenda and Li Longyuan, Nonlinear analysis of reinforced concrete shells, Chinese Journal of Applied Mechanics, 7(1990), 31-36.

[35] Lu Wenda et al., Proc. of the 3nd Int. Symp. of Natural Draught Cooling Towers, IASS-Paris, Springer-Verleg Berlin Heidelberg, France (1989), 259-269.

[36] A. K. Gupta, et al., Investigation on hyperbolic cooling tower ultimate behavior, Eng. Struct., 8(1986), 87-92.

[37] Li Longyuan, Science in China (Series A), 32(1989), 564-569.

[38] Viggo Tvergaard, Effect of plasticity on post-buckling behavior, Lecture Notes in Physics -- Buckling and Post- Buckling, Springer-Verlag, (1985), 144-180.

[39] Su Xuming and Lu Wenda, Post-buckling and imperfection sensitivity analysis of structure in the plastic range, PartI: Modal analysis, Thin-Walled Struct. 10(1990), 263-275.

[40] Xu Sunming and Lu Wenda, Post-bifurcation and imperfection sensitivity analysis of plastic clamped circular plates, Int. J. Solids Struct., 27(1991), 769-782.

RESEARCH ON UNREINFORCED HOLLOW CLAY TILE WALLS AND DEVELOPMENTS OF COMPUTATIONAL MECHANICS

by

James E. Beavers,[1] Richard M. Bennett,[2] and Roger D. Flanagan[3]

[1]Corporate Fellow and Director, Center for Natural Phenomena Engineering, Martin Marietta Energy Systems, Inc., Oak Ridge, Tennessee, USA
[2]Associate Professor of Civil Engineering, University of Tennessee, Knoxville, Tennessee, USA
[3]Senior Engineer, Martin Marietta Energy Systems, Inc., Oak Ridge, Tennessee, USA

ABSTRACT

This paper discusses various aspects of an unreinforced hollow clay tile wall (HCTW) research program being conducted by Martin Marietta Energy Systems, Inc.,* for the U.S. Department of Energy. The purpose of the program is to assist in determining the response of infilled HCTW buildings to earthquake forces.

INTRODUCTION

During the early to middle 1900s, hollow clay tile (HCT) was used extensively as a building construction material throughout the United States. Many buildings were constructed of steel and/or concrete frames with HCT infilled walls. In the central and eastern United States, seismic building codes did not exist at the time and the buildings were not designed for seismic loads.

This paper focuses on hollow clay tile wall (HCTW) buildings located in zones of low-to-moderate seismicity, although the results are directly applicable to high seismicity zones. If infilled, unreinforced HCTW building construction can be shown to have inherent strength above that normally allowed in design by codes, retrofit may not be required in low-to-moderate seismicity zones if a risk-based approach is adopted. A risk-based approach is based on the principle that while the building may not meet today's codes and standards requirements, the expected seismic performance is reasonably understood and has sufficient margins of safety. To help accomplish these goals testing and improved computational methods are required.

*Martin Marietta Energy Systems, Inc., is a managing contractor for the U.S. Department of Energy under contract DE-AC05-840R21400.

A second principle that has an impact on the issue of "to retrofit or not to retrofit" in low-to-moderate seismicity zones is the cost increment of retrofit. In many cases, the cost increment to retrofit a building to a higher seismic capacity rather than a lower seismic capacity is much less than the cost to retrofit to the lower seismic capacity. For example, the cost to install a one inch diameter bolt required for the retrofit of a building in a moderate or high seismic zone is basically the same cost to install a one-half inch diameter bolt in the same building for retrofit in a low seismic zone. Thus, the real cost impact for retrofit of buildings in low-to-moderate seismic zones is not the level of retrofit but the decision, or requirement, to retrofit.

Building codes in the United States generally consider the limit state for unreinforced masonry to be cracking due to its brittle nature. However, the overall seismic resistance of an infilled masonry building panel can be an order of magnitude higher than that when the first crack occurs. This leads to a key proposition that must be adopted in the risk-based approach to a "no retrofit philosophy" for existing construction in low-to-moderate seismic zones. That proposition is that the HCTWs will crack when the earthquakes occur, but overall stability must be maintained. Thus, significant cracking of infilled walls is being described by the authors as acceptable damage for buildings in low-to-moderate seismic zones. Therefore, the principal basis of the research and development program on HCTWs described in this paper is to investigate and develop computational methods for determining post-cracking behavior, including ultimate load, of infilled HCTW building construction.

PREVIOUS HOLLOW CLAY TILE RESEARCH

Most of the work completed during the last decade in the United States has focused on the seismic performance of reinforced concrete block and brick masonry. The U.S. Technical Coordinating Committee for Masonry Research program [1] funded by the National Science Foundation has done some work on unreinforced masonry, but focused on reinforced masonry. More recently, as a result of the unreinforced masonry retrofit ordinance [2] in Los Angeles, California, USA, and the reauthorization of the United States National Earthquake Hazards Reduction Program [3], increased activity for testing and development of computational methods for unreinforced masonry construction is occurring in the United States. The National Earthquake Hazards Reduction Program requires the President of the United States to adopt standards by December 1, 1994, for assessing and enhancing the seismic safety of all federally owned and leased buildings not designed for seismic loads.

For the HCTW configurations that are being investigated, no previous research work has been found other than studies of material properties, such as Plummer [4] and Johnson and Matthys [5]. However, unreinforced masonry research on concrete block and brick have application to the HCTW program. Dawe and Seah [6 and 7] performed experimental and analytical work on the out-of-plane response of unreinforced concrete block infilled walls. Lawrence [8] published an extensive annotated bibliography on the lateral loading of masonry infilled panels. Dawe and Seah [9] and Dawe and Hatzinikolas [10] summarize the findings of a major test program on unreinforced infilled concrete block walls subjected to in-plane loads. Dawe et al. [11] tested unreinforced masonry infilled walls on a shake table. The work of Page, Kleeman, and Dhanasekar [12] in developing a nonlinear constitutive relationship and biaxial failure surface for unreinforced masonry subjected to in-plane loads is also applicable to the HCTW program. This model was used in the evaluation of test results of infilled frames [13].

BUILDING TYPES

The building types being studied in this program are steel frame with infilled HCTWs and were constructed in the 1940s and 1950s. Typically the buildings are industrial type, are one- to five-stories in height, and have steel framing with simple framing connections and very little or no cross-bracing. This results in the infilled HCTWs becoming the major lateral load resisting elements. Although the buildings were designed for the HCTWs to be nonload bearing, significant load bearing can exist due to the method of construction and building additions that have occurred. The walls are constructed with HCT cores running horizontally, resulting in a full bedding joint, and are of either eight or thirteen inches in thickness. The eight-inch walls are constructed with a single eight-inch tile. The thirteen-inch walls have an eight-inch and a four-inch tile laid in a staggered fashion, Figure 1. All exterior walls are thirteen inches thick.

The potential collapse mechanism's of these buildings are loss of lateral load capacity by (1) out-of-plane failure of the infilled HCTWs as a result of cracking caused by out-of-plane inertia loads, (2) out-of-plane failure as a result of cracking caused by in-plane shear loads, (3) in-plane corner crushing or diagonal cracking causing loss of shear capacity and (4) out-of-plane failure as a result of cracking caused by building drift.

Figure 1. End view of hollow clay tile wall.

The seismic performance of this type construction is highly dependent upon the true boundary or interface conditions. Typical HCTWs being investigated rest on a running bond supported by a concrete floor. The vertical interface condition is the HCT inserted between the flanges of the framing steel. Several variations exist including the wall being slightly offset from the steel centerline. The top is grouted against the bottom flange of the beam.

HCTW TESTING PROGRAM

Fricke and Jones [14 and 15] published elements of the HCTW test program as it was being developed and included some early test results. Since that time, the program has expanded into a total of approximately nineteen types of tests with multiple tests of each type, resulting in 93 in situ tests and over 198 laboratory tests. The major in situ tests are out-of-plane air bag tests while major laboratory tests are out-of-plane deformational and in-plane tests. A complete listing of the tests is shown in Table 1. Shake table testing is also being planned but is not fully developed at this stage.

TABLE 1
HCTW in situ and laboratory tests

TEST	NUMBER Insitu	Lab	TEST	NUMBER Insitu	Lab
1. Push Test	18	0	11. Racking	1	
2. Normal Stress	6		12. In-plane	4	16
3. Block Compression Normal		20	13. Coupon Tensile	6	12
4. Block Compression Parallel		10	14. Bond Wrench	12	18
5. Prism Compression Normal	6	10	15. Non-Dest. Exam	4	32
6. Prism Compression Parallel	6	10	16. Block Tensile		12
7. Flex. Bond Normal	6	10	17. Wall-Col. Bending		2
8. Flex. Bond Parallel	6	10	18. Mortar Characteristic		20
9. Deformability Normal	12		19. Out Building	2	___
10. Airbag	4	16	TOTAL	93	198

The purpose of the tests is threefold. Some of the tests are designed to determine existing conditions. These include deformability test, push test, and bond wrench test. Some of the results of these tests will be correlated to assemblage tests. For example, comparison of laboratory prism compression tests and the in situ deformability test will help determine the feasibility of using the deformability test to assess the strength of a wall. Figure 2 shows a typical prism compression test on a staggered thirteen-inch wall while Figure 3 show a typical in situ deformability test using the "flat jack" method. An extensive nondestructive evaluation (NDE) research program is also being conducted to develop NDE methods for assessing wall conditions.

The second purpose of the testing is to develop an appropriate constitutive model. Both the elastic pre-cracking behavior and the post-cracking performance are examined. These tests include prism compression tests, flexural bond tests, prism split tensile tests, and biaxial tests.

"flat jacks."

The third purpose of the tests is to identify system behavior, and thus appropriate idealizations for computational analysis. The emphasis is on post-cracking behavior. These tests include air bag tests, in-plane racking tests and various other laboratory tests.

Currently one full-scale insitu air bag test is in progress. This test is being conducted on the first floor of a five-story steel-frame building shown in Figure 4. The infilled panel being tested is an 8-inch HCTW 29 feet long and 12 feet high with the bottom of the panel grouted to a concrete floor, the top grouted to a W30 x 108 beam, and the ends encased partially between W14 x 142 columns. Figure 5 shows an isometric drawing of the test setup and Figure 6 shows a photograph of the actual wall with the load frame installed.

Figure 4. Five-story hollow clay tile wall building.

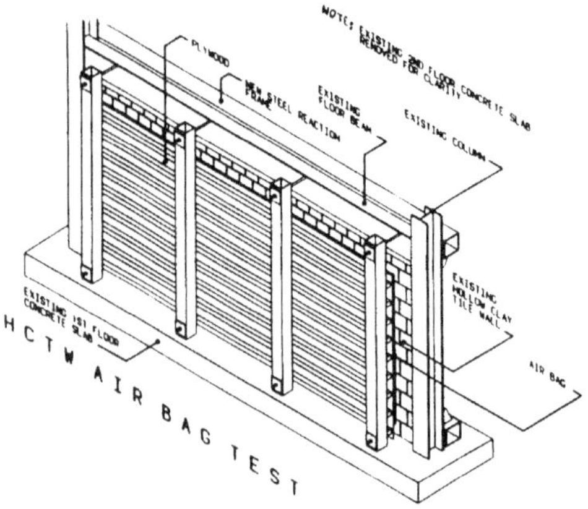

Figure 5. Isometric drawing of air bag test.

Figure 6. Air bag test wall with load frame.

One of the principal purposes of the laboratory static in-plane and out-of-plane tests is to gain knowledge of the in-plane performance following post-cracking caused by out-of-plane deformations. The out-of-plane loads will be supplied by point loads on the steel frame and will allow for the study of the impact of interstory drift. Frame flexibility and interface conditions will also be examined. Drawing schematics of these tests are shown in Figure 7.

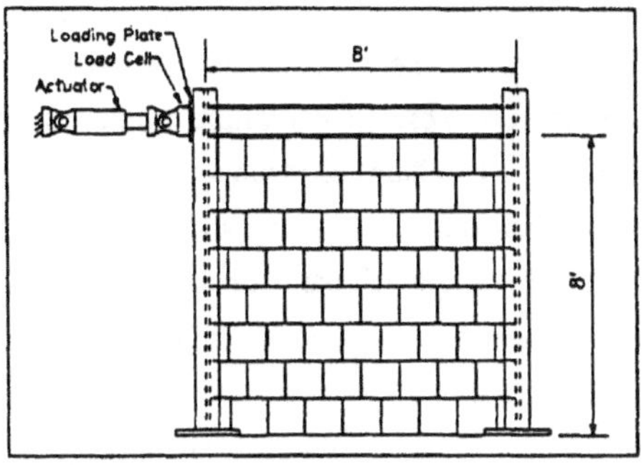

Figure 7a. In-plane test.

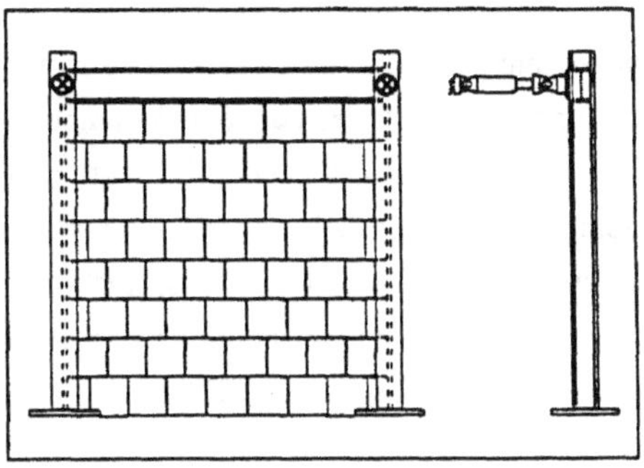

Figure 7b. Out-of-plane test.

The HCTW test program is quite extensive and has taken about twelve months in planning. Although some testing has occurred, the main stream of testing is just now beginning and will continue over the next two years.

COMPUTATIONAL MECHANICS

Current computational methods available to practicing engineers usually require ignoring the infilled walls as a structural member or assuming the infilled as a shear wall limited to elastic properties. The latter technique has been used, with limited success, for determining the seismic vulnerability of HCTW building types being examined in this study. Out-of-plane analyses typically are performed considering only one-way action with allowable material strength properties specified in a code. These computational methods are quite adequate for design. However, they can significantly underestimate the strength of a building when used for analysis of existing facilities. More refined computational methods are needed, particularly for the risk-based approach adopted herein. Computational methods for seismic evaluations of structures must be able to predict post-cracking performance in-plane and out-of-plane. They must include the complex interfaces that occur between post-cracking distortion of the infilled wall and flexibility of the surrounding frame. They must take into consideration the effects of interstory drift.

Based on what has been done to-date, it can be concluded that orthotopic linear elastic models are generally sufficient for predicting the out-of-plane inertial response [16 and 17] prior to cracking. Work has been done by a number of researchers to establish a failure criterion for first cracking [18, 19, 20, 21, and 22]. Based on recent work by Chua [23], supported by this HCTW program, the authors believe the failure criterion established by Thurlimann and Guggisberg [22] represents the best estimate for first cracking of a single infilled wall panel under a state of static stress. Using a commercial nonlinear finite element program [24] and failure criteria of Thurlimann and Guggisberg [22] for first panel cracking, Chua [23] produced some reasonable results for single panel out-of-plane behavior, as shown in Table 2. Chua also conducted parametric studies and found that for static out-of-plane loads, the modulus of rupture and modulus of elasticity had the most significant effect on first panel cracking while Poisson's ratio and shear moduli had little effect.

TABLE 2

First Panel Cracking Predictions

(After Chua, 1991)

UNB Specimen	Experimental Results (KPa)	Elliptical Interaction Method (KPa)
WE2	3.8	3.4
WE4	2.4	2.7
WE6	5.2	5.4
WE8	4.8	5.4

Considerable work has also been done examining the post-cracking out-of-plane behavior and ultimate load. One of the earliest accounts for out-of-plane behavior that found arching action occurs and should be considered in computational methods was McDowell et al. [25]. Tests by Gabrielsen and Kaplan [26] showed that arching action can be significant enough for a wall to resist load reversals equivalent to accelerations greater than 1g. Seah [27] developed a computational method based on concrete yield line theory modified for concrete block masonry. A key feature is that the method accounts for compressive membrane stresses, or arching action, which are induced by lateral restraint. The method accounts for panel frame flexibility and was able to produce results that compared well with Seah's laboratory air bag tests for post-cracking behavior and ultimate load. In a series of parametric studies, Chua [23] found that compressive strength and panel frame flexibility had the most significant effect on ultimate load of infills.

Little research has been found that addresses out-of-plane behavior of infilled unreinforced masonry wall panels caused by interstory drift. In his literature review, Lawrence [8] sited no work on interstory drift. However, Benedetti and Benzoni [28] indicated that interstory drift is a more severe out-of-plane loading than inertial loads. Recently, for a building of the type described herein, Butala, Jones, and Beavers [29] have confirmed through a progressive crack analysis that interstory drift, rather than inertial loads, is the primary failure cause for first panel cracking.

In tests on masonry walls that simulated interstory drift, Kariotis, Ewing, and Johnson [30 and 31] and Adham [32] found that ultimate load or stability of a wall panel was a function of the wall velocities. They developed an empirical relationship for wall survivability in an earthquake, thus allowing for retrofit decisions.

In-plane analyses of infilled structures have generally been conducted by replacing infill with an equivalent strut [33]. Computational methods have been considered for nonlinear degrading behavior of equivalent struts [34 and 35]. Plastic collapse methods have been proposed for obtaining ultimate strength of infilled panels [36 and 37]. Dawe and McBride [38] showed that Liauw and Kwan's method [36], with a penalty factor proposed by Wood [37], accurately predicted the capacity of several experimental frames.

Finite element modeling of infilled frames was attempted as early as 1985 [39]. Early models used linear elastic material properties, but recognized the importance of interface. Some various types of a gap/interface element would be used to account for debonding that would occur at the masonry/steel frame junction. Several nonlinear finite element models have been proposed [40, 41, and 42]. These have met with some success, but have the limitation that they only consider in-plane effects.

In the HCTW program, development of computational methods for analysis is divided into three phases. The first phase was an extensive literature search and review. The second phase has been the establishment of a baseline for computational development based on ability to repeat and study the works of others. Work is still ongoing in this phase and currently HCTW program staff are examining in-plane behavior. Attempts are being made to model extensive in-plane infilled frame tests conducted at the University of New Brunswick. A nonlinear concrete model in a commercial code [24] and a specialized masonry nonlinear finite element code [43] are currently being used.

The third phase of the program will be the development of computational methods to evaluate seismic response of steel frame buildings with infilled HCTW panels as structural members. These will take into consideration constitutive relationships and empirical methods developed from the test program. The goal is for computational methods to represent the effects of interstory drift, out-of-plane inertia loads, in-plane response including frame flexibility, panel-to-frame interfaces and progressive failure of the HCTW panel. Computational methods developed will be verified by the test program and tests conducted by others.

SUMMARY

Significant progress of the HCTW program has occurred to date. Tests have been conducted and computational methods are being developed and have shown reasonable comparison with other work. A large portion of the test program has been developed and tests include static, dynamic, and shake table. The shake table test program is now under development.

Results to date have shown that inherent strength does exist in infilled unreinforced HCTWs and they should be considered as a structural element when evaluating buildings for seismic loads. Results also tend to indicate that out-of-plane response is not a weak link in systems response and wall panels will not fail out-of-plane at the onset of earthquake loading, resulting in a loss of lateral load building stability.

Computational methods are being studied and improvements are being developed to better predict both building and component response. Due to the complexity of large structures, the development approach is toward simplified models using both equivalent strut formulations and finite element methods.

As the HCTW program progress continues and test results become available, new computational methods will be developed. The final result may well show that many buildings of the type described herein would (may) not require major structural retrofit in low-to-moderate seismic zones.

LIST OF REFERENCES

1. Noland, J. L., 1990 status report: U. S. coordinated program for masonry building research. Proceedings, Fifth North American Masonry Conference, University of Illinois at Urbana - Champaign, June 3-6, 1990.

2. Los Angeles Building Code Division 88, 1985 edition.

3. 101st Congress, "National Earthquake Hazards Reduction Program Reauthorization Act," Public Law 101-614, November 16, 1990.

4. Plummer, H.C., Brick and Tile Engineering, Brick Institute of America, 1962.

5. Johnson, F.B. and Matthys, J.H. Structural strength of hollow clay tile assemblages. Journal of the Structural Division, ASCE, 1973, **99**, 259-75.

6. Dawe, J. L. and Seah, C. K., Lateral load resistance of masonry panels in flexible steel frames. Proceedings, 8th International Brick and Block Masonry Conference, Dublin, Ireland, 1988, 606-16.

7. Dawe, J. L. and Seah, C. K., Out-of-plane resistance of concrete masonry infilled panels. Canadian Journal of Civil Engineering, 1989, **16**, 854-64.

8. Lawrence, S. J., Lateral loading of masonry infilled panels: a literature review. Technical Record 454, Experimental Building Station, Sydney, Australia, 1979.

9. Dawe, J.L. and Seah, C.K. Behavior of masonry infilled steel frames. Canadian Journal of Civil Engineering, 1989, **16**, 865-76.

10. Dawe, J.L. and Hatzinikolas, M.A., Behavior of masonry infilled panels in steel frames. Proceedings, 8th International Brick and Block Masonry Conference, Dublin, Ireland, 1988, 516-27.

11. Dawe, J.L. Schriver, A.B., and Sofocleous, C., Masonry infilled steel frames subjected to dynamic load. Canadian Journal of Civil Engineering, 1989, **16**, 877-85.

12. Page, A.W., Kleeman, P.W. and Dhanasekar, M., An in-plane finite element model for brick masonry. New Analysis Techniques for Structural Masonry, ASCE,1985, pp. 1-18.

13. Dhanasekar, M. and Page, A. W., Influence of brick masonry infilled properties on the behavior of infilled frames. Proceedings, Institution of Civil Engineers, 1986, **81**, 593-605.

14. Fricke, K. E. and Jones, W. D., Seismic analysis and testing of clay tile walls at the Oak Ridge Y-12 plant. Proceedings, Second DOE Natural Phenomena Hazards Mitigation Conference, Knoxville, Tennessee, 1989, 162-71.

15. Fricke, K. E. and Jones, W. D., A test program to determine the structural properties of unreinforced hollow clay tile masonry walls of the DOE Oak Ridge plants. The Masonry Society Journal, 1991, **9**, 56-63.

16. Haseltine, B. A., West, H. W. H. and Tutt, J. N., Part 2: design of walls to resist lateral loads. The Structural Engineer, 1977, **55**, 422-30.

17. Essawy, A. S. and Drysdale, R. G., Evaluation of available design methods for masonry walls subject to out-of-plane loading. Proceedings, Fourth North American Masonry Conference, 1987, 32:1-32:13.

18. Essawy, A. and Drysdale, R., Macroscopic failure criterion for masonry assemblages. Proceedings, Fourth Canadian Masonry Symposium, Fredericton, N. B., Canada, 1986, 263-77.

19. Fried, A., Anderson, C. and Smith, D., Predicting the transverse lateral strength of masonry walls. Proceedings, Eight International Brick and Block Masonry Conference, Dublin, Ireland, 2, 1988, 1171-83.

20. Gazzola, E., Drysdale, R. and Essawy, A., Bending of concrete masonry wallettes at different angles to the bed joints. Proceedings, Third North American Masonry Conference, University of Texas, Arlington, 1985, 27:1 - 27:14.

21. Hendry, A. W., The lateral strength of unreinforced brickwork. The Structural Engineer,1973, **51**(2), 43-50.

22. Thurlimann, B. and Guggisberg, R., Failure criterion for laterally loaded masonry walls: experimental investigations. Proceedings, 8th International Brick and Block Masonry Conference, Dublin, Ireland, 1, 1988, 699-706.

23. Chua, L. S., Out-of-plane behavior of unreinforced hollow clay tile masonry infilled wall panels. Thesis presented for the Master of Science Degree, University of Tennessee, Knoxville, 1991.

24. Abaqus User's Manual, Version 4.8. Hibbit, Karlsson, and Sorensen, Inc., 1989.

25. McDowell, E. L., McKee, K. E. and Sevin, E., Arching action theory of masonry walls. Journal of Structural Division, ASCE, 1956, **82**, 915:1- 915:18.

26. Gabrielsen, B. L. and Kaplan, K. , Arching in masonry walls subjected to out-of-plane forces. Earthquake Resistant Masonry Construction, NBS Building Science Series 106, 1977, 283-313.

27. Seah, C. K., Out-of-plane behavior of concrete masonry infilled panels. Thesis submitted in partial fulfillment of the requirements for the degree of Master of Science, University of New Brunswick, Canada, 1988.

28. Benedetti, D., and Benzoni, G. M., A numerical model for seismic analysis of masonry buildings: experimental correlations. Earthquake Engineering and Structural Dynamics, **12**. 1984, 817-31.

29. Butala, M. B., Jones, W. D. and Beavers, J. E., Out-of-plane behavior of hollow clay tile walls infilled between steel frames. Proceedings, Third Department of Energy Natural Phenomena Hazards Mitigation Conference, October 15-18, 1991.

30. Kariotis, J. C., Ewing, R. D., and Johnson, A. W. Predictions of stability for unreinforced brick masonry walls shaken by earthquakes. Proceedings, 7th International Brick and Block Masonry Conference, Australia, 1985, 1175-96.

31. Kariotis, J. C., Ewing, R. D. and Johnson, A. W., Predictions of stability for unreinforced brick masonry walls shaken by earthquakes. Proceedings, Third North American Masonry Conference, 1985, 49:1-49:11.

32. Adham, S. A., Out-of-plane response of masonry walls shaken by earthquakes. Proceedings of the Third North American Masonry Conference, pp. 47:1-47:14, 1985.

33. Stafford-Smith, B., Lateral stiffness of infilled frames. Journal of the Structural Division, ASCE, 1962, **88**, 183-99.

34. Chrysostomou, C.Z., Gergely, P. and Abel, J.F. Preliminary studies of the effects of degrading infilled walls on the nonlinear seismic response of steel frames National Center for Earthquake Engineering Research. NCEER-88-0046, 1988.

35. Klingner, R.E., and Bertero, V.V. Earthquake resistance of infilled frames. Journal of the Structural Division, ASCE, 1978, **104**, 973-89.

36. Liauw, T.C. and Kwan K.H., Unified plastic analysis of infilled frames. Journal of the Structural Division, ASCE, 1985, **111**, 1427-48.

37. Wood, R.H., Plasticity, composite action and collapse design of unreinforced shear wall panels in frames. Proceedings, Institution of Civil Engineers, 1978, pp.381-411.

38. Dawe, J.L. and McBride, R.T., Experimental investigation of the shear resistance of masonry panels in steel frames. Proceedings, Seventh International Brick Masonry Conference, Melbourne, Australia, 1985, pp. 791-801.

39. Mallick, D.V. and Severn, R.T. The behavior of infilled frames under static loading. Proceedings, Institute of Civil Engineers, 1967, **38**.

40. Dhanasekar, M., Page, A.W. and Kleeman, P.W., The behavior of brick masonry under biaxial stress with particular reference to infilled frames. Proceedings, Seventh International Brick Masonry Conference, 1985, pp. 814-24.

41. Liauw, T.C. and Kwan, K.H., Non-linear analysis of multistory infilled frames. Proceedings, Institution of Civil Engineers, 73, Part 2, 1982, pp. 441-54.

42. Rivero, C.E., and Walker, W.H., A model to study the interaction of frames and infilled masonry wall. Proceedings, Third North American Masonry Conference, 1985, pp. 76:1-76:14.

43. Ewing, R.D., El-Mustapha, A.M., Kariotis, J.C., FEM/I finite element computer program for the nonlinear static analysis of reinforced masonry building components. U.S.-Japan Coordinated Program for Masonry Building Research, Report No. 2.2-1, 1990.

THE ANALYSIS OF THE ENVIRONMENTS AND STRUCTURES INTERACTION——SOIL-PLATE AND SHELLS INTERACTION

Huang-Yih
(Xian Institute of Metallurgy and Construction Engineering, Xian China)

Ⅰ. Introduction

The interaction between environment and structure includes the mutual influence of environmental factors such as wind, sea wave, temperature, earthquake, etc. At present, study of the subject has attracted attention in China and abroad, and the interaction of plate and shell–structure with foundation is one of the important aspects of the subject studied. Analysis of the interaction between plate and shell–structure with foundation is of practical importance in engineering, such as in aeronautical, transportation, architectural, hydraulic, and harbour engineering as well as in other structures.In the theory of plate on the elastic foundation, the exact solution of thick plate with rectangular free boundary on the two–parameter elastic foundation is especially difficult to deal with. The problem has been investigated by a number of authors in China and other countries, while up to now the solution is still open. So far, only a few of numerical and approximate solutions are available. Since accuracy of the numerical solution at the plate singularities (corners) is rather poor, the exact solution to the problem is crucial for structural analysis and structure designers. The investigation work of the problem has been conducted in China as early as the sixties. Professor Chen Shutao secured the approximate solution to a thin plate of rectangular free boundary, employing Galerkin's method with Rayleigh function as the trial function. Since then, no essential progress has been made with the problem, during the past two decades. Until the early 1980s, a few results achieved for the problem have been still numerical. In 1984, the pioneer work was first carried out by Professor Zhang Fufan who derived, with the method of superpositon, the exact solution to the problem of a rectanguar thin plate in the Winkler's foundation (single parameter) with four edges. The author and co–workers put forward a mathematical modelling of the generalized slip boundary condition and with this model, the general solution to a rectangular plate (thin or thick) with four free edges on the Winkler's foundation was derived. Besides, the general solution to a thin rectangular plate with free edges on the two–parameter elastic foundation was obtained.This paper presents the general solution of the bending problem of thick rectangular plate with free boundary and of conical shell on the two–parameter elastic foundation and corresponding numerical results are also given.

Ⅱ. The problem of Thick plate

By the method of superposition, a rigorous solution of the completely free rectangular thick plate on the V.Z.Vlasov two–parameter elastic foundation is presented. We first derive some basic solutions of plate under the slip boundary condition, generalized slip boundary condition, simply supported boundary condition and generalized simply supported boundary condition, then by superposing these basic solutions, the most generally rigorous solution of a rectangular thick plate with four free edges on the two–parameter elastic foundation can be obtained. Some numerical results are given.

The method in this paper is of universal significance. It can be used to solve the problem of the rectangular thick plate with arbitrary complicated boundary conditions.

1. The basic equation.

Let us consider a rectangular thick plate on the two–parameter elastic foundation shown in Fig.1. we have the following governing differential equation of thick plate on the two–parameter elastic foundation, i.e.,

$$D\Delta^4 F - 2r^2\Delta^2 F + s^4 F = \bar{q} \qquad (2-1a)$$

$$\Delta^2 f - 2d^2 f = 0 \qquad (2-1b)$$

where $d^2 = c/\mu^* D \qquad \mu^* = 1-\mu$

$D = Eh^3/12(1-\mu)$

$2r^2 = (Dk + 2ct - \beta CD)/D, \quad S^4 = ck/D$

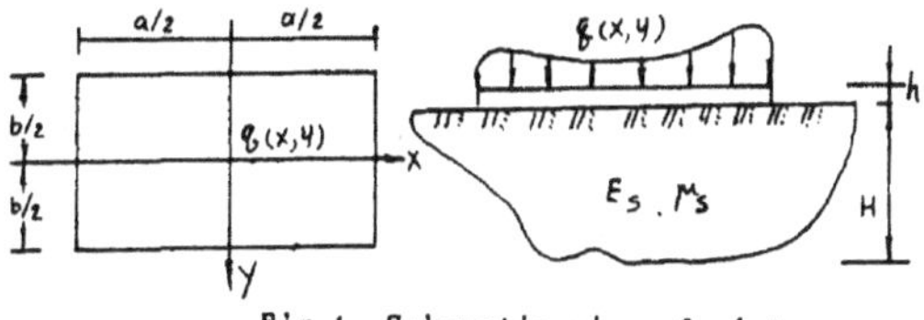

Fig. 1. Schematic view of plate

$$D = D(C + 2t - 2ct\beta) \qquad \Delta^2 = \frac{\partial^2}{\partial X^2} + \frac{\partial^2}{\partial Y^2} \qquad (2-2)$$

$$\bar{q} = (ck/D)\psi - \frac{2ct + c^2}{D}\Delta^2\psi - q/D$$

$\psi(x, y)$ is the function concerned with loading $q(x, y)$. It is the particular solution of the following equation:

$$2\frac{D}{C}\beta t\Delta^2\psi - \left(1 + \frac{D}{C}\beta k\right)\psi = -\frac{D}{C}\beta q \qquad (2-3)$$

According to V.Z.Vlasov mode [3], the parameter

$$k = \frac{E_0\gamma}{2a(1-\mu_0^2)}\psi_k, \quad t = \frac{aE_0}{8\gamma(1+\mu_0)}\psi_t, \quad E_0 = E_s/1-\mu_s$$

$$\mu_0 = \mu_s/1-\mu_s, \qquad (2-4)$$

$$\begin{matrix}\psi_k \\ \psi_t\end{matrix} = \frac{1}{\mathrm{sh}^2\left(\frac{\gamma H}{a}\right)}\left[\mathrm{sh}\left(\frac{\gamma H}{a}\right)\mathrm{ch}\left(\frac{\gamma H}{a}\right) \pm \frac{\gamma H}{a}\right]$$

γ is parameter concerned with the foundation feature; E_s μ_s are the elastic constants of foundation.

The expressions of displacement and internal forces of plate are as follows:

$$w = \frac{D}{C}\Delta^2 F - F + \psi, \qquad \psi_x = \frac{\partial F}{\partial x} + \frac{\partial f}{\partial x} + \beta\frac{D}{C}\Delta_s\frac{\partial F}{\partial x}$$

$$\psi_y = \frac{\partial F}{\partial y} - \frac{\partial f}{\partial x} + \beta\frac{D}{C}\Delta_s\frac{\partial F}{\partial y}, \qquad \Delta_s(\ ) = (k - 2t\Delta^2)\ (:)$$

$$M_x = -D\left[\frac{\partial^2 F}{\partial x^2} + \mu\frac{\partial^2 F}{\partial y^2} + \mu'\frac{\partial^2 f}{\partial x\partial y} + \beta\frac{D}{C}\Delta_s\left(\frac{\partial^2 F}{\partial x^2} + \mu\frac{\partial^2 F}{\partial y^2}\right) + \frac{B}{D}\left(\frac{D}{C}\Delta^2 F - F + \psi\right) - \frac{\beta}{D}q\right] \qquad (2-5)$$

$$M_y = -D\left[\mu\frac{\partial^2 F}{\partial x^2} + \mu\frac{\partial^2 F}{\partial y^2} - \mu'\frac{\partial^2 f}{\partial x\partial y} + \beta\frac{D}{C}\Delta_s\left(\mu\frac{\partial^2 F}{\partial x^2} + \mu\frac{\partial^2 F}{\partial y^2}\right) + \frac{B}{D}\left(\frac{D}{C}\Delta^2 F - F + \psi\right) - \frac{\beta}{D}q\right]$$

$$M_{xy} = -D\mu'\left[\frac{\partial^2 F}{\partial x\partial y} + \frac{1}{2}\left(\frac{\partial^2 f}{\partial y^2} - \frac{\partial^2 f}{\partial x^2}\right) + \beta\frac{D}{C}\Delta_s\frac{\partial^2 F}{\partial x\partial y}\right]$$

$$Q_x = D\left(\frac{\partial}{\partial x}\Delta^2 F - 2\frac{C}{D}\frac{\partial F}{\partial x} - \frac{C}{D}\frac{\partial f}{\partial y} - \beta\Delta_s\frac{\partial F}{\partial x} + \frac{C}{D}\frac{\partial\psi}{\partial X}\right)$$

$$Q_y = D\left(\frac{\partial}{\partial y}\Delta^2 F - 2\frac{C}{D}\frac{\partial F}{\partial y} + \frac{C}{D}\frac{\partial f}{\partial x} - \beta\Delta_s\frac{\partial F}{\partial y} + \frac{C}{D}\frac{\partial\psi}{\partial y}\right)$$

The boundary conditions of free edge (FE) are:

$$x=\pm a/2,\quad M_x=M_{xy}=0,\quad Q_x=Q_a^{*}$$
$$y=\pm b/2,\quad M_y=M_{xy}=0,\quad Q_y=Q_b^{*} \qquad (2-6)$$

Q_a^{*}、Q_b^{*} ——the given function

The solution of the governing equation (2–1) can be written in the following form:

$$F(x,\ y)=F_p(x,\ y)+\sum_{m=0}^{\infty}\left[A_m\varphi_{1m}(y)+B_m\varphi_{2m}(y)+C_m\varphi_{3m}(y)+D_m\varphi_{4m}(y)\right](\cos\alpha_m x \text{ or } \sin\alpha_m x) \qquad (2-7a)$$

$$f(x,\ y)=\sum_{m=0}^{\infty}(\alpha_m \mathrm{sh}\lambda_m y+b_m \mathrm{ch}v_m y)(\sin\alpha m_x \text{ or } \cos\alpha_m x) \qquad (2-7b)$$

$$F(x,\ y)=F_p(x,\ y)+\sum_{m=0}^{\infty}\left[A_n\varphi_{1n}(y)+B_n\varphi_{2n}(x)+C_n\varphi_{3n}(x)+D_n\varphi_{4n}(x)\right](\cos\alpha_n y\quad \sin\alpha_n y) \qquad (2-8a)$$

$$f(x,\ y)=\sum_{m=0}^{\infty}(\alpha_n \mathrm{sh}\lambda_n x+b_m \mathrm{ch}\lambda_n y)(\sin\alpha_m y \text{ or } \cos\alpha_m y) \qquad (2-8b)$$

in which, $\alpha m=2m\pi/a$, $\alpha n=2n\pi/b$, $F_p(x,\ y)$ is a particular solution, we can take the double trigonometric series as the particular solution. Therefore, we only discuss the homogeneous solution. A_m, B_m, C_m, D_m, a_m, b_m, A_n, B_n, C_n, D_n, a_n, b_n, are arbitrary constants. If $l^4=s^4-r^4>0$, the functions $\varphi_{im}(y)$ $(i=1,2,3,4)$ have the following form:

$$\varphi_{1m}(y)=\mathrm{ch}\tau_m y\cos\gamma_m y\qquad \varphi_{2m}(y)=\mathrm{sh}\tau_m y\cos\gamma_m y$$
$$\varphi_{3m}(y)=\mathrm{sh}\tau_m y\sin\gamma_m y\qquad \varphi_{4m}(y)=\mathrm{ch}\tau_m y\sin\gamma_m y \qquad (2-9)$$
$$\tau_m^2=\frac{1}{2}\left(t_m^2+\alpha_m^2+\gamma^2\right)\qquad \gamma_m^2=\frac{1}{2}\left(t_m^2-\alpha_m^2-\gamma^2\right)$$
$$t_m^2=\alpha_m^4+2r^2\alpha_m^2S^4,\qquad \lambda_m=\left(2d^2+\alpha_m^2\right)^{1/2}$$

In engineering, we often encounter $l^4=s^4-r^4>0$. Therefore, we only discuss the case below.

The quantities with subscript n are obtained from the corresponding quantities with subscript m by replacing m by n and by interchanging the dimension b with a. So, we will only define the quantities subscript m.

In order to obtain the solution of this problem, the following boundary conditions and corresponding solutions (basic solution) are of importance, for example, $x=a/2$;

simply supported edge(SSE):

$$W=M_{xy}=M_x=0 \qquad (2-10a)$$

generalized simply supported edge(GSSE):

$$M_x=M_{xy}=0,\ W=\overline{W} \qquad (2-10b)$$

Slip edge(SE):

$$\psi_x=M_{xy}=Q_x=0 \qquad (2-10c)$$

generalized slip edge(GSE):

$$M_{xy}=Q_x=0\qquad \psi_x=\overline{\psi}_x \qquad (2-10d)$$

in which, $\overline{W}$, $\overline{\psi}_x$ are given function on the boundary of plate.

2. **Basic solution**

For generalization, let us consider a free edge rectangular plate on the elastic foundation (Fig.1). We can always divide this problem in the combination of the different states of deformation of the plate: the double symmetric deformation (SS); the double antisymmetric deformation (AA); the symmetric — antisymmetric deformation (SA); the antisymmetric–symmetric deformation (AS). Under the symmetrical deformation, the antisymmetrical components of the internal forces and displacement of the plate equal to zero along the section of the symmetrical axes, and under the antisymmetrical deformation, the symmetrical components of the internal force and displacement of the plate equal to zero along the section of the antisymmetrical axes. Hence, we can respectively regard these sections along the symmetrical axes and antisymmetrical axes as the slip edge (SE) and simply supported edge (SSE). So, for the each of the deformation state of the plate, we need to focus our attention on only a quarter segment of plate, that has free edges and slip edges and simply supported edges, respectively, (Fig.2).

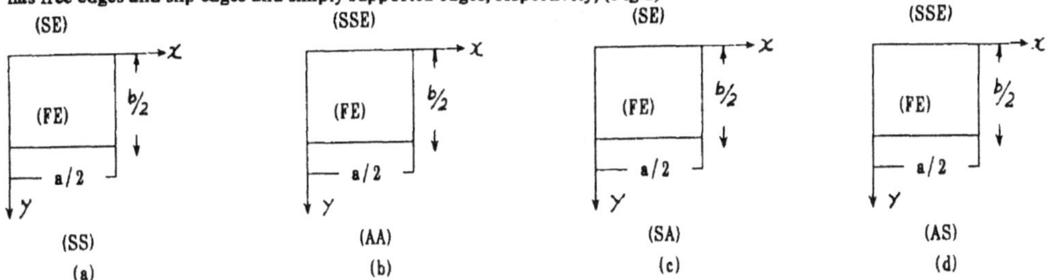

Fig. 2

If the solutions for each case (Fig.2) are obtained, we can obtain the general solution of this problem by superposing these results.

(1) For the double symmetrical deformation, we need to focus our attention on only a quarter plate shown in Fig.2a. In order to obtain its solution, we can superpose the basic solutions shown in Fig.3. i.e., the solution of this quarter plate is the superposition of each building block at right–hand side shown in Fig.3

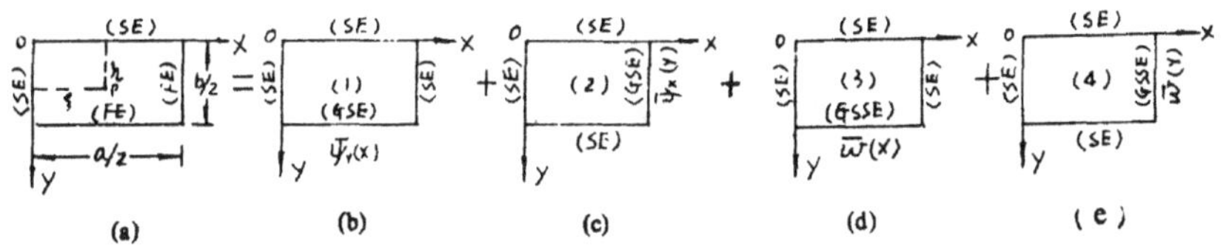

Fig.3 Building blocks for analyzing the double symmetric case

In Fig.3,

$$\bar{\psi}_y(x)=\sum_{m=0}^{\infty}E_m^{(1)}\cos\alpha_m x \qquad \bar{w}(x)=\sum_{m=0}^{\infty}G_m^{(1)}\cos\alpha_m x \qquad \bar{\psi}_x(y)=\sum_{n=0}^{\infty}F_n^{(1)}\cos d_n y,$$

$$\bar{w}(y)=\sum_{n=0}^{\infty}H_n^{(1)}\cos\alpha_n y \qquad \alpha_m=\frac{2m\pi}{a} \qquad \alpha_n=\frac{2n\pi}{b} \tag{2-11}$$

in which, $E_m^{(1)}$、$G_m^{(1)}$、$F_m^{(1)}$、$H_m^{(1)}$ are undetermined coefficients. They can be determined from the satisfaction of free edge conditions.

We turn now to the first building block (Fig.3b) where the boundary conditions are as follows:

$$y=b/2 \qquad M_{xy}=Q_y=0 \qquad \psi_y=\bar{\psi}_y(x) \tag{2-12}$$

Using expressions (2–7a) (2–7b) and considering the symmetrical condition we have the homogeneous solution, then called the basic solution,

$$F_1''(x,\ y)=\sum_{m=0}^{\infty}\frac{E_m^{(1)}}{\theta'_m}\left[\theta_{1m}\varphi_{1m}(y)+\theta_{3m}\varphi_{3m}(y)\right]\cos\alpha_m x$$

$$f_1''(x,\ y)=\sum_{m=0}^{\infty}\eta_{2m}E_m^{(1)}\mathrm{sh}\lambda_m y\sin\alpha_m x \tag{2-13}$$

We turn next to the second building block (Fig.3c) where the boundary conditions are as follows:

$$x=a/2 \qquad M_{xy}=Q_x=0 \qquad \psi_x=\bar{\psi}_x(y) \tag{2-14}$$

Using expressions (2–8a), (2–8b) and noting the symmetrical condition, we find

$$F_2''(x,\ y)=\sum_{n=0}^{\infty}\frac{F_n^{(1)}}{\theta_n'}\Big[\theta_{1n}\varphi_{1n}(x)+\theta_{3n}\varphi_{3n}(x)\Big]\cos\alpha_n y$$

$$f_2''(x,\ y)=-\sum_{n=0}^{\infty}\eta_{2n}F_n^{(1)}\mathrm{sh}\lambda_n x\sin\alpha_n y \tag{2-15}$$

We turn next to to the third building block (Fig.3d), the corresponding boundary conditions are as follows:

$$y=b/2 \qquad M_{xy}=M_y=0 \qquad W=\overline{W}(x) \tag{2-16}$$

Using condition (2–16), we find homogeneous solution

$$F_4''(x,\ y)=\sum_{m=0}^{\infty}\frac{G_n^{(1)m}}{\varepsilon_m'}\Big[\varepsilon_{1m}\varphi_{1m}(y)+\varepsilon_{3m}\varphi_{3m}(y)\Big]\cos\alpha_m x$$

$$f_3''(x,\ y)=\sum_{m=1}^{\infty}\frac{\xi_{1m}}{\varepsilon_m^3}G_m^{(1)}\mathrm{sh}\lambda_m y\sin\alpha_m x \tag{2-17}$$

For the fourth building block (Fig.3e), the boundary conditions are following form:

$$x=a/2 \qquad M_{xy}=M_x=0 \qquad W=\overline{W}(y) \tag{2-18}$$

Using condition (2–18), we obtain the corresponding homogeneous solution

$$F_4''(x,\ y)=\sum_{n=0}^{\infty}\frac{H_n^{(1)n}}{\varepsilon_n'}\Big[\varepsilon_{1n}\varphi_{1n}(x)+\varepsilon_{3n}\varphi_{3n}(x)\Big]\cos\alpha_n y$$

$$f_4''(x,\ y)=-\sum_{n=1}^{\infty}\frac{\xi_{3n}}{\varepsilon_n'}H_n^{(1)}\mathrm{sh}\lambda_n x\sin\alpha_n y \tag{2-19}$$

Adding the solution (2–13)、(2–15)、(2–17)、(2–19), we can obtain the the general solution

$$F''(x,\ y)=\sum_{m=0}^{\infty}\Bigg[\left(-\frac{E_n^{(1)}}{\theta_m^3}\theta_{1m}+\frac{G_m^{(1)}}{\varepsilon_m^3}\varepsilon_{1m}\right)\varphi_{1m}(y)+\left(\frac{E_m^{(1)}}{\theta_m^3}+\frac{G_m^{(1)}}{\varepsilon_m'}\varphi_{3m}(y)\right)\Bigg]\cos\alpha_m x$$

$$+\sum_{n=0}^{\infty}\Bigg[\left(\frac{F_n^{(1)}}{\theta_n^3}\theta_{1n}+\frac{H_n^{(1)}}{\varepsilon_n^1}\varepsilon_{1n}\right)\varphi_{1n}(x)+\left(\frac{F_n^{(1)}}{\theta_m^3}\theta_m+\frac{H_n^{(1)}}{\varepsilon_n^3}\varepsilon_{m}\right)\varphi_m(x)\Bigg]\cos\alpha_n y+F_p''(x,\ y)$$

$$f''(x,\ y)=\sum_{m=1}^{\infty}\left(\eta_{2m}E_m^{(1)}+\frac{\xi_{sm}}{\varepsilon_m'}G_m^{(1)}\right)\mathrm{sh}\lambda_m y\sin\alpha_m x-\sum_{n=1}^{\infty}\left(\eta_{2n}F_n^{(1)}+\frac{\xi_{sn}}{\varepsilon_n'}H_n^{(1)}\right)\mathrm{sh}\lambda_n x\sin\alpha_n y \tag{2-20}$$

Substituting solution (2–20) in (2–5) and using the boundary conditions (2–6) while developing the results in the same kind of trigonometric series, we obtain equations which can determine the undetermined coefficients $E_m^{(1)}$, $F_n^{(1)}$, $H_n^{(1)}$, $G_m^{(1)}$.

(2) For the doubly antisymmetric deformation, we turn to calculate the next quarter plate shown in Fig.2(b). In like manner, the general solution of quarter–plate is： (Fig.4):

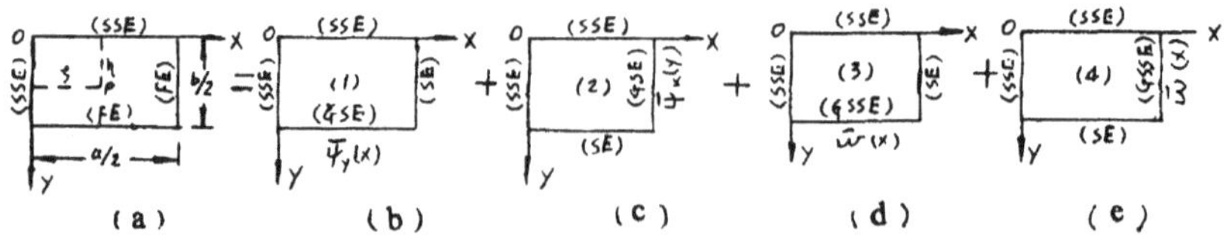

Fig.4 building blocks for analyyzing double antisymmetric

$$F^{aa}(x,y)=\sum_{m=1,3}^{\infty}\left[\left(\frac{E_m^{(2)}}{\theta_m^a}\theta_{2m}+\frac{G_m^{(2)}}{\varepsilon_m^a}\varepsilon_{2m}\right)\varphi_{2m}(y)+\left(\frac{E_m^{(2)}}{\theta_{4m}^a}\theta_{4m}+\frac{G_m^{(2)}}{\varepsilon_m^a}\varepsilon_{4m}\right)\varphi_{4m}(y)\right]\sin\alpha_m X$$

$$+\sum_{n=1,3}^{\infty}\left[\left(\frac{F_n^{(2)}}{\theta_n^a}\theta_{2n}+\frac{H_n^{(2)}}{\varepsilon_n^a}\varepsilon_{2n}\right)\varphi_{2n}(x)+\left(\frac{F_n^{(2)}}{\theta_n^a}\theta_{4n}+\frac{H_n^{(2)}}{\varepsilon_n^a}\varepsilon_{4n}\right)\varphi_{4n}(x)\right]\sin\alpha_n y+F_p^{aa}(x,y)$$

$$f^{aa}(x,y)=\sum_{m=1,3}^{\infty}\left(\eta_{1m}E_m^{(2)}+\frac{\xi_{am}}{\varepsilon_m^a}G_m^{(2)}\right)\mathrm{ch}\lambda_m y\cos\alpha_m x+\sum_{n=1,3}^{\infty}\left(\eta_{1n}F_n^{(2)}+\frac{\xi_{an}}{\varepsilon_n^a}H_n^{(2)}\right)\mathrm{ch}\lambda_n x\cos\alpha_n y \qquad (2-21)$$

In like manner, form boundary condition (2–6) the undetermined coefficients, $E_m^{(2)}$, $G_m^{(2)}$, $H_n^{(2)}$, $F_n^{(2)}$, are calculated.

In above expressions, the following notations are introduced, i.e.,

$$e=d^2l^2\iota_m^2,\quad m_0=d^2+\alpha_m^2,\quad e_1=\tau_m l^2+\gamma_m\gamma^2,\quad e_2=\tau_m r^2-\gamma_m l^2,\quad e_3=\mu\cdot d^2\alpha_m^2\gamma_m$$

$$e_4=1-2\frac{D}{C}r^2-\frac{D}{c}\mu\cdot\alpha_m^2,\quad e_5=\mu\cdot\lambda_m\alpha_m^2,\quad m_1=(d^2+\alpha_m^2)l^2$$

$$e_6=\tau_m\frac{D}{C}l^2-\gamma_m\left(1-\frac{D}{C}\gamma^2\right)\quad e_7=\tau_m\left(1-\frac{D}{C}r^2\right)-\gamma_m\frac{D}{C}l^2\quad \xi_{sm}=\alpha_m e_{10}'e_s'-\alpha_m e_q'e_s' \qquad (s\to a)$$

$$e_s'=\left(\alpha_m^2\mu^{\cdot}+r^2\right)\varphi_{3m}\left(\frac{b}{2}\right)+l^2\varphi_{1m}\left(\frac{b}{2}\right) \qquad (S\to a,\ \varphi_{1m}\to\varphi_{2m},\ \varphi_{3m}\to\varphi_{4m})$$

$$\bar{e}_s'=\left(\alpha_m^2\mu+r^2\right)\varphi_{1m}\left(\frac{b}{2}\right)-l^2\varphi_{3m}\left(\frac{b}{2}\right) \qquad (S\to a,\ \varphi_{1m}\to\varphi_{2m},\ \varphi_{3m}\to\varphi_{4m}) \qquad (2-22)$$

$$e_9'=\gamma_m\varphi_{2m}\left(\frac{b}{2}\right)+\tau_m\varphi_{4m}\left(\frac{b}{2}\right) \qquad (S\to a,\ \varphi_{2m}\to\varphi_{1m},\ \varphi_{4m}\to\varphi_{3m})$$

$$e_{10}'=\tau_m\varphi_{2m}\left(\frac{b}{2}\right)-\gamma_m\varphi_{4m}\left(\frac{b}{2}\right) \qquad (S\to a,\ \varphi_{2m}\to\varphi_{1m},\ \varphi_{4m}\to\varphi_{3m})$$

$$\varphi_1=\varphi_{2m}^2\left(\frac{b}{2}\right)+\varphi_{4m}^2\left(\frac{b}{2}\right) \qquad (1\to 2,\ \varphi_{2m}\to\varphi_{1m},\ \varphi_{4m}\to\varphi_{3m})$$

$$\varphi_{as}=\varphi_{1m}\left(\frac{b}{2}\right)\varphi_{2m}\left(\frac{b}{2}\right)+\varphi_{3m}\left(\frac{b}{2}\right)\varphi_{4m}\left(\frac{b}{2}\right) \qquad (a\to b,\ \varphi_{1m}\to\varphi_{3m})$$

$$\theta_m^s=e\varphi_1\quad (S\to a,\ 1\to 2)\qquad \eta_{1m}=a_m/d^2\mathrm{ch}\lambda_m\frac{b}{2}\quad \eta_{2m}=a_m/d^2\mathrm{sh}\lambda_m\frac{b}{2}$$

$$\theta_{1m}=(e_1m_0-e_3)\varphi_{2m}\left(\frac{b}{2}\right)+(e_2m_0-e_3)\varphi_{4m}\left(\frac{b}{2}\right) \qquad (1\to 2,\ \varphi_{2m}\to\varphi_{1m},\ \varphi_{4m}\to\varphi_{3m})$$

$$\theta_{3m}=(e_1m_0-e_3)\varphi_{4m}\left(\frac{b}{2}\right)-(e_2m_0-e_3)\varphi_{2m}\left(\frac{b}{2}\right) \qquad (3\to 4,\ \varphi_{4m}\to\varphi_{3m},\ \varphi_{2m}\to\varphi_{1m})$$

$$\varepsilon_m^s=m_1e_4\varphi_s\mathrm{sh}\lambda_m\frac{b}{2}+e_5e_6\varphi_{as}\mathrm{ch}\lambda_m\frac{b}{2}+e_6e_7\varphi_{ba}\mathrm{ch}\lambda_m\frac{b}{2} \qquad (s\to a,\ \mathrm{sh}\to\mathrm{ch},\ b\to a)$$

$$\varepsilon_{1m}=m_0e_s^a\mathrm{ch}\lambda_m\frac{b}{2}-m_0-\bar{e}_s^a\mathrm{sh}\lambda_m\frac{b}{2} \qquad (S\to a,\ 1\to 2,\ \mathrm{sh}\to\mathrm{ch})$$

$$\varepsilon_{3m}=e_5e_{10}^a\mathrm{ch}\lambda_m\frac{b}{2}-m_0e_s^a\mathrm{sh}\lambda_m\frac{b}{2} \qquad (3\to 4,\ s\to a,\ \mathrm{ch}\to\mathrm{sh})$$

In expression (2–22), exchanging a for b and m for n, we obtain the corresponding quantities with the subscript n.

(3) For the case of the symmetric–antisymmtrical deformation, we can in the same way obtain the solution of the quarter plate shown in Fig.2(c). The general solution is:

$$F^{sa}(x,y)=\sum_{m=1,3}^{\infty}\left[\left(\frac{E_m^{(3)}}{E_m^1}\theta_{1m}+\frac{G_m^{(3)}}{\varepsilon_m^1}\varepsilon_{1m}\right)\varphi_{1m}(y)+\left(\frac{E_m^{(3)}}{\theta_m^3}\theta_{3m}+\frac{G_m^{(3)}}{\varepsilon_m^3}\varepsilon_{sm}\right)\varphi_{sm}(y)\right]\sin\alpha_m X$$

$$+\sum_{n=0}^{\infty}\left[\left(\frac{F_n^{(1)}}{\theta_n^a}\theta_{2n}+\frac{H_n^{(3)}}{\varepsilon_n^a}\varepsilon_{2n}\right)\varphi_{2n}(x)+\left(\frac{F_n^{(3)}}{\theta_n^m}\theta_{4n}+\frac{H_n^{(3)}}{\varepsilon_n^a}\varepsilon_{4n}\right)\varphi_{4n}(x)\right]\cos\alpha_n y+F_p^{sa}(x,y)$$

$$f^{sa}(x,y)=\sum_{m=1,3}^{\infty}\left(\eta_{2m}E_m^{(3)}+\frac{\xi_{2m}}{\varepsilon_m^3}G_m^{(3)}\right)\mathrm{sh}\lambda_m y\cos\lambda_m x \qquad (2-23)$$

$$+\sum_{n=1}^{\infty}\left(\eta_{1n}F_n^{(3)}+\frac{\xi_{1n}}{\varepsilon_n^a}H_n^{(3)}\right)\mathrm{ch}\lambda_n x\sin\alpha_n y$$

From boundary condition (2–6), the $E_m^{(3)}$, $E_n^{(3)}$, $G_m^{(1)}$, $H_n^{(1)}$, are calculated.

The general solution $F^{sa}(x, y)$, $f^{sa}(x, y)$ are extracted from $F^{as}(x, y)$, $f^{as}(x, y)$ by interchanging the variables X and Y.

The final result should be:

$$F(x, y) = F^{ss}(x, y) + F^{aa}(x, y) + F^{as}(x, y) + F^{sa}(x, y)$$

$$f(x, y) = f^{ss}(x, y) + f^{aa}(x, y) + f^{as}(x, y) + f^{sa}(x, y) \qquad (2-24)$$

3. Numerical Results

As the application of this theory, we give an example of a square plate, shown in Fig.6. $E = 2 \times 10^6 T/m^2$, $\mu = 0.167$, $h = 0.2M$, $\mu_s = 0.4$, $E_s = 4 \times 10^6 T/M^2$, $\gamma = 1.55$, $a = 1M$, locally, uniform load, $g_0 = 100T/M^2$
This problem is double symmetric deformation, the calculation results are given in Table 1. and Table 2. In the table 1, the first lines are Winkler's results and the second lines are the V.Z.Vlasov's.

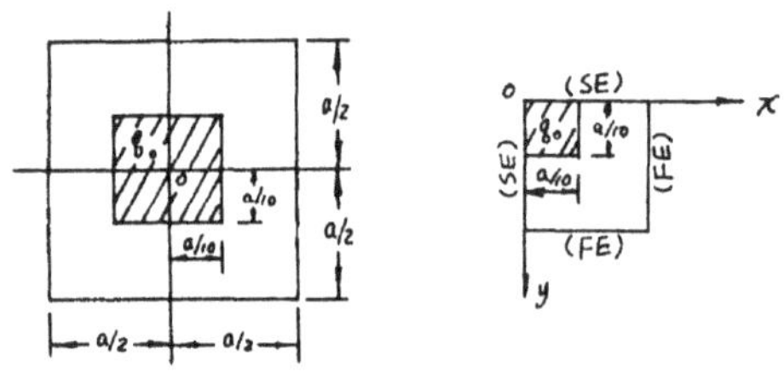

Fig.5 View of the completely free square plate

Table 1 Deflection values for plate when H = 0.6M

y x	0	a/8	a/4	3a/8	a/2
0	0.2586 0.2586	0.2538 0.2538	0.2470 0.2470	0.2425 0.2425	0.1824 0.1824
a/8	0.2438 0.2438	0.2058 0.2057	0.2455 0.2455	0.2417 0.2417	0.1775 0.1768
a/4	0.2470 0.2470	0.2455 0.2455	0.2424 0.2424	0.2396 0.2396	0.1614 0.1614
3a/8	0.2426 0.2425	0.2417 0.2417	0.2396 0.2396	0.2377 0.2377	0.1429 0.1430
a/2	0.1833	0.1775	0.1618	0.1429	0.0314

Table 2 Deflection values for plate when H = ∞

y x	0	a/8	a/4	3a/8	a/2
0	0.3167	0.3119	0.3051	0.3006	0.2317
a/8	0.3119	0.3088	0.3036	0.2998	0.2257
a/4	0.3051	0.3036	0.3005	0.2911	0.2095
3a/8	0.3006	0.2998	0.2977	0.2958	0.1900
a/2	0.2317	0.2259	0.2095	0.1900	0.0671

The results show that the difference between the Winkler's and Vlazov's gets bigger and bigger along with the increase of depth of foundation.

III. The problem of conical shell.

The general bending problem of conical shells on the two–parameter elastic foundation (V.Z.Vlasov model) is not solved. Here, the displacement solution method is presented. Under the arbitrary loads and boundary conditions, the general solution of the governing differential equation is obtained in series form. As a valuable application of this theory, the combination construction of shells (M–type combined shell) is calculated. The exact solution of this problem has never been obtained before.

1. The basic equation of conical shells

If the foundation reaction in u, v direction is omitted, the basic equations in terms of displacement for the general bending problems of conical
shells on the elastic foundation may be written as follows:

$$L^{11}(u)+L^{12}(v)+L^{13}(w)+q_1/k=0$$
$$L^{21}(u)+L^{22}(v)+L^{23}(w)+q_2/k=0 \qquad (3-1)$$
$$L^{31}(u)+L^{32}(v)+L^{33}(w)+q_n/k=0$$

where, material forms of the operator $L^{ij}(i,\ j=1,\ 2,\ 3)$ are as follows:

$$L^{11}=\frac{\partial^2}{\partial s^2}+\frac{1\partial}{s\partial s}-\frac{1}{s}+\frac{(1-\mu)\sec^2\varphi}{2s^2}\frac{\partial^2}{\partial\theta^2} \quad L^{12}=\sec\varphi\left(\frac{1+\mu}{2}\frac{1}{s}\frac{\partial^2}{\partial s\partial\theta}-\frac{3-\mu}{2}\frac{1\partial}{s^2\partial\theta}\right),$$

$$L^{21}=\sec\varphi\left(\frac{1+\mu}{2}\frac{1}{s}\frac{\partial^2}{\partial s\partial\theta}+\frac{3-\mu}{2}\frac{1}{s}\frac{\partial}{\partial\theta}\right),\quad L^{22}=\frac{1-\mu}{2}\left(\frac{\partial^2}{\partial s^2}+\frac{1}{s}\frac{\partial}{\partial s}-\frac{1}{s^2}\right)+\frac{\mathrm{Sec}^2\varphi}{s^2}\frac{\partial^2}{\partial\theta^2}$$

$$L^{23}=\frac{1}{S^2}\sec^2\varphi\sin\varphi\frac{\partial}{\partial\theta}\quad L^{31}=\frac{K}{D}\mathrm{tg}\varphi\left(\mu\frac{1}{s}\frac{\partial}{\partial s}+\frac{1}{s^2}\right)\quad L^{32}=\frac{K}{D}\mathrm{tg}\varphi\mathrm{Sec}\varphi\frac{\partial}{\partial\theta} \qquad (3-2)$$

$$L^{33}=\Delta_s^2\Delta_s^2+(K/D)/s^2\mathrm{tg}^2\varphi+\frac{1}{D}(k-2t\Delta_s^2)\quad \Delta_s^2=\frac{\partial^2}{\partial s^2}+\frac{1}{s}\frac{\partial}{\partial s}+\frac{\sec^2\varphi\partial^2}{s^2\partial\theta^2}\quad L^{13}=\mathrm{tg}\varphi\left(\mu\frac{1}{s}\frac{\partial}{\partial s}-\frac{1}{s^2}\right)$$

in which, $K=Eh/(1-\mu^2)$, $D=Eh^3/12(1-\mu^2)$, E, μ is elastic modules and Poisson's ratio of the shell material respectively, h is thickness of conical shells, k is Winkler modules, t is shear module of foundation. The other quantities have been shown in Fig.6. The internal force components expressed by displacement components of conical shells are written[6, 7].

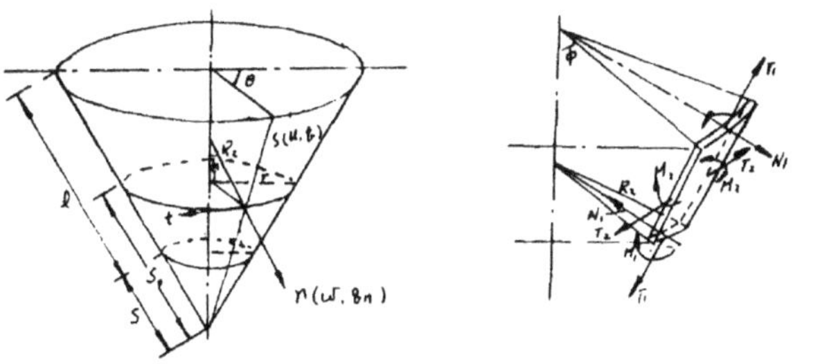

Fig.6

When the displacement. components of conical shell and the displacement function, $U(s,\ \theta)$, satisfy the following relations [6,7]

$$u=-\sin\varphi\quad \cos^3\varphi L'_u\quad (U)-\cos^2\varphi L'_{up1}(\psi_1)+\cos^2\varphi L'_{up2}(\psi_2)$$
$$v=-\sin\varphi\quad \cos^3\varphi L'_v\quad (U)+\cos^2\varphi L'_{up1}(\psi_1)-\cos^2\varphi L'_{up2}(\psi_2) \qquad (3-3)$$
$$w=\cos^4\varphi L'_w(U)$$

the first two of Eqs.(3–1) are satisfied automatically. $\psi_1(s,\ \theta)$, $\psi_2(b,\ \theta)$ are the functions concerned with q_1, q_2, respectively. They are the solution of the following equations:

$$L'_w(\psi_i)=-2q_is^2\sec^2\varphi/(1-\mu)k \qquad (i=1,\ 2) \qquad (3-4)$$

Generally, the general solution of (3–4) is not needed. It is enough to get a particular solution, because even if we take only the particular solution of it, the first two equations of Eqs. (3–1) can be satisfied.

The solution of equation (3–4) can be found in the reference (2).

Substituting (3–3) into the third one of Eq.(3–1), we get the governing equation expressed by the displacement function $U(s,\ \theta)$:

$$L'_1L'_2(U)+(1-\mu^2)\frac{K}{D}s^2\mathrm{tg}^2\varphi L'_3(U)+\frac{s^4}{D}(k-2t\Delta_s^2)L'_2(U)=Q \qquad (3-5)$$

where

$$L_1' = \left[(D_s - 2)^2 + \sec^2\varphi\frac{\partial^2}{\partial\theta^2}\right]\left[D_s^2 + \sec^2\varphi\frac{\partial^2}{\partial\theta^2}\right], \quad L_2' = L_w', \quad L_3' = (D_s^2 - 1)D_s^2 \tag{3-6}$$

and the free term

$$Q(s, \theta) = q_n s^4 \sec^4\varphi / D + \frac{1-\mu}{2}\frac{k}{D} s^2 \sec^2\varphi \sin\varphi \left\{\left[\mu(D_s^3 - D_s) + D_s^2 - 1 - \frac{\partial^2}{\partial\theta^2}(D_s + \sec^2\varphi)\right]\psi_1 + \sec\varphi\left[\frac{\partial^2}{\partial\theta^2} + (2-\mu)D_s^2 - (1-\mu)D_s + 1\right]\frac{\partial}{\partial\theta}\psi_2\right\}$$

Expression (3–3) is general solution expressed in displacement function of Eq.(3–1) of the general bending problems of conical shells on the elastic foundation.

Substituting (3–3) into (2–3)[7], the internal force components can be represented by displacement function $U(s, \theta)$. So, the general bending problem of conical shells on the two–parameter elastic foundation is reduced to solve the governing equation (3–5) under certain boundary conditions.

2. The exact solution of basic equation

Expanding loads into Fourier's series, i.e.

$$q_i(s, \theta) = q_{i0} + \sum_{n=1}^{\infty}\left[q_{inc}(s)\cos n\theta + q_{ins}(s)\sin n\theta\right] \qquad (i = 1, 2, n) \tag{3-7}$$

the solution of Eq.(3–5) can be taken in the following form:

$$U(s, \theta) = U_0(s) + \sum_{n=1}^{\infty} U_{nc}(S)\cos n\theta + \sum_{n=1}^{\infty} U_{ns}(S)\sin n\theta \tag{3-8}$$

substituting (3–7) and (3–8) into (3–5) and comparing the coefficients of each term of triangular series and at the same time introducing operator $\delta_s = sd / ds$, we have basic equations for two kinds of problems as follows:

1′ $\quad L_s(U_{nc}) = Q_{nc} \qquad n = 0, 1, 2......$ (3 – 9)

2′ $\quad L_s(U_{nc}) = Q_{ns} \qquad n = 1, 2,......$ (3 – 10)

where operator

$$L_s = \prod_{i=1}^{8}(\delta_s - \alpha_i \quad + (1-\mu^2)(k/D)\mathrm{tg}^2\varphi s^2 \prod_{i=1}^{4}(\delta_s - \beta_i) + (k/D)S^4\prod_{i=1}^{4}(\delta_s - \gamma_i) - \frac{2t}{D}s^2(\delta_s^2 - m^2)\prod_{i=1}^{4}(\delta_s - \gamma_i) \tag{3-11}$$

and each parameter $\quad (m = n\sec\varphi)$

$\alpha_1 = m + 2,\ \alpha_2 = -m + 2,\ \alpha_3 = m + 1,\ \alpha_4 = -m + 1,$

$\alpha_5 = m,\ \alpha_6 = -m,\ \alpha_7 = m - 1,\ \alpha_8 = -(m + 1);$

$\beta_1 = 0,\ \beta_2 = 0,\ \beta_3 = 1,\ \beta_4 = -1,\ \gamma_1 = m - 1,\ \gamma_2 = -m + 1,\ \gamma_3 = m + 1,\ \gamma_4 = -(m + 1)$ (3 – 12)

free term

$$Q_{nc}(S) = S^4\sec^4\varphi q_{nc} / D + \frac{1-\mu}{2}(k/D)S^2\sec^3\psi\sin\psi\left\{(\delta_s - 1)[(\delta_s - 1)(\mu\delta_s - 1) + m^2]\psi_{lnc} + m\left[(2+\mu)\delta_s^2 + m^2 + (1-\mu)\delta_s + 1\right]\varphi_{2ns}\right\} \tag{3-13}$$

$$Q_{ns}(S) = S^4\sec^4\varphi q_{nc} / D + \frac{1-\mu}{2}(k/D)S^2\sec^3\varphi\sin\varphi\left\{(\delta_s - 1)[(\delta_s - 1)(\mu\delta_s - 1) + m^2]\varphi_{lns} + m\left[(2+\mu)\delta_s^2 + m^2 + (1-\mu)\delta_s + 1\right]\varphi_{2nc}\right\}$$

Inasmuch as Eq.(3–9) is analogous in structure to the Eq.(3–10), we only solve one of them, for example (3–9). Making a transformation, put $\xi = (s/1)^2$, then Eqs.(3–9) becomes

$$L_\xi(U_{mc}) = Q_{mc}(\xi) \tag{3-14}$$

in which, the operator L_ξ is

$$L_\xi = \prod_{i=1}^{8}(\delta_\xi - \alpha_i/2) + A_1\xi\prod_{i=1}^{4}(\delta_\xi - \beta_i/2) + A_2\xi^2\prod_{i=1}^{4}(\delta_\xi - \gamma_i/2) - A_3\xi\left(\delta_\xi^2 - \frac{m^2}{4}\right)\prod_{i=1}^{4}(\delta_\xi - \gamma_i/2) \tag{3-15}$$

where, the operator, $\delta_\xi = \xi\frac{d}{d\xi}$, and the constants

$$A_1 = (1-\mu^2)Kl^2\text{th}^2\varphi/16D, \quad A_2kl^4/16D, \quad A_32tl^2/4D$$

the free term

$$Q_{mc} = (\xi) = (1/16)\Big[l^4\xi^2\sec^2\varphi q_{mc}/D + \frac{1-\mu}{2}(K/D)l^2\xi\sec^3\varphi\sin\varphi\Big\{(2\delta_\xi - 1)\Big[(2\delta_s - 1) + m^2\Big]\psi_{1mc}$$

$$+ m\Big[4(2+\mu)\delta_\xi^2 + m^2 + 2(1-\mu)\delta_\xi + 1\Big]\psi_{2ms}\Big\} \tag{3-16}$$

The total solution of Eqs. (3–14) can be written as

$$U_{mc} = U_{mc}^H + U_{mc}^P \tag{3-17}$$

where U_{mc}^H is homogenous solution, U_{mc}^P is the particular solution. They satisfy the following equations, respectively

$$L_\xi(U_{mc}^H) = 0 \tag{3-18a}$$

$$L_\xi(U_{mc}^P) = Q_{mc}(\xi) \tag{3-18b}$$

we take the solution of equation (3–18a) in following form:

$$U_{mc}^H = \sum_{\gamma=0}^{\infty}\alpha_\gamma\xi^{\sigma+\gamma} \qquad (\alpha_0 \neq 0) \tag{3-19}$$

Substituting (3–19) into (3–18a), we get a recurrence relation as follows:

$$\sum_{e=0}^{2} a_{r-e}f_e(\sigma + r - e) = 0 \tag{3-20}$$

where

$$f_1(\delta+\gamma-1) = A_1\prod_{i=1}^{4}[(\delta+\gamma-1) - \beta_i/2]$$

$$-A_3\left(\delta+\gamma-1+\frac{m}{2}\right)\left(\delta+\gamma-1-\frac{m}{2}\right)\prod_{i=1}^{4}[(\delta+\gamma-1) - \gamma_i/2] \tag{3-21}$$

$$f_0 = (\sigma+\gamma) = \prod_{i=1}^{8}[(\sigma+\gamma) - \alpha_i/2], \qquad f_2(\sigma+\gamma-2) = A_2\prod_{i=1}^{4}[(\sigma+\gamma-2) - \gamma_i/2]$$

and corresponding indicial equation is (3–22)

$$f_0(\sigma) = 0 \tag{3-22}$$

from it, the eight indices of equation (3–18a) are obtained, i.e.

$$\sigma_1 = (m+2)/2, \quad \sigma_2 = (-m+2)/2, \quad \sigma_3 = (m+1)/2, \quad \sigma_4 = (-m+1)/2,$$

$$\sigma_5 = m/2, \quad \sigma_6 = -m/2, \quad \sigma_7 = (m-1)/2, \quad \sigma_8 = -(m+1)/2 \tag{3-23}$$

1′. When $m \neq 2$, or $m/2$ is greater than 2 integer, the four regular solution corresponding to the homogeneous Eq. (3–18a) is

$$U_{mc}^{H(j)} = \sum_{\gamma=0}^{\infty}\alpha_\gamma^j\xi^{\sigma j+\gamma}, \qquad \sum_{e=0}^{2}a_{r-e}^j f_e(\delta_j + r - e) = 0 \qquad (j = 1, 2, 3, 4) \tag{3-24}$$

Since the difference between, as δ_1, δ_6 and δ_2, δ_6 and δ_3, δ_7 and δ_4 and δ_8 is l, the solutions corresponding to δ_5, δ_6, δ_7, δ_8 contain logarithmic term, thus

$$U_{mc}^{H(j+4)} = U_{mc}^{H(j)}\ln\xi + \sum_{\gamma=0}^{\infty}\left(b_\gamma^{j+4}\right)'\xi_{j+4}^{\sigma}+\gamma \qquad (j = 1, 2, 3, 4) \tag{3-25}$$

$$\sum_{e=0}^{2} b_{r-e}^{j+4} f_e(\sigma_{j+4}+r-e)+\sum_{e=0}^{2} a_{r-e-1}^{j} f_e(\sigma_{j+4}+r-e)=0$$

$$f'_0(\sigma+r)=\prod_{i=1}^{8}[(\sigma+r)-\alpha_i/2]\sum_{i=1}^{8}\left(\frac{1}{\sigma+r-\alpha_i/2}\right),$$

$$f'_1(\sigma+r-1)=A_1\prod_{i=1}^{4}[(\sigma+r-1)-\beta_{i/2}]\sum_{i=1}^{4}\left(\frac{1}{\sigma+r-1-\beta_{i/2}}\right)-A_3(\sigma+r-1-\frac{m}{2})(\sigma+r-1+\frac{m}{2})\cdot$$

$$\prod_{i=1}^{4}[(\sigma+r-1)-r_i/2]\left[\frac{1}{\sigma+r-1-m/2}+\frac{1}{\sigma+r-1+m/2}+\sum_{i=1}^{4}\left(\frac{1}{\sigma+r-1-r_i/2}\right)\right]$$

2'. $m=2$, if denote

$$\tau_1=2,\quad \tau_2=0,\quad \tau_3=-1/2,\quad \tau_4=3/2,\quad \tau_6=1,\quad \tau_7=-1,\quad \tau_8=-3/2 \tag{3-26}$$

we have

$$U_{nc}^{H(j)}=\sum_{\gamma=0}^{\infty}\alpha_\gamma^j\xi_j^\tau+\gamma \qquad \sum_{e=0}^{2}a_{r-e}^j f_e(\tau_j+r-e)=0 \quad (j=1,\ 2,\ 3,\ 4,\ 5) \tag{3-27}$$

and

$$U_{nc}^{H(j+5)}=U_{nc}^{H(j)}\ln\xi+\sum_{\gamma=0}^{\infty}\left(b_\gamma^{j+5}\right)\delta_{j+6}^{\tau}+\gamma \qquad (j=1,\ 2,\ 3)$$

$$\sum_{e=0}^{2} b_{r-e}^{j+5} f_e(\tau_{j+5}+r-e)+\sum_{e=0}^{2} a_{r-e-1}^{j} f'_e(\tau_{j+5}+r-e)=0 \tag{3-28}$$

The particular solution of Eq.(3–18a) is dependent on free term $Q_{nc}(\xi)$. For the distribution loads considered in this paper, free term can always be represented as

$$Q_{nc}(\xi)=\sum_{n'=0}^{\infty}Q_{ncn'}\xi^{\sigma n'} \tag{3-29}$$

in which, $Q_{ncn'}$ is known constant, N' is arbitrary integer. For each term $Q_{ncn'}\xi^{\sigma n'}$ of (3–29), Eq.(3–18b) becomes

$$L_\xi(U_{ncn'}^p)=Q_{ncn'}\xi^{\sigma n'} \qquad (n'=1,\ N') \tag{3-30}$$

the solution of Eq.(3–20) can be taken in the following form, i.e.

$$U_{ncn'}^p=Q_{ncn'}\sum_{\gamma=0}^{\infty}b_\gamma^{n'}\xi^{\sigma n'+\gamma} \tag{3-31}$$

where

$$b_0^{n'}=1/g_0(\sigma_{n'}),$$

$$\left[b_\gamma^{n'}g_0(\sigma_{n'}+\gamma)+A_1g_1(\sigma_{n'}+\gamma-1)-A_3g_3(\sigma_{n'}+\gamma-1)\right]b_{\gamma-1}^{n'}+A_2b_\gamma^{n'}-2g_2(\sigma_{n'}+\gamma-2)=0,$$

$$g_0(\sigma_{n'}+\gamma)=\prod_{i=1}[(\sigma_{n'}+\gamma)-\alpha_i/2],\quad g_1(\sigma_{n'}+\gamma-1)=\prod_{i=1}^{4}[(\sigma_{n'}+\gamma-1)-\beta_i/2],$$

$$g_2(\sigma_{n'}+\gamma-2)=\prod_{i=1}^{4}[(\sigma_{n'}+\gamma-2)-\gamma_i/2] \tag{3-32}$$

$$g_3(\sigma_{n'}+\gamma-1)=(\sigma_{n'}+\gamma-1+\frac{m}{2})(\sigma_{n'}+\gamma-1-\frac{m}{2})\prod_{i=1}^{4}[(\sigma_{n'}+\gamma-1)-\gamma_i/2]$$

So, the particular solution of Eq.(3–14) can be written in the following form:

$$U_{nc}^p=\sum_{n'=0}^{\infty}U_{ncn'}^p \tag{3-33}$$

the total solution of Eq.(3–14) is

$$U_{nc}=\sum_{\mu'=1}^{\infty}C_{\mu'}U_{nc}^{H(\mu')}+U_{nc}^p \tag{3-34}$$

Where C_μ, $(\mu'=1\sim8)$ are undetermined constants which are determined from boundary conditions.

For n=0 (or m=0), i.e. it coincides with axisymmetric deformation state. Omitting the common operator $\cos^2\varphi$ (σ_e^2-1), the expression (3–3) and basic Eq.(3–4), (3–5) can be reduced to the following form:

$$u_0 = -\sin\varphi\cos\varphi(\mu\sigma_s - 1)U_0 + (1-\mu)/2\psi_1,\quad V_0 = \psi_2,\quad w_0 = \cos^2\varphi(\delta_s^2 - 1)U_0 \tag{3-35}$$

$$(\delta_s^2 - 1)\psi_i = -2q_i s^2 / k(1-\mu) \tag{3-36}$$

$$L_1^s L_2^s(U_0) + (1-\mu^2)\frac{K}{D}\mathrm{tg}^2\varphi s^2 L_3^s(U_0) + S^4\frac{1}{D}(K - 2t\Delta_s^2)L_2^s(U_0) = Q(s) \tag{3-37}$$

in which, $L_1^s = (\sigma_s - 2)^2\sigma_s^2,\quad L_s^2 = \sigma_s^2 - 1,\quad L_3^s = \delta_s^2$

$$Q(S) = q_n S^4\sec^2\varphi / D - \frac{1-\mu}{2}(k/D)S^2\mathrm{tg}\varphi\sec^2\varphi(\mu\delta_s + 1)\psi_1 \tag{3-38}$$

The indices of Eq.(3–37) are respectively,

$$\rho_1 = 1,\quad \rho_2 = 0,\quad \rho_3 = 1/2,\quad \rho_4 = 1,\quad \rho_5 = 0,\ \rho_6 = -1/2$$

repeating the above–mentioned process, the homogeneous solution is

$$U_0^{H(j)} = \sum_{r=0}^{\infty} C_r^j \zeta^{\sigma_j + r},\quad \sum_{e=0}^{2} C_{r-e}^j f(\sigma_j + r - e) = 0 \qquad (j = 1,\ 2,\ 3) \tag{3-39}$$

$$U_0^{H(j+3)} = U_0^{H(j)}\ln\zeta + \sum_{r=0}^{\infty}(C_r^{j+3})'\xi\rho_{j+3} + \gamma \qquad (j = 1,\ 2)$$

$$\sum_{e=0}^{2} d_{r-e}^{j+3} f_e(\rho_{j+3} + r - e) + \sum_{e=0}^{2} C_{r-e}^j f'_e(\rho_{j+3} + r - e) = 0 \tag{3-40}$$

$$U_0^{H(6)} = U_0^{H(6)}\ln\xi + \sum_{\gamma=0}^{\infty}(C_\gamma^6)'\xi^{\gamma-1}$$

$$\sum_{e=0}^{2} d_{r-e}^6 f_e(\rho_6 + r - e) + \sum_{e=0}^{2} C_{r-e-1}^j f_e(\rho_6 + r - e) = 0 \tag{3-41}$$

where

$$f_0(\rho+\gamma) = (\rho+\gamma)^2(\rho+\gamma+1/2)(\rho+\gamma-1)^2(\rho+\gamma-1/2)\quad f_2(\rho+\gamma-2) = (\rho+\gamma-3/2)(\rho+\gamma-5/2)A_2$$

$$f_1(\rho+\gamma) = (\rho+\gamma-1)^2 A_1 - (\rho+\gamma-1)^2(\rho+\gamma-3/2)(\rho+\gamma-1/2)A_3$$

$$f'_0(\rho+\gamma) = f_0(\rho+\gamma)\left[\frac{2}{\rho+\gamma} + \frac{1}{\rho+\gamma+1/2} + \frac{2}{\rho+\gamma-1} + \frac{1}{\rho+\gamma-1/2}\right]$$

$$f'_1(\rho+\gamma-1) = 2A_1(\rho+\gamma-1)$$

$$-A_3(\rho+\gamma-1)^2(\rho+\gamma-3/2)(\rho+\gamma-1/2)\left(\frac{2}{\rho+\gamma-1} + \frac{1}{\rho+\gamma-3/2} + \frac{1}{\rho+\gamma-1/2}\right)$$

$$f'_2(\rho+\gamma-2) = A_2(e+\gamma-3/2)(\rho+\gamma-5/2)\left[\frac{1}{\rho+\gamma-3/2} + \frac{1}{\rho+\gamma-5/2}\right] \tag{3-42}$$

The solving method for the particular solution of Eq.(3–37) is similar to the particular solution of Equ.(3–14) From the Cauchy's criterion, the above series that we have obtained is convergent and calculations show that the convergence of series is very fast.

3. Numerical Results.

As a valuable application of the theory presented in this paper, let us consider the M–type combined shells foundation shown in Fig.7. The numerical results have been carried out. All calculations are divided in four parts, i.e., inner shell, ring beam, outer shell and exterior base, finally, the continuity conditions between the inner shell and ring beam, continuity conditions between outer shell and ring beam and the boundary condition of outer shell are used to determine the integral constants. The computation results are shown in Fig.9 and Fig.10.

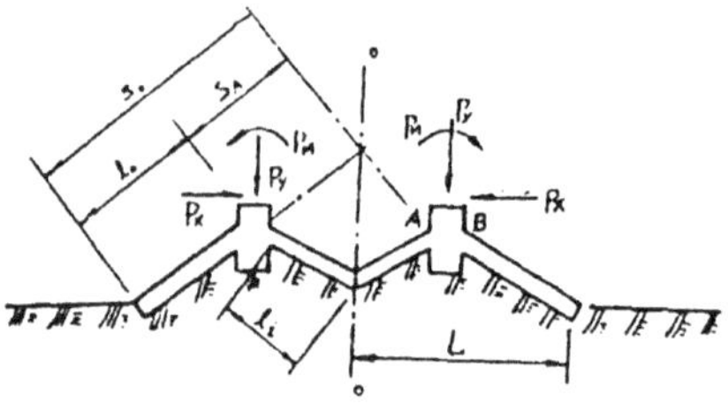

Fig. 7

Data: E_a=10Mpa μ_a=0, 3
μ=0. 167, E=11480Mpa
h=0. 05, γ=1, 55

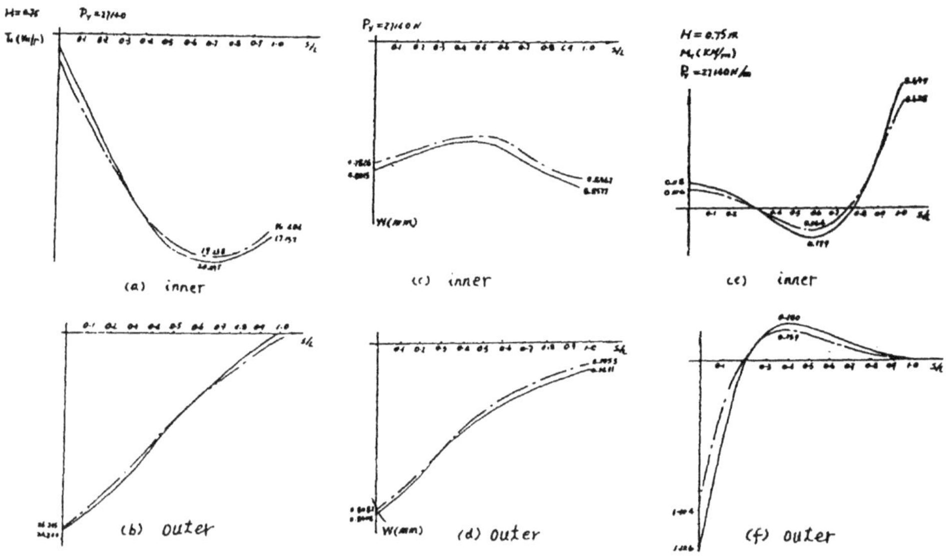

Fig. 8

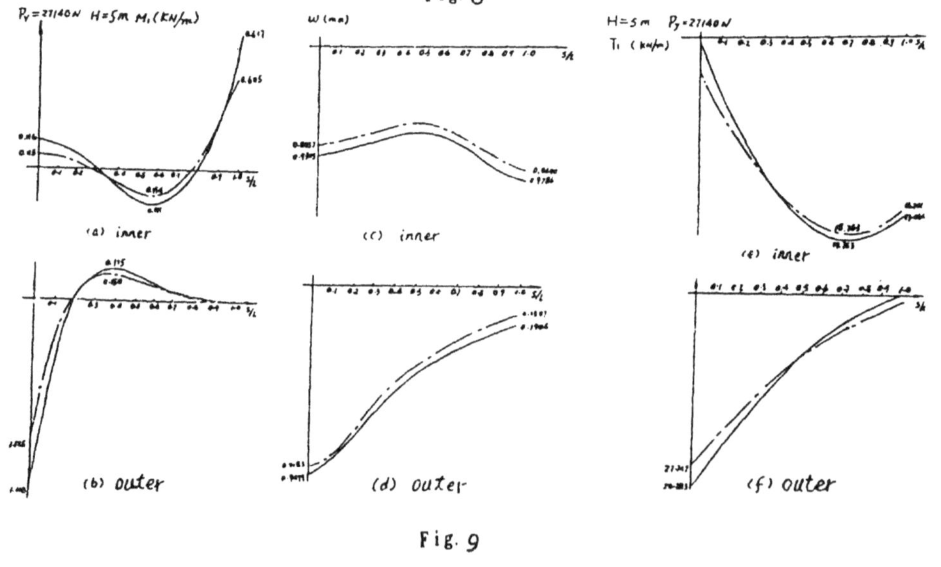

Fig. 9

By means of the above analyses, it is shown that when the compressible layer of foundation is relatively thin. The Winkler model is available. With the increasing of the thickness of foundation layer, the error resulting from using the Winkler model will increase.

Ⅳ. The Directions of cooperative Research.

In view of the wide application of plate, and shell structures in engineering and the development of environments and plate、shell structure interaction analyses, the following researches are very valuable and neceesary:

1. So far as the theory of the plates and shell is concerned, it is valuable to discuss the following directions:

(1) The nonlinear analyses of reinforced concrete plates and shells. The reinforced concrete is a special material.

(2) The problems of the dynamic buckling (the location and the whole buckling) and dynamic–buckling of the plate and shell structures.

(3) The analyses of the large scale combined structures of plates and shells (for examples, grain bins).

2. So far as the analyses of the environments and plates、shells interaction, the following researches are very necessary:

(1) The analyses of visco–elastic model of foundation and plates、shells interaction, the analyses of visco–plastic model of foundation and plates、shells interaction and the analyses of the variable modulus soil(non–uniform mediums) and plate、shell structure interaction.

(2) the analyses of the large scale combined structures of plates and shells under the action of the earthquake.

(3) The researches of the large combined structure vibration under the actions of wind.

(4) The dynamic analyses of the liquid and plate–shell structures interaction (including the earthquake).

References:

1. Hu Hai–Chang, ACTA Mechanic Sinica, No.6, 1963.
2. Huang–Yih, ACTA Mechanica Solida Sinica, No.1, 1986.
3. Selvadurai, A.P., Elastic Analysis of Soil–Foundation Interaction, Elsevier Scientific Publishing Co. 1979.
4. Vlasov, V.Z., The General Theory of shells and its Applications (in Russian) Gostekhizdat, 1949.
5. Luo–En, China Civil Engineering Journal. No.4, 1981.
6. Huang–Yih, Proceedings of the Fifth Engin. Mech. Conf. U.S.S. 1984.
7. Huang–Yih. J. of Appl. Math. and Mech. No.2, 1991.
8. Ince, Ordinary Differential Equations, London. 1927.

ACTIVE STRUCTURAL CONTROL FOR NATURAL HAZARD MITIGATION

T.T. SOONG
Department of Civil Engineering
State University of New York at Buffalo
Buffalo, NY 14260, USA

ABSTRACT

In order to safeguard structures against severe environmental loads, passive control systems have been incorporated into new as well as existing structures. These systems include tuned mass dampers and viscoelastic dampers against wind loads, and base isolation systems against earthquakes. More recently, the possibility of using active control systems has also received considerable attention. An active system generally consists of feedback control while relying upon the availability of structure response information and external energy supply. Examples of possible active control systems are active appendages, active mass dampers and active bracing systems.

This paper addresses the design, fabrication and performance evaluations of two active systems which have been installed in a full-scale structure. Important implementational issues are discussed.

INTRODUCTION AND BASIC PRINCIPLES

The possible use of active control systems as a means of structural protection against environmental loads, such as wind and earthquakes, has received considerable attention in recent years. It has now reached the stage where active systems have been installed in full-scale structures [1]. As an example, an active mass driver system was recently installed atop the eleven-story Kyobashi Seiwa Building in Tokyo. This system, designed primarily for comfort control of building occupants, has performed well under several hurricanes and moderate earthquakes encountered since installation.

Another active mass damper system was recently installed in a dedicated full-scale 600-ton test structure on the grounds of the Takenaka Research Laboratory. It has also performed well during a recent moderate earthquake. In addition, a full-scale active bracing system has been installed in the same structure for performance evaluation purposes. The advantage of having the performance of two active systems evaluated using the same structure is obvious. In addition to providing the same base parameters for performance comparisons, this arrangement allows the calibration of both systems by using one of the systems as motion inducer and the other as motion controller. Even without actual environmental disturbances, much of the performance characteristics of these systems can be assessed.

To provide a focus, this paper addresses design, fabrication, and performance issues related to the development of the active mass damper and the active bracing system. These discussions provide a realistic assessment of the potential benefit which can be derived from an active system on the one hand, and system design and capability requirements on the other.

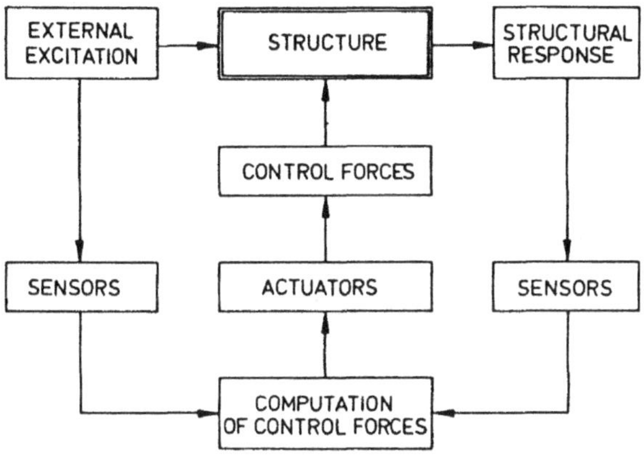

Fig. 1 Block Diagram of Active Control

In terms of basic principles, an active structural control system has the basic configuration as shown schematically in Fig. 1. It consists of (a) sensors located about the structure to measure either external excitations, or structural response variables, or both; (b) devices to process the measured information and to compute necessary control forces needed based on a given control algorithm; and (c) actuators, usually powered by external energy sources, to produce the required forces. When only the structural response variables are measured, the control configuration is referred to as *closed-loop control* since the structural response is continually monitored and this information is used to make continual corrections to the applied control forces. An *open-loop control* results when the control forces are regulated only by the measured excitations. In the case where the information on both the response quantities and excitation are utilized for control design, the term *open-closed loop control* is used.

To see the effect of applying such control forces to a structure under ideal conditions, consider a structure modeled by an n-degree-of-freedom lumped mass-spring-dashpot system. The matrix equation of motion of the structural system can be written as

$$M\ddot{\mathbf{x}}(t) + C\dot{\mathbf{x}}(t) + K\mathbf{x}(t) = D\mathbf{u}(t) + E\mathbf{f}(t) \tag{1}$$

where M, C and K are the $n \times n$ mass, damping and stiffness matrices, respectively, $\mathbf{x}(t)$ is the n-dimensional displacement vector, the r-vector $\mathbf{f}(t)$ represents the applied load or external excitation, and the m-vector u is the applied control force vector. The $n \times m$ matrix D and the $n \times r$ matrix E define the locations of the control force vector and the excitation, respectively.

Suppose that the open-closed loop configuration is used in which the control force $\mathbf{u}(t)$ is designed to be a linear function of the measured displacement vector $\mathbf{x}(t)$, the velocity vector $\dot{\mathbf{x}}(t)$ and the excitation $\mathbf{f}(t)$. The control force vector takes the form

$$\mathbf{u}(t) = K_1\mathbf{x}(t) + C_1\dot{\mathbf{x}}(t) + E_1\mathbf{f}(t) \tag{2}$$

where K_1, C_1, and E_1 are respective control gains which can be time-dependent.

The substitution of equation (2) into equation (1) yields

$$M\ddot{\mathbf{x}}(t) + (C - DC_1)\dot{\mathbf{x}}(t) + (K - DK_1)\mathbf{x}(t) = (E + DE_1)\mathbf{f}(t) \tag{3}$$

Comparing equation (3) with equation (1) in the absence of control, it is seen that the effect of open-closed loop control is to modify the structural parameters (stiffness and damping) so that it can respond more favorably to the external excitation. The effect of the open-loop component is a modification (reduction or total elimination) of the excitation.

It is seen that the concept of active control is immediately appealing and exciting. On the one hand, it is capable of modifying properties of a structure in such a way as to react to external excitations in the most favorable manner. On the other hand, direct reduction of the level of excitation transmitted to the structure is also possible through active control.

The choice of the control gain matrices K_1, C_1, and E_1 in equation (2) depends on the control algorithm selected. A number of control strategies for structural applications have been developed, some of which are based on the classical optimal control theory and some are proposed for meeting specific structural performance requirements. The reader is referred to [1] for discussions of some commonly used structural control algorithms.

ACTIVE BRACING SYSTEM (ABS)

Comprehensive experimental as well as analytical studies of active tendon or bracing control have been carried out. A comprehensive experimental study was designed and carried out in order to study the feasibility of active bracing control using a series of carefully calibrated structural models. As Fig. 2 shows, the experiments progressed from stage 1 with a simple structural model to stage 4 where a full-scale dedicated test structure is being used for testing and performance verification.

The model structures increased in weight and complexity as the experiments progressed from stage 1 to stage 3 so that more control features could be incorporated into the experiments. The model structure studied during the first stage was a three-story steel frame modeling a shear building by the method of mass simulation, whose top two floors were rigidly braced to simulate a single-degree-of-freedom system. The model was mounted on a shaking table which supplied the external load. The control force was transmitted to the structure through two sets of diagonal prestressed tendons mounted on the side frame.

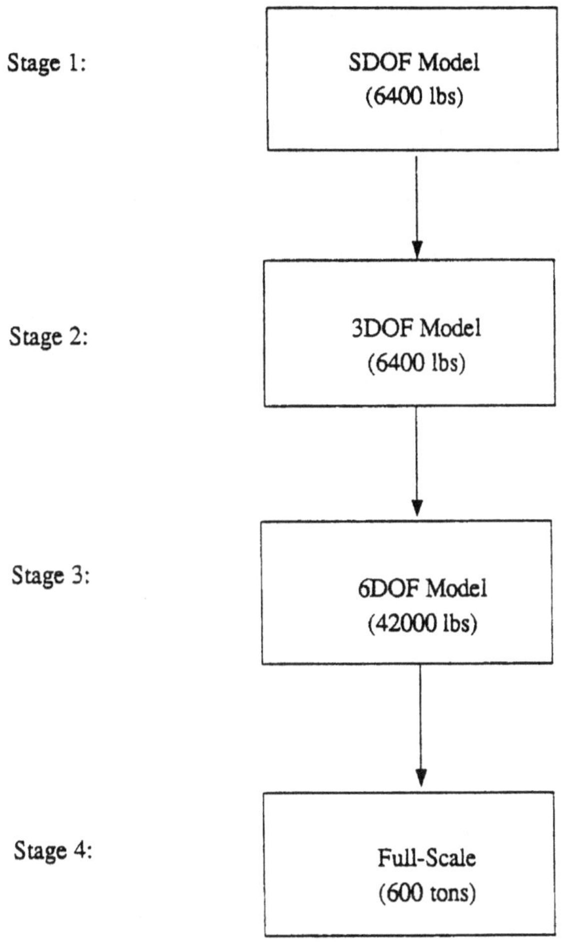

Fig. 2 Stages of Structural Tests

Several significant features of these experiments are noteworthy. First, they were carefully designed in order that a realistic structural control situation could be investigated. Efforts made towards this goal included making the model structure dynamically similar to a real structure, working with a carefully calibrated model, using realistic base excitation, and requiring more realistic control forces. Secondly, these experiments permitted a realistic comparison between analytical and experimental results, which made it possible to perform extrapolation to real structural behavior. Furthermore, important practical considerations such as time delay, robustness of control algorithms, modeling errors and structure-control system interactions could be identified and realistically assessed [2,3].

Experimental results show significant reduction of structural motion under the action of the simple tendon system. For example, a reduction of over 50% of the first-floor maximum relative displacement could be achieved. This is due to the fact that the control system was able to induce damping in the system from a damping ratio of 1.24% in the uncontrolled case to 34.0% in the controlled case [2].

At stage 2, rigid bracings on the top two floors of the model structure were removed in order to simulate a three-degree-of-freedom system. This multi-degree-of-freedom model provided opportunities for study and verification of a number of control features which were not possible in stage 1 study. These included modal control, time delay in the modal space, and control and observation spillover compensation. Moreover, further verification of simulation procedures could be carried out, giving added confidence in the use of simulation for extrapolating active control results to more complex situations. Experimental results compared favorably with analytical results obtained under the same conditions and showed that the motion of all three floors can be effectively controlled using a single actuator when control design is carefully carried out.

As a further step in this direction, a substantially larger and heavier six-story model structure was fabricated for stage 3 of this experimental undertaking. It is also a welded space frame utilizing artificial mass simulation.

Multiple tendon control was possible in this case and the following arrangements were included in this phase of the experiments:

(a) A single actuator is placed at the base with diagonal tendons connected to a single floor.

(b) A single actuator is placed at the base with tendons connected simultaneously to two floors, thus applying proportional control to the structure.

(c) Two actuators are placed at different locations of the structure with two sets of tendons acting independently.

Several typical actuator-tendon arrangements are shown in Fig. 3.

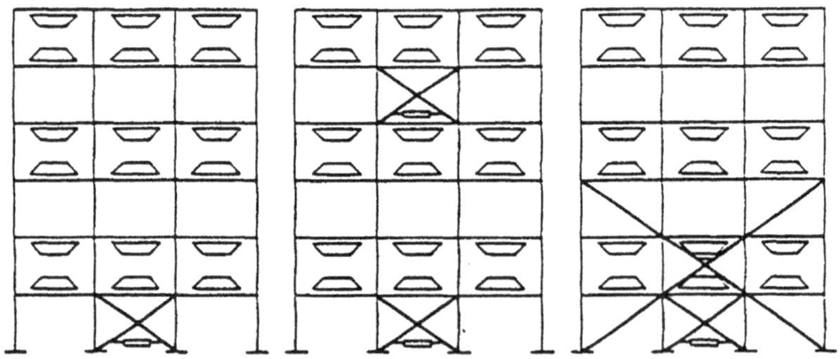

Fig. 3 Examples of Actuator-Tendon Arrangements at Stage 3

Another added feature at this stage was the testing of a second control system, an active mass damper, on the same model structure, thus allowing a performance comparison of these two systems. The active mass damper will be discussed in more detail in the next section.

For the active tendon systems, experimental as well as simulation results have been obtained based upon the tendon configurations stipulated above. Using the N-S component of the El Centro acceleration record as input, but scaled to 25% of its actual intensity, control effectiveness was demonstrated. For example, in terms of reduction of maximum relative displacements, results under all actuator-tendon arrangements tended to cluster within a narrow range. At the top floor, a reduction of 45% could be achieved. Control force and power requirements were also found to be well within practical limits when extrapolated to the full-scale situation [4].

Results obtained in the laboratory provided invaluable information in the development of the full-scale active bracing system [5], which was fabricated and installed in a full-scale 600-ton dedicated test structure at stage 4. The structure was erected in Tokyo by the Takenaka Corporation.

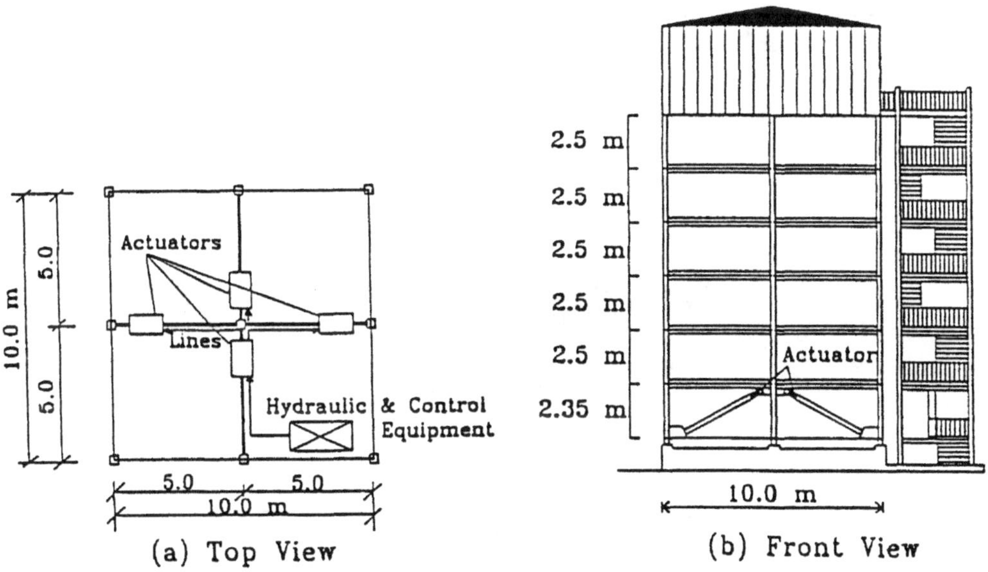

Fig. 4 Test Structure at Stage 4 and Active Bracing System

As shown in Fig. 4, the system consists of four actuators attached to bracings on the first floor and it is designed to provide motion control biaxially. Initial tests were performed using an active mass damper at the top as an exciter and the active bracing system as the controller. Figure 5(a) shows the uncontrolled structural response in both directions. The decay rate of the uncontrolled response is very slow because the system is only slightly damped (1% in the strong direction and 0.5% in the weak direction). Figure 5(b) shows the build-up of the structural vibration and then the influence of control applied at the peak of excitation. The increase in damping is evident, showing, for example, an increase from 0.5% to 3.5% in the weak direction. The influence of control during forced vibration is shown in Fig. 5(c). The response builds up in the transient stage and reaches its steady state, which is evidently smaller than the previous tendency as shown in Fig. 5(b).

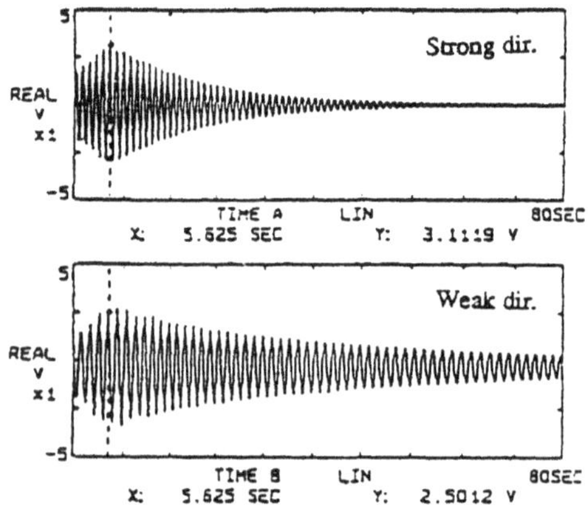

(a) Uncontrolled Free Vibration

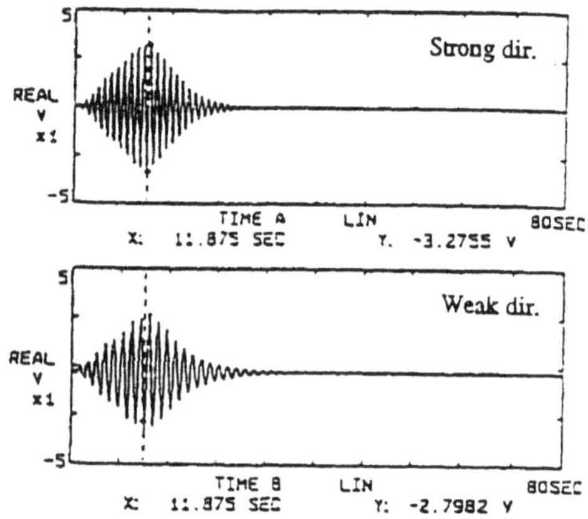

(b) Controlled Free Vibration

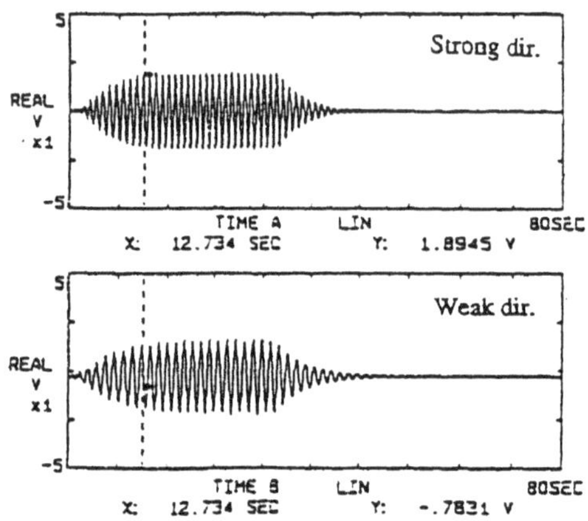

(c) Controlled Forced Vibration

Fig. 5 Preliminary Test Results of Full-Scale Structures

ACTIVE MASS DAMPER (AMD)

During stage 3 of the experimental series, an active mass damper fabricated by the Takenaka Corporation of Japan was placed on top of the structure, which could be operated under different conditions by changing its added mass, its stiffness and the state of the regulator. A total of 12 cases were performed in the experiment.

Extensive experimental results were obtained under various simulated earthquake excitations [4]. A summary of results obtained under the 25% intensity El Centro excitation is given below:

Percent Reduction of Maximum Top-floor Relative Displacement:	43.3-57.2
Percent Reduction of Maximum Top-floor Acceleration:	5.5-30.7
Percent Reduction of Maximum Base Shear:	31.4-44.4
Maximum Control Force Required (kips):	0.68-2.56
Maximum Mass Peak-to-Peak Stroke (in):	3.23-10.1
Maximum Control Power Required (Kw):	0.82-5.73

One of the advantages of testing two different active systems using the same model structure is that their performance characteristics can be realistically compared. Extensive simulation and experimental results obtained based on the six-story, 42,000-lb model structure show that both AMD and ABS display similar control effectiveness in terms of reduction in maximum top-floor relative displacement, in maximum top-floor absolute acceleration, and in maximum base shear. They also have similar control requirements such as maximum control force and maximum power. Other information which may shed more light on their relative merits but is not considered here includes cost, space utilization, maintenance and other practical observations.

Following this test series, a full-scale active mass damper was fabricated and installed by Takenaka atop the stage-4 structure. The biaxial AMD, shown in Fig. 6, is of the pendulum type with a fail-safe regulator. It weighs 6 tons, approximately 1/100 of the structural weight, and has a maximum stroke of ± 1.0 m with a maximum control force of 10 tons. The system has been performing as expected. Figure 7 shows that, during a recent earthquake in Tokyo, the maximum relative displacement at the top floor was observed to be 0.63 cm as compared with 2.16 cm, which is the estimated maximum value had the system not been activated.

Again, the advantage of having the performance of two active systems evaluated using the same structure is obvious. In addition to providing the same base parameters for performance comparison, this arrangement allows the calibration of the ABS and AMD systems by using one of the systems as motion inducer and the other as motion controller. Even without actual seismic motion, much of the performance characteristics can be assessed during this calibration method. During the calibration period, several feasible control algorithms can be evaluated and control parameters refined.

Performance observations of these systems under actual ground motions are being carried out by deactivating one of the systems for a period of six months in order to allow performance assessment of the other system. A total three-year observation period is planned under this activation-deactivation scheme.

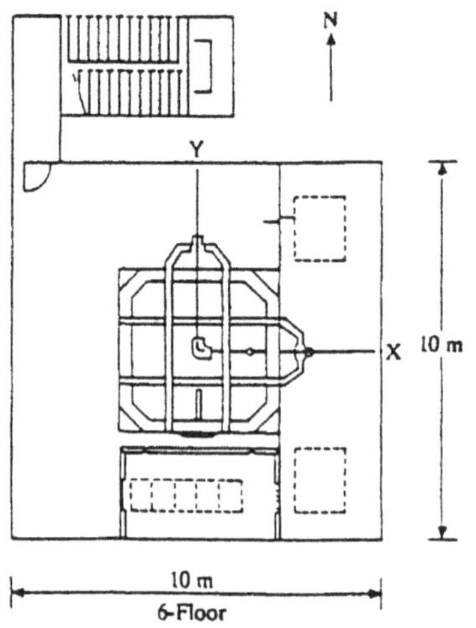

Fig. 6 Active Mass Damper at Stage 4

CONCLUDING REMARKS

With extensive experimental work and full-scale testing underway, active structural control research has entered an exciting phase. Faced with increasing demands on reliability and safety, active structural control can be an eminently logical alternative in ensuring structural integrity and safety to more traditional approaches.

At the same time, however, a large number of serious challenges remain and they must be addressed before active structural control can gain general acceptance by the civil engineering and construction professions at large. One of the important issues is the fact that, since the concept of an active-controlled structure is a significant departure from traditional structural concepts, obstacles exist with respect to its acceptance. This is particularly true when structural safety is to rely upon an active control system. More full-scale demonstration projects are thus needed for purposes of concept verification and education. Other issues, no less important, include capital cost and maintenance, reliance on external power, system robustness, and system reliability.

ACKNOWLEDGEMENTS

The work described in this paper has been sponsored in part by the National Science Foundation and by the National Center for Earthquake Engineering Research, Buffalo, New York. The

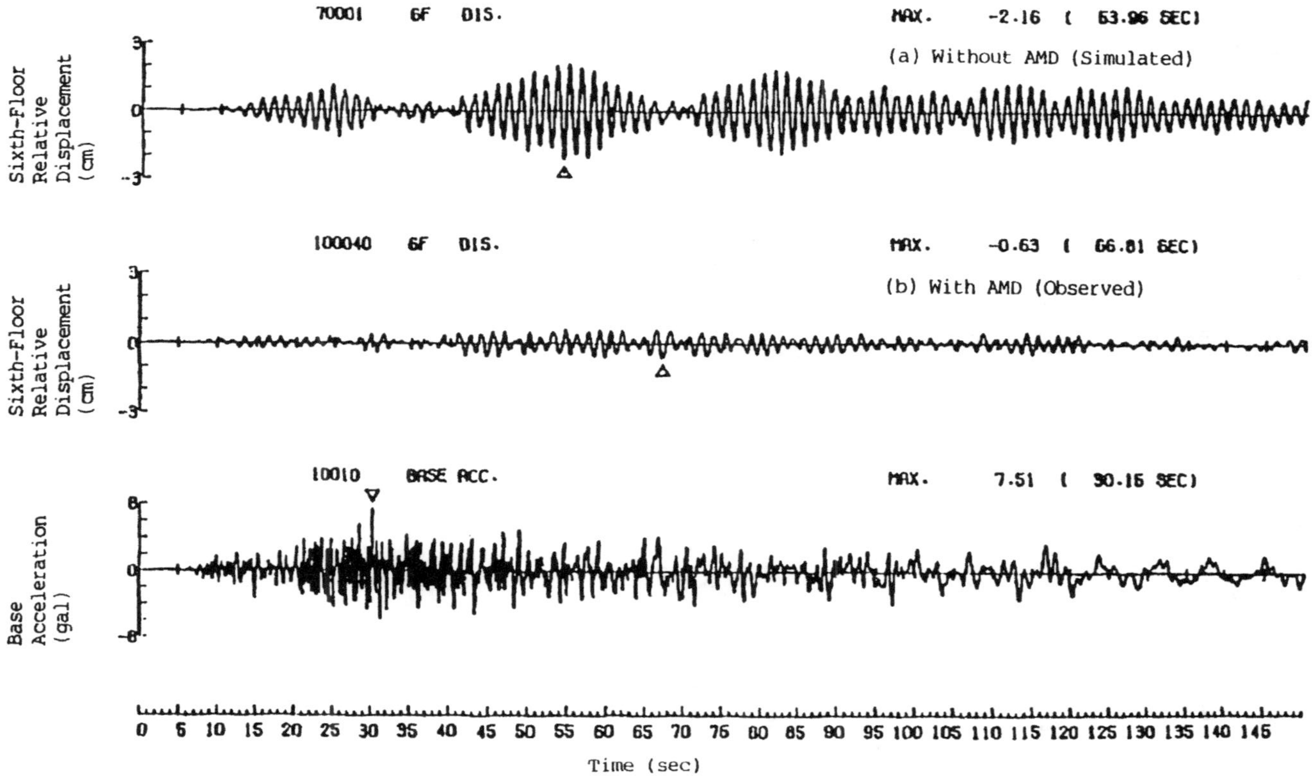

Fig. 7. Performance of AMD During 2/20/90 Earthquake in Tokyo, Japan

author is indebted to the following firms for their technical assistance, consultation and financial support: MTS Systems Corporation, Takenaka Corporation and Kayaba Industry Co., Ltd. Finally, the author is indebted to his colleague, Professor A.M. Reinhorn, and his graduate students who worked on the projects described in this paper.

REFERENCES

1. Soong, T.T., Active Structural Control: Theory and Practice, Longman, London and Wiley, New York, 1990.

2. Chung, L.L., Reinhorn, A.M. and Soong, T.T., Experiments on active control of seismic structures, ASCE J. Eng. Mech., 1988, **114**, 241-256.

3. Chung, L.L., Lin, R.C., Soong, T.T. and Reinhorn, A.M., Experimental study of active control of MDOF seismic structures, ASCE J. Eng. Mech., 1989, **115**, 1609-1627.

4. Reinhorn, A.M., Soong, T.T., et al, 1:4 scale model studies of active tendon systems and active mass dampers for seismic protection, Technical Report NCEER-89-0026, National Center for Earthquake Engineering Research, Buffalo, NY, 1989.

5. Soong, T.T. and Reinhorn, A.M., et al., Full-scale implementation of active control, I. Design and simulation, ASCE J. Str. Eng., 1991, in press,.

STUDY ON STRUCTURAL CONTROL

Li Guiqing Cao Hong
Institute of Aseismic Engineering Structures
Wuhan University of Technology
No.14, Luoshi Road, 430070, Wuhan, China

Key words: Structural Control, Controlled Structure, Prediction, Real Time

Abstract

In this paper, authors briefly outline the study of structural control, introduce partial results obtained by authors and their colleagues in this field, such as fuzzy state control, prediction of real time control algorithms, control of structural whipping effect by means of mass pump. Finally, a forecast is given for the study of structural control in China and overseas.

I. Outline of study of structural control in China

The study of structural control in China began several decades ago, but the progress has not been notable until recent years.

In the early 1950s, Li Li proposed the idea of using roller to isolate foundations from vibration (this concept was proposed first by Bechtold of Germany in 1906), and some tests of buildings isolated from vibration (including using sand as the isolation foundation) were made. Professor Chen Dan and the Institute of Building Science of Liao Ning Province engaged in some testing and theoretical research about vibration isolation by friction, and made the conclusion that this kind of foundation is effective. Now, there are over 30 bridges in China where rubber was used as an isolation foundation.

Professor He Yuao proposed some control design concepts in order to optimize distribution of structural stiffness and mass. Professor Ding Dajun proposed a design method of dissipated energy control in the view of structural construction. Professor Lin Gao proposed a method to control structural vibration by means of collisions between two adjacent parts of the structure or structures, and some experiments were made. Recently, Professor Liu Ji made several experiments using oil pressure to increase the damping force of structures, and some important achievements have been made.

Authors and their colleagues have been engaged in research on active and passive control of structural random vibration since 1984. We published a "review of structural vibration control[1]" in 1987, where we introduced this theory by its fundamental concepts, principles ,history and questions need to be investigated further, which open a whole new field of study within civil engineering in China. The treatise written by authors in same year, "Structural Control and Controlled Structure[2]" , first proposed a new concept "controlled structure". As pointed out in that paper, "controlled structure" is a kind of structural engineering with control devices, which may be classified as a new discipline because the control devices and the structure should be treated as an entirety in order to make optimal design according to a given control condition. Not only can the response of structures (such as displacement, speed or acceleration) be controlled within a given range, but also a structural model of failure, for example, the location and order of plastic hinges, if the structure is designed according to theory of "controlled structure". There are close connections and also distinctions between "controlled structure" and "structural control". "Controlled structure" investigates and designs structures in the view of control and conditions. Control objectives may be reached not only by means of active or passive control devices but also by changing the distribution of structural stiffness, mass, steel bar and construction measurement. For example, with a properly adjusted mass and stiffness, a projecting object on top of a structure will reduce greatly the earthquake response of main structure. "Structural control" is the means to investigate and design control devices from view of structures. In recent years, we have made great progress in the field of "controlled structure", including reliability analysis of controlled structures. We presented the treatise "Optimal control of random vibration of tall buildings" at the 9th World Conference on Earthquake Engineering in 1988. In this paper, we investigated the control of critical mode shape of tall buildings and structures by using a separation theorem and optimal filter theory, and also analysed the effect of non–critical mode shape on structural response. We also studied the working principle of the mass mump, the effect of changing parameters of structure, optimal theory for the parameter of the mass pump[3] and the control of structural whipping effect by using of mass pump[4]. Then, based on an equivalent linearization procedure, we investigated the problem of parameter optimization of the system of the isolation foundation from vibration[5], and proposed control methods for reducing the vibration of tall framed structures by *installing artificial hysteresis hinges at the* ends of beams [6]. This device can make structural response within the elastic range when structure is subjected to earthquakes of small intensity, and artificial hysteresis hinges enter the plastic range but columns still within the elastic range when subjected to a strong earthquake. Last year, we proposed three kinds of control methods for wind–resistant tall buildings[7]. The authors have written those achivements in a book "Theory and its application of structural control", which will be published soon.

II. Fuzzy state control

1. Structures with a known mathematic model

The control equation of fuzzy system is:

$$\underset{\sim}{X}_{k+1} = R(\underset{\sim}{X}_k, \underset{\sim}{U}_k) = A\underset{\sim}{X}_k + B\underset{\sim}{U}_k \tag{1}$$

Eq.(1) is an apparent equation of the fuzzy system F={X, U, Y, δ, β}, which is based on fuzzy logic. $\underset{\sim}{X}_k, \underset{\sim}{U}_k$ in Eq.(1) are fuzzy parameters, and there are control parameters and external loads in $\underset{\sim}{U}_k$, $\delta: X\times U\times X \to [0,1]$, β: $X\times Y \to [0,1]$, $X_0 \in X$ is given. The fuzzy aim function is:

$$J = \bigwedge_{i=0}^{N} \mu_{L(x_i, u_i, i)}(X,U) \tag{2}$$

$L(x_i, u_i, i)$: $X\times U \to [0, 1]$, $u_i \in U$, N is fixed; we have to find a $u \in U$ in order to meet: $J(u) = \sup_{u\in U} J(u)$. Limited by the condition of Eq.(1), a fuzzy dynamic program is used to solve this equation:

$$M(\hat{x}_k, k) = \sup_{u_k, u_{k+1}, \cdots, u_N} [\bigwedge_{j=k}^{N} \mu_{L(\hat{x}_j u_j, j)}(X,U)] \qquad k = 0,1,2,\cdots,N, \qquad \hat{x}_0 = X_0 \tag{3}$$

Using the Bellman optimization equation, we obtain:

$$\begin{aligned} M(\hat{x}_k, k) &= \sup_{u_k, u_{k+1}, \cdots, u_N} \sup \{\mu_{L(\hat{x}_k, u_k, k)}(X,U)[\bigwedge_{j=k}^{N} \mu_{L(\hat{x}_j u_j, j)}(X,U)]\} \\ &= \sup_{u_k} \{\mu_{L(\hat{x}_k, u_k, k)}(X,U) \sup_{u_{k+1}, \cdots, u_N} [\bigwedge_{j=k}^{N} \mu_{L(\hat{x}_j u_j, j)}(X,U)]\} \\ &= \sup_{u_k} \{\mu_{L(\hat{x}_k, u_k, k)}(X,U) \wedge M[R(\hat{x}_k, u_k), k+1]\} \end{aligned} \tag{4}$$

When K equal to N:

$$M(\hat{x}_N, N) = \sup_{u_N \in U} \mu_{L(\hat{x}_N, u_N, N)}(X,U) \tag{5}$$

Then, we obtain a recursive formula for fuzzy optimal control:

$$\hat{x}_0 = X_0$$

$$M(\hat{x}_N, N) = \sup_{u_N \in N} \mu_{L(\hat{x}_N, u_N, N)}(X,U)$$

$$\hat{x}_{k+1} = R(\hat{x}_k, u_k)$$

$$M(\hat{x}_k, k) = \sup_{u_k \in U} \{\mu_{L(\hat{x}_k, u_k, k)}(X,U) \wedge M[R(\hat{x}_k, u_k, k+1]\} \quad k = N-1, N-2, \ldots, 1, 0 \tag{6}$$

Using Eq.(6) recursively, we can obtain optimal control $\hat{u} = \{\hat{u}_1, \ldots, \hat{u}_N\}$ and optimal orbit $\hat{x}_k (k = 0,1,\ldots,N)$.

2. Structure with unknown mathematic model

Gernerally speaking, an exact mathematic model of a structure can not be predicted. The error induced by the idealization of model, the rounding off of all kinds of parameters and the approximations of the computational method, may make the structure deviate from design objectives or fail. The aim of re-

searcher is to pursue a kind of design method which coincides with real structure as closely as possible. Fuzzy state control can reach this aim, andits frame is shown in Fig.1.

Fuzzy state control is a language control, the control law in a fuzzy control machine is determined by the experiences of manual control, and manual control can only observe the output of object controlled and its rate of change. Because the magnitude of structural response is so small that it can only be measured by instruments, this measurement is an indirect measurement. Therefore, fuzzy state control will not be carried on until the measurement is dealt with by fuzziness. Fuzzy state control consists of three procedures:

(1) Transform deterministic parameter into fuzzy one, that is to say, deal with state variables by fuzziness;
(2) Analyze, determine and establish fuzzy control rules, that is , set up a kind of fuzzy algorithm by experience;
(3) Make a fuzzy decision, and transform this decision into deterministic parameters in order to carry it out.

(1). Transform deterministic parameters into fuzzy ones

The step of transforming deterministic parameters into fuzzy ones is:

In order to make fuzzy control at the ith point of structure, the displacement and speed at this point, x_i and $\dot{x}_i$, need to be measured. Assume the range of displacement and speed are $[a_i, b_i]$ and $[\alpha_i, \beta_i]$ respectively. Then, those two ranges can be transformed into the range [−6, 6] by following equation:

$$y_i = \frac{6[2x_i - (a_i + b_i)]}{b_i - a_i} \qquad \dot{y}_i = \frac{6[2\dot{x}_i - (\alpha_i + \beta_i)]}{\beta_i - \alpha_i} \tag{7}$$

The fuzzy set can be established at this range, and its number depends on the particular situation and the accuracy needed. Here seven fuzzy sets are established for displacement and speed, $\underset{\sim}{E}_j$, $\underset{\sim}{C}_j$(j = 1, 2, ⋯, 7). $\underset{\sim}{E}_1$, $\underset{\sim}{C}_1$ express "positive large" or PL, its range is near to +6, $\underset{\sim}{E}_2$, $\underset{\sim}{C}_2$ "positive middle" or PM, near to +4, $\underset{\sim}{E}_3$, $\underset{\sim}{C}_3$ "positive small" or PS and near to +2, $\underset{\sim}{E}_4$, $\underset{\sim}{C}_4$ "zero" or 0, $\underset{\sim}{E}_5$, $\underset{\sim}{C}_5$ "negative small" or NS near to −2, $\underset{\sim}{E}_6$, $\underset{\sim}{C}_6$ "negative middle" or NM near to −4, $\underset{\sim}{E}_7$, $\underset{\sim}{C}_7$ "negative large" or NL near to −6. The membership of every parameter at this range to fuzzy set can be expressed as a continuous or discrete function. If the continuous function is used, the Gaussian distribution $F(x) = \exp\left\{-\frac{[x - E(x_i)]^2}{Var(x_i)}\right\}$ will usuallly be used. But if the number of fuzzy sets is large, the calculation will be very complicated. Therefore a discrete function will be adopted in this case, that is to say, there are 13 elements at that range. The displacement, speed and control force will be thought of as unchanged for the duration of the sample. The membership of every parameter to fuzzy set is listed in table 1, which can be determined by fuzzy statistics.

Table 1 is the value of fuzzy set $\underset{\sim}{E}_i$, $\underset{\sim}{C}_i$ at the range. If y_i, $\dot{y}_i$ measured are not integer, they are rounded off to integer. Thereby, the continuous parameter is transformed into a discrete one. Similarly, seven fuzzy sets can be established for control force $\underset{\sim}{u}_i$, (i = 1, 2, ···, 7).

Table 1.

	E_1PL	E_2PM	E_3PS	$E_4$0	E_5NS	E_6NM	E_7NL
−6	0	0	0	0	0.2	0.8	1.0
−5	0	0	0	0	0.4	0.9	0.9
−4	0	0	0	0	0.7	1.0	0.8
−3	0	0	0	0.2	0.9	0.9	0.6
−2	0	0	0	0.6	1.0	0.7	0.4
−1	0	0	0.1	0.9	0.9	0.3	0.1
0	0	0.1	0.6	1.0	0.6	0.1	0
1	0.1	0.3	0.9	0.9	0.1	0	0
2	0.4	0.7	1.0	0.6	0	0	0
3	0.6	0.9	0.9	0.2	0	0	0
4	0.8	1.0	0.7	0	0	0	0
5	0.9	0.9	0.4	0	0	0	0
6	1.0	0.8	0.2	0	0	0	0

(2). Set up fuzzy law

There are two inputs and one output on fuzzy control system here, which are shown in Fig. 2.

Language inference of fuzzy control device is:

$$\text{if } \underset{\sim}{E} \text{ and } \underset{\sim}{C} \text{ then } \underset{\sim}{U} \tag{8}$$

which can be expressed further as:

$$\text{if } E = \underset{\sim}{E}_i \text{ and } C = \underset{\sim}{C}_j \text{ then } U = \underset{\sim}{U}_{ij} \qquad i = 1,2,\cdots,m, \qquad j = 1,2,\cdots,n \tag{9}$$

Fuzzy relationship $\underset{\sim}{R}$ is:

$$\underset{\sim}{R} = \bigcup_{i,j} (\underset{\sim}{E}_i \times \underset{\sim}{C}_j) \times \underset{\sim}{U}_{ij} \tag{10}$$

$$\mu_{\underset{\sim}{R}}(x,y,z) = \bigvee_{i,j=1}^{m,n} [\mu_{\underset{\sim}{E}_i}(x) \wedge \mu_{\underset{\sim}{C}_j}(y)] \wedge \mu_{\underset{\sim}{U}_{ij}}(z) \quad \forall x \in X, \quad \forall y \in Y, \quad \forall z \in Z, \tag{11}$$

IF $\underset{\sim}{E}$ *and* $\underset{\sim}{C}$ are known by fuzzy inference, then the control force of output $\underset{\sim}{U}$ will meet:

$$\underset{\sim}{U} = (\underset{\sim}{E} \times \underset{\sim}{C}) \cdot \underset{\sim}{R} \tag{12}$$

in which: $\underset{\sim}{E} \times \underset{\sim}{C} = \{(x, y) | \forall x \in X, \forall y \in Y\}$, and :

$$\mu_{\underset{\sim}{U}}(z) = \bigvee_{\forall x \in X, \forall y \in Y} \mu_{\underset{\sim}{R}}(x,y,z) \wedge [\mu_{\underset{\sim}{E}}(x) \wedge \mu_{\underset{\sim}{C}}(y)] \tag{13}$$

Therefore, a set of inference laws can be set up:

$$\text{if } y_i = (PL)_{y_i} \text{ and } \dot{y}_i = (PL)_{\dot{y}_i} \text{ then } u_i = (NL)_{u_i}$$

$$\text{if } y_i = (PL)_{y_i} \text{ and } \dot{y}_i = (PM)_{\dot{y}_i} \text{ then } u_i = (NL)_{u_i}$$

$$if\ y_i = (PL)_{y_i}\ \ and\ \dot{y}_i = (PS)_{\dot{y}_i}\ \ then\ u_i = (NL)_{u_i}$$

$$if\ y_i = (PL)_{y_i}\ \ and\ \dot{y}_i = (\ 0)_{\dot{y}_i}\ \ then\ u_i = (NL)_{u_i}$$

$$if\ y_i = (PL)_{y_i}\ \ and\ \dot{y}_i = (NS)_{\dot{y}_i}\ \ then\ u_i = (NL)_{u_i}$$

....

$$if\ y_i = (NL)_{y_i}\ \ and\ \dot{y}_i = (PM)_{\dot{y}_i}\ \ then\ u_i = (PL)_{u_i}$$

$$if\ y_i = (NL)_{y_i}\ \ and\ \dot{y}_i = (PL)_{\dot{y}_i}\ \ then\ u_i = (PL)_{u_i}$$

The number of inference laws totals to 49, which are shown in table 2.

Table 2.

E / U / C	PL	PM	PS	0	NS	NM	NL
PL	NL	NL	NL	NM	PM	PL	PL
PM	NL	NM	NS	NS	PS	PM	PL
PS	NL	NM	NS	NS	PS	PM	PL
0	NL	NM	NS	0	PS	PM	PL
NS	NL	NS	NS	0	PM	PL	PL
NM	NM	NM	NS	PS	PM	PL	PL
NL	NM	NM	NS	PS	PM	PL	PL

(3). Fuzzy decision

The fuzzy control parameters obtained above need to be transformed into deterministic ones, so that control device can operate. There are several methods for this transformation, among which we used the maximum membership method. If u_i has maximum membership at the range discussed:

$$\left.\begin{aligned} U_{max} &= sup\mu_{ui}(u), \quad \forall u \in U, \\ or\quad \mu_{ui}(U_{max}) &\geqslant \mu_{ui}(u), \quad \forall u \in U \end{aligned}\right\} \tag{14}$$

If the weighted power method is used, then we have:

$$U_{max} = \frac{\sum_{i=1}^{N} \mu_{u_i}(u_i)u_i}{\sum_{i=1}^{N} \mu_{u_i}(u_i)} \tag{15}$$

When U_{max} is obtained, transform it into a real operation parameter U_z by using the following equation:

$$U_{iz} = \frac{(\varphi_i - \gamma_i)U_{imax}}{12} + \frac{1}{2}(\varphi_i + \gamma_i) \tag{16}$$

where U_{iz} is real control force at the ith point of structure. If a tendon device is used, then every force of the tendon can be determined:

$$U_n = B^{-1}U \tag{17}$$

in which $U_n = (u_1, u_2, \cdots, u_n)^T$, $u = (u_{1z}, u_{2z}, \cdots, u_{nz})^T$ is the horizontal control force and B is force vector of the tendon:

$$B=\begin{bmatrix} cos\theta_1 & -cos\theta_2 & 0 & \cdots & 0 \\ \cdots & \cdots & \cdots & \cdots & \cdots \\ 0 & 0 & \cdots & cos\theta_{n-1} & -cos\theta_n \\ 0 & 0 & 0 & \cdots & cos\theta_n \end{bmatrix} \tag{18}$$

Control laws are listed in table 3. Therefore, control force can be obtained from that table directly if y_i and $\dot{y}_i$ are known. The most important merit of fuzzy control is that it controls directly, yields notable results and need not know the model of the structure. The weakness of this method is that it may cause shaking at some points of structure. The table of total control should be made elaborately to avoid shaking. One of the measures is to choose more fuzzy elements to represent deterministic parameters in order to guarantee the structure against losing effectiveness.

Table 3.

y_i \ u_i \ $\dot{y}_i$	−6	−5	−4	−3	−2	−1	0	1	2	3	4	5	6
−6	6	6	6	5	4	3	2	1	1	1	1	1	1
−5	6	6	5	5	4	4	3	2	1	1	1	1	1
−4	5	5	4	4	3	3	2	1	1	1	1	1	1
−3	5	4	4	3	3	2	2	1	1	1	1	1	1
−2	3	3	3	2	2	2	1	1	1	1	1	1	1
−1	2	2	2	2	2	2	1	−1	−1	−1	−1	−1	−1
0	1	1	1	1	1	1	0	−1	−1	−1	−1	−1	−1
1	−2	−1	−1	−1	−1	−1	−1	−2	−2	−3	−3	−4	−4
2	−3	−2	−2	−1	−1	−2	−2	−2	−3	−3	−4	−4	−5
3	−4	−3	−2	−1	−1	−1	−2	−3	−4	−4	−4	−5	−6
4	−5	−3	−2	−2	−1	−1	−2	−3	−4	−4	−5	−5	−6
5	−6	−3	−2	−1	−1	−1	−2	−3	−4	−5	−5	−6	−6
6	−6	−5	−4	−3	−2	−1	−2	−3	−4	−5	−6	−6	−6

Ⅲ. Prediction of real time control law

It will take time to compute the control force as well as to apply those forces to the structure, which known as the hysteresis problem. Therefore, the prediction of real time control law needs to be investigated. Here, we present a method proposed by authors.

Assuming the state equation of the structure is:

$$\dot{X}(t)=AX(t)+BU(t)+FW(t) \tag{19}$$

where A is the matrix of system, U(t) and W(t) are the vector of the control force and the random excitation respectively; B and F are the matrix of the location of the control force and the random excitation.

Let T be a model matrix composed of the characteristic vector of matrix A, then T is a complex matrix. Assume:

$$X(t) = TY_1(t)$$

and inserting it into Eq.(19), we obtain:

$$\dot{Y}_1(t) = \theta Y_1(t) + P(t) \qquad Y_1(0) = 0 \tag{20}$$

in which θ is a diagonal matrix composed of the characteristic root of $A(\omega_1^2, \omega_2^2, \cdots, \omega_{2n}^2)$, and P(t) is expressed as:

$$P(t) = T_L^{-1}[BU(t) + FW(t)] \tag{21}$$

where T_L^{-1} meet: $T_L^{-1}T = I$. The solution of Eq(21) can be expressed as following form:

$$Y_1(t_{k+1}) = exp(\theta\Delta t_k)Y_1(t_k) + \int_{t_k}^{t_{k+1}} exp[\theta(t_{k+1} - \tau)]P(\tau)d\tau$$

where $\Delta t_k = t_{k+1} - t_k$. Assume that U(t) and W(t) are constants$U(t_k)$, $W(t_k)$ in $[t_k, t_{k+1}]$; then we have:

$$Y_1(t_{k+1}) = exp(\theta\Delta t_k)Y_1(t_k) + exp(\theta t_{k+1}) \int_{t_k}^{t_{k+1}} exp(-\theta\tau)P(\tau)d\tau$$

and obtain following equation:

$$x(t_{k+1}) = A(t_k)X(t_k) + B(t_k)U(t_k) + F(t_k)W(t_k)$$

in which:

$$A(t_k) = Texp(\theta\Delta t_k)T_L^{-1} \qquad B(t_k) = T\theta^{-1}[exp(\theta\Delta t_k) - I]T_L^{-1}B$$

$$F(t_k) = T\theta^{-1}[exp(\theta\Delta t_k) - I]T_L^{-1}F$$

Let $\Delta t_k = \Delta t$ for convenience sake, and express $X(t_k)$ as X(k). Then, the state equation of continuous system made discrete by:

$$X(k+1) = A(k)X(k) + B(k)U(k) + F(k)W(k)$$

Assume also that the measurement equation of the structural state is:

$$Y(k) = CX(k) \tag{22}$$

Because the excitation sequence W(k) can not be predicted, following equation is used to predict X(k+1) if Y(k) is known:

$$\hat{X}(k+1) = A(k)C^{-1}Y(k) + B(k)U(k) \tag{23}$$

The error $\Delta X(kH)$ caused by Eq.(23) can be expressed as:

$$\Delta X(kH) = X(k+1) - \hat{X}(k+1) = F(k)W(k) \tag{24}$$

If closed loop control is used, then the control force predicted is:

$$\hat{U}(k) = -\frac{\Delta t}{2} R^{-1}B^T Q\hat{X}(k) \tag{25}$$

where R and Q are weight power matrices. We will discuss how to choose matrix Q.

If the vibration equation of system with m degrees of freedom is:

$$M\ddot{Y}(t) + C\dot{Y}(t) + KY(t) = B_0U(t) + F_0W(t)$$

then the parameters X(k) of system with m degrees of freedom is:

$$X(k) = \{Y(t);\ \dot{Y}(t)\}^T, \qquad B = \{\ 0\ ; M^{-1}B_0\}^T$$

Let:

$$Q = \begin{bmatrix} Q_{11} & Q_{12} \\ Q_{21} & Q_{22} \end{bmatrix}$$

then we have:

$$U(k) = -\frac{\Delta t}{2} R^{-1} B_0^T (M^{-1})^T [Q_{21} \hat{Y}(k) + Q_{22} \hat{\dot{Y}}(k)]$$

We can know from the above equation that U(k) is not a function of Q_{11}, Q_{12}, therefore let them be zero. Q_{21} relates the weight matrix of the control force to the displacement vector X(k), and Q_{22} relates the weight matrix of the control force to the speed vector $\dot{X}(k)$. Therefore, the relevant weight factor will be much greater than other weight factors when displacement or speed at a given position is the main control object.

We discuss an eight-story building with the present method. Assume the mass,stiffness and damping of every story is the same: $m_i = 345.6 \times 10^3$Kg, $K_i = 340.4 \times 10^3$KN / m, $C_i = 29.37$KN • s / m.

Suppose that active TMD is used, and relevant parameters are: $m_d = 29.63 \times 10^3$Kg, $K_d = 9.387 \times 10^3$KN / m, $C_d = 250$KN • s / m.

Active TMD is placed at the top of building, as shown in Fig.3. Suppose that the structure is subjected to an earthquake, and the acceleration of the ground motion is expressed by:

$$\ddot{X} = g(t)\ddot{\zeta}(t) \tag{26}$$

where g(t) is deterministic function which represent nonstationary property of earthquake:

$$g(t) = \begin{cases} \left(\frac{t}{t_1}\right)^2 & 0 \leqslant t \leqslant t_1 \\ 1 & t_1 < t \leqslant t_2 \\ exp[-e(t-t_2)] & t > t_2 \end{cases} \tag{27}$$

in which $t_1 = 3$s, $t_2 = 13$s, $e = 2.71828$.

The spectrum of the acceleration of the ground motion can be expressed as:

$$S_{\ddot{x}}(\omega,t) = \frac{1 + 4\xi_g^2\left(\frac{\omega}{\omega_g}\right)^2}{\left[1 - \left(\frac{\omega}{\omega_g}\right)^2\right]^2 + 4\xi_g^2\left(\frac{\omega}{\omega_g}\right)^2} S_0 g^2(t)$$

where $\omega_g = 18.85$, $\xi_g = 0.65$, $S_0 = 4.5 \times 10^{-4}$.

Several methods are used for analysis. The results of those methods for the maximum response of the structure and the control force are listed in table 4.

Table 4.

Control Law	Displacement At the Top (cm)	Shear Force At the Bottom (KN)	Active Control Force (KN)
No Control Force	4.10	2506	
Yang's Method	1.61	1075	250
Present Method	1.82	1247	282

It is easy to see that the results obtained by the present method for prediction of a real time control law and the method proposed by J. N. Yang are very close . Both of them control seismic–response effectively.

Ⅳ. Control structural whipping effect by mass pump

We analyse the result of control of structural whipping effect by using mass pump. The shear type structure is shown in Fig.4; its parameters are given as follows:

$$k_1 = 116.9 \times 10^6 N/m, \quad m_1 = 43316.0Kg,$$
$$k_2 = 93.6 \times 10^6 N/m, \quad m_2 = 44002.0Kg,$$
$$k_3 = 70.17 \times 10^6 N/m, \quad m_3 = 42924.0Kg,$$
$$k_4 = 46.84 \times 10^6 N/m, \quad m_4 = 38024.0Kg,$$
$$k_5 = k_6 = 0.347 \times 10^6 N/m, \quad m_5 = m_6 = 490.0Kg.$$

The first natural frequency of the main structure $\omega_1 = 16.44$, and the natural frequencies of the independent upper structure (that is to say, fix the main structure) $\omega_1 = 16.44$, $\omega_2 = 43.04$. Therefore, the first natural frequencies of the main structure and the indepedent upper structure is the same, which means structure prone to manifest the effect of whipping when subjected to the earthquake.

The first three natural frequencies and corresponding mode shapes are:

$$\omega_1 = 15.53, \ \omega_2 = 17.32, \ \omega_3 = 40.70,$$
$$\{A\}_1 = \{0.021; 0.046; 0.071; \ 0.094; \ 0.659; \ 1.000\}^T,$$
$$\{A\}_2 = \{0.023; 0.049; 0.074; \ 0.091; -0.576; -1.000\}^T,$$
$$\{A\}_3 = \{0.331; 0.491; 0.194; -0.545; -1.339; \ 1.000\}^T,$$

Suppose the damping ratio of every mode is 0.02, then damping of each story will be $c_i = 49299.0Kg/s$ $(i = 1,...4)$, and $c_5 = c_6 = 466.3Kg/s$[9].

The mass pump is supported on the 4th and 6th floors, as shown in Fig.5. The state equation of structure subjected to earthquake is:

$$\dot{X}(t) = AX(t) + W(t)$$

Using Eq.(26) to represent the earthquake, the seismic response of the structure can be computed by numerical methods. We compute four cases, which are shown in table 5.

1. Without mass pump
(1) parameters of the structure are shown in Fig. 4.
(2) increase the stiffness of the upper structure by 40 percent, that is $k_5 = k_6 = 0.486 \times 10^6 N/m$.
(3) decrease the stiffness of the upper structure by 20 percent, that is $k_5 = k_6 = 0.278 \times 10^6 N/m$.

2. Mass pump is supported at the 4th and 6th floor as shown in fig.5
Let $\alpha = 45^0$, length of tube $L = 16.0m$, area of tube $a = \pi(0.03)^2 m^2$.
(1) $\beta = 100$; (2) $\beta = 200$; (3) $\beta = 50$;

3. Mass pump is supported at the 4th and 5th floor as shown in fig.5

(1) $\beta=100;\ \alpha=45^0,\ L=10.0m,\ a=\pi(0.03)^2m^2$.
(2) $\beta=100;\ \alpha=60^0,\ L=14.4m,\ a=\pi(0.03)^2m^2$.
(3) $\beta=200;\ \alpha=45^0,\ L=10.0m,\ a=\pi(0.03)^2m^2$.

Table 5.

floor response parameters		1		2		3		4		5		6	
		MD	SD	MD	SD	MD	SD	MD	SD	MD	SD	MD	SD
case 1	(1)	0.410	0.138	0.446	0.155	0.434	0.158	0.359	0.134	7.248	3.536	4.678	2.182
	(2)	0.509	0.192	0.563	0.217	0.582	0.224	0.502	0.187	4.258	1.499	2.617	0.884
	(3)	0.504	0.182	0.543	0.202	0.527	0.204	0.436	0.158	7.257	2.799	4.761	1.817
case 2	(1)	0.424	0.176	0.476	0.197	0.486	0.202	0.405	0.164	0.709	0.241	0.681	0.235
	(2)	0.424	0.176	0.480	0.197	0.490	0.200	0.406	0.164	0.697	0.241	0.703	0.239
	(3)	0.424	0.176	0.476	0.197	0.486	0.202	0.405	0.164	0.746	0.241	0.648	0.235
case 3	(1)	0.411	0.179	0.468	0.200	0.486	0.202	0.391	0.167	0.110	0.045	1.251	0.488
	(2)	0.411	0.179	0.468	0.200	0.486	0.202	0.391	0.167	0.131	0.055	1.239	0.489
	(3)	0.411	0.179	0.468	0.200	0.486	0.202	0.391	0.167	0.125	0.055	1.230	0.488
case 4		0.459	0.182	0.507	0.202	0.515	0.207	0.432	0.167				

4. Without upper structure and mass pump

We can obtain important conclusions from the computational results mentioned above.

(1) The structural whipping effect is obvious; the maximum displacement of the upper structure (MD) and the standard deviation (SD) are over ten times of those of the main structure. Though this result was obtained based on an assumption of the elastic response, the structure would enter the plastic state and failure when subjected to the earthquake with same intensity. Even if one increases the stiffness of the upper structure by 40 percent or decreases it by 20%, the whipping effect of structure subjected to earthquake is only reduced a certain degree, and there is some possibility of an increase in the structural response. Meanwhile, the maximum displacement of a story of the main structure will increase, that is to say, this change reduces the effect of reducing vibration of the upper structure on the main structure.

(2) If a mass pump is placed on the upper structure, structural whipping effect will be reduced greatly in any case. In case 2, which shown in Fig.5, the whipping effect almost disappears, and the effect on the main structure is very small. Therefore, it is very effective to use a mass pump to reduce structural whipping effect, especially for those structures with projected upper structure.

(3) The seismic response of the main structure without an upper structure is larger than that of the main structure with an upper structure, which means an upper structure has the effect of reducing the vibration of the main structure. The installation of a mass pump will not reduce this effect.

(4) The effect of reducing vibration of the mass pump supported at the 4th and 6th floor will be better than that supported at the 4th and 5th floor.

(5) When a mass pump is installed in the upper structure, the structural

response will change slightly if some parameters change slightly, which shown in table 5 (case 2 and case 3).

V. Forecast of research of structural control

There are more and more countries which have applied device for the isolation, absorption of vibration and dissipation of energy to multi–storey buildings and bridges during recent years. There are, at least, over ten countries using devices of isolation on earthquake–resistant structures, and over 200 engineerings of this kind have been built according to imperfect statistics. New Zealand has collected item of isolation in its code of structural design. France is engaged in standardization of rubber and its use in earthquake–resistant structures as well as formulation of an improved criterion for design of nuclear power station.

The positive control device such as the device of isolation, absorption of vibration and dissipation of energy is expected to be used more widely in many countries in near future as well as a period of considerable length. And more and more countries will formulate their own criteria in this field.

Many countries will continue their research on positive control, and begin with research on active and semi–active control. Some active and positive control devices will be installed in the buildings located in regions of earthquakes or strong winds. Many simulation experiments will be carried out in wind tunnels or shake tables.

National Natural Science Foundation has entrusted us to compile a research outline about "Structural Control and Controlled Structure", and has entrusted us to convene a symposium on this subject. We suggest that this research project be divided into several items, including active control, positive control, combination of active, positive and semi–active control, optimal design of controlled structures subjected to earthquake or strong gustiness, and expert system for structural control and controlled structures.

We will concentrate on research of "structural control and controlled structure", and make some experimental buildings in cooperation with some construction units. We think it is better to investigate the research of positive, semi–active control and combinations of positive and active control which need little energy. In any case, the positive control may be better for multi–storey buildings, which are widely used in engineering. But, the combination of positive and active control may be suitable for tall buildings, high–rise structures and large span structures.

References

[1]. Li Guiqing, Zhou Zujun, "Review of Structural Vibration Control", Earthquake Engineering and Engineering Vibration, No.1, 1987.

[2]. Li Guiqing, "Structural Control and Controlled Structure", Journal of Wuhan University of Technology, No.4, 1987.

[3]. Li Giuqing, Zhou Zujun, "Analysis of Vibration of Shear–type Structures with control device of the mass pump", Journal of Wuhan University of Technology, No.4, 1990.

[4]. Li Guiqing, Xiong Huoqing, "Control of structural Whipping effect by the mass pump", Earthquake engineering and engineeering vibration, No.4, 1987.

[5]. Li Guiqing, Zhou Zujun,"Analysis of Vibration Control of Frame with device of isolation", Journal of Wuhan Institute of Urban Construction, No.2, 1989.

[6]. Qu Weilian, Li Guiqing, "Design of reducing vibration for tall framed buildings installed with artificial hysteresis hinges", the proceedings of 10th national symposium on tall buildings and structures, June, 1986.

[7]. Qu Weilian, Li Guiqing, "Design method of wind–resistant tall buildings using several control devices", Journal of Buildings and Structures, No.1,1990.

[8]. Li Guiqing, 《Reliability of Structures》, Press of Wuhan University of Technology, 1989.

[9]. Li Guiqing, 《 Computational Theory and Methods of Earthquake –Resistant Structures》 , Press of Earthquake, 1985.

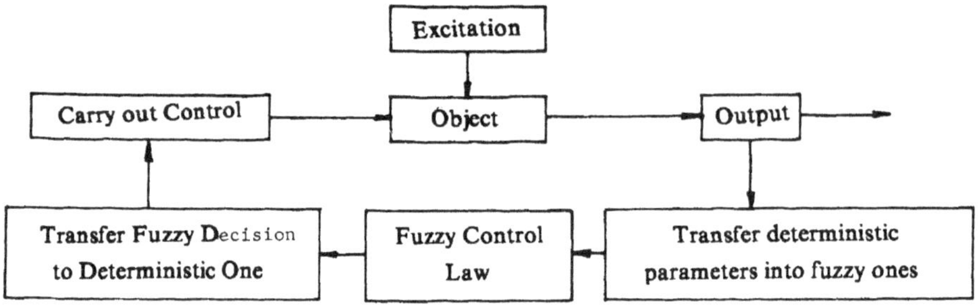

Fig.1

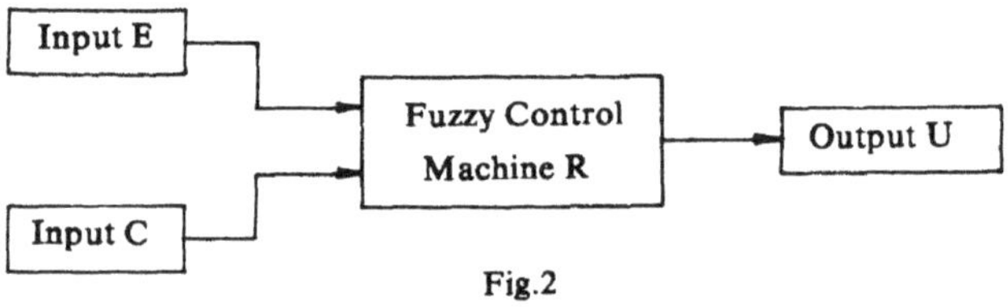

Fig.2

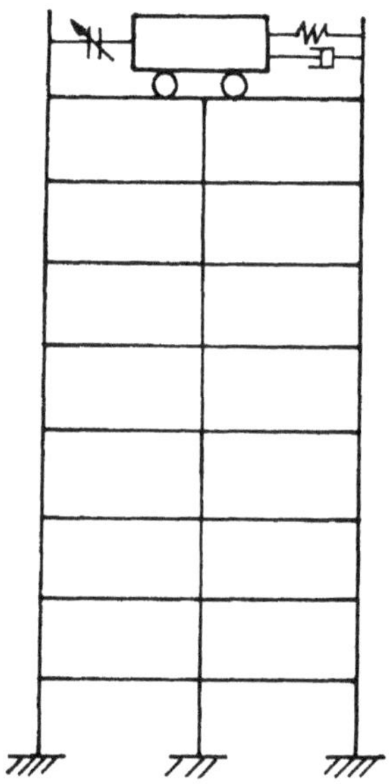

Fig.3

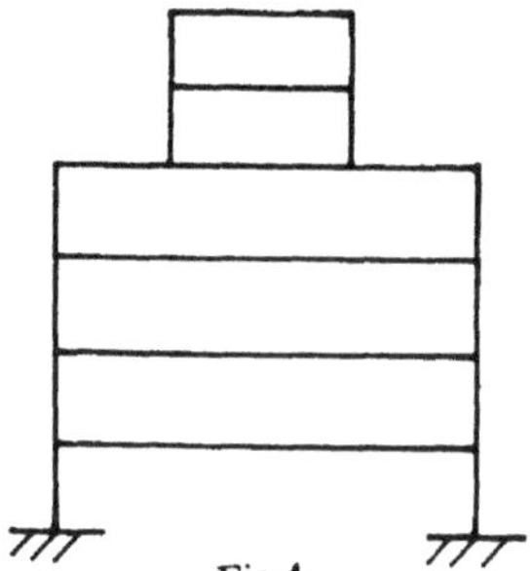

Fig.4

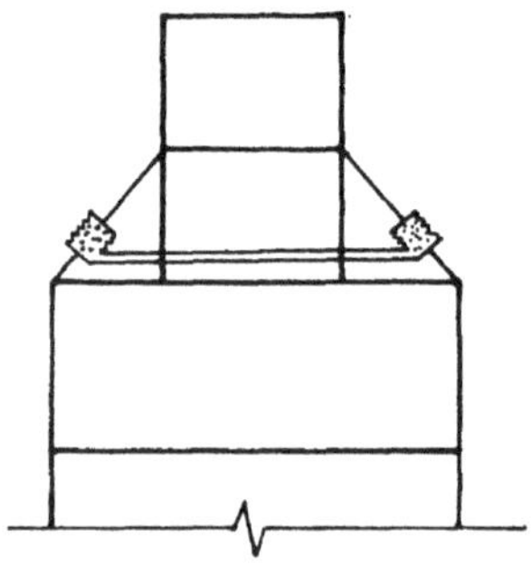

Fig.6

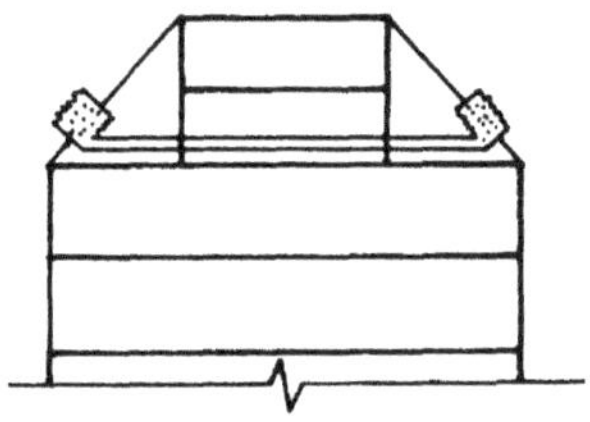

Fig.5

CONTROL OF COUPLED BENDING AND TORSIONAL VIBRATION AND MOTION OF BEAMS WITH SHEAR EFFECT

Xu Bohou and Yu Yimei
Department of Mechanics, Zhejiang University, Hangzhou 310027, China

ABSTRACT

Some problems of control of vibration and motion of flexible beams are considered in this paper; mode analysis, control scheme and observation spillover. The beam equation is Timoshenko's, while the visco—elasticity constitutive relationship obeys an integral equation with exponential kernel instead of Voight's equation. The velocity at the beam tip is observed as a feedback variable to reduce the observation spillover effect.

1. A SHORT SURVEY OF CONTROL OF ELASTIC BODIES

Control of elastic bodies is to control positions and suppress vibrations of elastic systems by controlling means. Since the middle of 60's, when the active control was introduced into elastic structures, some progress has been achieved in this field.

1. The main aim is to suppress vibrations. There are a lot of examples in mechanical engineering, civil engineering, aeronautical engineering, etc., for example, to suppress the vibration of a bridge caused by any exciting. Because the aim of this kind of control is rather simple, the problems are often simplified to be some finite dimensional ones, then solved as the control problems of finite dimensional systems. The main work of these problems is focused at two aspects; one is to increase the effect of suppressing vibrations by modifying the parameters and structures of the control system, the other is to reduce the active control energy to enlarge the application range. Chinese scientists have given some results about this field [1]-[3].

2. It is needed in control of elastic bodies not only the suppression of vibrations but also high location demand including position, attitude and shape along with the research need of large space structures and flexible manipulators. Now we could not first simplify our problems to finite dimensional ones; we must di-

rectly deal with the elastic bodies themselves instead (i. e. , we must deal with the partial differential equations describing the deformation of elastic bodies), because from the view of control, finite and infinite dimensional systems have essential difference. For example, for finite dimensional controllable system, we can design a state feedback which makes the closed—loop system with preassigned degree of steadiness; but for an infinite dimensional system, this is usually impossible [5]. In another aspect, the control and observation in practice are always finite dimensional; thus, we have to deal with the coupled finite and infinite dimensional problems. In addition, we often meet with special problems such as "frequency stacking", "rigidity having wide gap",etc. which are rather difficult not only in mechanics but also in control theory. In China, a group of mathematicians began their work in the middle of 70's and gained some good results in the suppression of elastic beams[6]-[8]. Because of complexity of these problems, it is necessary to understand the dynamic properties of elastic bodies in order to obtain some essential results of the control of elastic bodies, which needs the participation of scientists in mechanics.

Recently we have done work about manipulator control as following.

II. DYNAMIC MODEL OF FLEXIBLE BEAMS

We consider a flexible beam which lies in a horizontal plane, one end of the beam fixed at the shaft of a motor, the other end is free and stuck with a rigid body. When the motor rotates, the beam and the rigid body rotate in the plane. Because the mass center of the rigid body is not on the centroidal line of the beam, the beam has not only bending, but also torsional effects. Our purpose is to make the rigid body moving according to the preassigned demand. Similar problem has been discussed in [9], and the main difference from that paper is:

(1) We will take the shearing effect in account, namely use the Timoshenko's model, while the authors of [9] use the Euler — Bernoulli's one.

(2) We will consider a kind of exponential viscous model, instead of the Voight's model.

(3) We will use the velocity feedback at the tip of beam ($x=l$) instead of the curvature feedback at the joint of beam and motor.

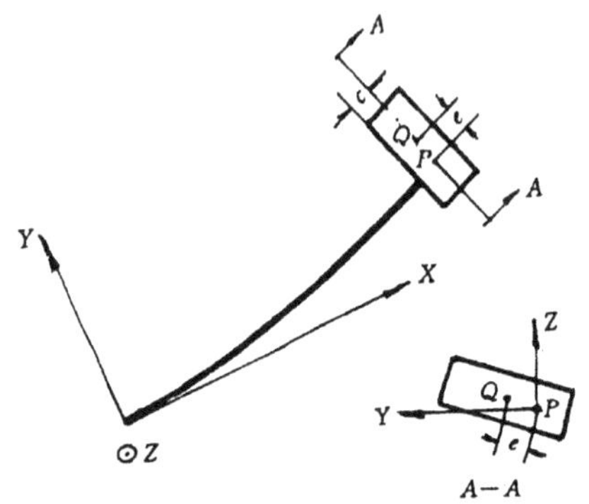

Figure 1 Flexible beam with a tip body

One knows that the Euler — Bernoulli's beam model deduced based upon the Kirchhoff's hypothesis is only suitable to the case of low vibration frequency of slender flexible beam; but on the other hand, the exact beam equation inferred by the Hamiltonian principle contains $\dot{\theta}^2$(here $\dot{\theta}$ is the angle velocity of the motor)[10], and the condition of neglecting the terms including $\dot{\theta}^2$ is that $\dot{\theta}$ is far below the vibration frequency of beam; therefore, the equation in [9] can be only used in the case of very small $\dot{\theta}$. In practice, when $\dot{\theta}$ is fairly large, engineers usually stiffen the beam, and it is needed to use more precise beam model. This is the reason for using the Timoshenko's model.

In vibration of elastic bodies, damping is rather important because in practice, one can only adopt finite dimensional control for elastic system (in this paper we only control the rotation moment of the motor); if the damping effect is neglected, it cannot be stabilized uniformly for the system[5]. Usually, people are used to the Voight's model (as in [9]), of which viscous force is proportional to the velocity; but when the viscosity of material is large (such as some high polymer material) or the frequ-ency is rather high, we will deduce overdamping vibration by the Voight's model , which is never observed in experiments. So many control experts develop some new visco-elastic model to modify the Voight's one. In this paper we suggest a kind of exponential model and the reason is the following:

The linear visco-elastic constitutive equation for infinite small stretching [11] is

$$\sigma(t) = E\varepsilon(t) + \int_{-\infty}^{t} \Phi(t-\tau)\dot{\varepsilon}(\tau)d\tau \tag{2.1}$$

where $\Phi(t)$ is the relaxing function which has the feature as in Fig. 2. Take $\Phi(t)$ to be δ-function, i. e.,

$$\Phi(t) = 2\eta E\delta(t)$$

then (2. 1) becomes

$$\sigma(t) = E(\varepsilon(t) + 2\eta\dot{\varepsilon}(t)) \tag{2.2}$$

that is the Voight's model. If one takes

$$\Phi(t) = 2\eta\alpha E e^{-\alpha t} \tag{2.3}$$

which can describe the relaxing function better, then

$$\sigma(t) = E(\varepsilon(t) + 2\eta\alpha\int_{-\infty}^{t} e^{-\alpha(t-\tau)}\dot{\varepsilon}(\tau)d\tau) \tag{2.4}$$

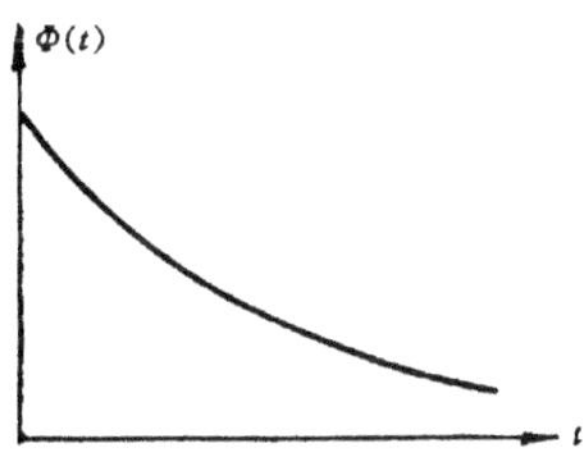

Figure 2 Relaxing function

This is the exponential model we adopt. It is obvious that Eq. (2. 4) becomes Eq. (2. 2) as α goes to infinity. It is easy to be proved that at the low frequency these two models coincide with each other well; at the frequency high (i. e. $\omega \gg \eta^{-1}$), the over-damping result can be deduced from the Voight's model, but not from the exponential one.

Similarly, the constitutive equation for shearing is

$$\tau(t) = G(\gamma(t) + 2\eta'\alpha\int_{-\infty}^{t} e^{-\alpha(t-\tau)}\dot{\gamma}(\tau)d\tau) \tag{2.5}$$

Usually $\eta \neq \eta'$; but at not very high frequency, the difference between them is about 10%, so we take $\eta = \eta'$ for convenience.

The main reason why we use the velocity feedback at $x = l$ instead of the deformation curvature is that our aim is to control the motion at $x = l$, and that we want to eliminate the spillover besides.

Now we deduce the equation without viscous effect. For convenience, we assume that the beam has uniform cross section, and that the bending center and torsional center lie on the centroidal axis. Assume $w(t,x)$ and $\psi(t,x)$ to be the transverse deformation function and normal turning function of the beam in the rotating frame XYZ, $\varphi(t,x)$ be the twisting angle (Fig. 1), then

$$\rho\frac{\partial^2 w}{\partial t^2} - C\frac{\partial}{\partial x}(\frac{\partial w}{\partial x} - \psi) = -\rho x\ddot{\theta} \tag{2.6}$$

$$\rho r_{\cdot}^2\frac{\partial\psi}{\partial t^2} - C(\frac{\partial w}{\partial x} - \psi) - D\frac{\partial^2\psi}{\partial x^2} = 0 \tag{2.7}$$

$$\rho x^2\frac{\partial\varphi}{\partial t^2} - G_J\frac{\partial^2\varphi}{\partial x^2} = 0 \tag{2.8}$$

where ρ is the mass density per unit length of the beam, C, D and G_J are the shear, bending and torsional rigidities respectively, ρr_0^2 is the moment of inertia, $\rho\kappa^2$ is the polar moment of inertia, $\ddot{\theta}$ is the angle acceleration of the motor. According to the previous discussion we have neglected the terms containing $\dot{\theta}^2$ in the equations.

The boundary condition at $x = 0$ is

$$w = \psi = \varphi = 0 \tag{2.9}$$

and the boundary condition at $x = l$ is (l is the length of the beam)

$$m\frac{\partial^2}{\partial t^2}[w + c\psi + e\varphi] + C(\frac{\partial w}{\partial x} - \psi) = -m(l + c)\ddot{\theta} \tag{2.10}$$

$$mc\frac{\partial^2}{\partial t^2}[w + c\psi + e\varphi] + J_0\frac{\partial^2 \psi}{\partial t^2} + D\frac{\partial \psi}{\partial x} = -mc(l + c)\ddot{\theta} \tag{2.11}$$

$$me\frac{\partial^2}{\partial t^2}[w + c\psi + e\varphi] + J_g\frac{\partial^2 \psi}{\partial t^2} + G_J\frac{\partial \psi}{\partial x} = -me(l + c)\ddot{\theta} \tag{2.12}$$

Eq. (2.10) indicates the force equilibrium in Y direction, and (2.11) and (2.12) indicate the moment equilibrium in Z and X directions respectively. m, J_0 and J_g are the mass of the rigid body, moments of inertia in Z and X directions respectively.

Now we express Eq. (2.6)—(2.8) and B. C. (2.10)—(2.12) in the form of matrix. Denote

$$\underline{u} = [u_1\ u_2\ u_3\ u_4\ u_5\ u_6]^T$$
$$= [\omega(t,x)\ \psi(t,x)\ \varphi(t,x)\ \omega(t,l)\ \psi(t,l)\ \varphi(t,l)]^T,$$

$$\Lambda = \begin{bmatrix} \Lambda_1 & \\ & \Lambda_2 \end{bmatrix}, \quad \Lambda_1 = \begin{bmatrix} \rho & & \\ & \rho r_0^2 & \\ & & \rho\kappa^2 \end{bmatrix}, \quad \Lambda_2 = \begin{bmatrix} m & mc & me \\ mc & mc^2 + J_0 & mce \\ me & mec & me^2 + J_g \end{bmatrix},$$

$$\Sigma = \begin{bmatrix} \Sigma_1 & 0 \\ \Sigma_2 & 0 \end{bmatrix},$$

$$\Sigma_1 = \begin{bmatrix} -C\frac{\partial^2}{\partial x^2} & C\frac{\partial}{\partial x} & 0 \\ -C\frac{\partial}{\partial x} & C - D\frac{\partial^2}{\partial x^2} & 0 \\ 0 & 0 & -G_J\frac{\partial^2}{\partial x^2} \end{bmatrix}, \Sigma_2 = \begin{bmatrix} C\frac{\partial}{\partial x} & -C & 0 \\ 0 & D\frac{\partial}{\partial x} & 0 \\ 0 & 0 & G_J\frac{\partial}{\partial x} \end{bmatrix}_{x=l},$$

$$\Omega = [-x\ 0\ 0\ -m(l + c)\ -mc(l + c)\ -me(l + c)]^T$$

The equation and boundary condition at $x = l$ are expressed as

$$\Lambda\ \ddot{\underline{u}} + \Sigma\ \underline{u} = \Omega\ddot{\theta} \tag{2.13}$$

where $\ddot{\underline{u}}$ denotes the second order derivative of $\underline{u}$ with respect to time t.

Obviously, matrix Λ is symmetric and positive definite; therefore, Eq. (2.13) becomes

$$\ddot{\underline{u}} + \Gamma\underline{u} = \Phi\ddot{\theta} \tag{2.14}$$

where $\Phi = \Lambda^{-1}\Omega, \Gamma = \Lambda^{-1}\Sigma$ are differential operators in Hilbert's space

$$\mathscr{H} = L^2(0,l) \times L^2(0,l) \times L^2(0,l) \times R^3$$

defined in a subspace of $\mathscr{H}$

$$H = \{\underline{u} \in \mathscr{H} \mid u_i \in H(0,l), u_{i+3} \in R, u_i(0) = 0,$$
$$u_{i+3} = u_i(l), i = 1,2,3\}. \tag{2.15}$$

The inner product of $\mathscr{H}$ is

$$\langle \underline{u}, \underline{v} \rangle_{\mathscr{H}} = \int_0^l \underline{u}_1^T \Lambda_1 \underline{v}_1 dx + \underline{u}_2^T \Lambda_2 \underline{v}_2 \tag{2.16}$$

where

$$\underline{u}_1 = [u_1\, u_2\, u_3]^T, \quad \underline{u}_2 = [u_4\, u_5\, u_6]^T, \quad \underline{v}_1 = [v_1\, v_2\, v_3]^T, \quad \underline{v}_2 = [v_4\, v_5\, v_6]^T.$$

III. MODE EXPRESSION OF OPERATOR Γ

It is easily proved that Γ is self-adjoint and positive definite and has compact inverse. From Hilbert-Schmidt theory, we have

Theorem 1. The eigenvalues and eigenfunctions of Γ have the following properties:

(1) The eigenvalues of Γ are positive, countable, and do not exist any finite limit points.

(2) For each eigenvalue there are only finite number of linearly independent eigenfunctions.

(3) The set of eigenfunctions forms a set of completed orthonormal bases of H.

We solve the eigenvalue problem

$$\Gamma \underline{u} = \lambda \underline{u} \tag{3. 1}$$

which contains two parts: equations corresponding Eq. (2. 6)—(2. 8) and boundary condition **corresponding** B. C. (2. 10)—(2. 12). The solution of the equations with B. C. (2. 9) is

$$w = C_1[a_1\beta_2 + a_2\beta_2^3)sin\beta_1 x\text{-}(a_1\beta_1\text{-}a_2\beta_1^3)sh\beta_2 x] + C_2[cos\beta_1 x\text{-}ch\beta_2 x] \tag{3. 2}$$

$$\psi = C_1[cos\beta_1 x\text{-}ch\beta_2 x]\text{-}C_2[(a_1\beta_1 + a_2\beta_1^3)sin\beta_1 x + (a_1\beta_2 + a_2\beta_2^3)sh\beta_2 x] \tag{3. 3}$$

$$\varphi = C_3 sin\beta_3 x \tag{3. 4}$$

where

$$a_1 = \frac{C^2 + \lambda\rho D}{C^2\text{-}\lambda\rho r_o^2 C}, \qquad a_2 = \frac{D}{C\text{-}\lambda\rho r_0^2}$$

$$\beta_{1.2} = (\frac{\lambda\rho}{2D})^{1/2}\{[\frac{D}{C} + r_0^2)^2 + \frac{4D(C\text{-}\lambda\rho r_0^2)}{\lambda\rho C}]^{1/2} \pm (\frac{D}{C} + r_0^2)^{1/2}\}^{1/2}$$

$$\beta_3 = (\lambda\rho\chi^2/G_J)^{1/2}$$

If we rewrite (3. 2)—(3. 4) in the form of matrix

$$\underline{u} = P(x,\lambda)\, \underline{C} \tag{3. 5}$$

where $C = [C_1\, C_2\, C_3]^T$, the part of B. C. at $x = l$ of (3. 1) is

$$[\Sigma_2 P(l,\lambda)\text{-}\lambda \wedge_2 P(l,\lambda)]C = 0 \tag{3. 6}$$

Thus, the algebraic equation of λ is obtained

$$det[\Sigma_2 P(l,\lambda)\text{-}\lambda \wedge_2 P(l,\lambda)] = 0 \tag{3. 7}$$

We can solve Eq. (3. 7) by numerical method. Therefore, a series of eigenvalues and eigenfunctions $\{\lambda_n, \underline{v}_n(x)\}$ are given.

IV. CONTROL SYSTEM

Now we introduce the visco-elastic model (2. 4)、(2. 5) to dynamic system (2. 14). Note that we have assumed $\eta = \eta'$; then

$$\ddot{\underline{u}} + 2\eta a\int_{-\infty}^{t} e^{-a(t-\tau)}\Gamma \dot{\underline{u}} d\tau + \Gamma \underline{u} = \Phi\ddot{\theta} \tag{4. 1}$$

If $\underline{u}$ expands the series of eigenfunctions $\underline{v}_n(x)$ in the previous section

$$\underline{u} = \sum_{n=1}^{\infty} \frac{1}{\omega_n^2} u_n(t)\, \underline{v}_n(x) \tag{4.2}$$

where $\omega_n^2 = \lambda_n, n = 1,2,\cdots$, Eq. (4.1) becomes a series of independent integral differential equations

$$\ddot{u} + 2\eta a\omega_n^2 \int_{-\infty}^{t} e^{-a(t-\tau)} u_n(\tau)d\tau + \omega_n^2 u_n(\tau) = -\omega_n^2 \bar{b}_n \ddot{\theta}, \qquad n = 1,2,3,\cdots\cdots, \tag{4.3}$$

where

$$\bar{b}_n = -\langle \Phi, \underline{v}_n \rangle$$
$$= \int_0^l x w_n(x)dx + m(l+c)[w_n(l) + c\psi_n(l) + e\varphi_n(l)]$$

To rewrite (4.3) in the form of state equation, let

$$s_n = \dot{u}_n, \quad t_n = \int_0^t e^{-a(t-\tau)} \dot{u}_n(\tau)d\tau$$

Eq. (4.3) becomes

$$\dot{\underline{x}}_n = M_n \underline{x}_n + N_n \ddot{\theta}, \; n = 1,2,\cdots\cdots \tag{4.4}$$

where

$$\underline{x}_n = [u_n\, s_n\, t_n]^T$$

$$M_n = \begin{bmatrix} 0 & 1 & 0 \\ -\omega_n^2 & 0 & -2\eta a\omega_n^2 \\ 0 & 1 & -a \end{bmatrix}, \quad N_n = \begin{bmatrix} 0 \\ -\omega_n^2 \bar{b}_n \\ 0 \end{bmatrix}$$

First, we consider the eigenvalues γ of M_n. Obviously, γ satisfies

$$\left(\frac{\gamma}{a}\right)^3 + \left(\frac{\gamma}{a}\right)^2 + (1 + 2\eta a)\left(\frac{\omega_n}{a}\right)^2\left(\frac{\gamma}{a}\right) + \left(\frac{\omega_n}{a}\right)^2 = 0 \tag{4.5}$$

By Sturm's theorem and the dissipation of damping, if $\eta a < 1$, Eq. (4.5) has one negative root and a couple of adjoint complex roots with negative real part. Therefore, each mode consists of three vibrations; one overdamping and two damping vibrations. Fig. 3 gives the figure that the degeneration constant -Re(γ/a) varies as ω_n/a varying.

Assume that the first m modes are taken into account only, i.e.,

$$\underline{u}(x,t) = \sum_{n=1}^{m} \frac{1}{\omega_n^2} u_n(t)\, \underline{v}_n(x) \tag{4.6}$$

The controllability matrix for the n-th mode is

$$W_n = [N_n \quad M_n N_n \quad M_n^2 N_n]$$
$$= \begin{bmatrix} 0 & -\omega_n^2 \bar{b}_n & 0 \\ -\omega_n^2 \bar{b}_n & 0 & \omega_n^4 (1 + 2\eta a) \\ 0 & -\omega_n^2 \bar{b}_n & \omega_n^2 \bar{b}_n a \end{bmatrix} \tag{4.7}$$

Obviously, the sufficient and necessary condition that the n'th mode is controllable is that $\bar{b}_n \neq 0$. We suppose that this condition always holds in practice. Furthermore, we have inversable T_n such that

$$T_n^{-1} M_n T_n = D_n = diag(\lambda_{n1}, \lambda_{n2}, \lambda_{n3}) \tag{4.8}$$

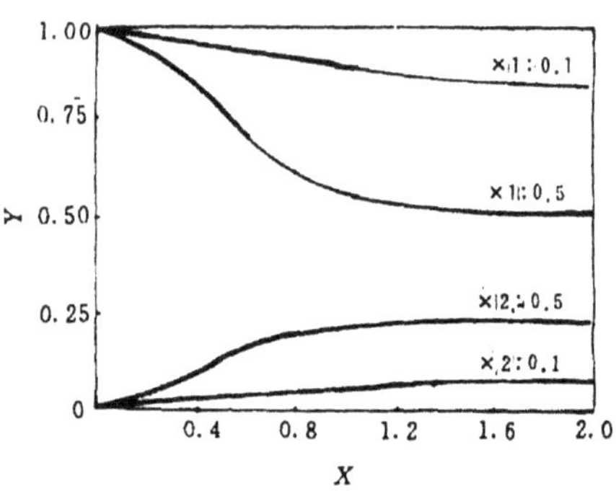

Figure 3 $X = \omega_n/a, Y = Y/a$, curves x1: 0.1, x2: 0.1 are the overdamping and damping vibrations with $\eta a = 0.1$.

and

$$B_n = T_n^{-1} N_n = [b_{n1}\, b_{n2}\, b_{n3}]^T$$

Because the controllability is invariant under similarity transformation, $b_{ni} \neq 0$, $i = 1,2,3$.

Now we consider the problem of controllability of combining the first m modes. Let

$$\underline{x} = [\underline{x}_1^T \cdots \underline{x}_m^T]^T$$

$$M = diag(M_1 \cdots M_m), \qquad N = [N_1^T \cdots N_m^T]^T$$

the control problem is

$$\dot{\underline{x}} = M\underline{x} + N\ddot{\theta} \tag{4.9}$$

With the transformation matrix $T = \text{diag}(T_1, \cdots, T_m)$, system matrix M becomes a diagonal one and control matrix N becomes

$$[B_1^T \cdots B_m^T]^T$$

Calculating the controllability matrix, it is found that the sufficient and necessary condition that (M, N) is controllable is that

(1)Every element of B is not zero;

(2)Every eigenvalue of the first m modes is different from each other.

It is easy to be proved that these conditions are satisfied. Therefore, we can suppress the system by pole allocation or quadratic regulator theory.

The aim of designing the controller is to make the motion at $x = l$ coincide with the preassigned $r(t)$, let

$$\tilde{\theta} = \theta(t) - r(t), \qquad \tilde{\gamma} = \dot{\theta}(t) - \dot{r}(t)$$

the state equation (4.9) expands into

$$\dot{\underline{y}} = A\underline{y} + B\ddot{\tilde{\theta}} + D\ddot{r} \tag{4.10}$$

where

$$\underline{y} = [\tilde{\theta} \quad \tilde{\gamma} \quad \underline{x}^T]^T$$

$$A = \begin{bmatrix} 0 & 1 & \\ 0 & 0 & \\ & & M \end{bmatrix}, \quad B = \begin{bmatrix} 0 \\ 1 \\ N \end{bmatrix}, \quad D = \begin{bmatrix} 0 \\ -1 \\ 0 \end{bmatrix} \tag{4.11}$$

Similarly, it can be proved that (A, B) is controllable. In fact, $\ddot{r}(t) = 0(\dot{r}(t)$ is regarded as step function); we can make $\underline{y}$ go to zero with a preassigned speed by controlling $\ddot{\tilde{\theta}}$ i. e. the rotating moment of motor.

We choose the velocity at $x = l$, $v(t)$, to observe, i. e.

$$v(t) = l\dot{\theta} + \dot{w}(l,t)$$

or

$$\begin{aligned} Z(t) &= v(t) - \dot{r}(t) \\ &= l\tilde{\gamma}(t) + \sum_{i=1}^{m} \frac{1}{\omega_i^2} s_i(t) w_i(l) \end{aligned} \tag{4.12}$$

because $r(t)$ is known. Besides, $\tilde{\theta}$ and $\tilde{\gamma}$ are known, the corresponding matrix of observation is

$$C = \begin{bmatrix} 1 & 0 & & & 0 & & \\ 0 & 1 & & & & & \\ 0 & l & 0 & w_1(l)/\omega_1^2 & 0 \cdots\cdots 0 & w_m(l)/\omega_m^2 & 0 \end{bmatrix} \tag{4.13}$$

With similar analysis in controllability, if and only if

$$w_i(l) \neq 0, \qquad i = 1, 2, \cdots, m \tag{4.14}$$

(A, C) is observable. We also suppose (4.14) hold in practice. Note that $\tilde{\theta}$, $\tilde{\gamma}$ are known; one only need to construct observer in the subspace expanded by $\underline{x}$ (see (4.11)).

V. SPILLOVER PROBLEM

Let $n > m$, and the first $3m + 2$ state variables (including two variables of rigid motion) be the control part, the other $3(n - m)$ variables be the residual part. Separate the rigid motion from others in the notation of (4.11); the state equations are

$$\dot{\underline{x}}_0 = M_0 \underline{x}_0 + N_0 \ddot{\theta} \tag{5.1}$$

$$\dot{\underline{x}}_C = M_C \underline{x}_C + N_C \ddot{\theta} \tag{5.2}$$

$$\dot{\underline{x}}_R = M_R \underline{x}_R + N_R \ddot{\theta} \tag{5.3}$$

$$z = C_C \underline{x}_C + C_R \underline{x}_R \tag{5.4}$$

where

$$\underline{x}_0 = [\theta \quad \dot{y}]^T, N_0 = [0 \quad 1]^T, \quad M_0 = \begin{bmatrix} 0 & 1 \\ 0 & 0 \end{bmatrix},$$

$$C_C = [0 \quad w_1(l)/\omega_1^2 \quad 0 \cdots 0 \quad w_m(l)/\omega_m^2 \quad 0]$$

$$C_R = [0 \quad w_{m+1}(l)/\omega_{m+1}^2 \quad 0 \cdots 0 \quad w_n(l)/\omega_n^2 \quad 0]$$

In the other notation of (5.1)—(5.4), the subscript C relates with the control part, while the subscript R with residual one. Here we assume that M_R has sufficient stabilization degree.

Construct the observer

$$\dot{\underline{w}}_C = (M_C - G_C C_C)\,\underline{w}_C + G_C(z - C_R \underline{w}_R) + N_C \ddot{\theta} \tag{5.5}$$

$$\dot{\underline{w}}_R = M_R \underline{w}_R + N_R \ddot{\theta} \tag{5.6}$$

Where G_C is a column vector such that $M_C - G_C C_C$ has sufficient degenerate ratio. Subtract (5.2), (5.3) from (5.5), (5.6) respectively and denote $\underline{e}_C = \underline{x}_C - \underline{w}_C, \underline{e}_R = \underline{x}_R - \underline{w}_R$, then

$$\dot{\underline{e}}_C = (M_C - G_C C_C)\,\underline{e}_C - G_C C_R \underline{e}_R \tag{5.7}$$

$$\dot{\underline{e}}_R = M_R \underline{e}_R \tag{5.8}$$

In fact we do the feedback proportional to $\underline{x}_0$ and $\underline{x}_C$, i. e.

$$\ddot{\theta} = -K_1 \underline{x}_0 - K_2 \underline{x}_C + K_2 \underline{e}_C \tag{5.9}$$

the equation containing $\underline{x}_0, \underline{x}_C, \underline{x}_R, \underline{e}_C, \underline{e}_R$ is

$$\begin{bmatrix} \dot{\underline{x}}_0 \; \dot{\underline{x}}_C \\ \dot{\underline{x}}_R \\ \dot{\underline{e}}_C \\ \dot{\underline{e}}_R \end{bmatrix} = \begin{bmatrix} M_0 - N_0 K_1 & -N_0 K_2 & 0 & N_0 K_2 & 0 \\ -N_C K_1 & M_C - N_C K_2 & 0 & N_C K_2 & 0 \\ -N_R K_1 & -N_R K_2 & M_R & N_R K_2 & 0 \\ 0 & 0 & 0 & M_C - G_C C_C & -G_C C_R \\ 0 & 0 & 0 & 0 & M_R \end{bmatrix} \begin{bmatrix} \underline{x}_0 \\ \underline{x}_C \\ \underline{x}_R \\ \underline{e}_C \\ \underline{e}_R \end{bmatrix} \tag{5.10}$$

The eigenvalues of system (5.10) consist of ones of the controller, observer and residual part, i. e. the choice of K_1, K_2, G_C does not change the stability of residual part.

But this conclusion is based on the observer (5.5) and (5.6). Because we often could not make M_R, N_R and C_R precise enough in fact, it is difficult to use the observer (5.5) and (5.6). We usually replace (5.5) by

$$\dot{\underline{w}}_C = (M_C - G_C C_C)\,\underline{w}_C + G_C z + N_C \ddot{\theta} \tag{5.11}$$

therefore, Eq. (5.10) becomes

$$\begin{bmatrix}\dot{\underline{x}}_0\\ \dot{\underline{x}}_C\\ \dot{\underline{x}}_R\\ \dot{\underline{e}}_C\\ \dot{\underline{e}}_R\end{bmatrix}=\begin{bmatrix}M_0-N_0K_1 & -N_0K_2 & 0 & N_0K_2 & 0\\ -N_CK_1 & M_C-N_CK_2 & 0 & N_CK_2 & 0\\ -N_RK_1 & -N_RK_2 & M_R & N_RK_2 & 0\\ 0 & 0 & -G_CC_R & M_C-G_CC_C & 0\\ 0 & 0 & 0 & 0 & M_R\end{bmatrix}\begin{bmatrix}\underline{x}_0\\ \underline{x}_C\\ \underline{x}_R\\ \underline{e}_C\\ \underline{e}_R\end{bmatrix} \tag{5. 12}$$

For the existence of C_R, the set of (5. 12) will be different from that of (5. 10); especially, the eigenvalues of the residual part will change, which may cause unstability[13]. Usually we will choose C_R as small as possible to reduce effect of the observer to the residual part, i. e. observer spillover.

VI. NUMERICAL EXAMPLES

There are two examples in this section; one is chosen from [9] to compare the result here with that of [9]; the other is obtained by stiffening the rigidity of the beam in Example 1.

Example 1. See Fig. 1, the parameters are

$$l = 0.825m, A = 0.05 \times 0.02, \rho = 0.777kg/m,$$
$$E = 2.06 \times 10^{11}N \cdot m^{-2}, \nu = 0.32, \eta = 2.5 \times 10^{-4}, \alpha = 400$$
$$c = 0.0m, e = 0.111m, m = 1.13kg, J_0 = 1.14 \times 10^{-4}kg \cdot m^2$$
$$J_R = 6.62 \times 10^{-5}kg \cdot m^2$$

where A is the area of the cross section of the beam. The shear and bending rigidities are calculated as

$$C = 6.634 \times 10^6 N, \qquad D = 6.87N \cdot m^2$$

so the effect of shearing is very small in the frequency range we consider. The lowest (circle) frequency $\omega_1 = 5.276s^{-1}$; the result of this example is limited in the case that $\dot{\theta} << \omega_1$, that means the speed of rotation far less than unit Hz.

We take $m = 2$, $n = 4$. From (4. 13) the nonzero elements of the observation matrix C are l and $w_i(l)/\omega_i^2$, $i = 1,\cdots,4$; in [9], the observed variable is the deformation curvature at x=0; the nonzero elements of the observation matrix C are $w_i''(0)/\omega_i^2$, $i = 1,\cdots,4$. Those two groups of values as below

	i=1	i=2	i=3	i=4
$wi(l)/\omega_i^2$:	3. 1711E-2	-4. 6352E-4	-1. 0407E-4	-7. 0193E-6
$w_i''(o)/\omega_i^2$:	1. 4245E-1	-5. 3022E-3	8. 6558E-3	-3. 4866E-3

It is easy to be found that $w_i(l)/\omega_i^2$ is much less than $w_i''(0)/\omega_i^2$, i. e. the velocity feedback can reduce the influence of the observation spillover.

We design the feedback matrix K of controller by the quadric controller theory with exponential function as weight[14], and the observer's one G_C by pole allocation which makes the observer have a little more degeneration than that of the controller. Take $\sigma = 1.5$; K and G_C are

$$K = [22.2, 15.5, -2.31, -0.943E\text{-}2, -0.169E\text{-}3, 1.85, 0.967E\text{-}1, -0.682E\text{-}1]$$
$$G_C = [-1100, 500, 1.3, 300, -6000, -35]$$

TABLE 1

Complex eigenvalues of the controller, observer and residual part

λ_C		λ_O		λ_R	
λ	$\bar{\lambda}$	λ	$\bar{\lambda}$	λ	$\bar{\lambda}$
−0. 496E1	−0. 497E1				
−0. 326E1	−0. 326E1				
−0. 354E1 ±0. 503E1i	−0. 355E1 ±0. 503E1i	−0. 439E1 ±0. 287E2i	−0. 430E1 ±0. 291E2i	−0. 134E1 ±0. 745E2i	−0. 683E0 ±0. 741E2i
−0. 400E3	−0. 400E3	−0. 400E3	−0. 400E3	−0. 397E3	−0. 397E3
−0. 323E1 ±0. 387E2i	−0. 321E1 ±0. 392E2i	−0. 530E1 ±0. 418E2i	−0. 605E1 ±0. 408E2i	−0. 963E1 ±0. 227E3i	−0. 964E1 ±0. 227E3i
−0. 399E3	−0. 399E3	−0. 399E3	−0. 399E3	−0. 380E3	−0. 381E3

In Tab. 1 the eigenvalues of Eq. (5. 10) and Eq. (5. 12) are listed, which denote ($\lambda_C, \lambda_O, \lambda_R$) and ($\bar{\lambda}_C, \bar{\lambda}_O, \bar{\lambda}_R$), respectively. From this table, one can see that the observer (5. 11) makes the degeneration coefficient of whole system decrease from 1. 34 to 0. 683.

Example 2. We replace the cross section A by 0. 004×0. 1 m^2 , the viscous constant η by 2. 5×10^{-5} only in Example 1.

The shearing and bending rigidities C, D are

$$C = 2.654 \times 10^7 N, \qquad D = 6.867 \times 10^3 N \cdot m^2$$

Obviously, the shearing effect is greater than in Example 1. In fact the first four frequency of the two beam models are

Euler-Bernoulli: 0. 4556E3, 0. 2885E4, 0. 4526E4, 0. 1184E5

Timoshenko: 0. 4522E3, 0. 2829E4, 0. 4322E4, 0. 1030E5

There is distinct difference between these two models.

Tab. 2 gives the first four complex frequency of the two viscous models;

TABLE 2

Complex Frequency of Different viscous models

Exponential	Voight'
−0. 2241E1±0. 4548E3i	−0. 5113E1±0. 4522E3i
−0. 3849E1±0. 2857E4i	−0. 2001E3±0. 2822E4i
−0. 3890E1±0. 4364E4i	−0. 4669E3±0. 4296E4i
−0. 3916E1±0. 1041E5i	−0. 2653E4±0. 9955E4i

There is obvious difference in the real part of two models; the degeneration ratio of (2. 4) is much smaller than that of (2. 2).

VII. RESEARCH SUBJECTS

We want to choose the following subjects between the mechanics and control theory:

1. Choice of new discretization methods

There are a lot of discretization methods such as finite element method, mode analysis method, etc., but they usually do not meet the requirement for two reasons. Firstly, we always wish that the control dimension is few, but the control precision is high. Secondly, we often want to avoid the troubles caused by control and observation spillover. The new discretization method should have a greater improvement in those two aspects.

2. The problem of frequency stacking

There are often very low and stacking natural frequency in a very flexible structure (i. e. there are many frequency in a very narrow range), which cause many problems to dynamic analysis and design of the control system. How to use some new control method and solve these problems is an interesting subject.

3. Analysis and control of the structure with wide gap of rigidity

This is a very realistic problem, which has caused some serious failure, for example, one of "Detector I" in 1958. The difficulty of this problem is ill-condition. How to analyse this structure and control it is an interesting problem.

4. Approximate modeling according to different confidence data

An approximate method has been developed in [4] to construct the model according to difference confidence observed frequency and modes. How to use this method in our problems is a subject.

REFERENCES

[1]Yue Wenbai and Ding Wenjing, Optimal Control of Random Vibration of Cantilever Beam, J. Building Structure (in Chinese), Vol. 7, No. 5(1987), pp. 49—58.

[2]Wang Shangxian, Discern the State Space of Machine Tool Based on Frequency Respond Test, J. Mech. Enging. (in Chinese), Vol. 23, No. 1(1987), pp. 42—53.

[3]Qu Weilian, Li Chaoyin and Zhong Aida, Random Fuzzy Mode Control for Very High Structure, Shanghai Mechanics (in Chinese), Vol. 9, No. 1(1988), pp. 33—46.

[4]Huang Ling, Chen Decheng and Luo Huogen, Approximate Modelling According to Different Confidence Data, Acta Mechanica Sinica (in Chinese), Vol. 20, No. 3(1988), pp. 236—242.

[5]J. G. Gibson, A note on stabilization of infinite dimensional linear oscillators by compact feedback, SIAM J. Contr. Optim, Vol. 18,No. 3(1980),pp. 311-316.

[6]Wang Kangning and Guo Zhaozhi, Control of Elastic Vibration, Scientia Sinica (in Chinese), No. 4 (1974), pp. 335—350.

[7]Fen Dexing and Zhu Gungtian, Problems on Suppress and Pole Allocation of Elastic Vibration Control, Scientia Sinica (in Chinese), (1982), No. 4,pp. 375—384.

[8]S. Chen and R. Triggiani, Proof of extensions of two conjectures on structural damping for elastic systems, Pacific J. of Math., Vol. 136, No. 1(1989),pp. 15-55.

[9]Y. Sakawa and H. Zheng, Modeling and Control of Coupled bending and torsional vibrations of flexible beams, IEEE Trans. Automat. Contr., Vol. AC-34, No. 9(1989), pp. 970-977.

[10]S. K. Biswas and R. D. Klafter, Dynamic modeling and optimal contorol of flexible robotic manipulators, Proc. of IEEE Inter. Conference on Robotics and Automation, Missouri, (1985), pp. 25-28.

[11]C. Trusdell and W. Noll, The Non-linear Theories of Mechanics, Springer-Verlag, Berlin, (1965).

[12]T. Kailath, Linear Systems, Prentice-Hall, Inc., Englewood Cliffs, N. J., (1980).

[13]M. J. Balas, Feedback control of flexible systems, IEEE Trans. Automat. Contr., Vol. AC-23 (1978), pp. 673-679.

[14]B. D. O. Anderson and J. B. Moore, Linear Optimal Control, Prentice-Hall, Englewood Cliffs, N. J., (1971).

EFFECTS OF FLEXIBLE FOUNDATION ON THE RESPONSES OF ACTIVELY CONTROLLED STRUCTURES

Juang, Der-Shin

Department of Civil Engineering
National Central University
Chung-Li, Taiwan, China

ABSTRACT

The objective of this paper is to investigate the effect of soil flexibility on the performance of active control systems for seismic-excited buildings. By using the critical-mode control algorithm, active tendon and active tuned mass damper were studied. Numerical examples are provided to illustrate the influence of foundation rigidity on the controlled response of the building and the required control forces. It is shown that the effectiveness of active tendon system and active tuned mass damper system is not degraded. However, the required maximum control force is getting large as foundation rigidity in rocking direction decreases. Time delay effect for the interaction system is also examined in this paper. It is found that, for both the fixed base system and the interaction system, the system response and control force become large as time delay increases.

INTRODUCTION

During the past two decades, the applications of active control mechanisms to improve the reliability and safety of civil engineering structures were studied extensively [3]. Usually, efficient control of the structural system requires accurate mathematical modeling of the system, such as sufficient number of degrees of freedom, complete sensing data to feedback, .. ,etc. For practical implementation, however, system modeling errors cannot be avoided. It is, therefore, an important issue to study the effects of system modeling errors in structural control. Many published papers have already discussed the influences of inaccurate system parameters, incomplete measured feedback data, and time delay on the control performance [3,5,7]. They all provided valuable information for practical implementation of active control system in large civil engineering structures.

A common assumption in most studies on active control of the dynamic response of structure is that soil-structure interaction effects are small and that the base motion is negligible. In reality, a building rested on a flexible foundation will introduce rocking and translational motions of the foundation. Usually, rocking motion will enlarge the inter-story relative displacements to a large extent. Since an objective in structural control is to use the active mechanism to reduce inter-story relative displacements, it is important to study the influence of foundation rigidity on the performance of active control systems.

In this paper, the model of the structural system is regarded as a shear beam type building with a foundation that can sway and rock in the soil. Based on the critical-mode control algorithm [6], the effects of foundation stiffness on the performances of active tendon system and active mass damper system are investigated. The effects of time delay for the interaction system are also examined in this paper.

FORMULATION

Figure 1 shows the configuration of the soil-structure interaction system. The structure is idealized by a n-degree of freedom system and is subjected to a one-dimensional earthquake ground acceleration $\ddot{x}_g$. The structural mass, damping and stiffness matrices are represented by $[M_s]$, $[C_s]$ and $[K_s]$, respectively. The soil is replaced by two springs and corresponding dashspots: one set corresponding to swaying (k_t and c_t), and the other corresponding to rocking (k_r and c_r). If the active mass damper is placed on the top floor and the active tendons are installed on every story unit (Figure 2(c)), the motion equation of the system can be written as :

$$[M]\{\ddot{X}\}+[C]\{\dot{X}\}+[K]\{X\}=-\{\underline{M}\}\ddot{x}_g+[H]\{U\} \tag{1}$$

where

$$\{X\}=\begin{Bmatrix} \theta \\ x_b \\ \cdots \\ \{X_s\} \\ x_d-x_n \end{Bmatrix}_{(n+3)x1} \qquad [M]=\begin{bmatrix} I_t & 0 & \vdots & 0 & \vdots & 0 \\ 0 & m_b & \vdots & 0 & \vdots & 0 \\ \cdots & \cdots & \cdots & \cdots & \cdots & \cdots \\ 0 & 0 & \vdots & [M_s] & \vdots & 0 \\ \cdots & \cdots & \vdots & \cdots & \vdots & \cdots \\ 0 & 0 & \vdots & 0 \cdots 0 \; m_d & \vdots & m_d \end{bmatrix}_{(n+3)x(n+3)}$$

$$[C]=\begin{bmatrix} c_r+\{h\}^T[C_s]\{h\} & \{1\}^T[C_s]\{h\} & \vdots & -\{h\}^T[C_s] & \vdots & 0 \\ \{1\}^T[C_s]\{h\} & c_t+\{1\}^T[C_s]\{1\} & \vdots & -\{1\}^T[C_s] & \vdots & 0 \\ \cdots & \cdots & \vdots & \cdots & \vdots & \cdots \\ -[C_s]\{h\} & -[C_s]\{1\} & \vdots & [C_s] & \vdots & 0 \\ \cdots & \cdots & \vdots & \cdots & \vdots & -c_d \\ 0 & 0 & \cdots & 0 & \cdots & c_d \end{bmatrix}_{(n+3)x(n+3)}$$

$$[K]=\begin{bmatrix} k_r+\{h\}^T[K_s]\{h\} & \{1\}^T[K_s]\{h\} & \vdots & -\{h\}^T[K_s] & \vdots & 0 \\ \{1\}^T[K_s]\{h\} & k_t+\{1\}^T[K_s]\{1\} & \vdots & -\{1\}^T[K_s] & \vdots & 0 \\ \cdots & \cdots & \vdots & \cdots & \vdots & \cdots \\ -[K_s]\{h\} & -[K_s]\{1\} & \vdots & [K_s] & \vdots & 0 \\ \cdots & \cdots & \vdots & \cdots & \vdots & -k_d \\ 0 & 0 & \cdots & 0 & \cdots & k_d \end{bmatrix}_{(n+3)x(n+3)}$$

$$[H]=\begin{bmatrix} 1 & -1 & & & \\ & 1 & & & \\ & & \ddots & & \\ & & & 1 & -1 \\ & & & & 1 \end{bmatrix}_{(n+3)x(n+3)}$$

$\{\underline{M}\}=[0\ \ m_b\ \ m_1 \ldots m_n\ \ m_d]^T$ is a (n+3)x1 matrix, m_b is the mass of the foundation;

$\{X_s\}=\{v\}+\{1\}x_b+\{h\}\theta$, x_b is the translational displacement of the foundation;

$\{v\}$ = the vector of relative displacements of each floor level to the ground;

θ = rocking angle of the foundation;

$\{1\} = [1\ \ 1 \ldots 1]^T$ is a nx1 matrix;

$\{h\} = [h_1\ \ h_2 \ldots h_n]^T$, h_i is the height of floor level i;

$\{U\} = [0\ \ 0\ \ U_1 \ldots U_n\ \ U_d]^T$, U_i is the control force provided by active tendon on story unit i, and U_d is the control force provided by active mass damper;

I_t = total mass moment of intertia of strucuture and foundation about the rocking axis at the foundation;

x_d = relative displacement of the TMD system to the ground;

m_d, k_d, c_d = the mass, stiffness, and damping coefficients of the TMD system, respectively.

Although Eq.(1) is developed according to the system as depicted in Figures 1 and 2(c), it is worthwhile to note that, by slightly modifying the equation, it can be used to study the control systems shown in Figures 2(a) and 2(b). In fact, Eq.(1) is a general form of the motion equations of all systems shown in Figure 2.

Parameters k_t, k_r, c_t and c_r are usually frequency dependent, and the interaction system does not possess classical normal modes. The structural responses can be rigorously computed only by direct step-by-step numerical integration. However, with appropriately specified modal damping values, the normal mode method can produce sufficiently accurate approximations. Several empirical methods are available for computing the composite modal damping for interaction system. The method adopted in this paper is the complex eigenvalue method as discussed in reference 1 and 8. Detailed procedures will not be discussed in this paper.

CONTROL ALGORITHM

Let the state vector be $\{Z\}=[\{X\}^T\ \{\dot{X}\}^T]^T$; then Eq.(1) can be rewritten as

$$\{\dot{Z}\} = [A]\{Z\} + [B]\{U\} + \{W\}\ddot{x}_g \tag{2}$$

in which [A] is the system matrix, [B] is the location matrix of controllers, and {W} is the location matrix of external excitations in the state space.

By solving the eigenvalues, $\mu_j+i\Omega_j$, and the corresponding eigenvectors, $\{a\}_j+i\{b\}_j$, of the system matrix [A], Eq.(2) can be transformed into n+3 pairs of independent equations:

$$\{\dot{\upsilon}\} = [\Lambda]\{\upsilon\} + [D]\{U\} + \{F\}\ddot{x}_g \tag{3}$$

where $\{Z\}=[\Gamma]\{\upsilon\}$, $[\Gamma]=[\{a\}_1\ \{b\}_1 \ldots \{a\}_{n+3}\ \{b\}_{n+3}]$, $[D]=[\Gamma]^{-1}[B]$, $\{F\}=[\Gamma]^{-1}\{W\}$, and

$$[\Lambda] = \begin{bmatrix} \mu_1 & \Omega_1 & & & \\ -\Omega_1 & \mu_1 & & & \\ & & \ddots & & \\ & & & \mu_{n+3} & \Omega_{n+3} \\ & & & -\Omega_{n+3} & \mu_{n+3} \end{bmatrix}$$

In Eq.(3), each pair of equations represents a vibration mode. The equation can be separated into critical modes as:

$$\{\dot{\upsilon}_c\} = [\Lambda_c]\{\upsilon_c\} + [D_c]\{U\} + \{F_c\}\ddot{x}_g \tag{4}$$

$$\{\dot{\upsilon}_r\} = [\Lambda_r]\{\upsilon_r\} + [D_r]\{U\} + \{F_r\}\ddot{x}_g \tag{5}$$

where $\{\upsilon\}=[\{\upsilon_c\}^T\ \{\upsilon_r\}^T]^T$, $[D]=[[D_c]^T\ [D_r]^T]^T$, $\{F\}=[\{F_c\}^T\ \{F_r\}^T]^T$, and $[\Lambda]=\mathrm{diag}[[\Lambda_c]\ [\Lambda_r]]$.

Using the critical-mode control algorithm, the performance index to be minimized can be written as [6]:

$$J = \frac{1}{2}\int_0^{t_f} (\{\upsilon_c\}^T[Q_c]\{\upsilon_c\} + \{U\}^T[R]\{U\})dt \tag{6}$$

in which t_f is a duration defined to be longer than that of earthquake, $[Q_c]$ is a positive semi-definite response weighting matrix corresponding to the critical modes, and $[R]$ is a positive definite control force weighting matrix. A minimization of the performance index J, given by Eq.(6), subjected to the constraint of Eq.(4) yields the feedback control force as:

$$\{U\}=-[R]^{-1}[D_c]^T[P]\{\upsilon_c\} \tag{7}$$

where $[P]$ is the Riccati matrix satisfies following Riccati equations:

$$0=[P][\Lambda_c]+[\Lambda_c]^T[P]+[Q_c]-[P][D_c][R]^{-1}[D_c]^T[P] \tag{8}$$

Substituting Eq.(7) into Eq.(4), yields

$$\{\dot{\upsilon}_c\} = ([\Lambda_c] - [D_c][R]^{-1}[D_c]^T[P])\{\upsilon_c\} + \{F_c\}\ddot{x}_g \tag{9}$$

Solve Eq.(9); we can obtain the controlled responses of critical modes of the structural system.

TIME DELAY

To apply the control algorithm to a real structural system we must take into account the time delay not only for digitizing the observed system response but for calculating the control force and for applying it to the structure. To account for time delay in the control algorithm, the optimal feedback control system is modified as [9]:

$$\text{Minimize} \quad J_1 = \frac{1}{2}\int_0^{t_f} (\{\upsilon_c\}^T[Q_c]\{\upsilon_c\} + \{U\}^T[R]\{U\})dt \tag{10}$$

$$\text{Subject to} \quad \{\dot{\upsilon}_c\} = [\Lambda_c]\{\upsilon_c\} + [D_c]\{U(t-\beta)\} + \{F_c\}\ddot{x}_g \tag{11}$$

in which β represents time delay to apply control force to the system, and $\{U(t-\beta)\}=0$ for $-\beta \leq t < 0$.

Let

$$\{\eta(t,s)\} = \{U(t-s\beta)\}, \qquad 0 \leq s \leq 1 \tag{12}$$

then

$$\frac{\partial\{\eta(t,s)\}}{\partial t} + \frac{1}{\beta}\frac{\partial\{\eta(t,s)\}}{\partial s} = 0, \qquad \text{for } t>0 \text{ and } 0 \leq s \leq 1 \tag{13}$$

If we separate the domain [0,1] into ε equally spaced regions, the first term in Eq.(13) may be approximated by

$$\frac{\partial\{\eta(t,s)\}}{\partial t} \cong -\frac{1}{\beta}\frac{\{\eta(t,s_j)\} - \{\eta(t,s_{j-1})\}}{1/\varepsilon}, \qquad \text{for } j = 1,2,..,\varepsilon \tag{14}$$

where s_j is the j-th separating point in the domain $0 \leq s \leq 1$. By defining $\{\eta_j(t)\} = \{\eta(t,s_j)\}$, we obtain

$$\begin{aligned} \frac{\partial\{\eta_1(t)\}}{\partial t} &\cong \frac{\varepsilon}{\beta}(\{U(t)\} - \{\eta_1(t)\}) \\ \frac{\partial\{\eta_j(t)\}}{\partial t} &\cong \frac{\varepsilon}{\beta}(\{\eta_{j-1}(t)\} - \{\eta_j(t)\}) \qquad \text{for } j = 2,3,....,\varepsilon - 1 \\ \frac{\partial\{\eta_\varepsilon(t)\}}{\partial t} &\cong \frac{\varepsilon}{\beta}(\{\eta_{\varepsilon-1}(t)\} - \{U(t-\beta)\}) \end{aligned} \tag{15}$$

Based on Eq.(15), we can redefine the optimal control system with time delay as:

$$\text{Minimize} \quad J_1 = \frac{1}{2}\int_0^{t_f} (\{\upsilon_c\}_d^T[Q_c]_d\{\upsilon_c\}_d + \{U\}^T[R]\{U\})dt \tag{16}$$

$$\text{Subject to} \quad \{\dot{\upsilon}_c\}_d = [\Lambda_c]_d\{\upsilon_c\}_d + [D_c]_d\{U(t)\} + \{F_c\}_d\ddot{x}_g \tag{17}$$

where

$$\{\upsilon_c\}_d = \begin{Bmatrix} \{\upsilon_c\} \\ \cdots\cdots \\ \{\eta_1(t)\} \\ \vdots \\ \{\eta_j(t)\} \\ \vdots \\ \{U(t-\beta)\} \end{Bmatrix} \qquad [Q_c]_d = \begin{bmatrix} [Q_c] & \vdots & [0] \\ \cdots & \cdots & \cdots \\ [0] & \vdots & [0] \end{bmatrix}$$

$$[\Lambda_c]_d = \begin{bmatrix} [\Lambda_c] & \vdots & 0 & 0 & \cdots & 0 & \vdots & [D_c] \\ \cdots & \cdots & \cdots & \cdots & \cdots & \cdots & \cdots & \cdots \\ 0 & \vdots & -\frac{\varepsilon}{\beta}[I] & 0 & \cdots & 0 & \vdots & 0 \\ 0 & \vdots & \frac{\varepsilon}{\beta}[I] & -\frac{\varepsilon}{\beta}[I] & \cdots & 0 & \vdots & 0 \\ \vdots & \vdots & 0 & \cdots & \ddots & & \vdots & \vdots \\ 0 & \vdots & 0 & \cdots & \frac{\varepsilon}{\beta}[I] & -\frac{\varepsilon}{\beta}[I] & \vdots & 0 \\ 0 & \vdots & 0 & 0 & \cdots & \frac{\varepsilon}{\beta}[I] & \vdots & -\frac{\varepsilon}{\beta}[I] \end{bmatrix}$$

$$[D_c]_d = \begin{bmatrix} [0] \\ \cdots\cdots \\ \frac{\varepsilon}{\beta}[I] \\ [0] \\ \vdots \\ [0] \end{bmatrix} \qquad \{F_c\}_d = \begin{Bmatrix} \{F_c\} \\ \cdots \\ \{0\} \\ \vdots \\ \{0\} \end{Bmatrix}$$

The optimal control system as described by Eqs.(16) and (17) can be solved by using a similar procedures dicussed in previous section. The optimal control force can be obtained as:

$$\{U\} = -[R]^{-1}[D_c]_d^T[P]_d\{\upsilon_c\}_d \tag{18}$$

in which the Riccati matrix $[P]_d$ satisfies the following Riccati equation:

$$0 = [P]_d[\Lambda_c]_d + [\Lambda_c]_d^T[P]_d + [Q_c]_d - [P]_d[D_c]_d[R]^{-1}[D_c]_d^T[[P]_d \tag{19}$$

Substituting Eq.(18) into Eq.(17), yields

$$\{\dot{\upsilon}_c\}_d = ([\Lambda_c]_d - [D_c]_d[R]^{-1}[D_c]_d^T[P]_d)\{\upsilon_c\}_d + \{F_c\}_d\ddot{x}_g \tag{20}$$

The vector $\{\upsilon_c\}_d$ can be solved using Eq.(20). As $\{\upsilon_c\}_d$ is solved, $\{\upsilon_c\}$ can be solved, too.

NUMERICAL DEMONSTRATION

To study the effects of soil-structure interaction on the performance of control systems we have considered an eight-story building founded on soil with different rigidities. The structural properties of each story unit are as follows: m_i = m = mass of each floor = 345.6 Tons, k_i = k = elastic stiffness of each story unit = 3.404x105 KN/m, $\xi_i = \xi$ = fixed base modal damping ratio = 2%. and h = height of each story unit = 3 m. The foundation was modeled as rigid circular base with radius R=15m and mass m_b=500 Tons. The Poisson's ratio of soil, ν, is set at 1/3. The stiffness and damping coefficients of the foundation are simply calculated by the following equations [1,8]:

$$k_t = \underline{k}_1 \frac{32(1-\nu)}{7-8\nu} Gr, \qquad k_r = \underline{k}_2 \frac{8}{3(1-\nu)} Gr^3$$

$$c_t = 2\xi_t \sqrt{k_t m_t}, \text{ and } \qquad c_r = 2\xi_r \sqrt{k_r I_o}$$

where $\underline{k}_1$ and $\underline{k}_2$ are dimensionless stiffness coefficients used to describe the dependency of foundation stiffness to forcing frequency, I_o is polar mass moment of inertia of structure and foundation, and ξ_t and ξ_r are damping ratios of soil in translational and rocking directions, respectively.

The control systems considered in this paper are active tendons and active tuned mass damper as given in Fig.2(a) and (b). The response weighting factor, Q_{ii}, and the control force weighting factor, R_{ii}, are taken as 10^5 and 10^{-4}, respectively. For the mass damper system, m_d, k_d, and c_d are assumed, respectively, to be 29.6 Ton, 957.2 KN/m and 7.3%. The 1940 El Centro earthquake acceleration records in N-S direction are employed in this study for which the maximum ground acceleration is set at 0.31g. Damping ratios of the soil in translational and rocking directions are assumed to be ξ_t = 7.3% and ξ_r = 40%, respectively.

To discuss the effects of foundation flexibility on the performance of control system, we first select different values of shear modulus of the soil: G=18000 KN/m^2, G=36000 KN/m^2, and G=infinite (fixed base). By changing G value, the characteristics of the structural system are changed, too. The computed first three modes natural frequencies and damping ratios of the structural system are listed in Table 1. It can be found that the frequencies are decreased and the damping ratios are increased as shear modulus of soil decreases.

If active tendons are installed on every story unit, the uncontrolled and controlled maximum inter-story displacements for different G values are shown in Figure 3(a). It shows that, for different G values, the maximum inter-story displacement in each story unit can be reduced to almost the same value. Because vibrational energy dissipated into the supporting medium of the interaction system, the maximum inter-story displacements of the lower story units are reduced. However, the maximum inter-story displacements of the upper story units and the required maximum control forces (Table 2) are increased as shear modulus decreases. It is because rocking motion of the foundation presented in the interaction system. Figure 1(b) is used to show the influence of time delay. It is found that time delay always degrades the control effectiveness of active control devices. But, with time delay, the difference of the required maximum control force between interaction system and fixed base system is very small (Table 2).

For active tuned mass damper system, the controlled maximum inter-story displacements of interaction system are also slightly different from that of fixed base system (Figure 4). It is interesting to note that the maximum inter-story displacements in the upper story units of interaction system are smaller than that of fixed base system. However, spill-over effect is always presented when active tuned mass damper system is used. The required maximum control force for active tuned mass system also increases as the shear modulus of soil decreases (Table 2).

Although different shear modulus will simultaneously change stiffness and damping of the structural system. Yet, in References 5 and 8, it has been verified that different damping values do not significantly affect the effectiveness of active control systems. More important factors that change the effectiveness of active control systems could be the stiffi ·ss coefficients of the foundation. To identify this phenomenon, we fixed the damping ratios of the foundation as used in the example discussed earlier and use different combinations of k_t and k_r values as follows:

(1) k_r=1x10^8, 2x10^8, and 3x10^8 KN/rad,

(2) k_t=6.5x10^5, 13x10^5, and 19.5x10^5 KN/m.

Results are given in Figures 5 through 8 and Tables 3 and 4. Figures 5 and 6 show that, as the values of k_t and k_r reduce, the maximum inter-story displacements are changed in a manner similar to that shown in the previous example. Figures 7 and 8 shows that the maximum rocking angle and swaying displacement of the foundation can be reduced significantly as we use active control systems. However, Tables 3 and 4 show that, as k_t and k_r reduce, the required maximum control force increases. That is, in order to control the inter-story displacements of the building, the control forces required for the interaction system are greater than that for the fixed base system. If we compare these results with that of previous example, it is not difficult to find that the both the required maximum control forces and the maximum inter-story displacements of this example are greater than that of previous example. It means that foundation stiffness coefficients are more important than foundation damping ratios in the active control of structures.

CONCLUSION

By using the critical-mode control algorithm, the influences of foundation flexibility on the performance of active tendon system and active mass damper system are presented in this paper. Numerical examples are provided to show that the active control systems work well for the interaction system. Yet, because rocking motion presented in the interaction system, inter-story displacements on the upper story units and the required maximum control forces of the interaction system are larger than that of the fixed base system, Consequently, it is necessary to pay more attention , if the active system is plan to be installed on the building for which the interactive effect is pronounced. The effects of foundation stiffness on the performance of active control systems were also proven to be more significant than that of foundation damping ratios. Furthermore, time delay degrades the effectiveness of active control devices in both fixed base system and interaction system.

In this paper, we assumed that complete response data can be sensed and provided to compute the required control force. In active tendon system, we also assumed that tendons are installed on every story unit. It is, however, impractical to allocate a sensor and an actuator for each degree of freedom [7], and the control efficiencies are always degraded as limited number of sensors and actuators can be installed in the structure. Under this circumstance, the effects of foundation rigidity might be more pronounced than the example we have discussed in this paper. To use active devices in practical engineering further studies in this area are needed.

ACKNOWLEDGEMENT

Financial support for this research from the National Science Council under Grant No. NSC-80-0410-P008-05B is gratefully acknowledged.

REFERENCES

1. Novak, M., and El Hifnawy, L., "Effect of Soil-Structure Interaction on Damping of Structures," Int. J. of Earthquake Engineering and Structural Dynamics, Vol.11, 1983, pp.595-621.
2. Soliman, M.A., and Ray, W.H., "Optimal Feedback Control for Linear-Quadratic Systems Having Time Delays," Int. J. of Control, Vol.15, No.4, 1972, pp.609-627.
3. Soong, T.T., "Active Structural Control in Civil Engineering," NCEER, Technical Report NCEER-87-0023, 1987.
4. Tsai, N.C., "Modal Damping for Soil-Structure Interaction," J. of Engineering Mechanics, ASCE, Vol.100, EM2, April, 1974, pp.323-341.
5. Yang, J.N., and Akbarpour, "Practical Considerations for Structural Control: System Uncertainty, System Time Delay, and Truncation of Small Control Forces," NCEER, Technical Report NCEER-87-0018, August 10, 1987.
6. Yang, J.N., and Lin, M.J., "Optimal Critical Mode Control of Building Under Seismic Load," J. of Engineering Mechanics, ASCE, Vol.108, EM6, Dec.1982, pp.1167-1185.
7. Yang, H.T.Y., Liaw, D.G., Hsu, D.S., and Fu, H.C., "A Simple Model for Optimal Control of High Rise Buildings," Proceedings of International Workshop on Intelligent Structures, Taiwan, 23-26 July 1990.
8. Juang, D.S., "Effects of Soil-Structure Interaction on the Actively Controlled Structures During Earthquakes," A Report to be Submitted to the National Science Council, 1991 August.
9. Soliman, M.A., and Ray, W.H., "Optimal Feedback Control for Linear-Quadratic Systems Having Time Delays," International Journal of Control, Vol.15, No.4, 1972, pp.609-627.

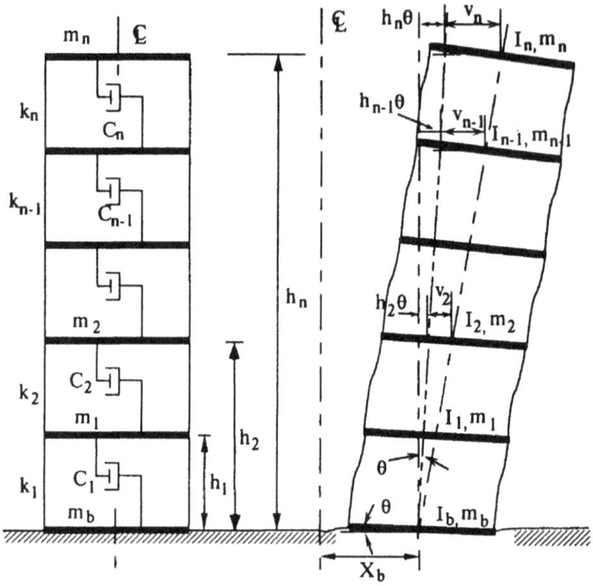

Figure 1. Modeling of Interaction System

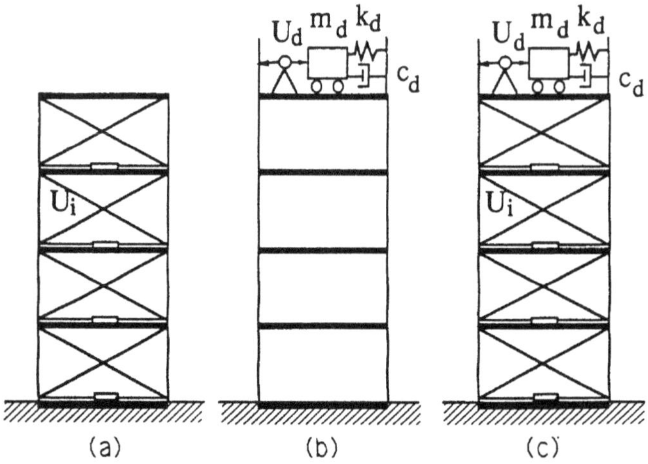

Fig 2 Control Systems: (a) Active Tendons (b) Active Tuned Mass Damper
(c) Combine Active Tendons and Active Tuned Mass Damper

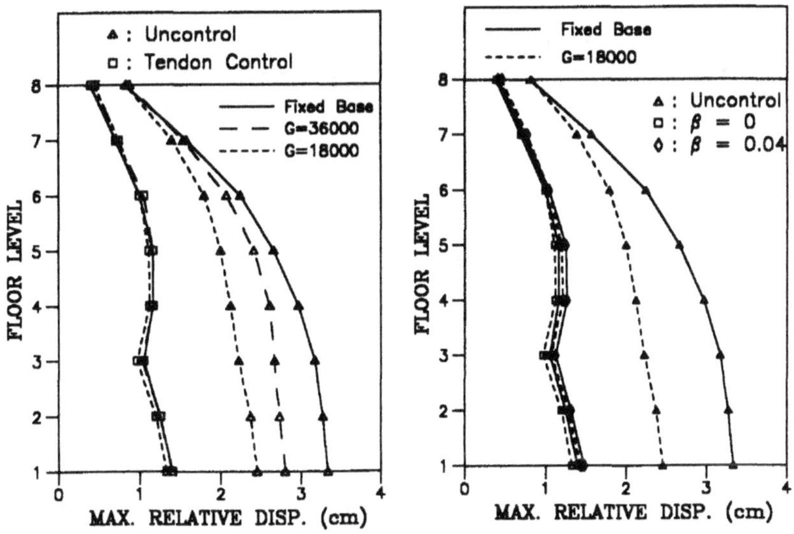

(a) (b)

Figure 3 Uncontrolled and Controlled Maximum Inter-Story Displacements By Using Active Tendons: (a) Effects of Shear Modulus G, (b) Effect of Time Delay

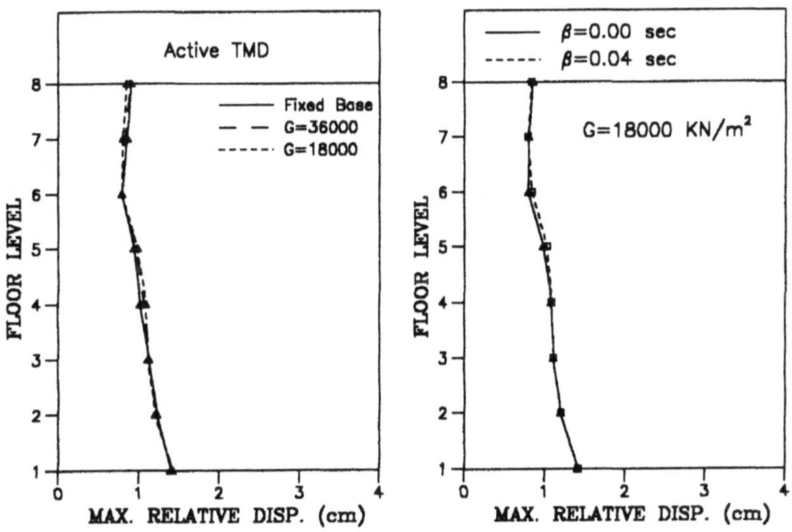

(a) (b)

Figure 4 Maximum Inter-Story Displacements Controlled by Using Active Tuned Mass Damper: (a) Effect of Shear Modulus G, (b) Effect of Time Delay

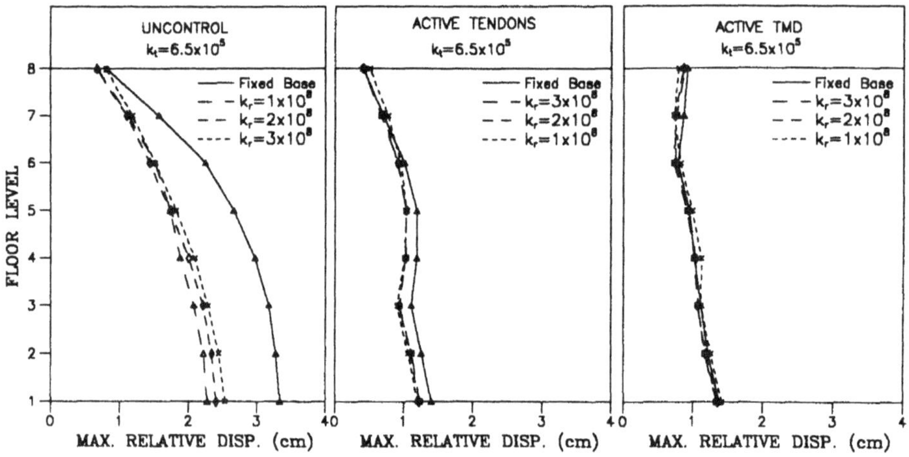

(a) (b) (c)

Figure 5. Influence of k_r on the Maximum Inter-Story Displacements: (a) Uncontrolled, (b) Controlled by Active Tendons, (c) Controlled by Active Tuned Mass Damper.

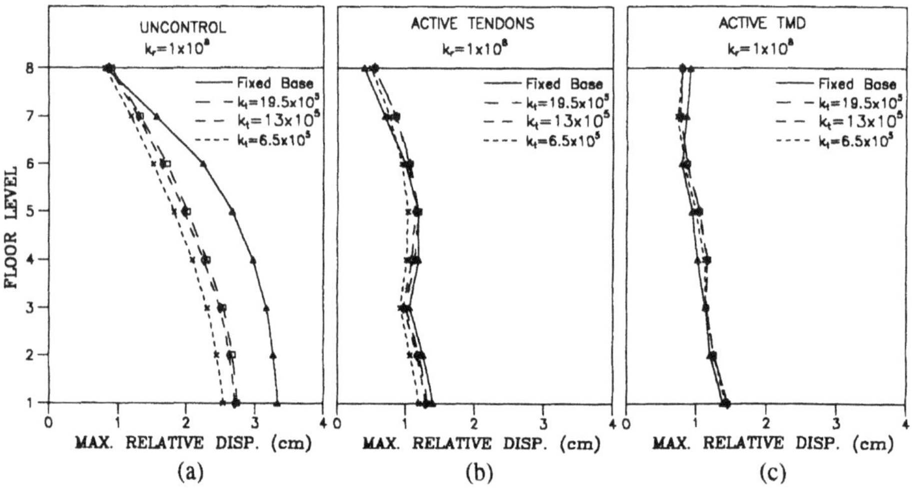

(a) (b) (c)

Figure 6. Influence of k_t on the Maximum Inter-Story Displacements: (a) Uncontrolled, (b) Controlled by Active Tendons, (c) Controlled by Active Tuned Mass Damper.

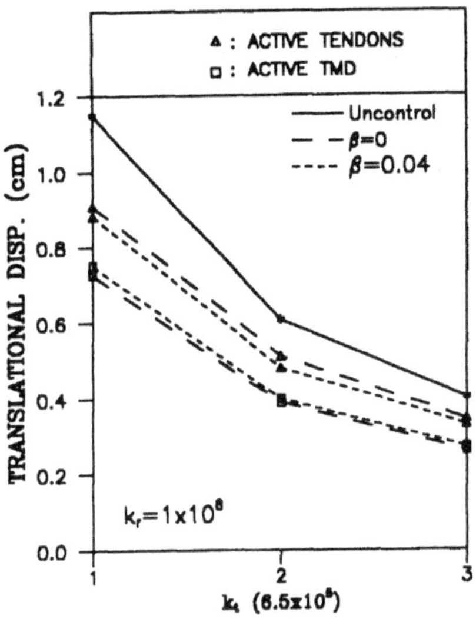

(a)

Figure 7. The Maximum Translational Displacement of the Foundation

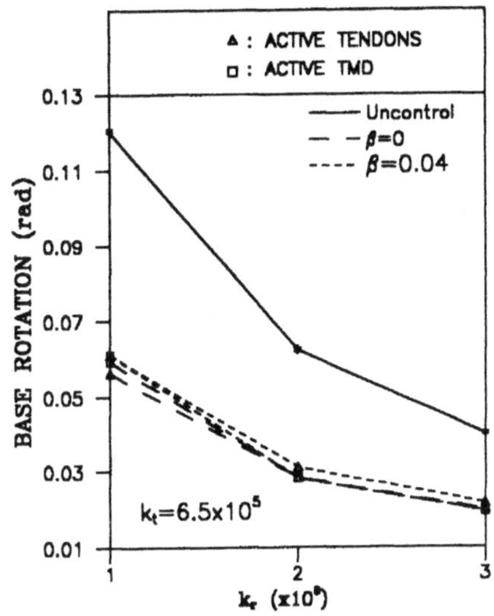

(b)

Figure 8. The Maximum Rocking Angle of the Foundation

TABLE 1. NATURAL FREQUENCIES (rad/sec) AND DAMPING RATIOS OF THE STRUCTURAL SYSTEM FOR DIFFERENT G VALUES

	ω_1	ω_2	ω_3	ξ_1	ξ_2	ξ_3
Fixed Base	5.79	17.21	27.91	2.000%	2.000%	2.000%
G=36000	5.56	16.79	27.61	2.084%	6.317%	10.12%
G=18000	5.31	16.61	27.41	2.371%	7.151%	10.83%

TABLE 2. THE MAXIMUM CONTROL FORCES (KN) PROVIDED BY DIFFERENT ACTIVE CONTROL SYSTEMS (KN)

Control Device	Time Delay	Fixed Base	G=36000	G=18000
Active	$\beta=0$	2352	2370(+0.8%)	2395(+1.8%)
Tendons	$\beta=0.04$	2464	2465(+0.04%)	2466 (+0.08%)
Active	$\beta=0$	1841	1891(+2.7%)	1924(+4.5%)
TMD	$\beta=0.04$	1902	1954(+2.7%)	1978(+4.0%)

TABLE 3. MAXIMUM CONTROL FORCE (KN) PROVIDED BY ACTIVE TENDONS FOR DIFFERENT VALUES OF k_t and k_r

k_r	$3x10^8$		$2x10^8$		$1x10^8$	
k_t	$\beta = 0$	$\beta = 0.04$	$\beta = 0$	$\beta = 0.04$	$\beta = 0$	$\beta = 0.04$
$19.5x10^5$	2350	2481	2357	2453	2392	2348
$13.0x10^5$	2378	2497	2396	2466	2432	2358
$6.5x10^5$	2484	2494	2500	2459	2531	2474

TABLE 4. MAXIMUM CONTROL FORCE (KN) PROVIDED BY ACTIVE TUNED MASS DAMPER FOR DIFFERENT VALUES OF k_t and k_r

k_r	$3x10^8$		$2x10^8$		$1x10^8$	
k_t	$\beta = 0$	$\beta = 0.04$	$\beta = 0$	$\beta = 0.04$	$\beta = 0$	$\beta = 0.04$
$19.5x10^5$	1903	1965	1906	1965	1990	2000
$13.0x10^5$	1919	1979	1924	1978	2027	2031
$6.5x10^5$	1976	2007	2009	2021	2124	2107

ACTIVE CONTROL OF BASE ISOLATED STRUCTURES

Jun-Ping Pu
Professor
Department of Civil Engineering
Feng Chia University, Taichung, Taiwan, China

and

James M. Kelly
Professor
Department of Civil Engineering
University of California at Berkeley, Berkeley, CA., USA

ABSTRACT

An active closed-open-loop control strategy that controls the absolute responses of the structure is established. Only relative displacement and relative velocity of the structure, and the absolute acceleration of the ground motion is needed to be measured. The gain of the closed-loop control is determined by pole assignment method, and that of the open-loop control is determined by the optimization of the dynamic spectrum. The application of this control algorithm to the base isolated structures is investigated and is shown to be effective in reducing the absolute response. The performance of closed-open-loop control is slightly better than that of closed-loop control system.

INTRODUCTION

Active control system is designed ordinarily to decrease the relative responses and the shear in the structure so as to guarantee the safety and comfort of the residents [1,4,5]. Although the relative responses of the structure can be reduced to be nearly zero at the expense of large control energy, it is not the best way since the absolute responses of the structure will then be the same as that of the excitation, and the equipment in the structure will be subjected to quite large an acceleration that is almost equal to the acceleration of the excitation.

Base isolation system is a kind of passive control system which can protect the structure from seismic excitation effectively. The low stiffness of the isolation bearings can decrease the natural frequency and change the mode shape of the dominate mode. The response of each floor of the superstructure will remain almost the same and the superstructure will vibrate as a whole under ground motion. The shear in each floor is quite small already; it is not necessary to introduce the active control system just for the sake of safety.

The disadvantage of the base isolation system is that it could cause a large displacement of the structure due to its low stiffness. The main goal of the control system established herein is to decrease the absolute responses of the structure so that the structure will remain almost "stationary" no matter how the seismic excitation is. Theoretically, an active control system can be designed to reduce the absolute response of the structure to be very small by sensing and feedbacking the absolute responses. However, this is impractical since the absolute displacement still can not be measured in the field up to this moment.

In this paper, we try to control and to reduce the absolute response of the structure through those measurements that can be sensed, such as relative responses of the structure, and the absolute acceleration of the ground motion. An efficient active control system is established. This control strategy is adopted to base-isolated structures and the control force is exerted only to the base of the structure [2,3]. Interesting result is obtained.

SYSTEM CONTROL UNDER INCOMPLETE MEASUREMENT

Assume that a building can be modeled as a n-degree-of-freedom structure. Let the mass, damping and stiffness at ith floor be m_i, c_i, and k_i respectively; then the equation of motion with control is

$$M \ddot{X} + C (\dot{X} - \dot{X}_g) + K (X - X_g) = U \tag{1}$$

where

M, C, K = $n \times n$ mass, damping, and stiffness matrices of the structure, X = n vector denoting the absolute displacement of the structure.

$X_g^t = [\ x_g \ \ x_g \ \cdot \ \cdot \ \ x_g\ ]$
x_g = seismic displacement, and
U = n vector denoting the control force.

Several control strategies can be used to reduce the

absolute response X by transforming Eq.1 from a second order differential equation to a first order state equation and measuring the absolute response. However, since we can not sense the absolute displacement of the structure yet, what we can measure on the field are the relative displacement $(X-X_g)$, the relative velocity $(\dot{X}-\dot{X}_g)$, and the ground acceleration $\ddot{x}_g$; we try to establish a closed-open-loop control strategy that feedbacks the relative responses of the structure in the closed-loop and the acceleration of excitation in the open-loop.

Let Y denotes the incomplete measurement of the system,

$$Y = S\ Z \tag{2}$$

where

$$Y^t = [\ x_g \quad D^t \quad D^t\ ]$$

in which

$$D = X - X_g$$

and

$$S = \begin{bmatrix} 1 & 0 & 0 & 0 & 0 \\ 0 & S_{22} & 0 & I & 0 \\ 0 & 0 & S_{33} & 0 & I \end{bmatrix}$$

in which

$$S_{22}{}^t = S_{33}{}^t = [\ -1 \quad -1 \quad . \quad . \quad -1\ ]$$

The observability matrix for the system has a rank of 2n+1, which is less than the dimension of the state.

We try to establish a linear control methodology that control force is proportional to the measurement, that is,

$$U = -\ G\ Y \tag{3}$$

where $G = [\ G_1 \quad G_2 \quad G_3\ ]$ is the gain matrix, in which G_1 is a $n \times 1$ open-loop gain matrix, G_2 and G_3 are $n \times n$ closed-loop gain matrices, respectively.

Substituting Eq.3 into Eq.1, it can be obtained that

$$\dot{V} = E\ V + H\ V_g \tag{4}$$

where

$$V^t = [\ \dot{X}^t \quad X^t\ ], \qquad V_g{}^t = [\ x_g \quad x_g \quad x_g\ ],$$

$$E = \begin{bmatrix} E_{11} & E_{12} \\ I & 0 \end{bmatrix} \qquad H = [\ H_1 \quad H_2 \quad H_3\ ]$$

in which

$E_{11} = - M^{-1}C - M^{-1}G_2,\qquad E_{12} = - M^{-1}K - M^{-1}G_3,$

$H_1 = - M^{-1}G_1,\qquad H_2 = \bar{H}_2 - M^{-1}G_2S_{22},\qquad H_3 = \bar{H}_3 - M^{-1}G_3S_{33}$

and

$\bar{H}_2^t = [\ c_1/m_1\quad 0\quad .\quad .\quad 0\ ]$

$\bar{H}_3^t = [\ k_1/m_1\quad 0\quad .\quad .\quad 0\ ]$

The first part on the right side of Eq.4 is the controllable part, and the second part is uncontrollable. It can be seen that by measuring the ground acceleration and relative responses of the structure, and designing the gain matrices G_1, G_2, and G_3 in Eq.3 properly, one may change the absolute responses of the structure in Eq.4 as desired.

CONTROL STRATEGY

It is seen from Eq.4 that the transient response of the state V depends on the closed-loop gains G_2 and G_3. The steady-state response depends on the open-loop gain G_1 in addition to G_2 and G_3. Through a proper choice of the closed-loop and open-loop gains, one may adjust both the transient and the steady-state of the absolute response of the structure.

Take Laplace transformation on both sides of Eq.4; we have

$$(sI - E)\ V = (s^2H_1 + sH_2 + H_3)\ x_g \qquad (5)$$

Closed-Loop Gain

The transient response of the structure is determined from the characteristic equation $\Delta(s)$

$$\Delta(s) = |\ sI - E\ | \qquad (6)$$

where |..| represents the determinant of a matrix.

The characteristic equation $\Delta(s)$ depends on the closed-loop gains G_2 and G_3 that can be determined by using the pole assignment method if the closed-loop poles are specified.

Open-Loop Gain

From Eq.5 we have the steady-state transfer function as

$$L(s) = (sI - E)^{-1} (s^2H_1 + sH_2 + H_3) \qquad (7)$$

Under sinusoidal excitation, set $s = i\omega$, we have

$$\bar{L}(i\omega) = (i\omega I - E)^{-1}(-\omega^2 H_1 + i\omega H_2 + H_3) \tag{8}$$

The magnitude of $\bar{L}(i\omega)$ is known as the dynamic magnification factor.

The transfer function depends on both closed-loop gains G_2, G_3, and open-loop gains G_1. Let us choose the cost function J of the transmissibility as

$$J = \int_{\omega_1}^{\omega_2} \bar{L}(i\omega)\, d\omega$$

where ω_1 and ω_2 are the lower and upper designed frequency limit of the excitation that the structure could be subjected to.

The controlled structure is expected to have minimum transmissibility under excitation. One may choose the open-loop gain G_1 by minimizing the cost J of transfer function after the closed-loop gains are chosen.

CONTROL OF ISOLATED STRUCTURES

Base isolation devices can filter the high frequency of the excitation and prevent it from transmitting to the superstructure. The natural frequency of the first mode is lowered and shifted from the main frequencies of the seismic excitation, so that the response of the superstructure of the base isolated structure is almost the same. Since the shear between floors is quite small, it has no need to introduce the active control system to the structure just for the sake of safety. However, active control system can be adopted to decrease the absolute response of the structure. The strategy proposed in the previous sections can be applied to control the base isolated structures conveniently.

If the control force is designed to exert to the base of the isolated structure only, the open-loop gain matrix and the closed-loop gain matrices in Eq.3 can be written as

$$G_1^t = [\, g_1 \quad 0 \quad . \quad . \quad 0 \,]$$

$$G_2 = \begin{bmatrix} g_{21} & g_{22} & . & . & g_{2n} \\ 0 & 0 & . & . & 0 \\ . & . & . & . & . \\ . & . & . & . & . \\ 0 & 0 & . & . & 0 \end{bmatrix} \qquad G_3 = \begin{bmatrix} g_{31} & g_{32} & . & . & g_{2n} \\ 0 & 0 & . & . & 0 \\ . & . & . & . & . \\ . & . & . & . & . \\ 0 & 0 & . & . & 0 \end{bmatrix}$$

The controllability matrix under this circumstance has a rank 2n, so that the absolute response can be controlled by this control force.

NUMERICAL EXAMPLE

Let us consider a two-degree-of-freedom structure as an illustration. Assume that the mass of each floor is 6.0 tons, the stiffness is 500 kN/m and 5000 kN/m, respectively. The natural frequencies of the structure is 6.3738 and 41.3446 rad/sec. The damping ratio is assumed to be 2%, so that the original poles is located at $-0.127 \pm 6.373i$ and $-0.827 \pm 41.33i$. The transfer function of the structure without controlling is shown in Figure 1(a). The uncontrolled acceleration in time domain under the excitation of El Centro earthquake (May 18, 1940) is shown in Figure 1(b).

If we try to lower the first mode frequency to be 2 rad /sec, and increase the damping ratio of the structure to be 10%, then the poles expected is on $-0.2 \pm 1.99i$ and $-4.134 \pm 41.336i$. The transfer function under closed-loop control is shown in Figure 2(a). More than 90% of the peak transmissibility is reduced. Time domain acceleration is shown in Figure 2(b). It can be seen that the response is much more sluggish. About 88% of the acceleration is diminished. The acceleration is less than 30% of that of the ground excitation, thus maintain the comfort of residents and the serviceability of the equipment of structures.

The transfer function can be decreased further through a proper choice of the open-loop gain g_1. The relationship between the open-loop gain g_1 and the cost function J is shown in Figure 3. The optimum choice of g_1 that minimize the cost J is -0.06. The transfer function with optimum closed-open-loop control is shown in Figure 4(a). The vibration response is shown in Figure 4(b). The efficiency is just slightly improved by introducing the open-loop control to the system; that is not unexpected since the optimum open-loop gain is nearly zero.

Although the frequency and the response of the structure are diminished to quite large an extent, large control force is needed in both closed-loop and closed-open-loop control system as shown in Figure 5(a) and 5(b), in which the maximum amplitude of the control force is about 81% of the total dead weight of the structure.

The control energy needed can be reduced through the proper choice of the location of the poles. Since the first modal frequency of the isolated structure is small already due to the installation of the base isolators, it has no need to change the damped frequencies of the isolation structure through active control system. Let the poles of the controlled system be $-0.637 \pm 6.373i$ and $-4.134 \pm 41.336i$ respectively, so that the damping ratio is increased to be 10% ; the closed-loop transfer function is then shown in Figure 6. The closed-open-loop transfer function is shown in Figure 7(a). The response of the structure under closed-open-loop control is shown in Figure 7(b). About 36% of the displacement, 51% of the velocity, and 55% of the acceleration were diminished. The maximum control force needed now less than 8% of the total weight of the structure as seen in Figure 5(c).

CONCLUSION

An active control algorithm that controls the absolute response of the structure is established. Only relative velocity and relative displacement of the structure, and the absolute ground acceleration are needed to be measured and feedbacked to the structure. This is quite practical. The closed-loop gains are determined by the pole assignment method, so as to obtain a transient response as we expected. The open-loop gains is chosen to minimize the area of the transfer function of the structure, so as to minimize the steady-state response.

The control strategy that controls the absolute response of the structure is especially recommendable to be applied on the base isolation structure which has small shear in the building. The control forces are exerted only to the base of the isolated structure. It is not necessary to adjust the damped frequencies of the structure through the installation of the control system, so that small control forces are needed.

The performance of the closed-loop control system is close to that of the closed-open-loop system with optimum open-loop gains. The responses can be reduced efficiently both in the frequency domain and in the time domain.

REFERENCES

1. Miller, R. K., Masri, S. F., Dehghanyar, T. J., and Caughey, T. K., Active vibration control of large civil structures, J. Engrg. Mech., ASCE, 1988, 9, 1542-70.

2. Pu, J. P. and Chu, T. H., Active control of space structures, Computer Application in Struc. Engrg., Struc. Congress '87 Proc., Struc. Div., ASCE, Orlando, FL , Aug. 1987.

3. Pu, J. P. and Kelly J. M., Active control and seismic isolation, J. Engrg. Mech., ASCE, 1991, 10.

4. Soong, T. T., State-of-the-art review: Active structural control in civil engineering, Eng. Struc. 1988, 10, 74-84.

5. Yang, J. N., Danielians, A., and Liu, S. C., Aseismic hybrid control systems for building structures, J. Engrg. Mech., ASCE, 1991, 4, 836-52.

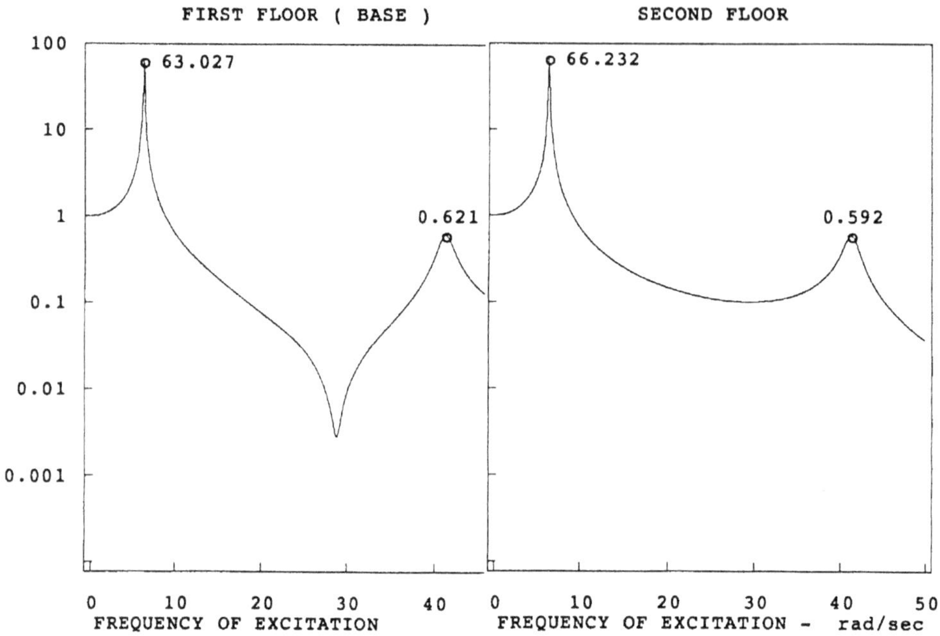

Figure 1(a)

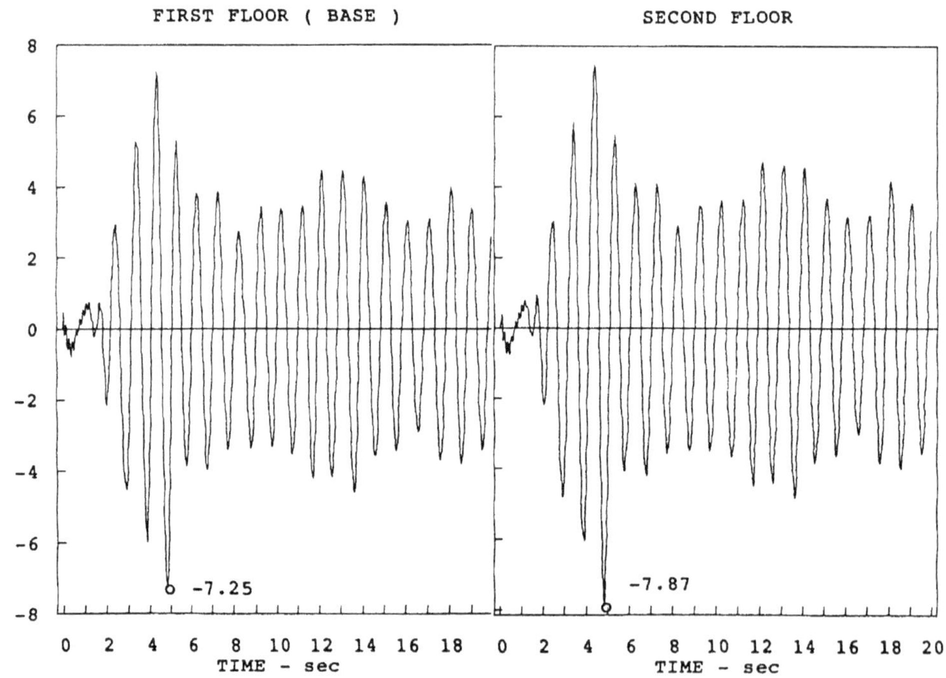

Figure 1(b)

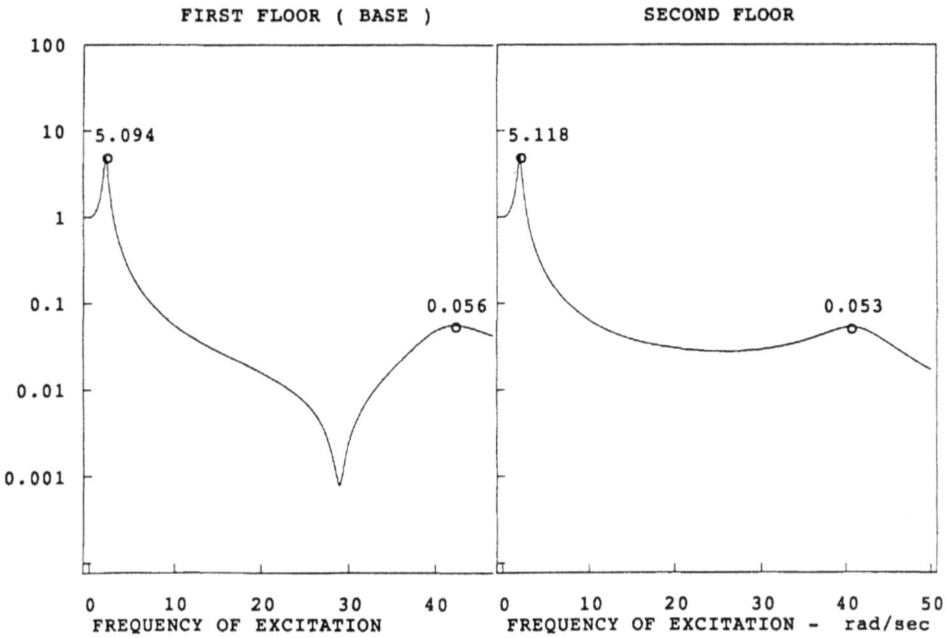

Figure 2(a)

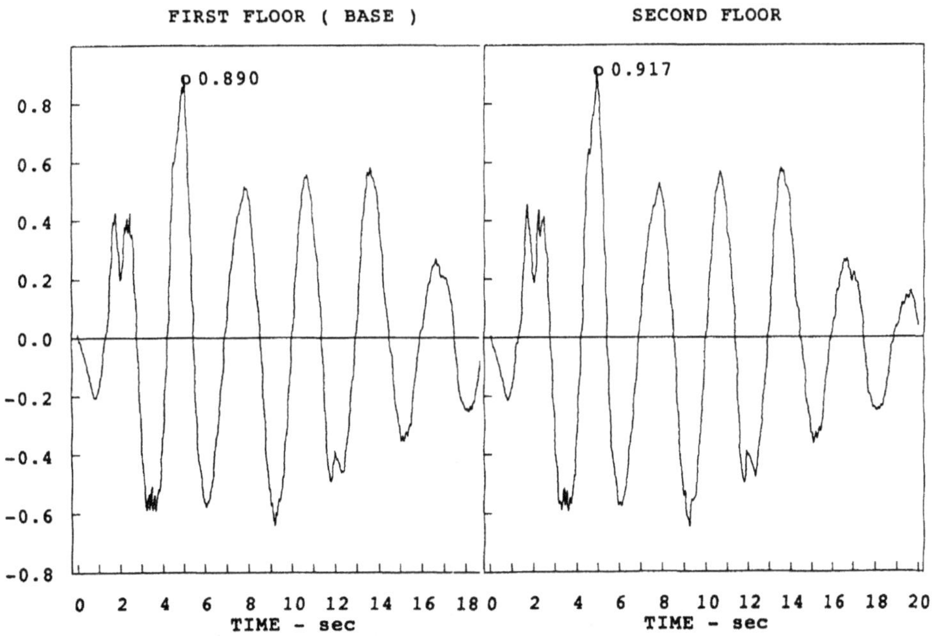

Figure 2(b)

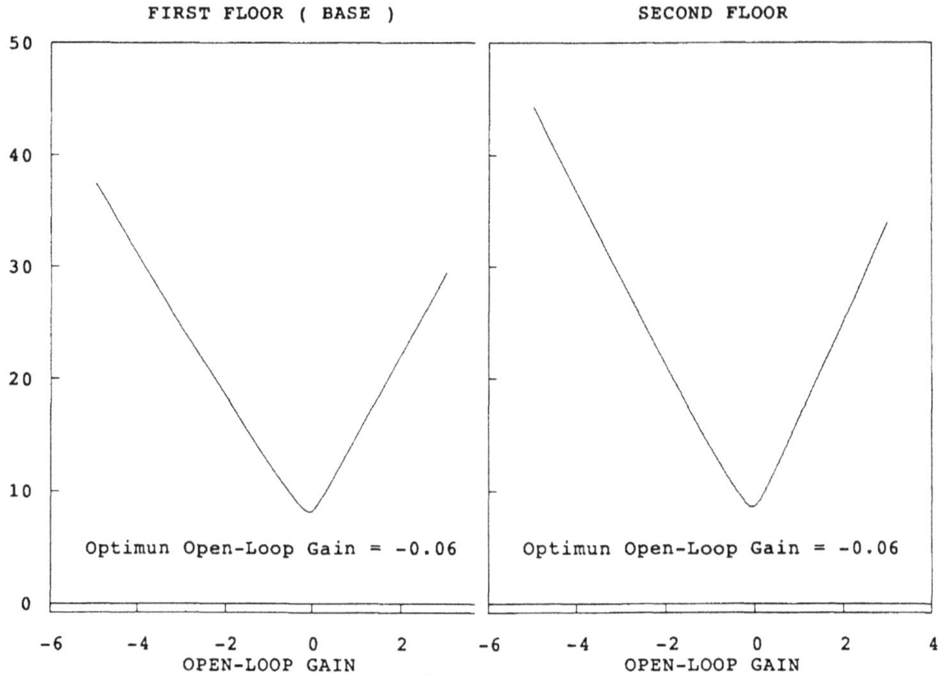

Figure 3

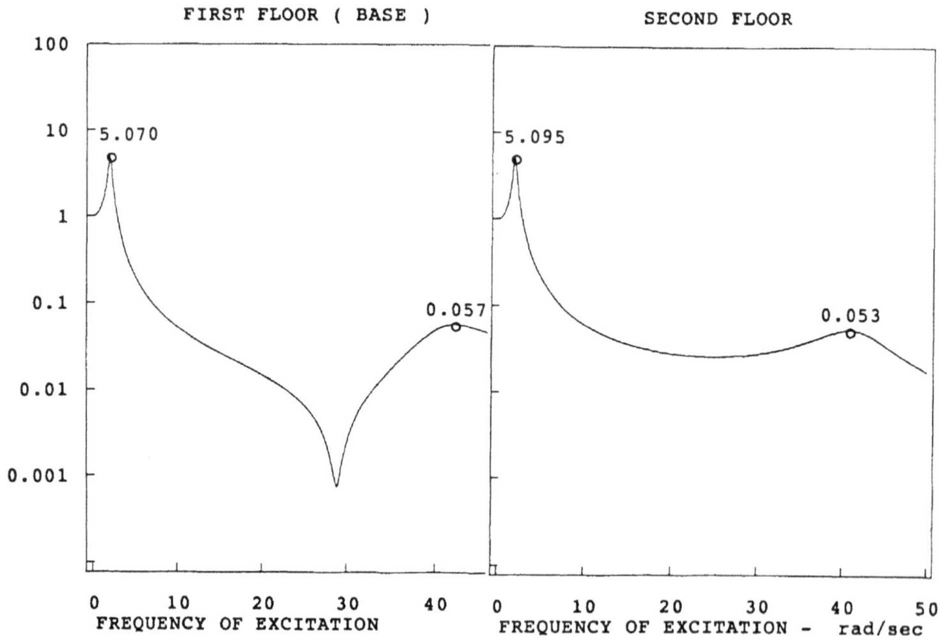

Figure 4(a)

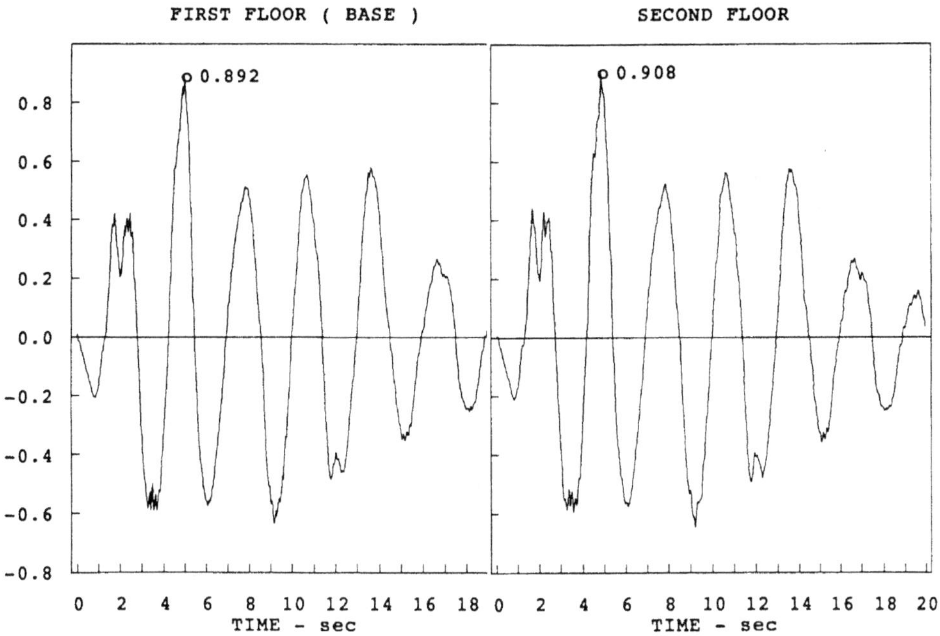

Figure 4(b)

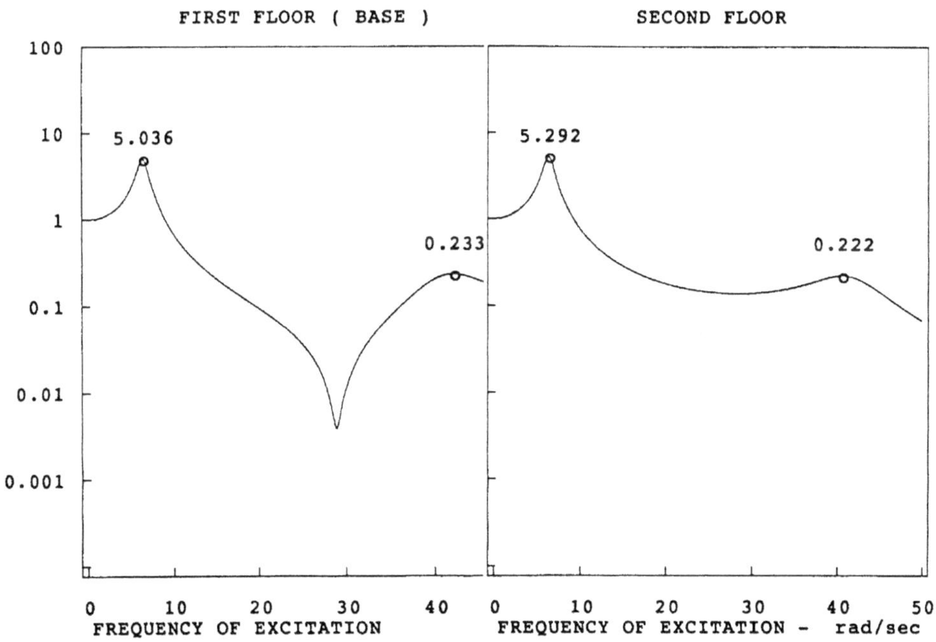

Figure 6

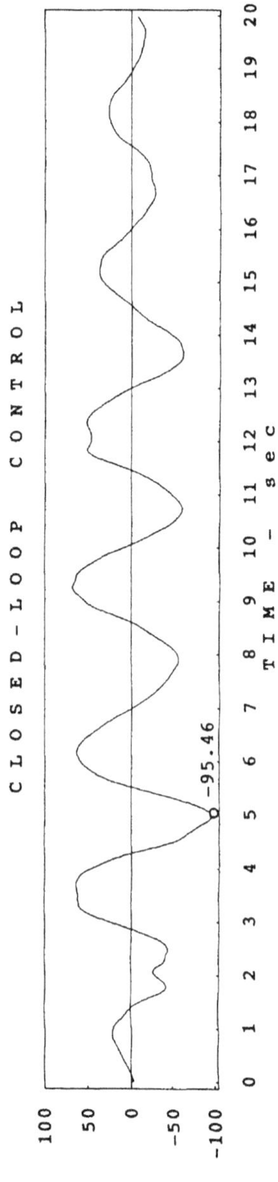

Figure 5(a)

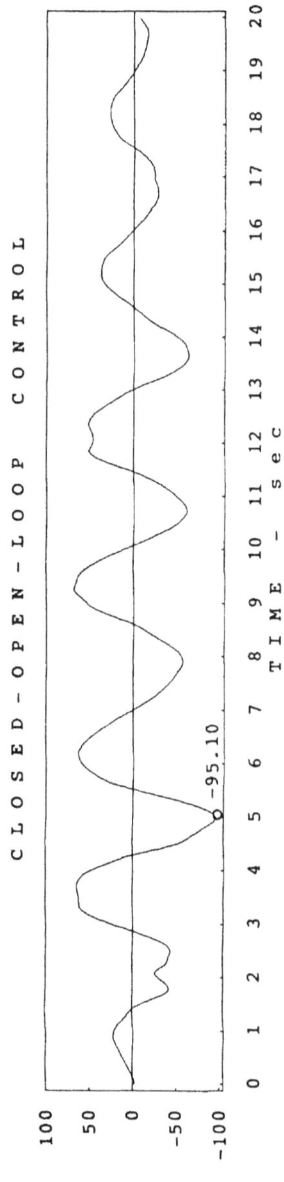

Figure 5(b)

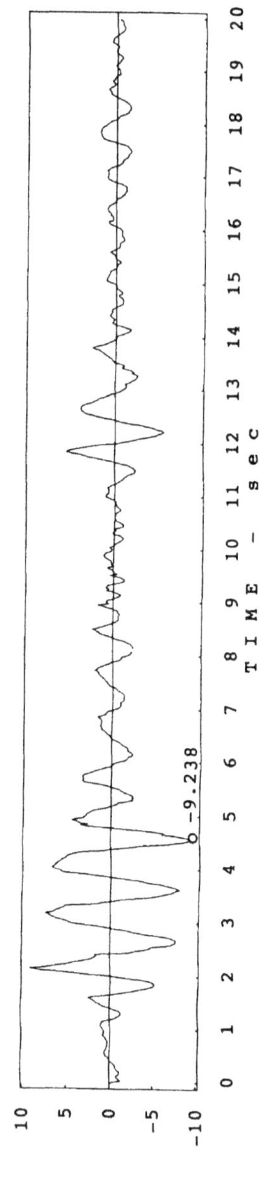

Figure 5(c)

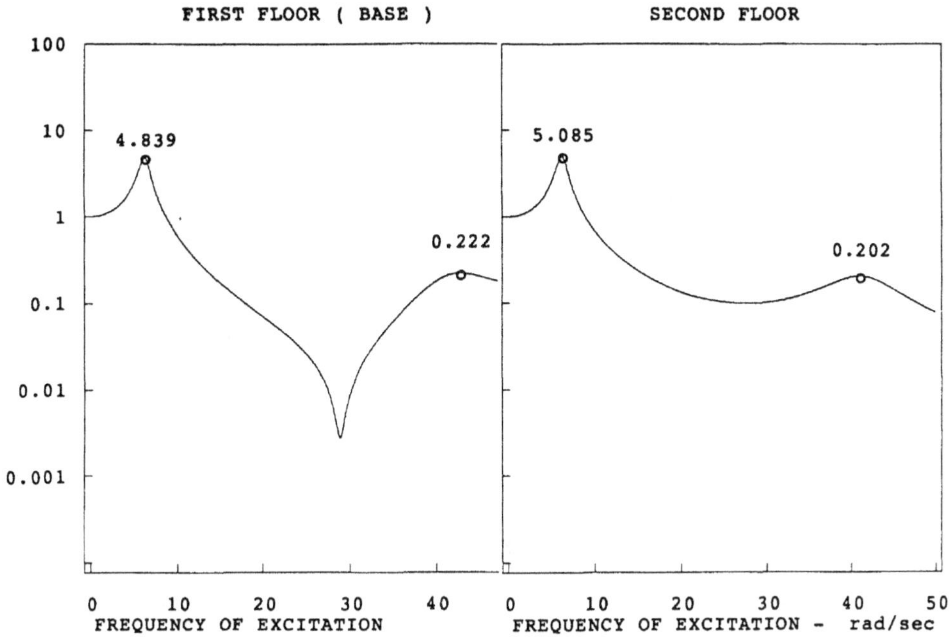

Figure 7(a)

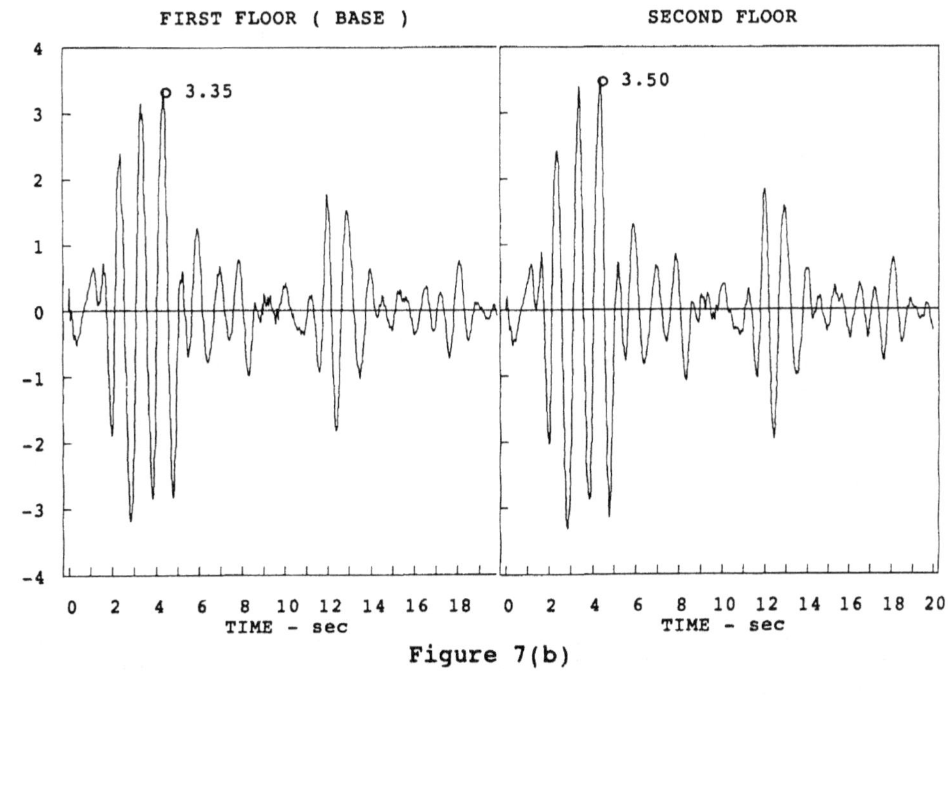

Figure 7(b)

SUBSTRUCTURED COMPUTER–ACTUATOR HYBRID ANALYSIS FOR INELASTIC EARTHQUAKE RESPONSE OF STRUCTURES

Yoshikazu YAMADA Hirokazu IEMURA William TANZO
Professor, Associate Professor, and Graduate Student, respectively
Earthquake Engineering Laboratory, Dept. of Civil Engineering
Faculty of Engineering, Kyoto University
Yoshida Honmachi, Sakyo-ku, Kyoto 606, Japan

ABSTRACT

A substructured on-line computer-actuator hybrid method is developed in which the advances of computational mechanics and experimental mechanics are combined in order to evaluate accurately and reliably the inelastic earthquake response of structures. A complete structural system is divided into analytical substructures and experimental substructures. Restoring forces for the experimental substructures modeling only the critical regions are directly measured from a specimen loaded based on its current deformation state. The remaining portions are taken as analytical substructures in which analytical models are used to describe their restoring-force characteristics. Using this substructured hybrid loading system, critical members with proper boundary conditions can be tested under realistic earthquake loads for accurate evaluation of their restoring-force characteristics. Consequently, inelastic earthquake response of the total structure can be accurately and reliably predicted. Using this hybrid analysis system, inelastic earthquake response behavior of high-damping rubber seismic isolators and steel box-section columns were evaluated.

INTRODUCTION

In modern seismic design philosophy, structures should not collapse even when heavily damaged by strong earthquakes. Certain critical elements (e.g., columns) should be ductile enough to sustain inelastic deformation while maintaining adequate strength. In recent years, seismic isolation concept is gaining wide interests for providing stiff structures with protection from earthquake damage. Both of these two diverse design methodologies (the conventional fixed-based design and the seismic isolation design) are means to protect structures from collapse through inelastic behavior of certain nonlinear elements. The complex behavior of these critical elements will affect the very nature of the induced seismic loads on a structure and thus affecting the response of the total structure. Understanding the inelastic behavior of members and their effect on the integrity of the total structure is an essential approach to evaluate reliability of structures against possible strong earthquakes.

Inelastic Seismic Behavior of Members and Seismic Response of Structures

With vast improvements in recent design, construction techniques and materials, light-weight and flexible structures such as high-rise buildings and long-span bridges have been designed and constructed in Japan. Recent observation of earthquake ground motion has suggested high possibility of structural response exceeding the yielding limits. Hence, it is becoming critically important to investigate earthquake safety of these structures in the inelastic range. Understanding of inelastic nonlinear behavior of members and total structures is a key to estimate safety of structures against possible strong earthquake ground motion.

In conventional earthquake engineering, earthquake ground acceleration has been considered as nonstationary random motion with broad band of frequency components. The El Centro Record of the 1940 Imperial Valley Earthquake is a typical case. This is still true for sites with good soil conditions. However, narrow frequency banded ground motion with long duration has been observed in recent earthquakes at soft soil sites, where modern structures have been constructed. When subjected to this type of earthquake ground motion, flexible and lightly-damped structures have shown resonant phenomena, which leads to high dynamic amplification of structural response. There has been some cases of structural damages caused by the high amplification effects due to some recent earthquakes. In designing flexible and lightly-damped structures, one has to accept high dynamic amplification factor of structural response exceeding yielding limits of structures. Under these situations, research on nonlinear behavior of structures and their members, as well as reasonable inelastic earthquake design methods are urgent.

Due to possible large earthquake ground motion, response of structures might go beyond yielding limits. Hence, their safety should be checked with inelastic design methods. Depending on the way inelastic earthquake response of structures is determined, the methods are divided into approximate and rigorous approaches. The most accurate procedure of estimating inelastic response of structures approaching failure under combined stresses is the step-by-step time history analysis. This method has become feasible with rapid developments in digital computers and modeling techniques. Because inelastic behavior of structures and members is totally estimated, damage and safety level of structures can be judged by many different failure criteria. It is emphasized that accuracy of inelastic modeling of members should be carefully examined when using the step-by-step time history analysis procedure. The results are only as good as the quality of the models assumed, especially on the hysteretic behavior of members.

Analytical and Experimental Evaluation of Inelastic Seismic Behavior

The development of computers and the dramatic increase in computational power have made possible the analysis of complex structures subjected to earthquake ground motions. In earthquake response analysis of structures, one has to deal with multi-degree-of-freedom system in which time-varying response characteristics are sought for. In addition, vast computational speed and storage requirements are needed when inelastic behavior is considered.

Although analytical simulation has progressed a lot through the use of powerful computers and accumulated theoretical and experimental data on structures, it is not yet possible to analyze and predict inelastic response of structures with total confidence, even on materials like steel and reinforced concrete which have been utilized for a long time now. Furthermore, more new materials and devices will be introduced in search of better protection against earthquake disasters. Extensive experimental tests are needed to study the complex behavior of these elements under realistic conditions that are likely to occur during earthquakes. Experimental tests are needed to verify and calibrate mathematical models, as well as to provide data for design.

Traditionally, quasi-static loading tests and shaking-table tests have been used in studies of inelastic behavior of structures. These methods have provided earthquake engineering with a wealth of information on inelastic behavior. In quasi-static static tests, arbitrary cyclic loading histories on an isolated specimen are usually prescribed; and moreover, behavior of total system is not considered. Since seismic loads are not applied per se on a structure but rather loads induced on a structure related to the structural vibrational response due to some ground disturbances, the loads on an element are affected by the response of the other elements of the system. On the other hand, overall response characteristics can be simulated on a shaking table by using a prototype of a complete system. However, size and scale of models mounted on shaking tables have to be compromised due to limitation in table capacities.

During the last two decades, on-line hybrid test method (or pseudo-dynamic test method) has become a very powerful tool to test for inelastic earthquake response of structures. The development has been so significant that it was the only feasible experimental technique available to test two full-scale multistory structural frames (a 6-story steel frame and a 7-story reinforced concrete frame) of a U.S.-Japan joint research project. In addition, it is finding its way into many structural testing laboratories around the world (notably Japan and U.S.) because of its efficiency, versatility, and reliability.

Both the shaking-table test and the original development of on-line hybrid test techniques involve fabrication of the complete structural propotypes for proper modeling. Thus, models are drastically simplified (e.g., into single degree-of-freedom model) or full-model specimens are drastically reduced in scale in order to accommodate them in currently available test facilities. Inelastic earthquake behavior are difficult to interpret from data obtained due to unrealistic earthquake loading used or lack of available similitude relationship for small-scale specimens.

Objectives of This Research

In most cases, severe inelastic deformations are likely occur only in certain critical elements. By incorporating substructuring concepts into on-line hybrid test technique, a substructured on-line hybrid test method is developed in which only the critical regions are tested experimentally while the other 'well-behaved' regions of the structure are modeled analytically. Thus, critical sub-assemblages and components can be tested economically under realistic load conditions considering proper boundary conditions. At the same time, inelastic earthquake response of the total structure can be reliably predicted.

However, in spite of its evident advantages, not much actual implementation of substructured on-line hybrid tests has been done. Most of the tests done have been conducted to demonstrate and verify its feasibilty and reliability using simple cases. There are two main impediments to the development of substructured on-line hybrid test technique towards its general application. One is the numerical stability and accuracy problems when applied to multi-degree-of-freedom (MDOF) structures using available numerical integration schemes currently available from finite-element analysis. Another major impediment is the lack of general loading system and the necessary control algorithms capable of testing structures under combined axial, shear, and bending loads simulating what a structure actually experiences in an earthquake.

In this paper, a substructured on-line computer-actuator hybrid method for inelastic earthquake response of structures is presented. To extend the applicability of substructured hybrid method, a three degree-of-freedom loading system capable of testing seismic isolators and columns under combined axial, shear, and bending loads has been developed and implemented. In addition, the usual quasi-static cyclic loading test and the basic hybrid method are incorporated under the same physical setup, thereby providing a complete testing program for inelastic behavior of structures. Finally, the developed system is applied to high-damping rubber seismic isolators and steel box-section columns.

ON-LINE COMPUTER-ACTUATOR HYBRID ANALYSIS/TEST METHOD

On-line hybrid test (pseudo-dynamic test) is a computer-controlled experimental technique in which direct numerical time integration is used to solve the equations of motion. The computed displacement at each step is statically imposed on a specimen through a computer-controlled load actuators in order to measure its restoring forces at the current deformation state. The measured restoring forces are then fed into the equations of motion to compute the next set of displacements. Reviews of the development and future prospects have been presented by Mahin et al. [1], Takanashi and Nakashima [2], and Iemura [3].

Basic Formulation

The equations of motion of a nonlinear structural system may be expressed as follows:

$$[M]\{\ddot{u}\}_t + [C]\{\dot{u}\}_t + \{R_f\}_t = \{P\}_t \tag{1}$$

where $\{u\}_t$, $\{\dot{u}\}_t$, $\{\ddot{u}\}_t$: displacement, velocity, and acceleration vector at time t;

$[M]$, $[C]$: mass and viscous damping matrices;

$\{R_f\}_t$: restoring forces to be measured experimentally from a specimen loaded based on its current deformation state $\{u\}_t$;

$\{P\}_t$: load vector at time t.

In an on-line hybrid test, the restoring force vector $\{R_f\}_t$ is measured from a specimen loaded based on its current deformation state. Using an explicit integration scheme (e.g., the central difference scheme), the equations of motion can be solved to obtain the displacement vector at the current step $\{u\}_{t+\Delta t}$ depending only on quantities computed in the previous steps.

$$\{u\}_{t+\Delta t} = \frac{\{P\}_t - \{R_f\}_t + \left(\frac{2}{\Delta t^2}[M]\right)\{u\}_t - \left(\frac{1}{\Delta t^2}[M] - \frac{1}{2\Delta t}[C]\right)\{u\}_{t-\Delta t}}{\frac{1}{\Delta t^2}[M] + \frac{1}{2\Delta t}[C]} \tag{2}$$

Substructured Hybrid Method

In the basic formulation of on-line hybrid test, specimen modeling the total structural system is fabricated and the total restoring force vector is obtained experimentally. However, inelastic deformations in structures during strong earthquake are often confined to some localized regions. By incorporating substructuring concepts into the basic form of on-line hybrid test, only the critical regions need to be fabricated and tested.

A complete structural system is considered to be divided into analytical substructures and experimental substructures. Portions of the structure that are as yet difficult to model mathematically are taken as experimental substructures. The remaining portions are taken as analytical substructures in which presently available analytical models are used to describe their restoring-force characteristics. For instance, the critical first-story columns of a two-story frame (Figure 1a) may be taken as the experimental substructures. As another example, the seismic isolation devices are taken as the experimental substructures in the seismically-isolated structures shown in Figure 1b.

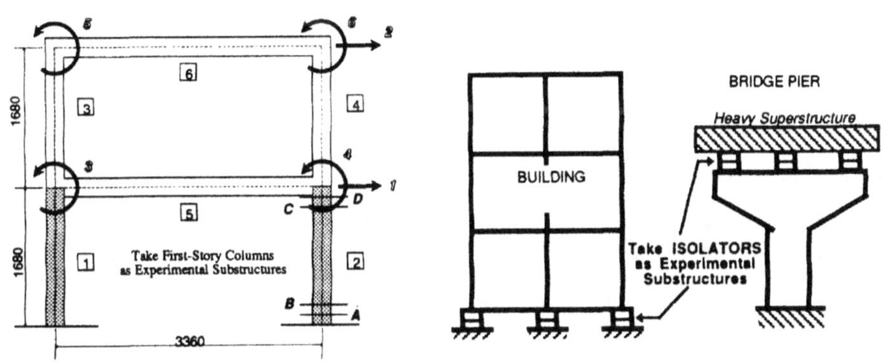

(a) Multi-story frame structure (b) Structures with seismic isolation

Figure 1. Analytical and Experimental Substructures

The equations of motion may be similarly expressed as in Eqn. 1. However, the contributions into the restoring-force vector $\{R_f\}_t$ come from both analytical models and measured restoring forces from the loaded specimen.

$$\{R_f\}_t = \{R_f\}_t^{experimental} + \{R_f\}_t^{analytical} \tag{3}$$

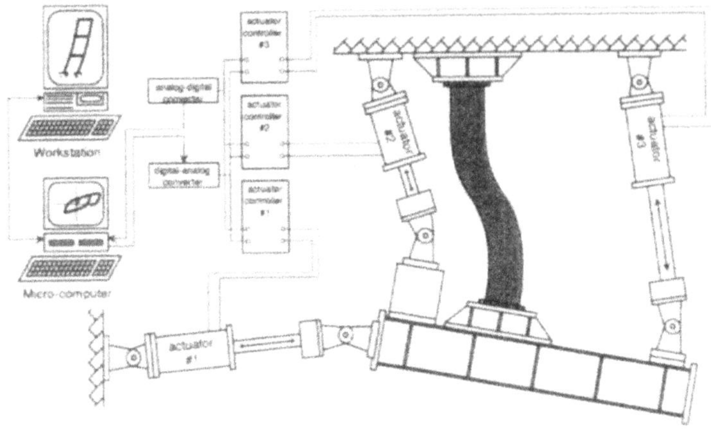

Figure 3. Set-up for testing column

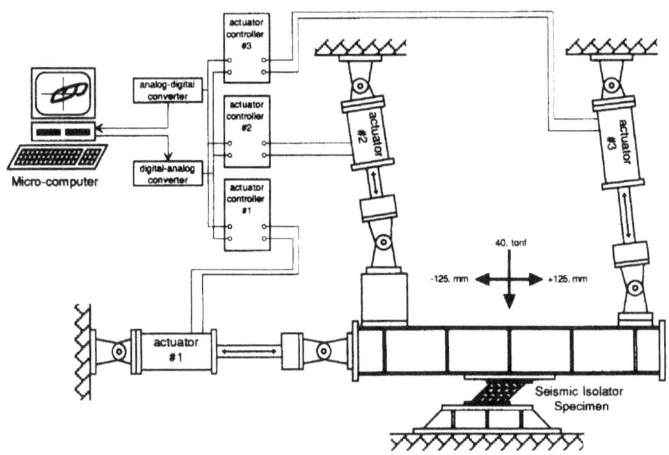

Figure 4. Set-up for testing seismic isolator

On-line Computer Control System

The general setup of the on-line computer control system is shown in Figure 5. The control computer used in these sets of experiments is a TEAC PS-9000 series 300 technical computer (Yokogawa Hewlett-Packard) that is equipped with sophisticated interface programming control. An internal GPIB (IEEE 488 standard) interface bus is used to connect the analog-digital (A/D) converter, hard disks and other storage media, printers, pen plotters, and other peripherals. Digital strain reading device is also attached through the GPIB bus. Three channels of 12-bit digital-analog (D/A) converters are multiplexed through a Serial Output I/F module to the GPIO (General-Purpose Input/Output has 16 input lines and 16 output lines) interface in the control computer. In substructured hybrid tests, a Sony NEWS workstation or a PC microcomputer is connected by RS-232C serial interface to handle all the computational tasks, while the TEAC computer is solely devoted to controlling the actuators and other instruments.

A brief summary of the procedure is illustrated by the flowchart in Figure 2.

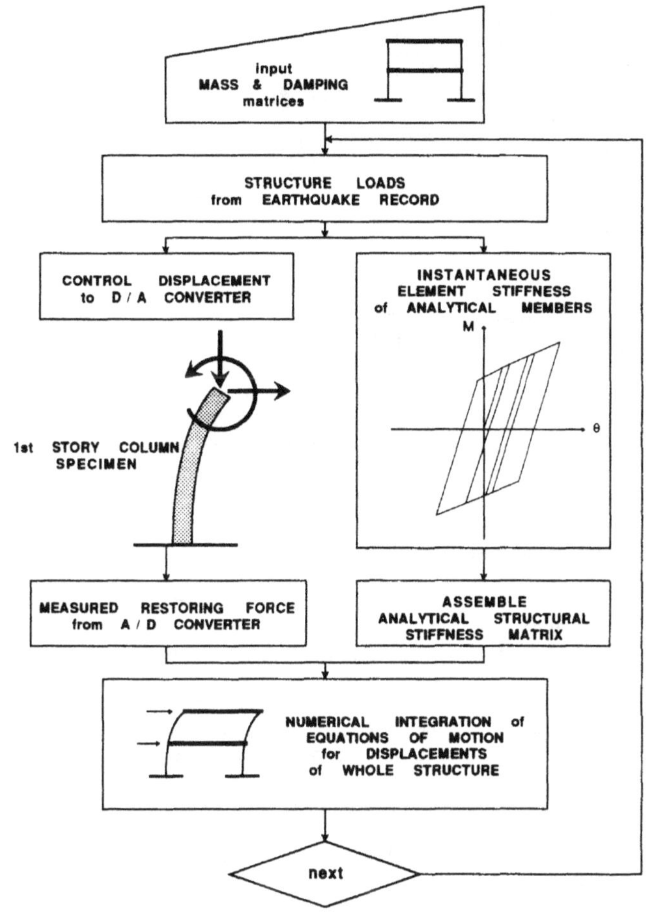

Figure 2. Substructured Hybrid Method for Inelastic Earthquake Response of Structure

HARDWARE AND SOFTWARE IMPLEMENTATION

A 3-DOF General In-plane Loading System

A 3-DOF general in-plane loading system capable of subjecting a specimen under combined axial, shear, and bending loads has been developed and implemented. The general set-up of the test bed and test rig is shown in Figure 3. One actuator is attached horizontally to the reaction wall, while two others are hanged vertically from the overhanging reaction girder. A rigid load-transfer beam was fabricated to transfer displacements and forces between the actuators and the specimen. Universal joints consisting of swivel heads and swivel bases are attached to both ends of each actuator. The loading system has been used to test RC and steel box-girder columns, and high-damping rubber seismic isolators (Figure 4).

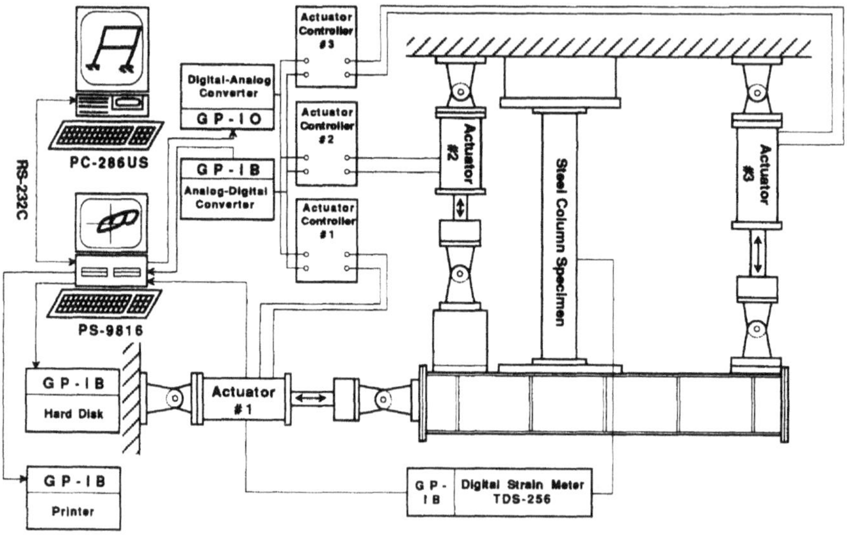

Figure 5. On-line computer control system

Inelastic Earthquake Analysis and Analytical Modeling

The procedure of computing inelastic earthquake response in a substructured hybrid test is very much similar to conventional analytical procedure, except for the contributions of the experimental restoring forces which may simply be treated as a subroutine for describing the hysteretic load-deformation characteristics of those experimental substructures. While very sophisticated modeling of structures and elements (e.g., finite element or fiber models) may be used in principle, there is a time constraint due to load relaxation problem in loaded specimen. Nevertheless, simple analytical modeling is sufficient using good hysteretic models for the analytical substructures.

In the present implementation, a structure is discretized into lumped-parameter system model and analytical members are described by member models (e.g., single-component member models). A wide variety of hysteretic models are available such as bilinear models and some degrading models (e.g., Takeda model for RC). In this present work on the steel columns, the analytical values of the yielding girders are varied to simulate strong-column weak-girder models and weak-column strong-girder models in which the inelastic response behavior of the critical first-story columns are investigated.

Numerical Integration Scheme

Using the conditionally stable explicit integration method in substructured on-line hybrid tests of a MDOF system would necessitate extremely small integration time intervals in order to satisfy the required numerical stability criterion. This requirement is especially stringent in systems such as structural frame where the axial and rotation D.O.F.s are very much stiffer than the lateral ones. A class of so-called mixed integration methods originally developed for finite-element analysis of structure-fluid interaction has been found to be suitable for substructured hybrid tests [4,5].

Belytschko and Mullen [6] proposed an implicit-explicit node-by-node partitioning, while Hughes and Liu [7] presented an element-by-element partitioning scheme. An explicit integration scheme is employed in the stiff subdomains, while an implicit integration scheme is used in the remaining flexible subdomains. Park [8] presented a unified partitioning procedure which can incorporate the above two pioneering schemes and introduced a D.O.F. by D.O.F. partitioning scheme which can be applied to separate computationally the D.O.F.s exhibiting low frequencies from those exhibiting high frequencies. Belytschko, Yen, and Mullen [9] presented a subcycling procedure in which different time steps are used in different partitions. All of the above procedures involve partitioning of the structural system in which appropriate procedures are used in each partitioned subsystem. Later, Plesha and Belytschko [10] introduced a constitutive operator splitting method for nonlinear transient analysis (based on an earlier proposition by Hughes, Pister, and Taylor) in which the material constitutive law is split into a constant history-independent relation (implicit portion) and a variable history-dependent relation (explicit portion).

While all of the above procedures have been directed toward the solution of very large coupled-field interaction problems, the partitioning schemes lend themselves very amenably to substructured hybrid test procedure. Whereas implicit integration is used mainly for the stiff partitions in numerical analysis of coupled-field interaction problem, the suitability of mixed methods for substructured hybrid tests is that explicit integration can be used for the determination of the current deformation state of the experimental substructure. Hughes-Liu member partitioning scheme had been used in several cases of substructured hybrid tests and studies [4,5] probably due to its natural adaptation to the member substructuring step in a substructuring hybrid test. The first mixed method by Belytschko and Mullen is also applicable to substructured on-line test and had been used in numerical studies [5]. Taking advantage of the popularity of the central difference integration scheme in hybrid tests, the subcycling technique may be worked to simplify programming and computational efforts. The DOF-by-DOF partition may be used to complement member partitioning scheme [ ] in which each element or subdomain is further partitioned into DOF's exhibiting low frequencies from high frequencies. The study of mixed integration methods is still a very active field of interest and it is expected that more useful schemes may be adapted for substructured hybrid tests.

Nakashima et al. [5] compared the 3 integration techniques (Belytschko-Mullen, Hughes-Liu, and OS methods) adapted for substructured hybrid test and has found that the constitutive operator splitting (OS) method is the most effective to substructured on-line hybrid tests in terms of both solution stability and accuracy. The OS method has been frequently used for substructured hybrid tests in Japan for the most recent tests based on the merits studied by Nakashima et al. [5]. It is adapted in the current implementation of this substructured hybrid loading test system used in the testing of steel box bridge piers presented here.

To implement the OS algorithm, Eqn. (1) is expressed as follows:

$$[M]\{\ddot{u}\}_{t+\Delta t} + [C]\{\dot{u}\}_{t+\Delta t} + \{R_f\}^{linear}_{t+\Delta t} + \{R_f\}^{nonlinear}_{t+\Delta t} = \{P\}_{t+\Delta t} \quad (4)$$

$$[M]\{\ddot{u}\}_{t+\Delta t} + [C]\{\dot{u}\}_{t+\Delta t} + [K]\{u\}_{t+\Delta t} + (\{R_f\}_{t+\Delta t} - [K]\{\tilde{u}\}_{t+\Delta t}) = \{P\}_{t+\Delta t}$$

An explicit trapezoidal rule for the predictive displacement vector is given by:

$$\{\tilde{u}\}_{t+\Delta t} = \{u\}_t + \Delta t\{\dot{u}\}_t + \left(\Delta t^2/4\right)\{\ddot{u}\}_t \quad (5)$$

On the other hand, an implicit trapezoidal rule (Newmark method with $\gamma = 0.5$ and $\beta = 0.25$)

$$\{u\}_{t+\Delta t} = \{\tilde{u}\}_{t+\Delta t} + \left(\Delta t^2/4\right)\{\ddot{u}\}_{t+\Delta t} \quad (6)$$

$$\{\dot{u}\}_{t+\Delta t} = \{\dot{u}\}_t + (\Delta t/2)(\{\ddot{u}\}_t + \{\ddot{u}\}_{t+\Delta t}) \quad (7)$$

Substituting the displacement and velocity expressions given above into Eqn. (2),

$$\{\ddot{u}\}_{t+\Delta t} = \frac{\{P\}_{t+\Delta t} - \{R_f\}_{t+\Delta t} - [C]\{\dot{u}\}_t - [C](\Delta t/2)\{\ddot{u}\}_t}{([M] + [C]\Delta t/2 + [K]\Delta t^2/4)} \quad (8)$$

It has been shown [10] that the OS integration method is unconditionally stable for materials with decreasing stiffness (hysteresis is of 'softening' type).

Loading and Instrumentation Control Algorithm

General displacement pattern of the experimental substructure is transformed into deformation of an equivalent cantilever model. For a desired deformation state in the cantilevered specimen tip defined by $\delta_x, \delta_y, \theta_z$, the restoring forces P_x, P_y, M_z are to be determined. The procedure is outlined in the following: (1) Initial lengths of the actuator pistons are computed at the initial set position with the specimen tip O taken as reference origin; (2) Compute constant distances and orientations within a rigid body between specimen tip O and the respective positions of the pivot hinge of the swivel heads; (3) For a desired deformation state defined by $[\delta_x, \delta_y, \theta_z]$, locate the new position of the pivot hinge of the respective swivel heads; (4) From the computed new positions of the swivel heads, compute the changes in lengths in order for the piston to reach these new positions; (5) Find the actuator displacement control signals $\Delta_1, \Delta_2, \Delta_3$ and send commands to D/A converter; (6) Load cells inside the actuators measure the forces induced in the pistons F_1, F_2, F_3 and readings taken by A/D converters; (7) Specimen restoring forces P_x, P_y, M_z are computed based on equilibrium equations of the rigid transfer beam; (8) And finally, cantilever restoring forces are transformed into the required restoring forces at the member ends.

ANALYSIS AND TEST RESULTS

High-Damping Rubber Seismic Isolators

Using the developed loading system described above, extensive tests were conducted on the HDR seismic isolators. Seismic performance of HDR-supported systems can be evaluated under different earthquake ground motions. Resonant response under frequency-sweeping sine input excitations can also be tested. In addition, the normal procedure of repeated cyclic loading test can be conducted using the same testing facility. Most recently, a substructured hybrid loading system for MDOF bridge structures with seismic isolators has been developed. Thus, the integrated testing system can test for: (a) fundamental mechanical properties, (b) resonant response, and (c) earthquake response. This offers a versatile testing system in the early development stage, as well as for final verification and proof tests.

A series of high-damping rubber (HDR) seismic isolators with equivalent damping rated at about 15% had been tested. The specimens are of 25-cm $\times$ 25-cm square in plan. A 3-layer configuration, a 4-layer configuration, and a 9-layer configuration were fabricated to determine differences in stiffness and damping properties due to different heights. Each rubber layer has a thickness of 12mm and 1mm thin steel plates are interleaved between the rubber layers to form laminated elastomers with high vertical stiffness but are laterally flexible.

The setup shown in Figure 4 was used in the test. A specimen is bolted to the underside of the rigid load-transfer beam and to a rigid platform that is attached to the rails of the strong reaction floor. Different heights of isolator specimens could be accommodated by inserting spacer plates between the specimen and the test platform. In this series of tests, the load-transfer beam was controlled to impose horizontal displacements while maintaining a constant axial load on the isolator specimen.

Cyclic load-deformation behavior of the HDR isolator subjected to cyclic displacements of $\pm$25%, $\pm$50%, ..., $\pm$150% of the specimen height with two repetitions of each cycle was first conducted. The specimen was finally loaded to $\pm$200% to check its behavior at extreme overload situations. Other tests under this mode include different levels of axial loads and effects of loading rate. Detailed results are presented in Iemura et al.[11].

A HDR-isolated structure is then modeled as a 2-sec SDOF system and subjected to sweeping sine excitation with frequency of sine waves ranging from 0.2 to 2.0 cps. Figure 6 shows the load-deformation response and the displacement and acceleration time history response (compared with analytical equivalent linear response). Different levels of acceleration amplitude are also conducted to establish resonant characteristics of the isolator at low loads as well as high loads.

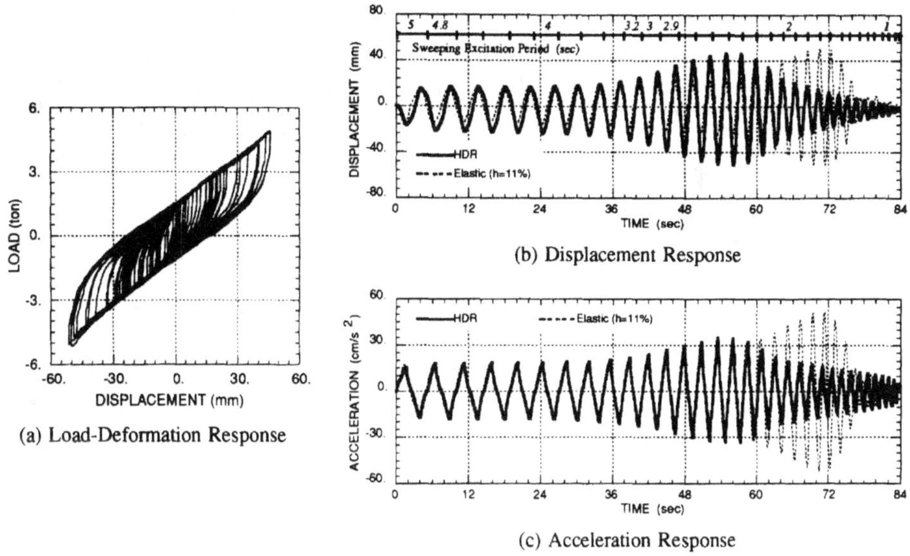

(a) Load-Deformation Response

(b) Displacement Response

(c) Acceleration Response

Figure 6. Resonant Response due to Sweeping-Sine Excitation for 100% Shear Strain

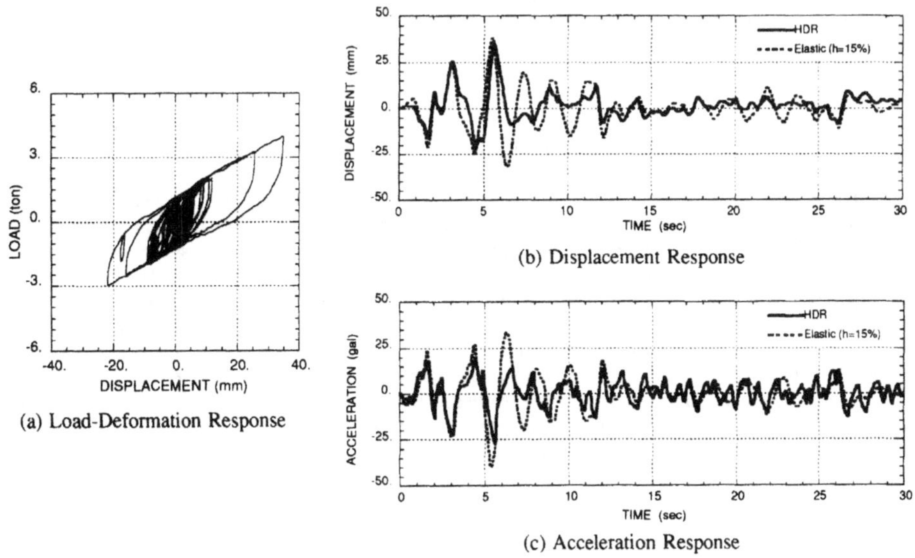

(a) Load-Deformation Response

(b) Displacement Response

(c) Acceleration Response

Figure 7. Earthquake Response due to El Centro Excitation for 100% Shear Strain

The 2-sec SDOF HDR system was also subjected to the El Centro earthquake which has short-period waves and to the Hachinohe earthquake which has long-period waves. The load-deformation response and displacement and acceleration response of the HDR isolator under 0.1g El Centro earthquake are shown in Figure 7. The load-deformation response and displacement and acceleration response of the HDR isolator under 0.1g Hachinohe earthquake are shown in Figure 8.

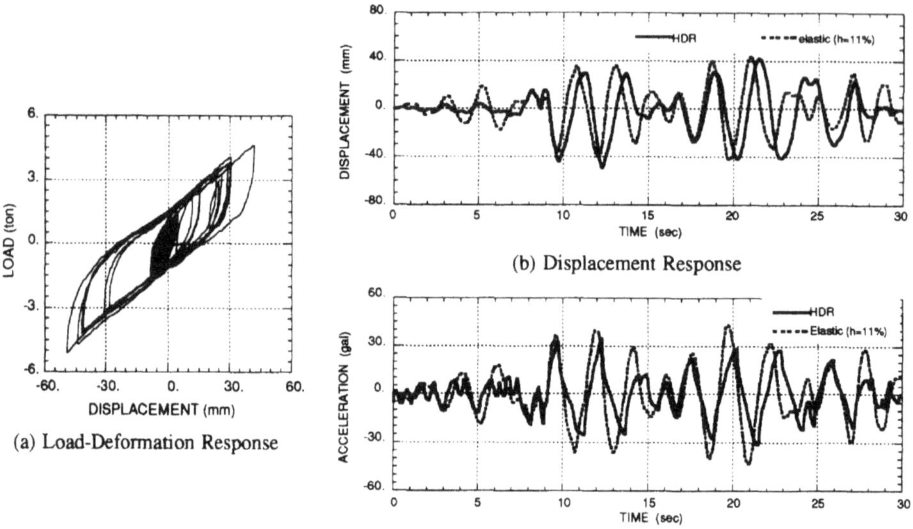

(a) Load-Deformation Response

(b) Displacement Response

(c) Acceleration Response

Figure 8. Earthquake Response due to Hachinohe Excitation for 100% Shear Strain

Most recently, a substructured hybrid loading system for MDOF bridge structures with seismic isolators has been developed. Figure 1b shows two examples of base-isolated building and a seismic isolated bridge pier. In the substructured loading test, the isolator elements are taken as experimental substructures and are tested. This test provides a proof test of the effectiveness of the isolator for a specific application. After the verification tests of the HDR isolators presented above, the current phase of the test program on the HDR isolators includes substructured hybrid loading test of the isolator for inelastic earthquake response of a MDOF bridge structural model.

The following main results have been obtained: (1) From cyclic loading tests, it has been confirmed that the HDR isolators behave as bilinear elements without strain hardening when strain is below 100%. Equivalent stiffness and damping decreases with increasing strain. Also, lateral stiffness decreases but damping increases with increasing axial bearing pressure. (2) From resonant tests using sweeping sine excitation, the resonant response of a 2.sec SDOF HDR system was obtained and has been found to possess sufficient damping and resistance (especially with increasing strain and strain hardening) to suppress large amplification. (3) Under El Centro earthquake, equivalent damping of about 15% has been observed for strain of about 100%, the same level obtained from static tests. However, equivalent damping of only 11% was observed when subjected to the Hachinohe earthquake which contains long-period waves. (4) Finally, the performance of HDR isolators as part of a MDOF bridge structures can be determined and verified by the newly developed substructured hybrid loading system.

Steel Box-Section Columns

A substructured hybrid loading system capable of subjecting a column specimen under combined axial, shear, and bending loads has been developed and used to test steel box-section columns. In addition, cyclic loading tests were also conducted. A series of steel box-section columns with different yield strength and stiffeners were fabricated.

In the first part of the testing program, cyclic loading tests were conducted on the steel column specimens in which patterns of combined axial, shear, and bending loads were imposed. The main parameters are different steel material properties and different ratios of stiffeners. Detailed results are presented in Yamada et al.[12].

In the second part of the testing program, inelastic earthquake response of two-story structural steel frames were determined by substructured hybrid loading tests. The critical first-story columns were taken as the experimental substructures, while the rest of the structure was modeled by analytical substructures. Inelastic behavior of the columns and the response of the total structure under earthquake ground motion was determined. The effect of varying the analytical stiffnesses of the yielding girders on the column behavior was investigated. Loading cases 3 and 4 are substructured hybrid loading cases. In Case 3 loading, the girders are modeled to have the same stiffness as the columns. In Case 4 loading, the girders are modeled to have twice the stiffness of the columns. The fundamental natural period of the model is set at 0.7 sec. The El Centro NS record (1940) is used as the input earthquake ground motion.

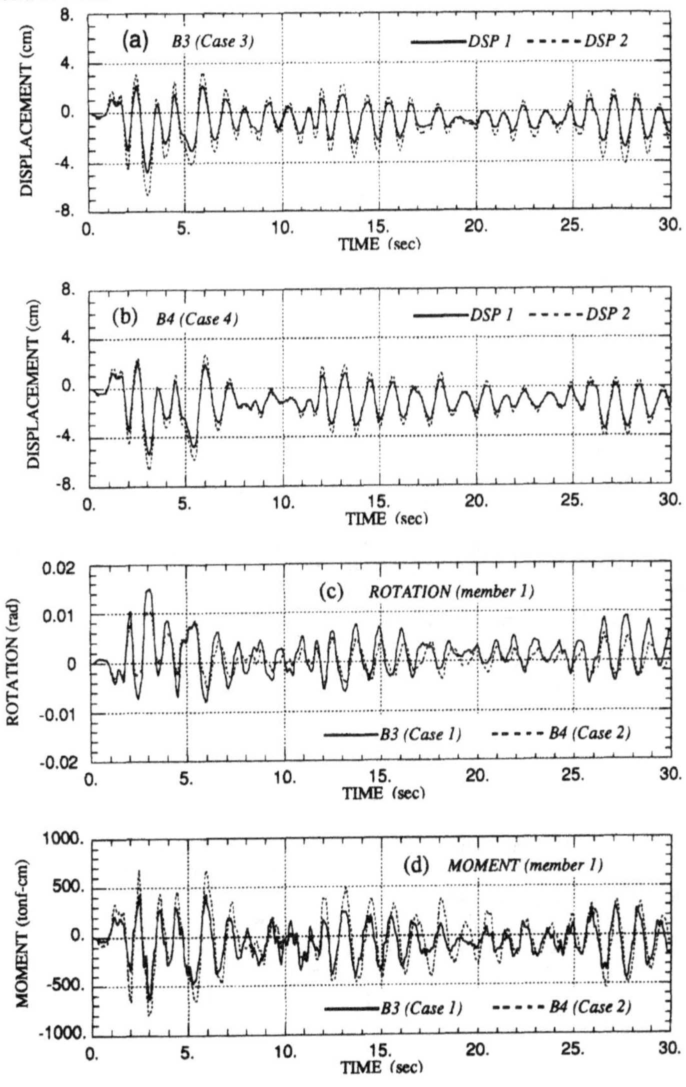

Figure 9. Time Histories of Earthquake Response

Results of Type B specimens are presented in this paper. Time history response is given in Figure 9. Figure 9a shows the story displacements for Case 3 substructured hybrid loading of the column in the frame with more flexible girder, while Figure 9b shows the story displacement response for Case 4 substructured hybrid loading test. Comparing (a) and (b), the upper story displacement is reduced due to stiffer structure (stiffer girders). The stiffer girders also reduced the rotation of the joints as shown in Figure 9c. Consequently, the first-story end of the tested column in Case 4 connected to a stiffer girder is displaced in a more constraining manner producing larger moment than in Case 3. Moment-curvature measurements taken at four sections indicated in Figure 1a are shown in Figure 10 for Case 3 frame and in Figure 11 for Case 4 frame.

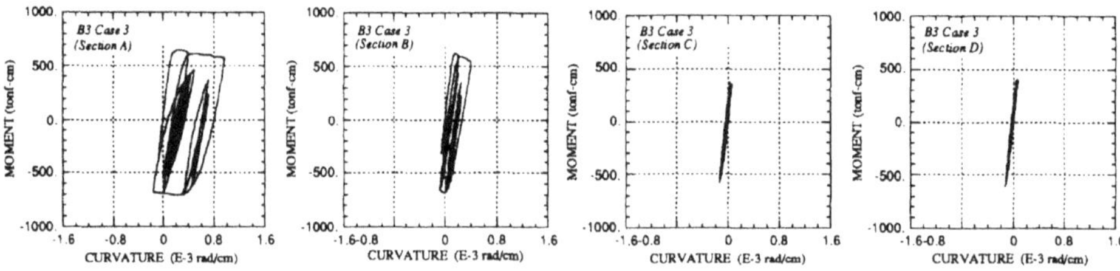

Figure 10. Moment-Curvature Response for Type B3 under Case 3 Substructured Hybrid Loading

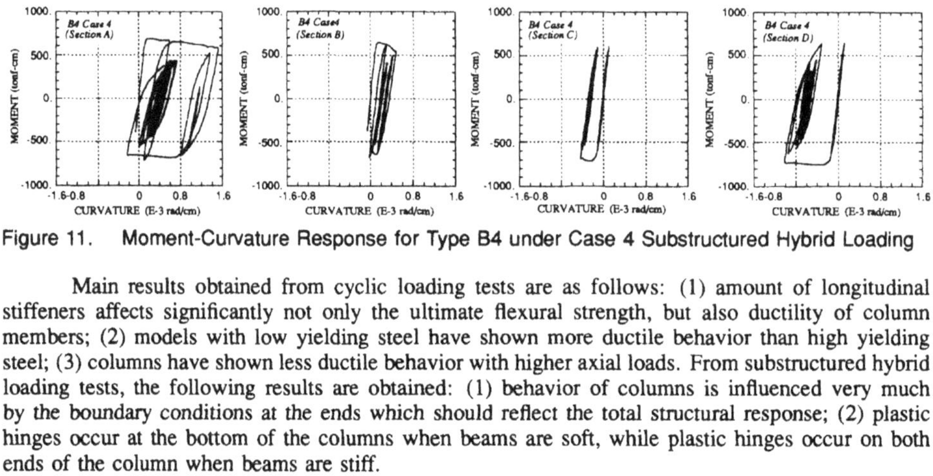

Figure 11. Moment-Curvature Response for Type B4 under Case 4 Substructured Hybrid Loading

Main results obtained from cyclic loading tests are as follows: (1) amount of longitudinal stiffeners affects significantly not only the ultimate flexural strength, but also ductility of column members; (2) models with low yielding steel have shown more ductile behavior than high yielding steel; (3) columns have shown less ductile behavior with higher axial loads. From substructured hybrid loading tests, the following results are obtained: (1) behavior of columns is influenced very much by the boundary conditions at the ends which should reflect the total structural response; (2) plastic hinges occur at the bottom of the columns when beams are soft, while plastic hinges occur on both ends of the column when beams are stiff.

CONCLUSIONS

Combining the best features of computational mechanics and experimental mechanics, an on-line computer-actuator substructured hybrid system for inelastic earthquake response of structures is developed. Using the 3-DOF general in-plane loading system, seismic isolators and columns had been tested and restoring forces are accurately determined for earthquake response analysis. Combined with substructured hybrid loading test procedure in which analytical models are used to characterize a total structural system, critical members can be tested under realistic loads and proper boundary conditions. Using restoring forces that are directly measured from a loaded specimen, inelastic earthquake response of the total structure can be reliably predicted.

ACKNOWLEDGEMENTS

The authors would like to acknowledge the support and cooperation of the following persons and organization: Mr. Kazuyuki Izuno (research associate), Mr. Shinji Nakanishi (laboratory engineer) of Kyoto University, Mr. Manabu Yasuda of Kajima Corp. (formerly graduate student), Katayama Ironworks Co., Toyo Tire & Rubber Co., Kawaguchi Metal Industries Co., and Nitto Engineering Consultants.

REFERENCES

1. S. Mahin, P. B. Shing, C. Thewalt, and R. Hanson, "Pseudodynamic Test Method—Current Status and Future Directions", *Journal of Structural Engineering, ASCE,* 115(8), 1989.

2. K. Takanashi and M. Nakashima, "Japanese Activities on On-line Testing", *Journal of Engineering Mechanics, ASCE,* 113(7), July.

3. H. Iemura, "Development and Future Prospects of Hybrid Experiments", *Transaction, Japan Society for Civil Engineers,* 356(I-3), 1985. (in Japanese)

4. S. Dermitzakis and S. Mahin, "Developement of Substructuring Techniques for On-line Computer Controlled Seismic Performance Testing", Report No. UCB/EERC–85/04, Earthquake Engg. Research Center, Univ. of California, Berkeley.

5. M. Nakashima, T. Kaminosono, M. Ishida, and K. Ando, "Integration Techniques for Substructure Pseudo Dynamic Test", *4th U.S. National Conference on Earthquake Engineering,* 1990.

6. T. Belytschko and R. Mullen, "Mesh Partitions of Explicit-Implicit Time Integration", *Formulations and Computational Algorithms in Finite Element Analysis: US-Germany Symposium,* (eds. Bathe, Oden, Wunderlich), pp. 673–690.

7. T. Hughes and W. K. Liu, "Implicit-Explicit Finite Elements in Transient Analysis: Implementation and Numerical Examples", *Journal of Applied Mechanics, ASME,* Vol. 45, pp. 375–378.

8. K. C. Park, "Partitioned Transient Analysis Procedures for Coupled-Field Problems: Stability Analysis", *ASME Journal of Applied Mechanics,* 47, 1980, pp. 370–376.

9. T. Belytschko, H. J. Yen and R. Mullen, "Mixed Methods for Time Integration", *Computer Methods in Applied Mechanics and Engineering,* Vol. 17/18, pp. 259–275.

10. M. Plesha and T. Belytschko, "A Constitutive Operator Splitting Methods for Nonlinear Transient Analysis", *Computers and Structures,* 20(4), pp. 767–777.

11. H. Iemura, Y. Yamada, W. Tanzo, Y. Uno, and S. Nakamura, "On-line Earthquake Response Tests of High-Damping Rubber Bearings for Seismic Isolation", *US-Japan Seminar on Seismic Isolation,* SUNY Buffalo, New York, Sept. 1991.

12. Y. Yamada, H. Iemura, W. Tanzo, M. Yasuda, and S. Oka, "Substructured Hybrid Earthquake Loading Test of Steel Box Pier Models", *US-Japan Seminar on Cyclic Buckling of Steel Structures and Structural Elements Under Dynamic Loading Conditions,* Osaka, Japan, July 1991.

RECENT DEVELOPMENT OF 3-DIMENSIONAL ANALYSIS OF TALL BUILDING STRUCTURES BY CONTINUUM METHOD

Liang Qizhi
(South China University of Technology, Guangzhou 510641, China)

Keywords: continuum method, 3-D analysis, tall building structures, second-order analysis.

ABSTRACT

The routine continuum method has proved itself to be extremely practical in structural analysis and design of tall buildings. But it can only be applied to coupled shear walls. Its recent development performed by the author and his graduate students has extended its application to all types of tall building structures, inclusive of 3-D analysis of static and dynamic problems, second-order analysis and overall stability of structures etc. In this paper, the governing equations as well as their theoretical (analytical) and numerical solutions for these problems are outlined with comment.

1. FORMULATION OF GOVERNING EQUATIONS

1.1 For Plane Multi-Bay Coupled Shear Walls

It is well known in literature[1][2] that the governing differential equation for single-bay coupled shear walls may be written in the form (Fig. 1):

$$t''(z) - bt(z) = -b_0V(z) \tag{1}$$

Where t(z) is the distributed shear force (per unit length) in the laminae; z — the ordinate along the height of building; V(z)—shear due to external load, and

$$b = \frac{12\tilde{I}_b}{hl^3}\left(\frac{a^2}{I_w} + \frac{1}{A_1} + \frac{1}{A_2}\right) \tag{2}$$

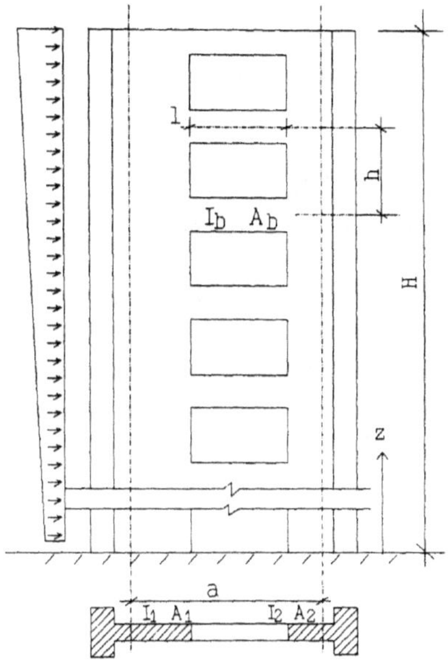

Figure 1

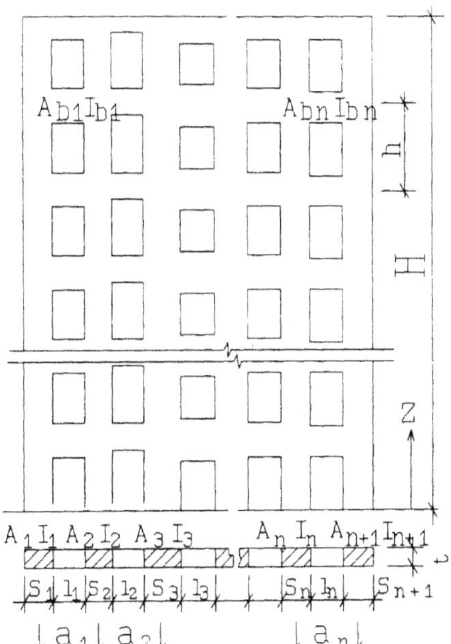

Figure 2

$$I_w = I_1 + I_2 \tag{3}$$

$$\tilde{I}_b = \frac{I_b}{1 + \dfrac{28\mu I_b}{A_b l^2}} \tag{4}$$

$$b_0 = \frac{12\tilde{I}_b a}{hl^3 I_w} \tag{5}$$

where μ is the shape factor for non - uniform distribution of shear stress in the cross section of a lintel beam.

J. Balaš et al. [3]attempted to present the governing differential equations and their theoretical solution for multi-bay coupled shear walls, but failed due to some errors both in establishing and in solving the governing equations. In Reference[4], the author of this paper first successfully constructed such equations and delivered their theoretical solution. The governing equations for multi-bay coupled shear walls may be written in the following matrix form (Fig. 2):

$$\{t\}''[B_p]\{t\} = -\{B_{op}\}V(z) \tag{6}$$

where

$$\{t\} = [t_1(z) \quad t_2(z) \quad \cdots \quad t_n(z)]^T \tag{7}$$

$t_i(z)$—shear force(per unit length) in the i-th band of laminae

$$\{B_{op}\} = [b_{01} \quad b_{02} \quad \cdots \quad b_{0n}]^T \tag{8}$$

$$b_{0i} = \frac{a_i}{\beta_i I_w} \tag{9}$$

$$I_w = \sum_{j=1}^{n+1} I_j \tag{10}$$

$$\beta_i = hl_i \left[\frac{l_i^2}{12I_{bi}} + \frac{\mu E}{GA_b}\right] \tag{11}$$

$$[B_p] = \begin{bmatrix} b_{11} & b_{12} & \cdots & b_{1n} \\ b_{12} & b_{22} & \cdots & b_{2n} \\ \vdots & \vdots & & \vdots \\ b_{n1} & b_{n2} & \cdots & b_{nn} \end{bmatrix} \tag{12}$$

$$b_{ij} = \left\{ \begin{array}{ll} \dfrac{1}{\beta_i}\left(\dfrac{a_i^2}{I_w} + \dfrac{1}{A_i} + \dfrac{1}{A_{i+1}}\right), & when\ i = j \\ \dfrac{1}{\beta_i}\left(\dfrac{a_i a_j}{I_w} - \dfrac{1}{A_j}\right), & when\ j - i = 1 \\ \dfrac{1}{\beta_i}\left(\dfrac{a_i a_j}{I_w} - \dfrac{1}{A_i}\right), & when\ i - j = 1 \\ \dfrac{1}{\beta_i}\left(\dfrac{a_i a_j}{I_w}\right) & when\ |i - j| \geqslant 2 \end{array} \right\} \tag{13}$$

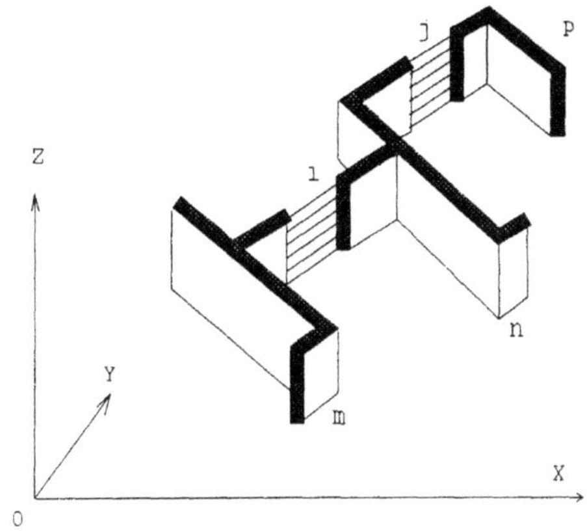

Figure 3

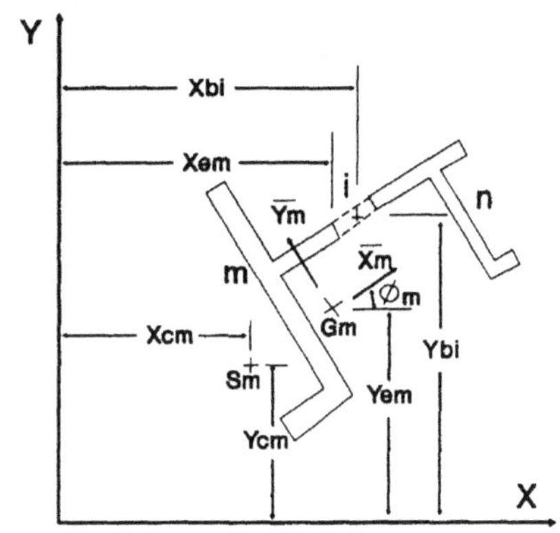

Figure 4

1.2 For Space Coupled Shear Walls

Regarding the shear walls as thin walled beams with open cross-sections, J. K. Biswas et al.[5] formulated the basic equations for space coupled shear walls and derived an approximate solution. In References[6][7], the author developed the governing equations for space coupled shear walls in matrix form and first carried out an exact (theoretical) solution for them. Such matrix form is

$$\{t\}'' - [B_s]\{t\} = [B_{os}]\{L\} \tag{14}$$

where

$$[B_s] = E[\beta]^{-1}([R][K_0]^{-1}[R]^T + [A]) \tag{15}$$

$$[B_{0s}] = E[\beta]^{-1}[R][K_0]^{-1} \tag{16}$$

$$\{L\} = [-V_x(z) \quad -V_y(z) \quad -M_t(z)]^T \tag{17}$$

$V_x(z), V_y(z)$—shear in direction X, Y respectively due to external load (Fig. 3);

$M_t(z)$—the torque about Z — axis due to external load;

$$[\beta] = diag\,[\beta_1 \quad \beta_2 \quad \cdots \quad \beta_n] \tag{18}$$

$$[R] = \begin{bmatrix} r_{x1} & r_{y1} & r_{\theta 1} \\ r_{x2} & r_{y2} & r_{\theta 2} \\ \vdots & \vdots & \vdots \\ r_{xn} & r_{yn} & r_{\theta n} \end{bmatrix} \tag{19}$$

$r_{xi} = X_{en} - X_{em}$ = the distance, measured along direction X, between the centroids of cross-sections of walls m and n which are connected by the i-th band of laminae (Fig. 4); (20a)

$$r_{yi} = Y_{en} - Y_{em} \tag{20b}$$

$$r_{\theta i} = X_{cm}(Y_{bi} - Y_{em}) - Y_{cm}(X_{bi} - X_{em}) - X_{cn}(Y_{bi} - Y_{en}) + Y_{cn}(X_{bi} - X_{en}) + \omega_{im} - \omega_{in} \tag{20c}$$

ω_{im}(or ω_{in})—the sectorial coordinate of midspan of the i-th bandof laminae with respect to the shear center of wall m (or n);

X_{cm}, Y_{cm} (or X_{cn}, Y_{cn})—the coordinates of shear center or wall m (or n);

$$[K_0] = E\begin{bmatrix} I_{yy}^* & I_{xy}^* & -I_{yc}^* \\ & I_{xx}^* & I_{xc}^* \\ \text{Symmetry} & & I_{\omega}^* \end{bmatrix} \tag{21}$$

$$I_{yy}^* = \sum_{m+1}^{M} I_{ym}, \quad I_{xx}^* = \sum_{m+1}^{M} I_{xm} \tag{22a}$$

$$I_{xy}^* = \sum_{m=1}^{M} I_{xym} \tag{22b}$$

$$I_{xc}^* = \sum_{m=1}^{M} I_{xcm} \tag{22c}$$

$$I_{yc}^* = \sum_{m=1}^{M} I_{ycm} \tag{22d}$$

$$I_{\omega}^* = \sum_{m=1}^{M} (I_{\omega m} + X_{cm}^2 I_{xm} + Y_{cm}^2 I_{ym} - 2X_{cm}Y_{cm}I_{xym}) \tag{22e}$$

M——the total mumber of walls;

I_{ym}, I_{xm}, I_{xym}——moments and product of inertia of the cross-section of wall m with respect to its own centroidal axes parallel to Y and X;

$$I_{ycm} = Y_{cm}I_{ym} - X_{cm}I_{xym} \tag{22f}$$

$$I_{xcm} = X_{cm}I_{xm} - Y_{cm}I_{xym} \tag{22g}$$

and $[A]$ is a symmetrical square matrix of order n, the elements of which $a_{ij}(i, j = 1, 2\cdots, n)$ can be determined as follows:

$$\left.\begin{array}{l} a_{ii} = \dfrac{1}{E}\left(\dfrac{1}{A_m} + \dfrac{1}{A_n}\right), \textit{with i-th band of laminae spanning between walls m and n} \\ a_{ij} = \begin{cases} \dfrac{u_{mp}}{EA_m}, \textit{with } \begin{matrix} i\text{-}th \\ j\text{-}th \end{matrix} \textit{ band of laminae spanning between walls m and } \begin{matrix} n \\ p \end{matrix} \\ -\dfrac{u_{nq}}{EA_n}, \textit{with } \begin{matrix} i\text{-}th \\ j\text{-}th \end{matrix} \textit{ band of laminae spanning between walls } \begin{matrix} m \\ n \end{matrix} \textit{ and } \begin{matrix} n \\ q \end{matrix} \\ 0, \textit{with } \begin{matrix} i\text{-}th \\ j\text{-}th \end{matrix} \textit{ band of laminae connecting } \begin{matrix} \textit{walls m and n} \\ \textit{neither wall m nor wall n} \end{matrix} \end{cases} \end{array}\right\} \tag{23}$$

$$u_{mp}(\text{or } u_{nq}) = \begin{Bmatrix} +1, \text{ when } m < p (\text{or } n < q) \\ -1, \text{ when } m > p (\text{or } n > q) \end{Bmatrix} \tag{24}$$

1.3 **For Tall Building Comprised of Various Types of Structures**

In a general case, all common types of tall building structures, namely, rigid frames, plane or space coupled shear walls, coupled system of walls and frames, framed-tube, tube-in-tube and bundled-tube structures etc., are comprised of vertical elements and connecting beams which are simulated by lamina system spanning between two vertical elements. A vertical element may be either a column or a wall; due to the probably great difference in flexural stiffness between vertical elements and the small stiffness of columns, the points of contraflexure are no longer located at the midspan of connecting beams. The hypothesis of the inflexion points located at midspan of connecting beams, which is applied in traditional continuum method for coupled shear walls, must be repudiated. For this purpose, not only the unknown distributed shears $t_k(z)$, but also the unknown distributed moments $m_k(z)$ are introduced at both sides of the cut at midspan of the lamina system (Fig. 5). Moreover, the effect of local bending of the columns must be taken into account. Based on the idea mentioned above, the author first elaborated the Generalized Continuum Method for tall building structures in References [8][9]. The governing equations may be written in the following matrix form:

$$\{t\}'' - [B]\{t\} = [B_0]\{L\} \tag{25}$$

where

$$[B] = [\beta'']^{-1}([R][K_0]^{-1}[R]^T + [A]) \tag{26}$$

$$[\beta''] = [\beta] + [U] - [U'][W]^{-1}[V] \tag{27}$$

$$[B_0] = [\beta'']^{-1}[R][K_0]^{-1} \tag{28}$$

$$[W] = [F] + [V'] \tag{29}$$

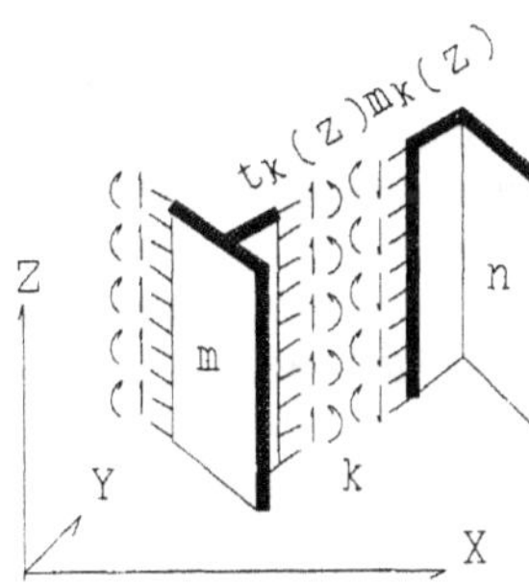

Figure 5

$[F]$—diagonal matrix of order n, the k-th element of which is h/i_{bk}, where $i_{bk} = EI_{bk}/l_k$ is the flexural stiffness of the k-th connecting beam;

$[U]$、$[U']$—square matrices of order n, their elements U_{ki}、U'_{ki} are the relative displacements at both sides of the cut at midspan of k-th band of laminae contributed by local bending of vertical elements due to $t_i = 1$、$m_i = 1$ respectively;

$[V]$、$[V']$—square matrices of order n, their elements V_{ki}、V'_{ki} are the relative rotations at both sides of the cut at midspan of k-th band of laminae contributed by local bending of vertical elements due to $t_i = 1$、$m_i = 1$ respectively.

We should notice that the unknown distributed moments $m_k(z)$ $(k = 1, 2, \cdots, n)$ are not involved in governing equations (25). In deriving Eqs. (25), the unknown distributed moments have been eliminated by condensation according to the following relationship:

$$\{m\} = -[W]^{-1}[V]\{t\} \tag{30}$$

where $$\{m\} = [m_1(z) \quad m_2(z) \quad \cdots \quad m_n(z)]^T \tag{31}$$

We should also notice that the types of all governing equations (25)(14)(6)and (1)are quite similar. In fact, if the deformation of local bending of vertical elements may be neglected, i. e. , the bending stiffness of vertical elements is very large relative to that of connecting beams, then we have $[U]=[U']=[V]=[V']\rightarrow[0]$, *and* $[\beta'']\rightarrow[\beta]$. As a result, Eqs. (25)degenerate
into Eqs. (14). Furthermore, when all shear walls are located parallel to YOZ plane, and only the actions and resistances parallel to this plane are considered, we have $I^*_{yy} = I^*_{zy} = I^*_{yc} = I^*_{xc} = I^*_{\omega} = 0, r_{xi} = r_{\theta i} = 0, r_{yi} = a_i, V_x(z) = M_t(z) = 0, V_y(z) = V(z), [B_s]\rightarrow[B_y], [B_{0s}]\rightarrow\{B_{0y}\}$, and Eqs. (14) degenerate into Eqs. (6). Finally, when n=1, Eqs. (6) degenerate into Eq. (1).

1.4 Using Folded Plate Model For Shear Walls

In many cases, it would be more convenient to simulate the shear walls with folded-plate model. Any shape of shear wall is subdivided into plane plates, and distributed shear $t_{vi}(z)$ should be inserted at both sides of the cut along each cant as shown in Fig. 6. All these $t_{vi}(z)(i=1,2\cdots)$ can be eliminated by condensation, and the governing equations consequently can be led to the same type of Eqs. (25), or the following equivalent type[10]:

$$\{S\}''-[B]\{S\} = -[B_0]\{M\} \tag{32}$$

where $$\{S\} = \int_z^H \{t\}dz \tag{33}$$

$$\{M\} = [M_{y0} \quad M_{x0} \quad B_{z0}]^T$$

M_{y0}、M_{x0}—applied moment due to external load about axis parallel to Y and X respectively;

$$B_{z0} = \int_z^H M_t(z)dz \tag{34}$$

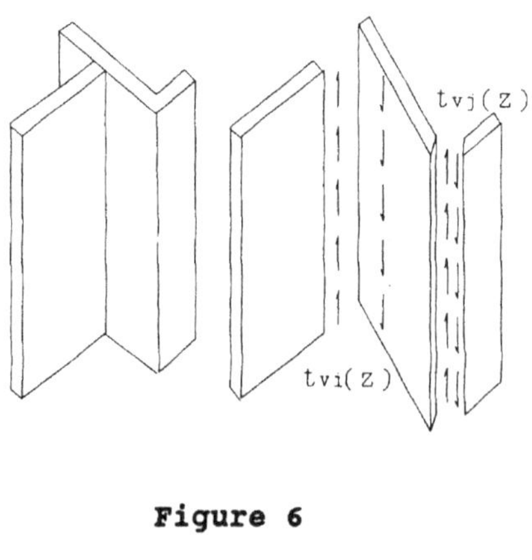

Figure 6

The order of matrix in the governing equations remains the same as that when using thin walled beam model to simulate the shear walls. The solution of stress resultants for plane plates is much more convenient than that for thin walled beams in designing the cross-sections of shear walls.

1.5 Using Strip Model For Shear Walls

When the shear walls are simulated by either thin walled beam models or folded - plate models, shear deformation in walls has been neglected and the distribution of normal stresses along the width of plane plates is linear, i. e. , the effect of shear lag has not been included. For wide plates of shear walls, the effect of shear lag cannot be ignored. In Ref. [11], we presented an alternative analytical model, in which shear walls were partitioned into planar strips and shear deformantion in each strip was considered according to Timoshenko' s beam theory. A set of governing differential equations established is of the same type as Eq. (25) or (32).

2. THEORETICAL SOLUTION OF GOVERNING EQUATIONS

The author first successfully presented the theoretical (analytical) solution for governing Eqs. (6) in Ref. [4], for Eqs. (14) in Refs. [6][7], and for Eqs (25) in Refs. [8][9]. Since governing Eqs. (6) & (14) may be regarded as special cases of Eqs. (25), only the theoretical solution of Eqs. (25) will be elucidated, for brevity, as follows.

From Theory of Differential Equations it is well known that the solution of Eqs. (25) consists of two parts — the homogeneous solution $\{t_h\}$ and the particular solution $\{t_p\}$:

$$\{t\} = \{t_h\} + \{t_p\} \tag{35}$$

The homogeneous solution is

$$\{t_h\} = [G](\{c_{cz}\} + \{d_{sz}\}) \tag{36}$$

in which

$$[G] = [G_1 \quad G_2 \quad \cdots \quad G_n] = \begin{bmatrix} g_{11} & g_{12} & \cdots & g_{1n} \\ g_{21} & g_{22} & \cdots & g_{2n} \\ \vdots & \vdots & \vdots & \\ g_{n1} & g_{n2} & \cdots & g_{nn} \end{bmatrix} \tag{37}$$

$$\{c_{cz}\} = [c_1 ch\lambda_1 z \quad c_2 ch\lambda_2 z \quad \cdots \quad c_n ch\lambda_n z]^T \tag{38}$$

$$\{d_{sz}\} = [d_1 sh\lambda_1 z \quad d_2 sh\lambda_2 z \quad \cdots \quad d_n sh\lambda_n z]^T \tag{39}$$

Each column of $[G]$, *i. e.* $\{G_j\} = [g_{1j} \quad g_{2j} \quad \cdots \quad g_{nj}]^T$, represents an eigenvector corresponding to the eigenvalue $\lambda_j (j = 1, 2, \cdots, n)$ of the characteristic equation;

$$|\lambda^2[I_n] - [B]| = 0 \tag{40}$$

where $[I_n]$— unit matrix of order n.

The column matrices of constants $\{c\} = \{c_{cz}\}_{z=0} = [c_1 \quad c_2 \quad \cdots \quad c_n]^T$ *and* $\{d\} =$

$[d_1 \quad d_2 \quad \cdots \quad d_n]^T$ will be determined by the boundary conditions.

The particular solution $\{t_p\}$ can be presented by

$$\{t_p\} = [C_0]\{L\} + \{D_0\} \tag{41}$$

where $[C_0]$—matrix of order $n \times 3$; $\{D_0\}$— column matrix of order n. It is sufficient for engineering practice to confine our discussion to that the lateral load is composed of concentrated forces and linearly distributed forces. Then $\{L\}''$ is independent of coordinate Z. Substituting Eq. (41) into Eq. (25) and comparing the terms on both sides of the equation, we obtain the following formulae for determining $[C_0]$ and $\{D_0\}$:

$$[C_0] = -[B]^{-1}[B_0] \tag{42}$$

$$\{D_0\} = [B]^{-1}[C_0]\{L\}'' = -[B]^{-1}[B]^{-1}[B_0]\{L\}'' \tag{43}$$

Substituting Eqs. (36) and (41) into Eq. (35), we have

$$\{t\} = [G](\{c_{cz}\} + \{d_{sz}\}) + [C_0]\{L\} + \{D_0\} \tag{44}$$

Substituting Eqs. (44) into the following boundary conditions

$$\{t\}_{z=0} = \{0\} \tag{45}$$

$$\{t\}'_{z=H} = \{0\} \tag{46}$$

we obtain the following formulae for determining $\{c\}$ and $\{d\}$:

$$\{c\} = -[G]^{-1}([C_0]\{L_0\} + \{D_0\}) \tag{47}$$

$$\{d\} = diag\,[\lambda_{\overline{cH}}]([G]^{-1}[C_0]\{L'_H\} + [\lambda_{SH}]\{c\}) \tag{48}$$

where $\{L_0\} = \{L\}_{z=0}; \qquad \{L'_H\} = \{L\}'_{z=H};$

$$diag[\lambda_{\overline{cH}}] = diag\left[\frac{1}{\lambda_1 ch\lambda_1 H} \quad \frac{1}{\lambda_2 ch\lambda_2 H} \quad \cdots \quad \frac{1}{\lambda_n ch\lambda_n H}\right]$$

$$diag[\lambda_{SH}] = diag[\lambda_1 sh\lambda_1 H \quad \lambda_2 sh\lambda_2 H \quad \cdots \lambda_n sh\lambda_n H]$$

The solution (44) has so far been determined; the stress resultants of lintel beams and vertical elements can easily be determined by statics. The displacements of the structural system can then be calculated by routine method of structural mechanics.

3. FINITE DIFFERENCE SOLUTION[10]

Governing Eq. (32) can be solved by finite difference method, with storey height h as step length. Using the formulae of central difference, Eq. (32) can be transformed into the following difference equation:

$$\begin{bmatrix} [C] & -2[I] & & & & 0 \\ -[I] & [C] & -[I] & & & \\ & \ddots & \ddots & \ddots & & \\ & & -[I] & [C] & -[I] & \\ & & & \ddots & \ddots & \ddots \\ 0 & & & -[I] & [C] & -[I] \\ & & & & -[I] & [C] \end{bmatrix} \begin{Bmatrix} \{S\}_0 \\ \{S\}_1 \\ \vdots \\ \{S\}_i \\ \vdots \\ \{S\}_{p-2} \\ \{S\}_{p-1} \end{Bmatrix} = \begin{Bmatrix} \{F\}_0 \\ \{F\}_1 \\ \vdots \\ \{F\}_i \\ \vdots \\ \{F\}_{p-2} \\ \{F\}_{p-1} \end{Bmatrix} \qquad (49)$$

where $\{S\}_i = \{S\}_{Z=Z_i=ih} \quad , i = 1,2,\cdots,p$

p=total storey number of the building

$$[C] = 2[I] + h^2[B]; \quad \{F\}_i = h^2[B_0]\{M\}_i$$

The boundary conditions are

$$\{S\}'_{Z=0} = \{S\}'_0 = \{0\}; \quad \{S\}_{Z=H} = \{S\}_P = \{0\} \qquad (50)$$

Expanding each row of Eq. (49), and taking Eq. (50) into account, the following recursive relationship results:

$$\{S\}_i = [C^*]_i(\{F^*\}_i + \{S\}_{i+1}), \quad i = 0,1,\cdots,p-1 \qquad (51)$$

where $$[C^*]_0 = \frac{1}{2}[C]; \quad [C^*]_i = [C] - [C^*]_{i-1}^{-1} \qquad (52)$$

$$[F^*]_0 = \frac{1}{2}\{F_0\}; \quad \{F^*\}_i = \{F\}_i - [C^*]_{i-1}^{-1}\{F^*\}_{i-1} \qquad (53)$$

Using Eqs. (50)(52)(53), we can solve the recursive Eq. (51) for $\{S\}_i$ in the order of $i=p-1, p-2, \cdots, 0$.

Numerical examples show that the finite difference solution requires much less calculation work as compared with the theoretical solution, and still has practically satisfactory accuracy.

4. ANALYSIS OF NONUNIFORM STRUCTURES[10]

When the geometrical and physical parameters of a structure have a stepwise change at some levels along the height, the structure can be divided into R segments, and its parameters remain unchanged within each segment. Supposing segments r and $r-1$ meet at the level $Z = Z_J$, the difference equation of the whole structure takes the following form:

$$\begin{bmatrix} [C^1] & -2[I] & & & & & & & \\ -[I] & [C^1] & -[I] & & & & & 0 & \\ & \ddots & \ddots & \ddots & & & & & \\ & & -[I] & [C^{r-1}] & -[I] & & & & \\ & & & -[I] & [C^J] & -[I] & & & \\ & & & & -[I] & [C^r] & -[I] & & \\ & 0 & & & & \ddots & \ddots & \ddots & \\ & & & & & & -[I] & [C^R] & -[I] \\ & & & & & & & -[I] & [C^R] \end{bmatrix} \begin{Bmatrix} \{S\}_0 \\ \{S\}_1 \\ \vdots \\ \{S\}_{J-1} \\ \{S\}_J \\ \{S\}_{J+1} \\ \vdots \\ \{S\}_{P-2} \\ \{S\}_{P-1} \end{Bmatrix} = \begin{Bmatrix} \{F\}_0 \\ \{F\}_1 \\ \vdots \\ \{F\}_{J-1} \\ \{F\}_J \\ \{F\}_{J+1} \\ \vdots \\ \{F\}_{P-2} \\ \{F\}_{P-2} \end{Bmatrix} \tag{54}$$

where

$$[C^J] = \frac{1}{2}([C^{r-1}] + [T^J][C^r]) \tag{55}$$

$$\{F\}_J = \frac{1}{2}(h_{r-1}^2[B_0^{r-1}] + h_r^2[T^J][B_0^r])\{M\}_J \tag{56}$$

$$[T^J] = \frac{E_{r-1}h_{r-1}}{E_r h_r}[\beta'']_{r-1}^{-1}[\beta'']_r \tag{57}$$

It is evident that the block tridiagonal character in Eq. (54) is maintained, and the solution for this difference equation can be obtained in the same manner as that for Eq. (49).

5. STRUCTURAL ANALYSIS FOR VERTICAL LOADS[12]

The effect of all transverse vertical loading on beams is treated in a similar manner as that in the routine matrix stiffness method of rigid frames. In establishing the compatibility equations, besides the axial deformation due to vertical load, the effect of bending moment in vertical elements due to eccentricity of vertical loads is also considered. Using folded-plate model for shear walls, the difference equation is of the type as Eq. (49) for uniform structures, or of the type as Eq. (54) for nonuniform structures.

Using this method of analysis and considering the construction process of structure, the effect of axial deformation of vertical elements due to vertical dead load has been studied[13].

6. THREE-DIMENSIONAL SEISMIC ANALYSIS[14]

The characteristic equation of free vibration of the structure is

$$[f][M]\{G\} = \lambda\{G\} \tag{58}$$

where $\{G\}$—*eigenvector of amplitude;*

$\lambda = \frac{1}{\omega^2}$; ω—*circular frequency;*

$[M]$—*mass matrix;*

$[f]$—*matrix of lateral flexibility.*

Using the method mentioned in § 2 or § 3, the elements of j—th column in [f] may be determined as the displacements due to unit force acted in the direction of j-th degree of freedom. Eq. (58) can be solved by iteration method or other numerical methods.

More generalized formulae for determining seismic response of tall buildings by response spectrum and modal analysis approach have been derived. Selecting different numbers of vibration modes for combination of seismic responses of different types of tall buildings, we have made comparisons between the SRSS and CQC methods of modal combination. The studies we have performed indicate that the SRSS method gives satisfactory result for tall buildings without torsion or with slight torsion, while CQC method is favourably applied to torsion-coupled tall buildings, for which 9 vibration modes should be taken into account in the calculation.

7. P-△ APPROXIMATIONS AND STABILITY OF STRUCTURES[15]

The additional lateral load {P} due to P-△ effect is $\{P\}=[W]\{\triangle\}$ (59)

where $\{P\} = [P_x \quad P_y \quad T_z]^T$

$\{P_x\} = [P_{x1} \quad P_{x2} \quad \cdots \quad P_{xp}]^T$

$\{P_y\} = [P_{y1} \quad P_{y2} \quad \cdots \quad P_{yp}]^T$

$\{T_z\} = [T_{z1} \quad T_{z2} \quad \cdots \quad T_{zp}]^T$

P_{xi}, P_{yi}, T_{zi}—the additional lateral forces in directions X, Y and torque about axis Z at i-th storey respectively;

$\{\triangle\} = [u \quad v \quad \theta]^T$

$\{u\} = [u_1 \quad u_2 \quad \cdots \quad u_p]^T$

$\{v\} = [v_1 \quad v_2 \quad \cdots \quad v_p]^T$

$\{\theta\} = [\theta_1 \quad \theta_2 \quad \cdots \quad p_\theta]^T$

u_i、v_i、θ_i—the displacements in directions X, Y and rotation about axis Z at i-th storey respectively.

The stability equation of structural system results in

$$\Phi[f][W]\{\triangle\} = \{\triangle\} \qquad (60)$$

where Φ—load parameter. Eq. (60) is of the same type as Eq. (58), and can be solved in similar manner.

8. SECOND-ORDER ANALYSIS[16][17]

For tall building structures with stepwise change of structural parameters at some levels along the height, as a general case, the basic equations of second-order analysis for k-th segment result in the following system of differential equations with variable coefficients:

$$\{S\}''_k + [A]_k\{S\}'_k + [B]_k\{S\}_k = \{P\}_k, \quad k = 1,2,\cdots,R \qquad (61)$$

where $[A]_k$、$[B]_k$ are nonsingular square matrices of order n. With known boundary conditions and interface conditions between two neighboring segments, numerical solution of Eqs. (61) can be obtained with high precision by using finite integral method, the transfer matrix approach and Riccati Transformation. Either linear finite integral method or quadratic finite integral method can be applied. Using the transfer matrix approach, the number of independent unknowns remains equal to the order of matrices in the original system of equations. Application of Riccati Transformation can guarantee the computation to be free from numerical instability which frequently takes place in transfer matrix approach.

9. APPLICATION IN ENGINEERING PRACTICE

Based on the computational method mentioned above, a computer program TBAD for analysis and design of tall building structures has been compiled. It has the function of structural analysis for horizontal loads, vertical loads, seismic reponses and their combinations (SRSS or CQC), effect of temperature changes, effect of foundation settlements, P-$\triangle$ effect, overall stability and second-order analysis, as well as the function of reinforcement design of all structural elements. Using a microcomputer with 512k on-board memory, it can carry out the 3-D analysis and design of a tall building structure with up to 80 storeys, 80 vertical elements and 80 bands of connecting beams. Many tall building structures in South China have been designed by using this program. The analysis of more than 100 numerical examples shows that the results obtained by this program are very close to those by finite element method, matrix stiffness method and structural test, although it requires much less computer storage and time than the other methods.

10. SUGGESTION OF FURTHER RESEARCH

There is room for further research of 3-D analysis of tall building structures by continuum method. The most pressing need is the analysis of behavior and response of structures in elasto-plastic range with consideration of the effect of geometrical nonlinearity of the deformed structures. It is highly promising to extend continuum method to the full range analysis up to collapse of structures, with the aim of meeting the requirements of optimal design and control of structures.

REFERENCES

[1] H. Beck: Contribution to the Analysis of Coupled Shear Walls, J. ACI, Vol. 59, No. 8, 1962.

[2] R. Rosman: Die statische Berechnung von Hochhauswänden mit öffnungsreihen, Bauingenieur-Praxis, H. 65, 1965.

[3] J. Balaš, V. Szabó: Ein Beitrag zur theoretischen Lösung von tragenden Querwänden mit öffnungen unter horizontaler Belastung, Bauingenieur 40(1965) 482-485.

[4] Q. Z. Liang: Analysis of Interconnected Shear Walls to Lateral Load, J.

South China Institute of Technology, vol. 7, No. 4, 1979. (in Chinese)

[5] J. K. Biswas, W. K. Tso: Three-Dimensional Analysis of Shear Wall Buildings to Lateral Load, J. Struct. Div., ASCE, vol. 100, No. ST5, 1974.

[6] Q. Z. Liang: Analysis of Interconnected Space Shear Walls to Lateral Load, J. South China Institute of Technology, vol. 8, no. 4, 1980(in Chinese).

[7] Q. Z. Liang: Eine allgemeine Theorie der Berechnung gegliederter Scheiben unter horizontaler Belastung im Hochbau, Bauingenieur 58(1983)19-26.

[8] Q. Z. Liang: Continuous Medium Technique for Calculating Tall Building Structures, J. of Building Structures, vol. 5, No. 4 & No. 5, 1984. (in Chinese)

[9] Q. Z. Liang: A Generalized Continuum Method for Structural Analysis of Tall Buildings to Lateral Load, Proc. 3rd International Conference on Tall Buildings, 1984

[10] Q. Z. Liang, S. Y. Li: Continuum-Discretization Method for the Analysis of Tall Building Structures, Proc. 3rd International Conference on Tall Buildings, 1984.

[11] Q. Z. Liang, L. Xie: Strip Model in Analysis of Tall Buildings by the Generalized Continuum Method, Proc. 4th International Conference on Tall Buildings, 1988.

[12] S. Y. Li, Q. Z. Liang: Continuum-Discretization Method for the Analysis of Tall Building Structures under Vertical Load, Proc. 9th National Conference on Tall Building Structures, pp. 385-394, 1986(in Chinese).

[13] Q. Z. Liang, X. Li, S. Y. Li: The Effect of axial deformation of Tall Building structures due to Vertical Dead Load, Proc. 9th National Conference on Tall Building Structures, pp. 405-412, 1986(in Chinese).

[14] Q. Z. Liang, S. Y. Li: Three-Dimensional Seismic Analysis of Tall Buildings, J. Building Structures, CAA, vol. 9, No. 1, pp. 24-33, 1988 (in Chinese)

[15] S. Y. Li, Q. Z. Liang: 3-D Analysis of Overall Elastic Stability of Tall Building Structures, Proc. 9th National Conference on Tall Building Structures, pp. 395-407, 1986 (in Chinese).

[16] Q. Z. Liang, L. Xie: Second-Order Analysis of Wall Frame Structures, J. Building Structures, CAA, vol. 7, No. 5, pp. 1-8, 1986 (in Chinese).

[17] Q. Z. Liang, L. Xie: On the Methods of Numerical Solution for the Boundary-Value Problems of Systems of Second Order Linear Differential Equations, J. South China University of Technology, vol. 16, No. 2, pp. 67-76, 1988 (in Chinese)

NONLINEAR MODELLING AND ANALYSIS OF REINFORCED CONCRETE FRAME STRUCTURES

Graham H. Powell
Professor of Civil Engineering

Filip Filippou
Associate Professor of Civil Engineering

Kirk Martini
Former Graduate Student

Scott Campbell
Graduate Student

University of California at Berkeley
Berkeley, CA 94720

ABSTRACT

This paper examines a number of issues related to the modelling and analysis of reinforced concrete frame structures, where the members resist axial force, bending, shear and possibly torsion. The emphasis in the paper is on how to capture, in an analysis model, the important aspects of reinforced concrete behavior.

The purpose of nonlinear analysis is first reviewed briefly, to place modelling needs in perspective. The basic requirements for a reinforced concrete frame member model are identified, particularly the need to model inelastic behavior both within a cross-section and along the member length. Beam-column models based on lumped and distributed plasticity assumptions, and on force (flexibility) and displacement (stiffness) formulations, are summarized. Cross-section models based on the analogy with plasticity theory are examined, and serious weaknesses in this approach are noted. Models based on "fiber" concepts are described, and it is argued that these models provide the best compromise between accuracy and computational effort. It is noted that although fiber models are superior to plastic hinge models, the basic fiber concept accounts only for axial and flexural effects, ignoring inelastic behavior in shear. An extension of the fiber concept to account for shear cracking and the yield of shear reinforcement is briefly described. The importance of deformation in the beam-to-column connection region is also noted, and inelastic connection models are briefly reviewed. The paper concludes with a discussion of practical modelling and computation issues.

INTRODUCTION

Because of the hazards posed by earthquakes and other extreme loads, there is increasing interest in computing the nonlinear static and dynamic response of civil engineering structures. In the past, nonlinear analyses have been performed mainly for research purposes and for critical structures such as nuclear power plants. As theories and computational techniques improve, however, and as computers become ever more powerful, it is becoming increasingly feasible to perform practical nonlinear analyses of structures such as buildings and bridges. Also, whereas nonlinear structural analyses in the past have tended to be based on simple nonlinear models to reduce the computational cost, it is becoming possible to consider much more complex, and hence more realistic, models. This is important for the investigation of reinforced concrete structures because reinforced concrete is an extremely complex material. Models which may be adequate for steel structural members may fail to capture the essential aspects of reinforced concrete members, and hence may not give meaningful results.

Until relatively recently, the design of reinforced concrete structures was based largely on strength considerations, assuming that the load magnitudes were well defined. Relatively little attention was given to behavior under extreme loading conditions such as earthquakes. As a result, many older reinforced concrete structures are brittle, and may not have sufficient ductility to survive strong earthquakes without serious damage. This is true not only of buildings but also of bridges. In future years it will be necessary to modify (retrofit) many structures to increase their level of safety. A major problem facing designers is that it is extremely difficult to compute the behavior of reinforced concrete structures under extreme loads, and hence it is extremely difficult to estimate the amount of damage which any particular structure might experience in a future earthquake. There is a strong need for better analysis tools for use in decisionmaking and in retrofit design.

It is important to note that structural analysis is used for two distinctly different reasons, as follows.

(1) Demand analysis: compute the demands on structures and structural members. For example, for a building frame which is to be designed to resist specified loadings, structural analysis can be used to compute the axial force, bending moment and shear force demands on the frame members. The members can then be designed with strength capacities which equal or exceed these demands. Demand analyses of this type can usually be performed with sufficient accuracy using elastic models. A demand analysis might, however, account for inelastic behavior, for example to compute the deformation demands on an inelastic base isolation system. The system can then be designed to accommodate these deformations without damage.

(2) Capacity analysis: compute the capacities of structures and structural members. For example, given an existing building frame designed to an old, and possibly inadequate, design code, a capacity analysis estimates its true strength. For this type af analysis it is essential to consider inelastic behavior. Also, if the analysis is to be a rational one, the analysis model must capture all those aspects of behavior which can influence strength. For example, if shear failure is a possible cause of collapse, the analysis model should ideally capture inelastic shear effects. Capacity analysis is thus much more difficult than demand analysis.

Capacity analysis of single members is often used as an aid in the development of design formulas for member strength. However, such analysis plays a secondary role, and most of the data is obtained from destructive tests of actual members, not from analysis. In producing a retrofit design for a building or bridge, a designer does not have the luxury of performing a destructive test, and must rely heavily on analysis. Designers do not yet have the analysis tools to do this effectively.

A reinforced concrete frame structure consists essentially of beam, column and wall members, subjected to axial forces, bending moments, shear forces, and possibly torsional moments. There may also be significant deformation in the beam-to-column connection regions. Three-dimensional behavior, of both the structure and the members, must often be taken into account. This paper is concerned with the modelling of reinforced concrete beam and column members for capacity analysis of such frames.

BASIC BEAM-COLUMN ELEMENT MODELS

Beams, columns and slender walls can be modelled using beam-column elements in which the effects of axial force, bending moment, shear force and possibly torsion are considered. Consider first the basic problem of modelling axial force and bending moment effects.

Figure 1 illustrates the effects of axial force and bending moment in a simple yielding member. For uniform bending the plastic zones extend along the full length and over part of the depth. For nonuniform bending the plastic zones extend partially along the length but are still symmetrical. For combined axial force and nonuniform bending the plastic zones are unsymmetrical, because of force-moment (P-M) interaction. Beam-column element models must capture these effects. For reinforced concrete they must also account for the fact that concrete is not a simple plastic material, but one which cracks and degrades.

Figure 2 shows alternative models for beam-column elements. Figure 2a shows a typical "lumped plasticity" model, in which the inelastic deformation is lumped in zero length plastic hinges at the element ends. Strain hardening effects can be modelled in the hinges, and/or provided by adding an elastic element in parallel with the inelastic one. Figure 2b shows a typical "distributed plasticity" element, in which the inelastic behavior is explicitly monitored at several discrete sections along the element length. Assumptions are then made about the variation of inelastic behavior at all other sections (e.g., linear variation of flexibility between monitored sections). Figure 2c shows a substructured element, divided into a number of subelements. The element as a whole is of distributed plasticity type, although the subelements could be of lumped type. Fundamentally, there is no difference between a model which uses a single substructured element and one which uses several non-substructured elements. However, it is easier to prepare computer program input data for a substructured element, since the user needs to specify only one element rather than several. There can also be computational advantages with substructured elements.

In general, distributed plasticity models are more accurate than lumped plasticity models. The reason is that the distribution of plasticity along the element length varies with the loading on the element, as indicated in Figure 1, and it is not generally possible to determine lumped hinge properties which are accurate for all loadings. For example, if hinge properties are assigned to give accurate results for uniform bending, they will generally not be accurate for nonuniform bending. Elements with lumped hinges can still be useful, but their limitations must be recognized.

The two most important steps in nonlinear analysis are (a) given the current state of a nonlinear element, compute its tangent stiffness matrix (linearization), and (b) given the current state and a displacement increment at the end nodes, determine the new state (state determination, or state update). For this paper it is sufficient to focus on the tangent stiffness computation.

For a lumped plasticity model the hinge is typically assumed to be rigid-plastic, and the main problem is to define its tangent stiffness in the plastic range. The element tangent stiffness can then be obtained using conventional analysis techniques. For a distributed plasticity model the main problem is to define the tangent stiffness or flexibility of a monitored section. The element tangent stiffness can then be obtained by interpolating stiffnesses or flexibilities at intermediate sections, and integrating over the element length. It may be noted that for the element in Figure 2b it will generally not be accurate to assume a shape function and use a displacement formulation, because the shape function changes as the element yields. For example, a cubic displacement assumption (linear curvature) will generally not be accurate. A force (flexibility) formulation is thus indicated. For the subelements in Figure 2c, however, a displacement formulation may be suitable, since the subelements are short.

If the inelastic behavior of a hinge or monitored section were affected only by bending moment, then the constitutive relationship would be uniaxial (moment-rotation for a hinge, moment-curvature for a section). These relationships tend to be complex for reinforced concrete members, because of steel yielding and concrete cracking and strength loss. It is possible, however, to develop equations or rules to describe such uniaxial behavior. This approach is <u>empirical</u> at the hinge or section level, since the equations or rules are chosen to match experimentally observed behavior. If, however, there is P-M interaction, then the constitutive relationship becomes multi-axial instead of uniaxial. The hinge or section stiffness is thus defined not by a single number but by a matrix (2x2 for two-dimensional P-M interaction, 3x3 for three-dimensional P-M-M interaction). The problem of defining empirical rules for the <u>diagonal</u> terms of the constitutive matrix is the same as for the uniaxial case. However, it can be extremely difficult to define empirical rules for the <u>off-diagonal</u> terms, because of the complexity of the underlying causes of the interaction. It is necessary, therefore, to base the interaction modelling on a rational theory of some kind rather than on empirical rules. A theory which has been commonly used is based on an analogy with the yield of plastic materials under multiaxial stress as considered in the next section.

INTERACTION BASED ON PLASTIC THEORY ANALOGY

Figure 3a shows a von Mises yield surface for a plastic material in two-dimensional principal stress space. If the normality rule is assumed for plastic flow, it is a simple matter to develop an expression for the tangent constitutive matrix after yield. Essentially, the von Mises yield criterion plus the normal flow assumption define a rational theory for interaction between stresses. By analogy, the same approach can be used for stress resultants, as indicated in Figure 3b for two-dimensional P-M space. Unfortunately, there are problems with this analogy.

Figure 4a shows yield surfaces for a linearly hardening plastic material, with an initial yield surface, a subsequent yield surface, and stress-strain curves for uniaxial stress along the x and y directions. With the assumptions which are usually made, the initial and subsequent yield surfaces have identical shapes, and the stress-strain curves are the same. Figure 4b shows the analogous case in P-M space for a reinforced concrete section. Now, the initial and subsequent yield surfaces do not have the same shape. Also, the action-deformation curves along the P and M directions are not the same. Further, for a complete theory it is necessary to provide empirical hardening rules, which define how the yield surfaces move, change size and possibly change shape as the material yields. Rules, such as kinematic hardening, can be defined fairly easily for yield of plastic materials, but not for P-M interaction.

In spite of these limitations, the plasticity analogy can capture some interaction effects, and it can serve as a basis for useful models. However, it is not rational enough for general use. To provide accurate models it is necessary to model the underlying causes of inelastic behavior and interaction, by moving one level lower before introducing empirical rules. This can be done using "fiber" concepts.

FIBER REPRESENTATION OF A SECTION

Figure 5 shows a simplified reinforced concrete column section and a fiber representation of the section. In the fiber representation the section is divided into a number of steel and concrete longitudinal fibers, and each is assigned a uniaxial stress-strain relationship. These relationships are empirical, based on observed behavior of steel and concrete. Note, in particular, that different stress-strain curves are assumed for confined and unconfined concrete. In order to assign the confined curve it is necessary to account for the amount of confinement and its effect on the stress-strain curve. A truly rational model would use a multiaxial model for concrete and consider the confining steel directly. However, such a model would be much more complex.

A major advantage of the fiber representation is that it allows P-M interaction to be considered without postulating multiaxial yield surfaces and flow rules. Figure 6a shows that a two-fiber model with simple yielding fibers provides a P-M interaction surface that is a reasonable approximation of that for a steel I section. A three-fiber model (not shown) gives a hexagonal yield surface. Figure 6b shows that a model with two yielding steel fibers combined with two yielding and cracking concrete fibers gives a P-M interaction surface of reinforced concrete type (the four-fiber interaction surface is actually somewhat more complex than that shown). As the number of fibers is increased, the interaction surfaces more closely approximate those of actual cross-sections. Hardening and softening behavior is also captured, without the need for complex hardening rules.

The fiber procedure can also be applied to hinges. In order to obtain the force-moment-strain-curvature behavior for a section, the fibers are assigned stress-strain curves. The force-moment-extension-rotation behavior of a lumped hinge can be obtained by assigning force-extension relationships to zero-length fibers in a hinge. This is more empirical, since the hinge seeks to capture inelastic behavior along the element length as well as over the cross-section, but the model can be useful.

Models of both hinges and cross-sections have also been developed with only small numbers of fibers (as few as 4 or 5 for a three-dimensional concrete element). Since the number of fibers is small, the fiber stress-strain curves are not those of the basic steel and concrete materials. Instead, they are empirically determined curves which combine steel and concrete properties into one fiber, and which also account for the small number of fibers.

Although these curves must be determined empirically, elements based on small numbers of fibers are computationally much less costly than more rational elements with large numbers of fibers. The issue of computational cost is addressed later in this paper.

EXTENSION OF FIBER CONCEPT TO SHEAR DEFORMATION

The controlling mode of inelastic deformation for a frame member is ideally flexural, since this provides the greatest ductility. Some frame members may, however, be controlled by shear, particularly in structures designed to old design codes. Inelastic shear deformation is less ductile than flexural deformation, and it could be argued that shear failure will immediately be followed by structural collapse. However, a structure may be able to redistribute load in such a way that local shear failure does not lead to overall collapse. Also, inelastic shear deformation is not entirely brittle, and there may be sufficient ductility to avoid major damage. It may be important, therefore, to model inelastic shear deformation and the interaction of shear forces with axial forces and bending moments. It is possible to extend the fiber modelling concept to include shear deformation. An outline of the procedure is as follows.

Figure 7a shows a column section, and Figure 7b shows its fiber representation. Figure 7b also shows a side view of a beam-column "slice" of infinitesimal length dx. This slice is a cuboid of dimension b by d by dx. Its behavior under axial force and bending moment is defined by the properties assigned to the longitudinal steel and concrete fibers, as already considered. In addition, the slice may deform in shear. Shear resistance is provided by the concrete and by transverse shear reinforcement. In Figure 7b the shear reinforcement is shown as a transverse fiber. The fiber area is the shear reinforcement area per unit length multipled by dx. Figure 7c indicates the way in which shear deformation is usually modelled in beam elements: the shear force produces shear deformation in the slice, based on the shear stiffness of the material. Since concrete has a large shear stiffness, the amount of shear deformation of this type is small and can usually be ignored. Figure 7d shows that shear deformation can also be caused by diagonal cracking. In this figure, the cracks are shown as opening with zero sliding parallel to the cracks, corresponding to perfect aggregate interlock. In fact, there will be sliding as well as crack opening, but the essential behavior is similar to that shown. In particular, the cracking produces not only shear deformation but also longitudinal and transverse extension. There is thus interaction between shear cracking and axial effects (axial force and bending moment). Shear cracking also deforms the transverse reinforcement in tension, so that the reinforcement resists shear. By adding cracking degrees of freedom to the slice, it is possible to account for shear cracking, to include the shear reinforcement fiber in the model, and to account for interaction among shear force, axial force and bending moment. The theory is presently under development by the authors. It is planned to extend it to incude torsional shear cracking in three-dimensional members, by allowing a spiral pattern of cracks.

CONNECTION DEFORMATION

In frame analysis it is usual to assume that beam-to-column connections are rigid, in the sense that beams and columns remain at right angles to each other as the frame deforms. It is well known, however, that there can be significant deformation in these connections. In steel frames this deformation is caused by high shear stresses in the "panel zone" of the connection. In concrete frames it is caused by shear cracking and by bond slip in the joint region. In some cases the connection may be the weakest part of the frame, and large deformations can be present. In order to develop a rational model for inelastic frame analysis it may be necessary to model these deformations.

Connection models which are similar in concept to a plastic hinge are the simplest. In these models the beam elements to the left and right of the connection are rigidly connected to each other, and similarly the column elements above and below the connection. However, the beams and columns are not joined rigidly but are connected by a deformable connection element. Any moments which are transferred from beams to columns thus pass through this element, causing it to deform. Inelastic connection behavior is modelled by assigning inelastic properties to the connection element. This type of element can be useful, but since its inelastic properties must be assigned empirically, it has the same weaknesses as other empirical elements. Reinforced concrete connections are complex, and it is desirable to capture the underlying behavior in the analysis model.

The main causes of deformation in a reinforced concrete connection are indicated in Figure 8. These causes are (a) yield of the main reinforcement, (b) bond slip, and (c) diagonal cracking. Reinforcement yield and bond slip can lead to significant crack opening in the beams at the column faces, and under cyclic loading cracks may open across the full beam depths. The authors are developing a connection element which directly models these effects. It is a complex substructured element which combines subelements for main reinforcement, column ties, concrete in the panel zone, bond links, and gaps at the column faces.

COMPUTATIONAL COST AND MODELLING ACCURACY

With today's computers, nonlinear analysis using rational, but elaborate, element models of the types described here is too expensive computationally for most structures. Cost can be reduced by using simpler, and hence more empirical, models. However, unless the correct empirical properties are specified for such models, the analysis results may not be meaningful. In academic studies, agreement with experimental results is often reported using simple models. However, such studies typically benefit from hindsight, and the models have often been adjusted, by trial and error, to obtain the agreement. For practical applications the engineer rarely has experimental data to support this "calibration" of empirical models.

A solution to this dilemma is to use rational, elaborate models to calibrate simpler, empirical models. The steps are essentially as follows.

(1) Using available experimental results, calibrate the rational model. If the model is truly rational, this should be relatively easy.

(2) Set up two analysis models of a single element, using (a) the rational model and (b) the empirical model with a best estimate of the empirical properties.

(3) Subject both analysis models to the same loadings, in the form of imposed end forces and/or displacements. The load magnitudes should be those expected to act on the element in the complete structure.

(4) Compare the results. If they agree within engineering accuracy, the empirical properties assigned in Step 2 are "correct". If not, revise these properties and repeat. That is,

calibrate the empirical model against the rational model. It might be possible to automate the process using formal system identification techniques.

(5) Repeat the process for a representative selection of members.

(6) Analyze the complete structure using the calibrated empirical models for the elements.

(7) From the analysis results, confirm that the loadings on the elements are similar to those used in Step 3, and hence that the calibration is correct.

With this approach, the overall computational cost can be minimized, and the engineer can have confidence in the analysis results. In the future, as computers get faster, it will be possible to use elaborate models directly in the analysis of the complete structure.

This procedure addresses an important preprocessing problem. Problems of computational cost can also arise in postprocessing. During a nonlinear analysis it is usual to save histories of node displacements and element response. These saved results are then used for postprocessing. For a large structure the amount of data to be saved can be immense, straining the capacity of even very large disks. This is especially true for dynamic analysis. Also, the element data to be saved must be specified before the analysis starts, and the engineer can not always be sure that all needed element response data has been saved.

A procedure which can be used to solve this problem is as follows.

(1) During the analysis, save time- or load-histories of node displacements, but not of element responses. The amount of data to be saved is thus greatly reduced.

(2) Identify those elements which are likely to sustain the most damage, by examining the maximum (envelope) values of element forces and deformations. These envelopes are computed during the analysis.

(3) Choose one of these elements. Extract, from the histories of node displacements, a history of displacements at the nodes to which the element connects.

(4) Set up an analysis model containing only this single element, and analyze it for these node displacements. During the analysis, calculate and output damage measures, and other pertinent results.

(5) Repeat for other potentially critical elements.

In Step 4 this process repeats calculations already performed during the main analysis. However, there can be a cost saving because histories of element response no longer need to be saved, and because single element calculations can be performed on low cost microcomputers. Also, decisions do not need to be made on which element responses to save during the main analysis. The postprocessing analysis re-creates all of the detailed time history information on the element state, and can process this information in ways which may not have been foreseen during the main analysis.

CONCLUSION

In order to provide realistic assessments of the safety of existing reinforced concrete structures, it is necessary to perform realistic capacity analyses. This requires more sophisticated inelastic modelling than is necessary for demand analyses. Unless the analysis model captures the important aspects of behavior, the analysis results may not be meaningful.

This paper has reviewed a number of concepts for modelling beams, columns and connections in reinforced concrete frame structures. The authors believe that in order to capture the important aspects of behavior it is necessary to use sophisticated, rational models for both members and connections, and that simple models based on empirical concepts are suspect unless used very carefully. We recognize, however, that even with today's high-powered computers, the computational cost of using fully rational models is too high. Semi-empirical fiber models with small numbers of fibers seem to offer the best compromise for the immediate future. The paper outlines a preprocessing "calibration" strategy to help the engineer choose properties for such empirical models. The paper also outlines a postprocessing strategy which can save computational cost by reducing the size of results files, and can also provide greater postprocessing flexibility.

The number of published papers and reports on frame element modelling is huge. Identifying the few most important ones is a tricky task, and we have not attempted it. Notable institutions at which work has been performed in the United States include Cornell University, the University of Illinois, Georgia Tech, SUNY Buffalo, the University of Nevada at Reno, and the University of California at Berkeley. Much excellent work has also been performed in Japan and in Europe.

It may finally be noted that an alternative approach to modelling reinforced concrete is to use solid finite elements for the concrete and embedded bar elements for the reinforcement, possibly with added bond link elements. Much excellent work has been done in this area, and the approach could be suitable for modelling single elements. However, it will be a long time before computers are fast enough to use this type of modelling for complete frames. The approach in this paper has been on elements which are specifically intended for frame analysis, and which could be practical in the near future.

ACKNOWLEDGEMENTS

The research described in this paper has been sponsored by Kajima Construction Co., Tokyo, under the CUREe research program, and by the California Department of Transportation. The support of these organizations is gratefully acknowledged. The opinions expressed are those of the authors, and not necessarily those of the sponsors.

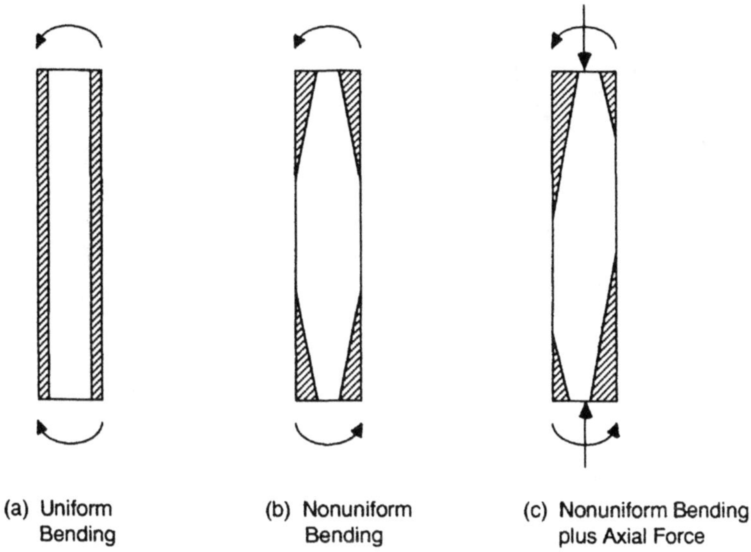

FIG. 1 PLASTIC ZONES FOR DIFFERENT LOADINGS

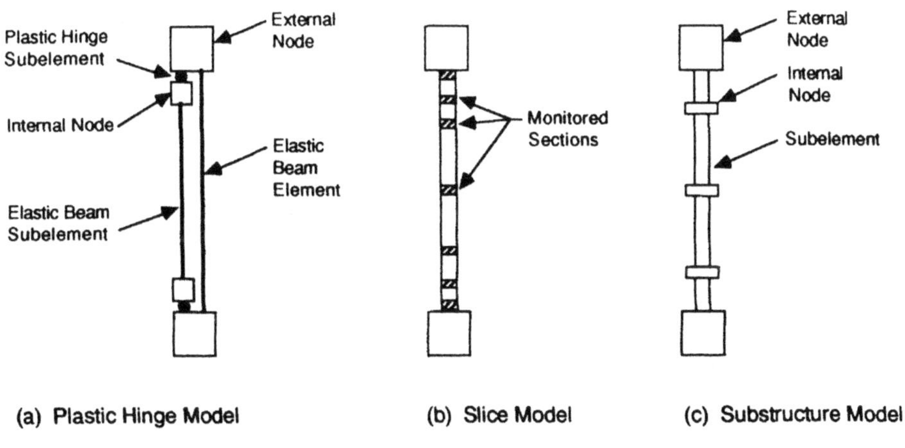

FIG. 2 ALTERNATIVE ELEMENT MODELS

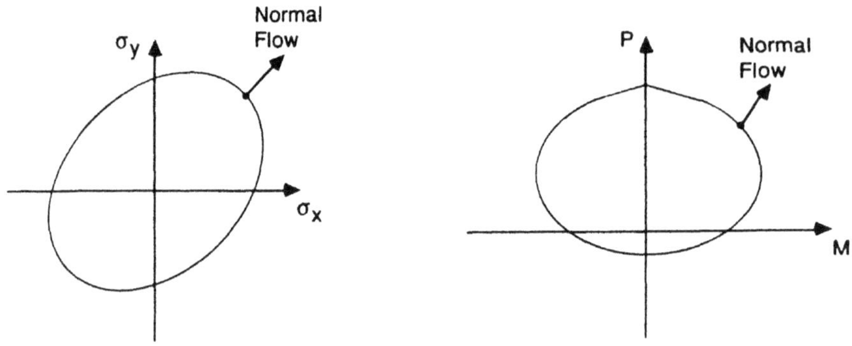

(a) Stresses

(b) Stress Resultants

FIG. 3 PLASTICITY THEORY ANALOGY

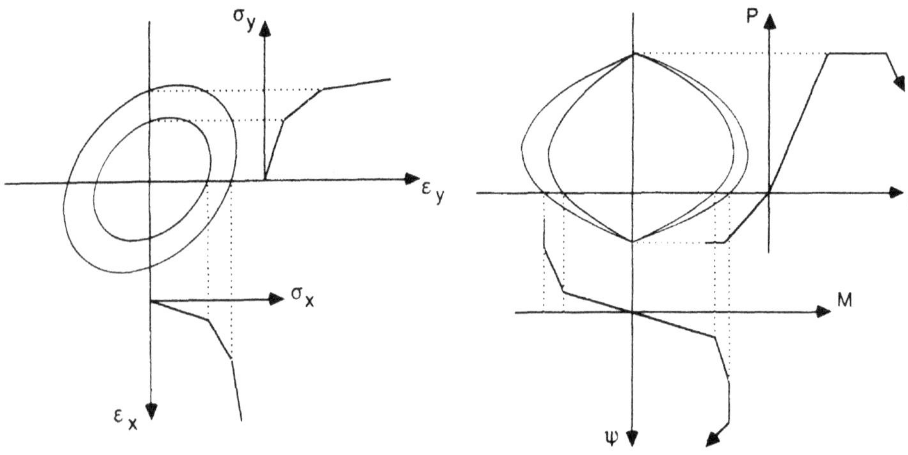

(a) Stresses and Strains

(b) Stress and Strain Resultants

FIG. 4 DIFFICULTIES WITH ANALOGY

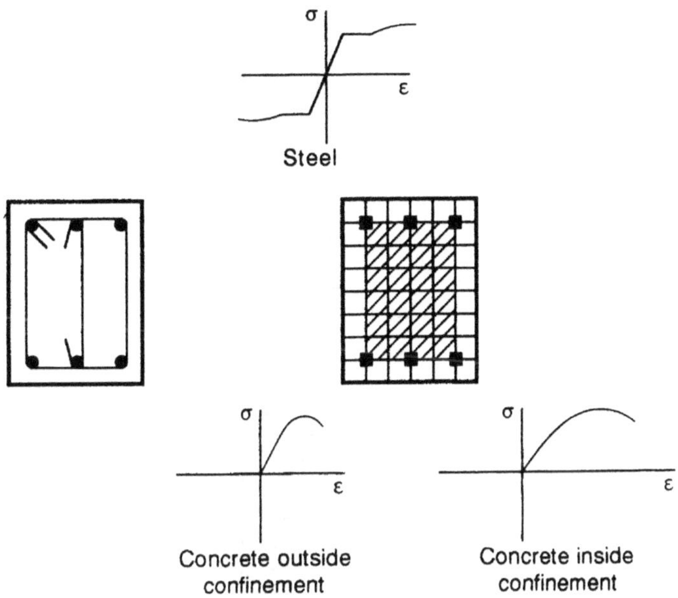

FIG. 5 FIBER REPRESENTATION OF A SECTION

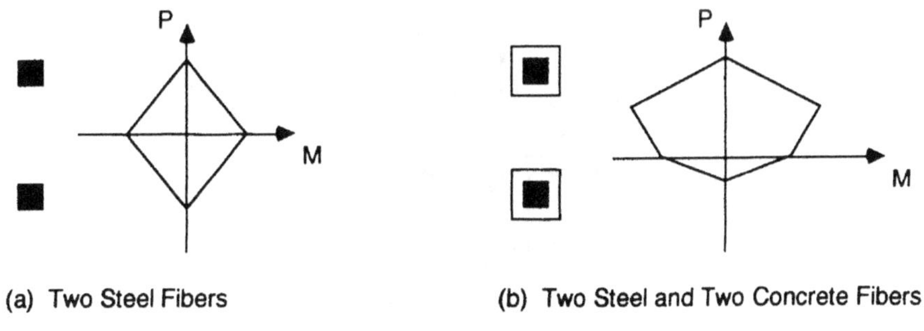

FIG. 6 P-M INTERACTION SURFACES FOR VERY SIMPLE SECTIONS

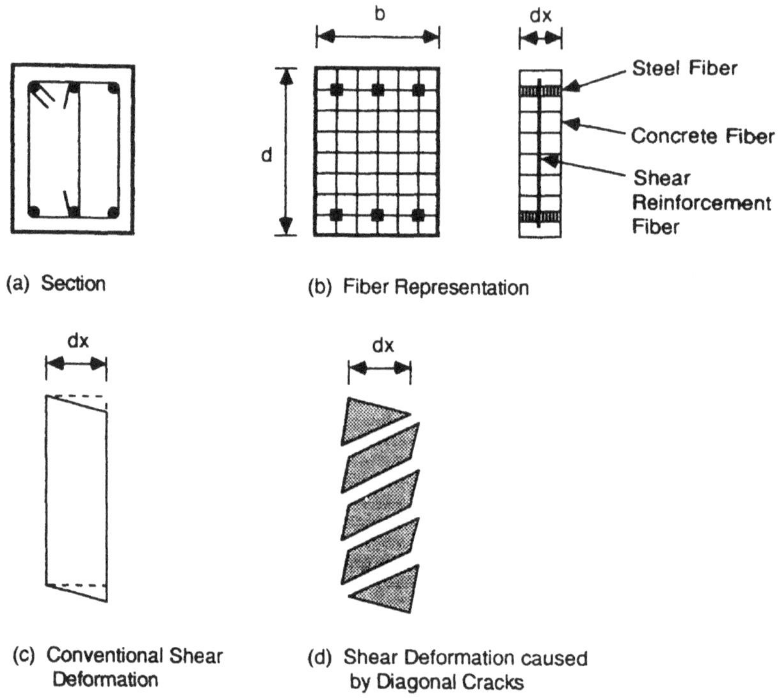

FIG. 7 EXTENSION OF FIBER MODEL TO INCLUDE SHEAR DEFORMATION

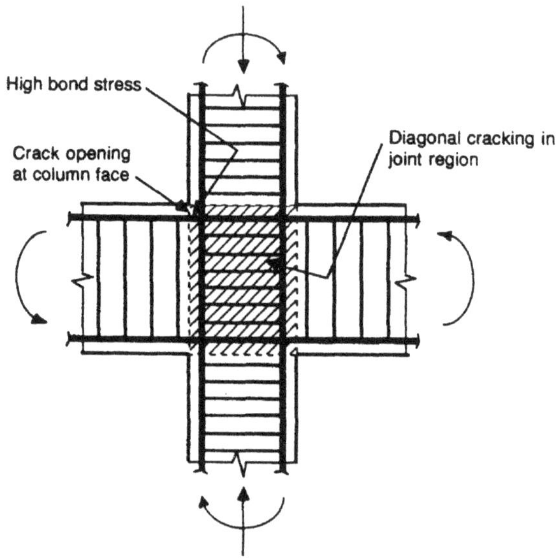

FIG. 8 SOURCES OF DEFORMATION IN BEAM-COLUMN CONNECTION

MECHANICAL PROPERTIES OF PLASTIC CONCRETE AND NONLINEAR STRUCTURAL ANALYSES OF PLASTIC CONCRETE CUT–OFF WALL

WANG QIZHENG WANG QINGYOU PU JIALIU
Department of Hydroelectric Engineering, Tsinghua University
100084, Beijing, CHINA

ABSTRACT

Plastic concrete has many advantages specially in mechanical properties compared with normal concrete. This paper presents the studies on plastic concrete and its application in hydraulic engineering in China.Influences of the proportioning of plastic concrete, surrounding pressure, and age of concrete on the constitutive relationship, failure status, and dynamic modulus, etc. of plastic concrete have been studied via unconfined and triaxial tests. Static and dynamic nonlinear structural analyses on plastic concrete cutoff wall were carried out. The influences of structure type and load conditions of the cutoff wall and the elastic moduli of concrete on stress states of the wall were analyzed. It shows that the stress state in plastic concrete cutoff wall is notably superior to that in normal concrete one and the plastic concrete is worth using widely in foundation seepage prevention works.

INTRODUCTION

Plastic concrete consists of cement, water, gravel, clay and bentonite.Sometimes it also mixed with fly ash and some additive, etc. Different from normal concrete, plastic concrete not only can save a lot cement economically, but has an advantage in mechanical properties over normal concrete as well. For instance, plastic concrete has lower deformation modulus, rather large limiting strain, larger plasticity as well as better impermeability. In addition, its mechanical properties can be controlled artificially to meet the requirements by changing its proportion of mixture.

When plastic concrete is used as the material for foundation cutoff wall, the stress state of the wall can be improved and the safety of the wall will be increased due to the fact that the deformation modulus and stress–strain relationship of plastic concrete can be modified to be consistent with those of the surrounding soil media. Therefore, the plastic concrete cutoff wall has been used more and more in hydraulic engineering.

In China, the plastic concrete has come into study and has been applied in hydraulic engineering since 80's. In recent years, plastic concrete cutoff wall has been adopted as the foundation diaphragm in many projects, and efforts are being made to use in large cofferdams of Xiaolangdi Project in Yellow River and Three Gorge Hydropower Project in Yangtze River. [1,2,3,4]

In coordination with the construction of above–mentioned projects, a great deal of research work has been carried out in China, which includes:

1. Study on static and dynamic characteristics of plastic concrete material. Influences of the proportioning of plastic concrete, surrounding pressure, age of concrete and loading speed on failure status, strength, initial and dynamic moduli, peak strain and damping, etc. of plastic concrete have been studied.

2. Structural static and dynamic analyses on plastic concrete cutoff wall. By means of linear and nonlinear finite element methods, the stress and strain of the wall under the actions of static and seismic loads have been studied. The influences of structure type and load conditions of the cutoff wall and the material mechanical parameters, especially the elastic moduli on stress states of the wall were analyzed.

3. Study on the theory of proportioning design for plastic concrete. By the analysis on the factors and regularities affecting strength, deformation modulus and plasticity index of plastic concrete, procedure has been studied to define the proportion of mixture as requested by optimum design.

4. Study on the construction technology and method for plastic concrete cutoff wall. The approaches have been studied to increase the construction speed and initial strength by using additives, and to control the quality of plastic concrete, etc.

5. In–situ tests and observations on plastic concrete cutoff wall. The permeability and the stress–strain states of the wall have resulted.

EXPERIMENTAL RESEARCH ON MECHANICAL PROPERTIES OF PLASTIC CONCRETE [5]

Unconfined compressive tests, consolidation drained triaxial tests and dynamic triaxial tests have been carried out on 57 groups of proportioning of plastic concrete samples with ages of 28 days, 3 months, 5 months and one year respectively. The surrounding pressure σ_3 was taken generally from 0.0 to 1.2 MPa.

Stress–strain Relationship and Failure Status

Typical shear stress $(\sigma_1-\sigma_3)$–axial strain ε_a–volume strain ε_v curves obtained from triaxial tests are given in Fig.1a. It can be seen that there is an obvious turning point like proportion limit for plastic concrete and the value is approximately equal to the peak strength of the sample during brittle failure under unconfined condition. The corresponding peak strain varies within the range of 0.4–0.7%, generally. Below this turning point, the stress–strain relationship is approximately a straight line, its slope the initial elastic modulus E_i. With the increasing of surrounding pressure σ_3, the stress–strain relationship turns to be of work hardening curve gradually, and the volume strain changes from shear expansion to shear compression slowly. It shows the obvious nonlinear character which is close to that of soils.

For plastic concrete samples, the failure pattern under triaxial condition is different from that under unconfined condition. Under unconfined condition the main failure type is compression, while under triaxial condition it is a typical shear failure with a distinct banded shear surface. Therefore it is suitable to describe the strength character of plastic concrete in terms of shear strength indexes—cohesion C and internal friction angle ϕ being used for soils. According to the results of triaxial tests, the cohesion is about 0.2–0.8 MPa and the internal friction angle varies within a range of 25.6 ° ~ 39.7 ° , both of which are related to the proportioning, material properties and water cement ratio of plastic concrete.

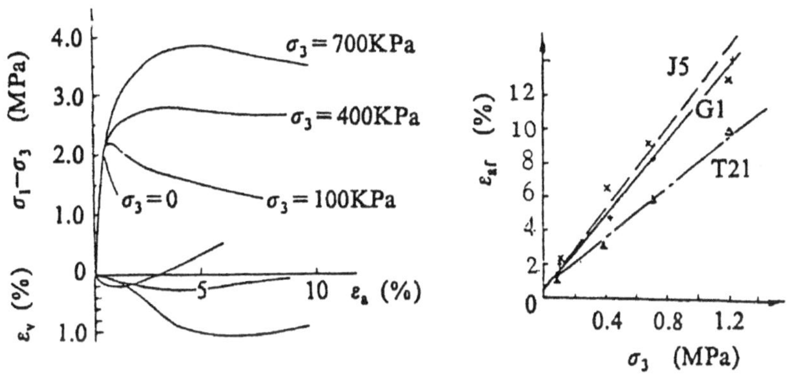

(a) $(\sigma_1-\sigma_3)\sim\varepsilon_a\sim\varepsilon_v$ relationship (b) Peak strain~confined pressure curves

Figure 1. Typical traixial tests stress–strain relationship

Influences of Surrounding Pressure on Mechanical Properties of Plastic Concrete

From Fig.1a, it can be seen that the strength and the peak strain ε_{af} of plastic concrete will rise remarkably with the increase of surrounding pressure. The relations between peak strain ε_{af} and surrounding pressure σ_3 from three series of triaxial tests are shown in Fig.1b. It can be seen that the increment of ε_{af} is nearly in line with that of σ_3. Fig.1 shows that when σ_3 is increased to 1.2 MPa ε_{af} comes to be 10–14% which is far greater than the ultimate strain of normal concrete and the axial failure stress of the sample is 3~ 8 times that of the unconfined strength. The initial modulus E_i, however, does not depend on the surrounding pressure for plastic concrete. This characteristic is completely different from that of common soils whose initial modulus E_i remarkably depend on the surrounding pressure σ_3. These mechanical properties of plastic concrete obviously will be helpful to improve the work condition and the safety of plastic concrete cutoff wall.

Influences of Age of Plastic Concrete on Its Mechanical Properties

Influences of age of plastic concrete on the strength q_u, initial modulus E_i and peak strain ε_{af} have been studied via the results of unconfined compressive tests for various ages of concrete. It indicates that q_u and E_i increase with the increasing of age . Strength of 90–day, 180–day and one–year is about 1.6, 2.0 and 2.7 times greater than that of 28–day respectively. The relations between the strength q_{ut} and initial modulus E_{it} versus age t after 28 days can be approximately described by the following formulas:

$$q_{ut} / q_{u28} = (t / 28)^n \qquad n = 0.379$$
$$E_{it} / E_{i28} = (t / 28)^m \qquad m = 0.349$$

where, the exponents m and n are very close to each other, illustrating that the strength of plastic concrete increased with age in step with the initial modulus roughly.

Peak strain ε_{af}, however, does not change with the age fairly and does not decrease with the increasing of the modulus. This feature will be beneficial to the load condition of plastic concrete as well.

Dynamic Properties of Plastic Concrete

Typical relationship between dynamic shear modulus G and dynamic shear strain γ has been studied. The results illustrate that the form of G~ γ curve is basically the same for the samples of various kinds of proportioning. When $\gamma < 2 \times 10^{-5}$, the dynamic shear modulus G basically does not vary with γ, which is called initial dynamic shear modulus G_0; when $\gamma > 2 \times 10^{-5}$, G decreases sharply with the increasing of γ. This is similar to the regularity of common soil. The relation of the variance of the ratio of G to G_0 with dynamic shear strain γ is illustrated in Fig. 2a. [6] This curve can be formulated as follows:

$$G / G_0 = 0.0834 + 1.759 \times 10^{-5} / \gamma \qquad \gamma > 2 \times 10^{-5}$$
$$G / G_0 = 1 \qquad \gamma \leqslant 2 \times 10^{-5}$$

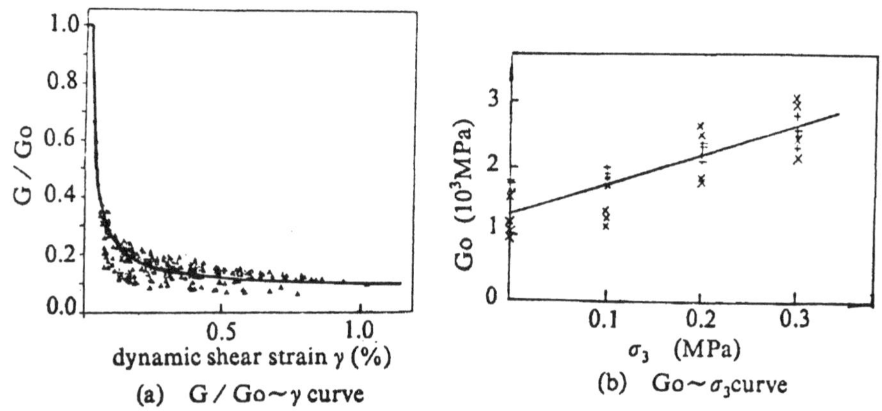

Figure 2. Dynamic shear modulus of plastic concrete

The dynamic shear modulus G also rises with the increasing of the surrounding pressure σ_3. Fig. 2b gives the relation of the variance of initial dynamic shear modulus G_0 versus σ_3 for plastic concrete [5]. It can be seen that G_0 increases nearly linearly with the increasing of σ_3.

Tests on damping characteristics have been completed on Pseudo–static Torsional Resonant Column Service. It shows that there seems to be a tendency for damping ratio to increase with γ within the range of strain tested, which is not very clear because the range of variance of γ is too little. In addition, damping ratio notably decreases with the increasing of age and dynamic shear modulus. For instance, the range of damping ratio's variance is 0.035~0.052 at age of 28–day and 0.016 ~ 0.021 at age of 90–day.

COMPARATIVE ANALYSES ON STRUCTURAL STRESSES OF PLASTIC AND NORMAL CONCRETE CUT–OFF WALL

In the light of the structure type and working conditions of concrete cutoff walls, there are three kinds of load–bearing state: the first is the cutoff wall in deep overburden underneath embankment dams with high central core whose main load–bearing characteristic is that the wall bears the great load transmitted from the core; the second is the cutoff wall under inclined core or in strengthening work of embankment dams, cofferdams, etc., where static water pressure leads the head place instead of the loads from upper soil media; the third is the situation that earthquake acts on the cutoff wall in above two cases. Comparative analyses have been carried out on static and dynamic structural stresses of cutoff walls of normal concrete and plastic concrete with various mechanical characteristics in above–mentioned three cases [7]. It is shown that the stress state in plastic concrete cutoff wall is notably superior to that in normal concrete one in all loading cases. And bases have been provided for the proportioning design of plastic concrete.

Analysis on Concrete Cutoff Wall underneath the Central Core of High Embankment Dam

The object analyzed is a central–core embankment dam with maximum height 188m, and the depth of overburden about 70m (Fig. 3). Two 1.4–meter wide concrete cutoff

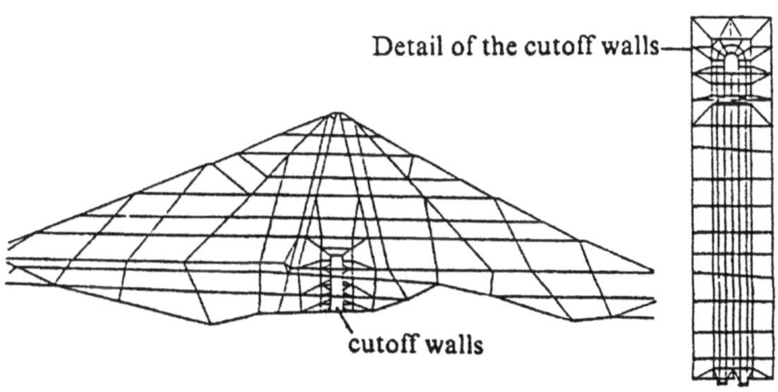

Figure 3. Finite element mesh of the central core embankment dam

walls will be adopted for seepage control in the foundation, their maximum height both being 70m and the clear distance between the walls is 3m. A gallery will be constructed on the top of the walls for connecting with the central core. Two alternatives of normal (rigid) and plastic concrete cutoff walls have been considered during the original design period. Hence comparative analyses of these two alternatives have been made as follows: [8]

1. Analysis Method: Program THEPD developed by Tsinghua University has been adopted in analysis and plane strain problem has been assumed. Elastic E–μ model has been used for the concrete cutoff wall and bedrock; for all kinds of soil and embankment materials in the alluvium and in the dam, hyperbolic nonlinear elastic E–μ model suggested by Duncan–Chang and quadrilateral iso–parametric element have been adopted. Four–node one–dimensional interface sliding elements have been also adopted in order to consider the interaction between the cutoff wall and the slurry layer as well as that between the cutoff wall and the material filled under the gallery bedding. The computational mesh is shown in Fig. 3.

An approach of incremental loading stage by stage has been used to simulate the process of construction and the storage of water. There are eight incremental loading stages divided as a total: the alluvium is taken as the first–stage load, and water storage stage is the last one; the middle six are construction stages. Each stage is divided into several minor increments with total number of 23.

In order to analyze comparatively, the materials of cutoff wall are rigid concrete ($E_W = 23.0 \times 10^3$MPa) and three kinds of plastic concrete respectively, two of which are linear elastic material ($E_W = 100$ MPa & 200 MPa) with elastic moduli 0.6~0.8 & 1.2~1.6 times those of the surrounding alluvium layer; and for the other, nonlinear E–μ model has been assumed having just the same mechanical characteristics as that of the alluvium. So the effect of the nonlinear characteristics of plastic concrete on the structural stresses can be studied.

2. Comparison of deformations of cutoff walls: Fig. 4a,4b and 4c give the comparison of displacements for plastic concrete cutoff wall and rigid one, and the variance of the ratio V_S / V_W of settlement in soil on the upstream side of the wall to that of the wall along the wall height in the completion stage respectively. It can be seen that the most

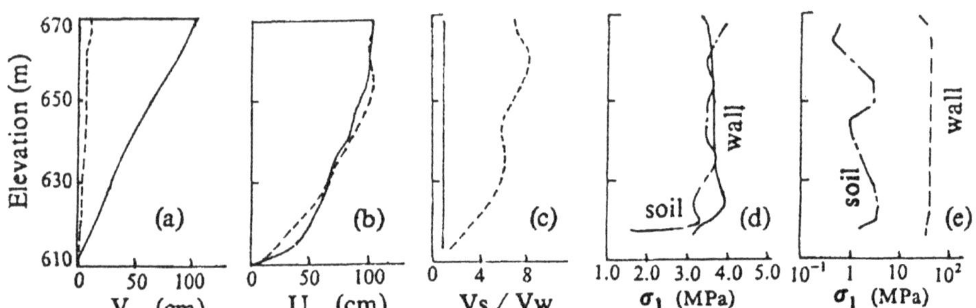

(a) Vertical settlement in completion stage (b) Horizontal displacement in impounding stage (c) Vs / Vw in completion stage (d) σ_{1w} and σ_{1s} for plastic wall (e) σ_{1w} and σ_{1s} for rigid wall ——— plastic wall - - - - rigid wall

Figure 4 Comparison of deformations and max. principal stresses between plastic and rigid cutoff wall difference between the deformations of plastic wall and rigid one occurs on the vertical settlement in completion stage. When $E_W = E_S$, the settlement of the plastic wall is almost the same with that of the surrounding soil, i.e. deformations are in coordination. The settlement of rigid wall, however, is far less than that of plastic wall, which is only about 1 / 10 of the latter. And the settlement of the surrounding soil is much greater than that of the wall, its maximum value being 8.5 times that of the latter. This great difference between deformations of the wall and surrounding soil will certainly bring about unfavourable influences on the stress state of normal concrete wall.

3. Comparison on stresses in cutoff wall: Fig. 4d and 4e give the comparison on maximum principal stresses σ_{1W} for the cutoff walls of plastic concrete($E_W = E_S$) and rigid one and σ_{1S} for soil on the upstream sides of the walls. It can be seen that the maximum principal stress σ_{1W} in plastic wall is the same as that of the soil σ_{1S} out of the wall on the whole when $E_W = E_S$. It shows that the wall and the surrounding soil share the loads transmitted from the upper body of the dam. The maximum σ_{1W} in the wall is 3.84 MPa at the lower portion of the wall. When the elastic modulus of plastic concrete has changed, the maximum σ_{1W} is 2.95 ~ 4.85 MPa, increasing with the elastic modulus.

For normal concrete cutoff wall, the stress of the soil layer outside the wall is far less than that for plastic one and the decrement is about 10~ 90%, averaging 54%. Therefore, the load borne by the surrounding soil is transmitted to the rigid wall by the shear stress between soil and the wall, due to the difference of the settlement between the wall and the soil; hence the wall stress increases greatly. The maximum stress reaches 27.9 MPa at the lower portion of the wall which is 7 times that of plastic wall at the same place.

In addition, it is valuable to point out that the minimum principal stress σ_3 of plastic wall is 0.8~ 1.2 MPa at the average and has a tendency to increase with increasing of the maximum principal stress σ_1. The average of σ_3 / σ_1 is about 1 / 4~ 1 / 3 and its maximum value occurs when $E_W = E_S$. This character will be helpful to increase the shear strength of plastic concrete. By our tests, the axial failure stress is 3~ 8 times the unconfined strength for plastic concrete when σ_3 is 1.2MPa. By the tests on plastic concrete of Converto Viejo Dam [9], when σ_3 increases from 0.1 MPa to 0.7 MPa, the ultimate deviation stress will increase from 0.22 MPa to above 3.87 MPa. This is beneficial to the safety of plastic concrete for its unconfined compressive strength is not high. For rigid wall, however, the average of σ_3 is about 1 MPa. This value is equal to that of plastic wall and the ratio σ_3 / σ_1 is less than 10% generally, its average value being only 4%, very close to the unconfined compressive state. It shows that it is hard to improve the strength of rigid concrete wall by the confined pressure of the foundation.

Analysis on Concrete Cutoff Wall by the Action of Water Pressure as Main Load

The object analyzed is a homogeneous earth dam [3] with a cutoff wall crossing it from dam crest through alluvium layer and finally inserted into the bedrock (Fig. 5). The wall is 32.5 m high and 0.8 m thick and the surrounding soil layers from top to bottom are clay in the dam, sandy loam, weathered basalt and sound basalt. The finite elemental mesh is shown in Fig. 5.

Duncan's Model is applied for clay in dam, sandy loam, weathered basalt and plastic concrete in computation, and elastic models for normal concrete and sound basalt. In order to optimize the stress in cutoff wall, the moduli of plastic concrete have been taken as 150, 300, 500, 1,000 MPa respectively and 20,000 MPa for normal (rigid) concrete.

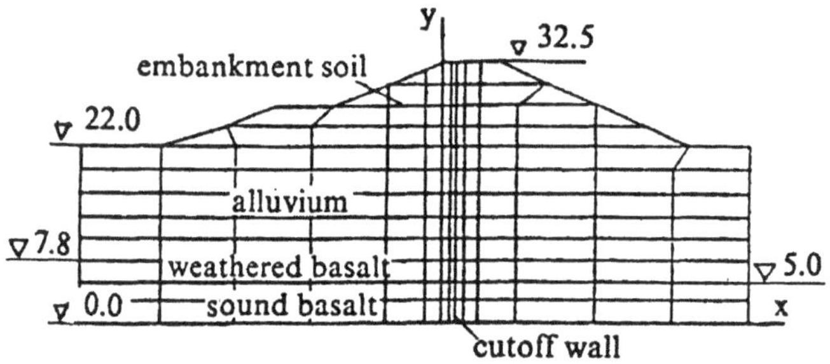

Figure 5. Finite element mesh

Fig. 6 gives the distribution of the vertical normal stress σ_y on the upstream and downstream surfaces of the wall along the height of the wall. It can be seen that by the actions of deadweight of the wall and upstream water pressure, the cutoff wall is basically in a deviation compressive state where the bottom part is the most unfavourable place, because of the existence of maximum bending moment over there.

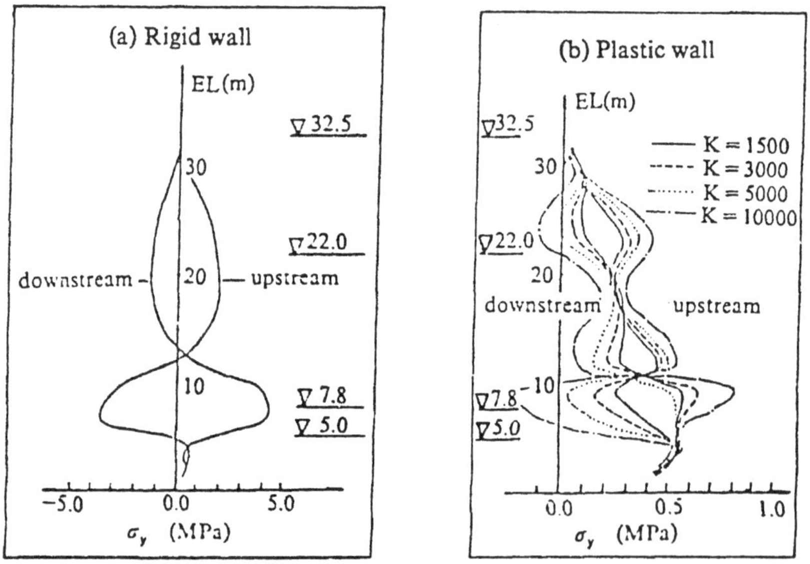

Figure 6. Distribution of the vertical normal stress along the height of the cutoff wall

It can be seen, for rigid wall, the maximum tensile stress occurring on upstream surface is as high as 3.6 MPa and the shear one 1.8 MPa. Obviously, these are beyond the ultimate strength of normal concrete and it is unsafe to adopt normal concrete cutoff wall here. With the decrement of the elastic modulus of concrete, the water pressure is gradually transferred to the soil media behind the wall, and the bending moment gradually decreases at the bottom. When the concrete is plastic, the working state comes gradually from bending action to axial compression as the leading roles. When elastic modulus E is less than 500 MPa, there is no tensile stress in the wall on the whole, shear stress is less than 0.15 MPa and the maximum compressive stress is less than 0.64 MPa. Now it is easier to design and obtain concrete whose modulus is less than 500 MPa and which can meet the requirements on strength and safety. When $E < 300$ MPa,improvement of stresses in the wall is not notable and the strength of concrete will decrease. So it is suggested to adopt plastic concrete with elastic modulus of 300 MPa and the 28–day strength should be greater than 1.5 MPa in the above–mentioned project.

In another study on plastic concrete cutoff wall [10], the influences of the Duncan's factors K, n, C and ϕ on the stress state of the cutoff wall have been systematically analyzed. The study indicates that it is beneficial to increase C and ϕ and decrease K and n for the improvement of the stress state of the cutoff wall, as well as to decrease the tensile stress and cut down the stress level. The tensile stress of the plastic concrete cutoff wall can be effectively reduced by the decrease of elastic modulus index n of plastic concrete. Therefore, it is better to make up a kind of plastic concrete with low values of K, n and with high values of C and ϕ.

Seismic Analysis on Concrete Cutoff Wall [3]

Dynamic analysis has been carried out by FEM for studying the antiseismic behavior of the cutoff wall shown in Fig. 5. The direct integration method has been adopted to evaluate the seismic stresses by using the time history of accelerations (shown in Fig. 7) recorded at Qianan during the 1976 Tangshan earthquake as base excitation. The ordinates of this time history have been adjusted to provide a maximum acceleration of 0.1g. The time step Δt is taken as 0.01 seconds. Both vertical and horizontal earthquake have been considered.

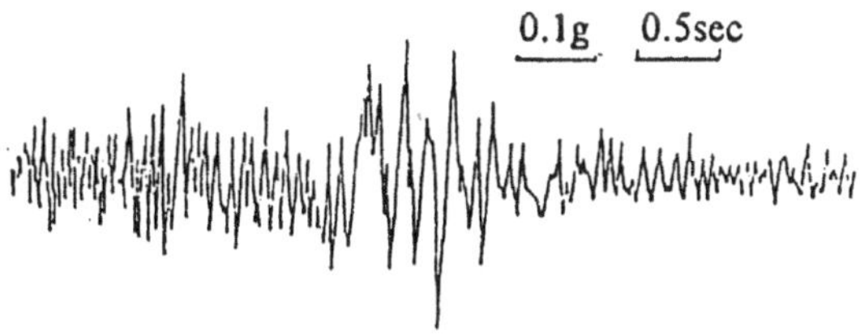

Figure 7. Qian–an earthquake record (1976. 8. 9)

Two cases have been considered in which the surrounding soil has linear or non–linear properties, and the equivalent linear method has been used for the nonlinear analysis. Four kinds of elastic modulus of concrete i.e. 20×10^3 MPa (as for normal or rigid concrete), 1×10^3, 300 and 100 MPa (as for plastic concrete) have been taken into account in the calculation.

Results of the seismic stresses of the wall obtained by linear and non–linear analysis are given in Fig. 8. It shows that the results evaluated by both methods are about the same for normal concrete wall. Same as the static stresses, the place of the most unfavorable seismic stress is still located at the bottom of the wall. The maximum vertical normal stresses are 1.17 MPa on upstream edge and 0.94 MPa on downstream edge of the wall, and the maximum shear stress is 0.5 MPa. When superposing these stresses on static ones, the total maximum tensile stress will increase by 17%, and reaches 5.8 MPa. The total maximum compressive stress will increase by 12%, and reaches 6.6 MPa and the total maximum shear stress reaches 0.86 MPa. Obviously, it aggravates the stress state of the wall so much and reduces the aseismatic safety of the cutoff wall.

For plastic concrete cutoff wall it can be seen from Fig. 8b that with the decreasing of the elastic modulus, the stress distribution along the height of the wall tends to be uniform and their values also decrease rapidly with E. When E is less than 300 MPa, the maximum seismic stress is not over 0.05 MPa. Considering the non–linear effect of surrounding soil, the stress can be reduced more. When E is down to 100 MPa, the value of stress drops down to 0.023 MPa which is only about 2% of that of the rigid wall. Because of the less values of the seismic stresses in plastic wall, there are still no unfavourable tensile stresses occurred when superposing them on the static stresses. It illustrates that plastic cutoff wall would be better for fitting in with the foundation deformation during earthquake, and is a better structure against earthquakes.

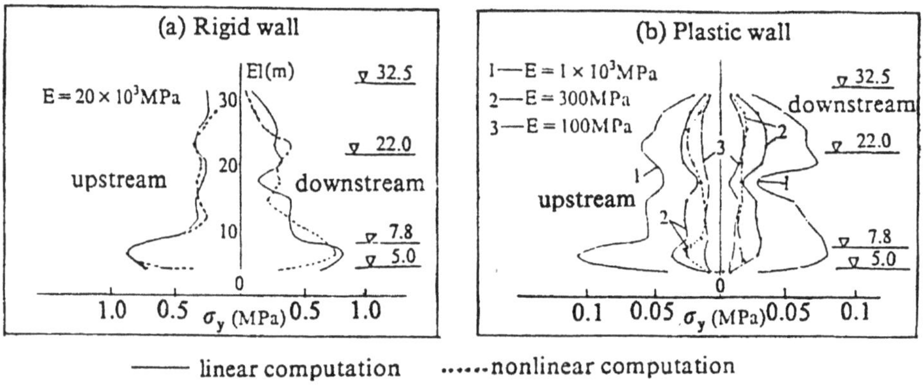

Figure 8. Envelope of the seismic stress in cutoff wall

EXPERIMENTAL RESEARCH ON PROPORTION OF MIXTURE OF PLASTIC CONCRETE

It has been indicated by the above analyses that the stress state of plastic concrete cutoff wall is determined by the ability for its deformation to coordinate with that of surrounding soil to a large extent. Therefore, the key to design plastic concrete cutoff wall is based on the principle of optimizing the stress state in the cutoff wall, and by means of the analysis of structural stress and according to the deformational character of surrounding soil layers and boundary condition to define the optimum structural form and dimension as well as the stress–strain relationship and other mechanical properties of the plastic concrete, and finally to define the optimum proportioning of plastic concrete materials. Hence, the question of design on plastic concrete cutoff wall is a question of "optimization and control of structural material" in nature. Compared with normal concrete, even more difficulty and complexity are added into the design of proportioning for plastic concrete, because of more kinds of raw materials in the composition of plastic concrete and more complicated factors influencing the mechanical characteristics.

In order to improve the safety and durability of plastic concrete, on one hand the deformation modulus should be decreased and nonlinear characteristics of plastic concrete should be assigned to fit in with that of surrounding soil layers so that the stress in the wall can be improved; on the other hand the strength and ultimate strain ε_{ult} of the concrete should be increased in the concrete so that the strength safety factor will be improved.

A great deal of test data [5] indicate that the higher the strength, the greater the modulus for plastic concrete. So in the design of proportion of mixture attention should also be paid to decrease the modulus–strength ratio $E_i/(\sigma_1-\sigma_3)_f$ i.e. the ratio of the initial modulus E_i to the strength $(\sigma_1-\sigma_3)_f$ of plastic concrete, along with controlling the strength and modulus. As for the nonlinear characteristics of plastic concrete another factor, nonlinearity index $\lambda=E_i/E_{0.5\%}$, to describe the type of the curves for stress–strain relationship has been taken, where $E_{0.5\%}$ is the secant modulus of plastic concrete when the corresponding axial strain is 0.5%.

Based on the study [11], main factors influencing the mechanical characteristics of plastic concrete are water cementitious ratio W/C_t (ratio of water to the total amount of cement, clay, bentonite and fly ash), consumption of bentonite C_b, dry unit weight r_d and consumption of aggregate C_a (sand and gravel). According to the analysis on these data, the influences of the increase of these factors on the mechanical characteristics of plastic concrete are illustrated in following table.

TABLE

Factors influencing the mechanical characteristics of plastic concrete

Factors	$E_i/(\sigma_1-\sigma_3)_f$	ε_{ult}	λ
W/C_t	up	down	up
C_a	up	down	up
C_b	up	down	up
r_d	up	down	up

After the deformation modulus, nonlinearity index λ, ultimate strain and strength or modulus – strength ratio of plastic concrete are defined through the optimization analysis on structural stress, orthogonal design and tests should be employed on water–cementitious ratio, consumption of aggregate, consumption of bentonite and other factors in the light of the regularities illustrated in the table. Then combining with the consideration on requirements for seepage–control, economy, construction condition, structural safety and durability, suitable proportioning of plastic concrete can finally be determined.

PROBLEMS FOR FURTHER STUDY

To sum up, plastic concrete is a kind of better construction material especially suitable to foundation cutoff wall. Plastic concrete cutoff wall has advantages such as good stress state, excellent capability against earthquake, low construction cost, constructional convenience, etc. In nature, design of plastic concrete and its structures is a problem of "optimization and control of structure and materials". Many new problems needed to be solved have been brought out to study on this subject, because of so many kinds of materials for composing plastic concrete, more complicated factors influencing and the nonlinearity of surrounding media and the plastic concrete itself. Although a great deal of test and research work has been completed on plastic concrete and cutoff wall by researchers in the world and many valuable experiences in construction practice have been gained, there are still questions needing study as follows:

1. Study on the mechanical characteristics of plastic concrete and its influencing factors especially on dynamic characteristics of this material including damping character, stress–strain relationship, failure type and dynamic strength at large strain, etc. This is the foundation for structural computation and analysis. In addition, further study is needed on the influences of every kind of material on the mechanical characteristics as basis for proportioning design.

2. Study on impermeability and durability of plastic concrete. The coefficient of permeability, critical seepage gradient, strength against erosion and the influencing factors and other problems should be determined, so that durability and safety will be improved for cutoff wall.

3. Study on nonlinear static and dynamic analysis on plastic concrete cutoff wall and the interaction between the wall and surrounding soil. Study on physical and mathematical model for numerical computation, particular type of element (e.g., element to simulate the action of the slurry layer outside the wall), and nonlinear computational program and technique and other aspects.

REFERENCES

〔1〕 The Yangtze River Research Institute, Study on embankment properties and flexible material of the cutoff wall in the cofferdam of Three Gorge Project, 1990.8.

〔2〕 Beijing Survey & Design Institute, Design on the plastic cutoff wall in the tailing cofferdam of Ming–tomb Pumped–storage Power Station, 1990.10.

〔3〕 Wang Qingyou, Wang Qizheng, Analysis on structural stress of the concrete cutoff wall in southern vice dam of Cetian Reservoir, Hydroelectric Department of Tsinghua University, 1990.7.

〔4〕 Feng Jiaji, Concrete face rockfill dams constructed on overburden, Proceedings of the Conference on Concrete Face Rockfill Dams, Hehai University publishing house, 1990.9.

〔5〕 Pu Jialiu, Yu Yuzhen, Wang Qingyou, Wang Qizheng, Experimental research on static & dynamic characteristics of plastic concrete, 1990.10.

〔6〕 Zhao Yong, Wang Qizheng, Dynamic interaction between cutoff wall and surrounding soil, Tsinghua University Report, 1991.

〔7〕 Wang Qizheng, Wang Qingyou, Pu Jialiu, Comparative analysis on structural stress of plastic and rigid concrete cutoff walls, 1990.10.

〔8〕 Wang Qizheng, Yu Yuzhen, Design & stress analysis on plastic concrete cutoff wall, Hydroelectric Department of Tsinghua University, 1989.7.

〔9〕 Luis Alvarez, Julio, Larenas, Alberto Berna, Juan Anlinio Marin, Characteristics of the plastic concrete of the diaphragm wall of Convento Viejo Dam, Fourteenth Congress on Large Dams. Q55R22P371–388.

〔10〕 Pu Jialiu, Yu Qionghua, Wang Shiting, Elastoplastic analysis & study on stress and displacement of the secondary cofferdam of Three Gorge Project, 1991.3.

〔11〕 Wang Qingyou, Pu Jialiu, Wang Qizheng, Yu Yuzhen, Experimental research on the proportion of mixture of plastic concrete. 1990.10.

ANALYSES OF TALL BUILDING STRUCTURES BY THE METHOD OF ANALYTIC ODE SOLVER

BAO SHIHUA
Department of Civil Engineering,
Tsinghua University, Beijing 100084, China

ABSTRACT

In this paper, by using the piecewise continuum technique for tall building structures, the differential equations for the static, dynamic and stability analyses of a tall building frame–shear wall–thin walled core cooperating system, in which frames, shear walls and cores are connected perpendicularly or skewly with each other in the building plan and the deformations of the floor slabs should be considered, have been derived. The solutions of these analytical differential equations are solved by Ordinary Differential Equations Solver COLSYS, which is a recently–developed general–purpose ordinary differential equations solver.

I. INTRODUCTION

In the late 70's, tall buildings had been rapidly developed in China, not only in its great numbers but also in the building heights. Till 1990, 75 tall buildings that exceed 100m in height had been built in the Mainland; the tallest one is the International Business Building with sixty–two storeys and 195m in height in Guangzhou. New structural systems have appeared more and more. Apart from the usually used structural systems frame, shear wall and frame–shear wall, there are the frame–tube, shear wall–tube, framed–tube, tube–in–tube, megaframe and so on. Most tall buildings in China are reinforced concrete structures, but more than 10 steel tall building structures have been erected in recent years.

The state of art of tall building structural analysis and the relevant references are given in [1]~[5], and will not be cited in detail here. A brief introduction is given as follows.

1. Various methods based on analytical solutions

These kinds of methods are usually differential equation solutions based on the assumptions of in–plane rigid floor slabs and continuum technique. Such as the frame–shear wall structure is treated as a cooperating system of the bending shear wall and the shear deformed frame to established 4–th (or 2nd) order differential equation, the coupled shear wall with flexible connecting beams is reduced to the similar type of differential equations as well. The latter method has been expanded to space shear walls.

Due to its clear physical concept and simple computational work, these kinds of methods had been widely used in engineering design. Due to the difficulty of solving large numbers of differential equations, these methods can only be used in symmetrical and regular structures.

2. The matrix displacement method based on member systems

Two dimensional analysis. Taking plane structures (frames or sheer walls) as substructures and under the assumption of infinite in–plane rigidity of the floor slabs, the matrix displacement method is used in combining the substructures together. Since 1974, it is the most popular method used in analyzing the normal frame, shear wall and frame–shear wall systems.

Three dimensional analysis. Entering 80′s, the complicated structural systems such as frame tube and tube–in–tube become more and more used. The characteristics of these types of structures are the space effects, and the space compatibilities of the resisting structures should be taken into consideration. Matrix displacement methods based on space member (including thin walled member) systems and the relevant computer programs have been developed.

The matrix displacement methods based on member systems are the widely adopted methods for tall building structural design in China.

3. Finite element methods

The tall building structure is divided into plane elements, plate bending elements, columns and beam elements which will form the global spatial structures. Various Finite Element techniques and computer programs have been developed and used in engineering design in China.

4. The semi–analytical methods

In which the finite strip method is a representation. Some significant improvements have been made in the finite strip techniques and relevant computer programs have been developed in China.

5. The methods of analytic and semi–analytic ODE Solver proposed by the author of this paper

In my opinion, the methods of tall building structural analysis should be developed in two ways. One is the discrete method, another is the analytical and semi–analytical method. Due to the very large computational work of the discrete method, large computer is needed and the expenses are large too. Analytical method for the analysis of tall building structure had been developed and widely used in simple structures. Due to the difficulty of solving large numbers of differential equations, analytical method had not been expanded to complicated structural systems. The recently–developed general–purpose ordinary differential equation (ODE) solvers have made it possible to renew or develop analytical method and semi–analytical method in analyzing various complicated tall building structures. In this paper, several typical problems of tall building structural analyses (including static, dynamic and stability analyses) are given as the examples of the method of analytic ODE Solver. The basic idea of the method of semi–analytic ODE Solver will be given later; the detailed analysis will be given in another paper.

II. STATIC, DYNAMIC AND STABILITY ANALYSES OF TALL BUILDING STRUCTURES

1. Static analysis of tall building structure with floor slab deformation considered

The assumption of in–plane rigid floor slab cannot be used in many tall building structural analyses, such as cases where length–width ratio of the building plan is very large, the building plan has 'bottle neck' reducing the rigidity of the floor slab, the structural stiffnesses change rapidly along the storey height and so on. The computational work for the analysis of tall building structure with the floor slab deformation taken into consideration by using the method of space structure is very large.[1,6]

In this paper the normally used continuous techniques in analyzing frame–shear wall structures are still used, but the assumption of in–plane rigid floor slab is released. The floor slabs are treated as deep beams in the floor plan with the shear deformation taken into consideration. The differential equations of equilibrium for each lateral resisting structure (including frame, shear wall and frame–shear wall) are obtained (Fig.1) as follows:[9]

$$\left[EI_i v_i'''' - \left(C_{fi} + \frac{D_i}{h}\right)v_i'' - \frac{GA_{i-1}}{\mu l_{i-1} h}(v_{i-1} - v_i) + \frac{GA_i}{\mu l_i h}(v_i - v_{i+1})\right]_n = q_{in} \tag{1}$$

$$i = 1,\cdots,s; \quad n = 1,2$$

in which, v_i represents the lateral displacements, symbol prime represents derivative to coordinate z; I_i represents the summation of the moment of inertia of all shear walls in the i–th resisting structure; C_{fi} represents the shear stiffness of the frame in the i–th resisting structure, i.e. the required shear force producing unit shear angle of the frame storey; D_i represents the summation of rotational stiffness of the connecting beams in the i–th resisting structure; A_i represents cross–sectional areas of the i–th floor slab; μ is a constant coefficient depending upon the shape of the cross–section,

$\mu = 1.2$ for rectangular cross–section.

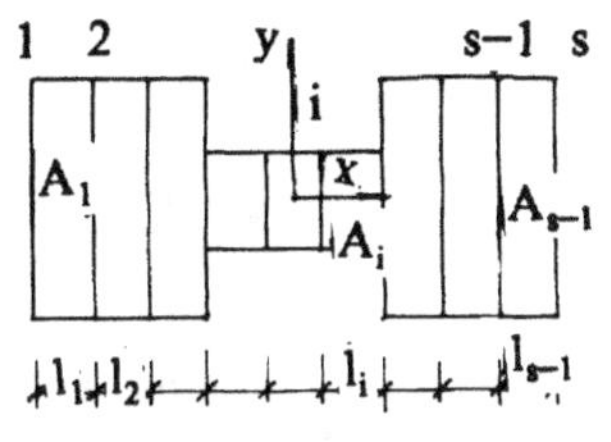

(a) Structural plan

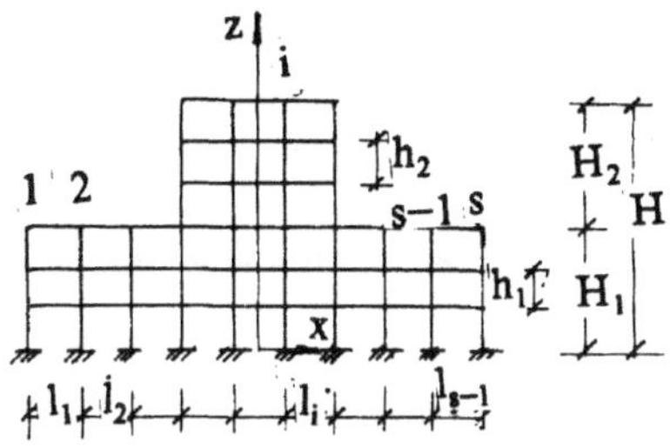

(b) Structural elevation

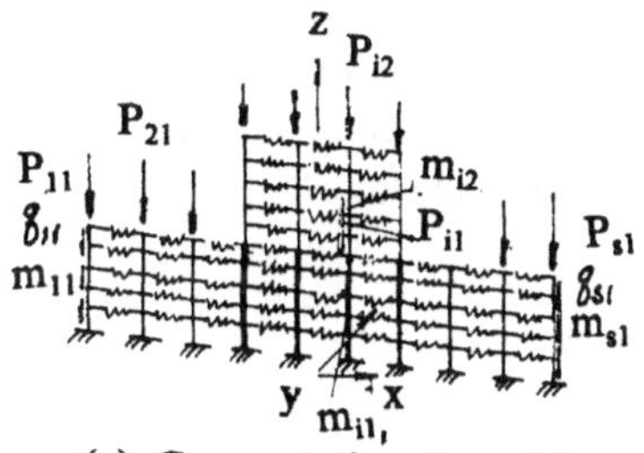

(c) Computational model

Figure 1

Eqs.(1) are the equilibrium equations in terms of lateral displacements, in which the first term represents the effect of the flexural deformation of the shear wall, the second term represents the effects of the elastic resistance of the frame and the restraint moment of the connecting beams on the shear walls, the 3rd and 4th terms represent the effects of floor slabs shear forces on the shear walls. The equations and expressions for the upper and lower parts of the structures are distinguished by subscript n; n = 1 denotes the lower part, n = 2 denotes the upper part.

The boundary conditions and connecting conditions are given as follows. The bottom ends of each resisting structure are clamped:

$$For \quad z = 0, \quad [v_i]_1 = 0 \quad and \quad \left[v_i'\right]_1 = 0, \quad i = 1, \cdots, s \tag{2}$$

The top ends of each resisting structure are free (or subjected to horizontal concentrated leads P_i):

$$For \quad z = H, \quad \left[v_i''\right]_2 = 0; \quad \left[-EI_i v_i''' + (C_{fi} + \frac{D_i}{h}) v_i'\right]_2 = 0 \quad (or P_i) \tag{3}$$

Connecting conditions between the upper and lower parts: for $z = H_1$;

$$[v_i]_1 = [v_i]_2; \quad \left[v_i'\right]_1 = \left[v_i'\right]_2; \quad \left[EI_i v_i''\right]_1 = \left[EI_i v_i''\right]_2;$$

$$\left[-EI_i v_i''' + (C_{fi} + \frac{D_i}{h})v_i'\right]_1 = \left[-EI_i v_i''' + (C_{fi} + \frac{D_i}{h})v_i'\right]_2 \qquad (4)$$

2. Horizontal vibrations of tall building structure with floor slab deformation considered

In the normally used method of multi–degree–of–freedom dynamic analysis, the masses are lumped in each floor slab, and the 'series– parallel lumped mass model' is used in the dynamic analysis of tall building structure with the floor slab deformation taken into consideration.[3,8] The computational works for this model are very large.

In this paper, the normally used continuous techniques are used and a 'piecewise continuous distributed mass parallel model' (Fig. 1c) is used. The vibrational differential equations are obtained as below:[10]

$$\left[EI_i V_i'''' - (C_{fi} + \frac{D_i}{h})V_i'' - \frac{GA_{i-1}}{\mu l_{i-1} h}(V_{i-1} - V_i) + \frac{GA_i}{\mu l_i h}(V_i - V_{i+1}) + m_i \ddot{V}_i\right]_n = 0, \qquad i = 1,\cdots,s; \quad n = 1,2 \qquad (5)$$

in which m_i represent distributed mass of the i–th resisting structure; symbol point represents derivative with respect to time t; symbol prime represents derivative with respect to coordinate z; other symbols are the same as before.

Assume $\quad V_i(z,t) = v_i(z) sin\omega t$

Substituting into Eq. (5), then

$$\left[EI_i v_i'''' - (C_{fi} + \frac{D_i}{h})v_i'' - \frac{GA_{i-1}}{\mu l_{i-1} h}(v_{i-1} - v_i) + \frac{GA_i}{\mu l_i h}(v_i - v_{i+1}) - m_i \omega^2 v_i\right]_n = 0, \qquad i = 1,\cdots,s; \quad n = 1,2 \qquad (6)$$

in which $v_i(z)$ are functions of z; ω is the natural frequency of the structure.

Eqs. (6) are the horizontal vibrational equations of the structure; the boundary conditions and connecting conditions are the same as those shown in Eqs (2)–(4).

3. Stability analysis of tall building structure with floor slab deformation considered

With the increase of building height or the ratio of building height to width, the lateral stiffness of the structure is decreased. In addition, the vertical loads of tall building are very large, and the stability analysis of tall building structure attracts more attention. The references of the overall stability analysis of tall building are few, especially through the simple and practical methods[7]

In this paper, the continuous technique is used and a continuous parallel model for stability analysis is used (Fig. 1c). The Stability differential equations are obtained as below[11]:

$$\left[EI_i v_i'''' - \frac{D_i}{h} v_i'' + \left\{[q_{i2}(H-z) + P_{i2}]v_i'\right\}' - \frac{GA_{i-1}}{\mu l_{i-1} h}(v_{i-1} - v_i) + \frac{GA_i}{\mu l_i h}(v_i - v_{i+1})\right]_2 = 0, \qquad i = 1,\cdots,s \tag{7}$$

$$\left[EI_i v_i'''' + \frac{D_i}{h} v_i'' + \left\{[q_{i1}(H_1 - z) + P_{i1} + P_{i2} + q_{i2}H_2]v_i'\right\}' - \frac{GA_{i-1}}{\mu l_{i-1} G}(v_{i-1} - v_i) + \frac{GA_i}{\mu l_i h}(v_i - v_{i+1})\right]_1 = 0$$

in which q_{in} represents the vertical load intensity, P_{in} represents the vertical concentrated load and other symbols are the same as before.

Eqs. (7) are the stability equations of the structure, and are two sets of variable coefficient differential equations, the boundary conditions are the same as those shown in Eqs. (2)~(4), except the last equation of Eqs.(4) which is replaced by

$$\left(-EI_i v_i''' + \frac{D_i}{h} v_i'\right)_1 - P_{i1} v_{i1}' = \left(-EI_i v_i''' + \frac{D_i}{h} v_i'\right)_2$$

4. Static analysis of frame–shear wall–thin walled core skewly combined system with floor slab deformation considered.

In recent years, one characteristic in the development of tall building in China is the complication of the structural systems due to architectural reasons. There are some common points: in structural types they are combined by frame, shear wall and thin walled core; in building plan they are arranged perpendicularly or skewly; in the height direction the geometrical dimensions of the system are uniform or stepped or with setback; for the effects of floor slabs in some regions the assumption of in–plane rigid floor slab can be adopted and in some regions (such as 'bottle neck' regions) the floor slab deformations should be taken into consideration. These points are shown in Fig.2 and which can be seen as the computational model of the system.

Substructure 1 consists of shear walls, frames and cores with stepped parameters which are connected by in–plane rigid floor slab, and has three displacement components u, v and θ. The substructures are connected by in–plane elastic floor slabs which are treated as deep beams in the floor plan and with the axial, bending and shear deformations taken into consideration. The equilibrium differential equations in terms of global displacements for each substructure are obtained as below:[12]

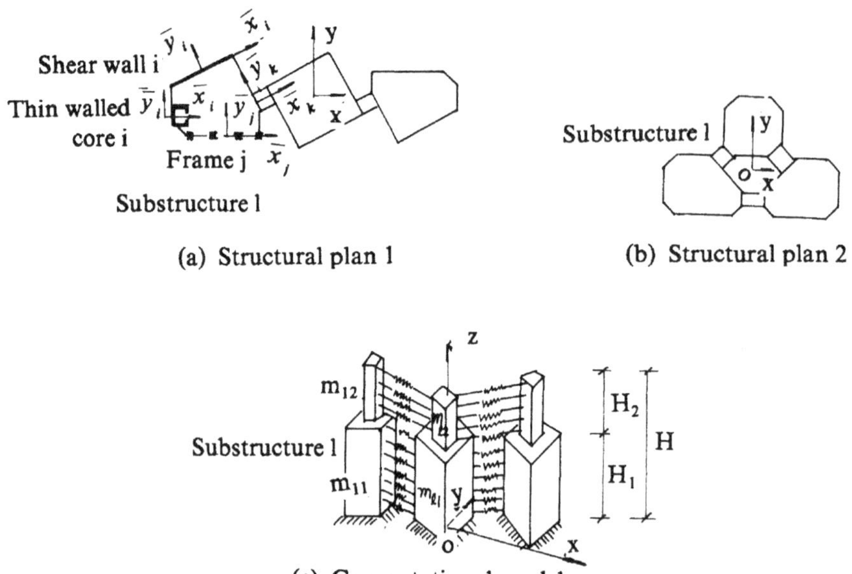

(a) Structural plan 1

(b) Structural plan 2

(c) Computational model

Figure 2

$$
\begin{aligned}
&[A]_{ln}\{U''''\}_{ln} - [B]_{ln}\{U''\}_{ln} + [C_1]_{ln}\{U\}_{ln} + [C_2]_{l-1,n}\{U\}_{l-1,n} \\
&\qquad + [C_3]_{l+1,n}\{U\}_{l+1,n} = \{p\}_{ln}, \qquad l = 1, \cdots, s; \; n = 1, 2
\end{aligned}
\tag{8}
$$

where $\{U\}_{ln} = [u \quad v \quad \theta]^T_{ln}, \quad \{p\}_{ln} = [p_x \quad p_y \quad m_z]^T_{ln}$

$$[A]_{ln} = \sum_i [T_i]_{ln}[D_i]_{ln}[T_i]^T_{ln}$$

$$[B]_{ln} = \sum_i [T_i]_{ln}[D_{ti}]_{ln}[T_i]^T_{ln} + \sum_j [T_j]_{ln}[D_{F_j}]_{ln}[T_j]^T_{ln}$$

$$[C_1]_{ln} = \frac{1}{h}[T_{k1}]_{ln}[D_{11}]_{kn}[T_{k1}]^T_{ln} + \frac{1}{h}[T_{k-1,2}]_{ln}[D_{22}]_{k-1,n}[T_{k-1,2}]^T_{ln}$$

$$[C_2]_{l-1,n} = \frac{1}{h}[T_{k-1,2}]_{ln}[D_{21}]_{k-1,n}[T_{k-1,1}]^T_{l-1,n}$$

$$[C_3]_{l+1,n} = \frac{1}{h}[T_{k1}]_{ln}[D_{12}]_{kn}[T_{k2}]^T_{l+1,n}$$

in which u, v and θ represent the displacements along x,y directions and rotation angle about z axis for each substructure; p_x, $p_y$and m_z represent the applied loads along x,y directions and torques about z axis for each substructure; $[D_i]$, $[D_{ti}]$, $[D_{Fj}] and [D_{lm}]_k$ represent the stiffness matrices in the local coordinates of shear wall i (including thin walled core), frame j and floor slab k in each substructure, and their expressions are shown below:

$$[D_i]_{ln} = diag(EI_{\bar{y}i} \quad EI_{\bar{x}i} \quad EI_{\bar{\omega}i})_{ln}, \quad [D_{ti}]_{ln} = diag(0 \quad 0 \quad GJ_{di})_{ln}$$

$$[D_{Fj}]_{ln} = diag(C_{fj} \quad 0 \quad 0)_{ln}, \quad a = \frac{EA}{l}, \quad b = \frac{E\bar{I}}{l}, \quad c = 6\frac{E\bar{I}}{l^2}$$

$$d = 12\frac{E\bar{I}}{l^3}, \qquad \beta = \frac{12\mu EI}{GAl^2}, \qquad \bar{I} = I/(1+\beta)$$

$$[D_{11}]_{kn} = \begin{bmatrix} a & 0 & 0 \\ 0 & d & c \\ 0 & c & (4+\beta)b \end{bmatrix}_{kn}, \quad [D_{12}]_{kn} = \begin{bmatrix} -a & 0 & 0 \\ 0 & -d & c \\ 0 & -c & (2-\beta)b \end{bmatrix}_{kn}$$

$$[D_{21}]_{kn} = \begin{bmatrix} -a & 0 & 0 \\ 0 & -d & -c \\ 0 & c & (2-\beta)b \end{bmatrix}_{kn}, \quad [D_{22}]_{kn} = \begin{bmatrix} a & 0 & 0 \\ 0 & d & -c \\ 0 & -c & (4+\beta)b \end{bmatrix}_{kn}$$

$[T_i]$, $[T_j]$ and $[T_{kl}]$ represent the coordinate transformation matrices of shear wall i (including thin walled core), frame j and floor slab k(subscript k1 denotes end 1, k2 denotes end 2) and having the similar form as

$$[T_i] = \begin{bmatrix} \cos\alpha_i & -\sin\alpha_i & 0 \\ \sin\alpha_i & \cos\alpha_i & 0 \\ x_i^0 \sin\alpha_i - y_i^0 \cos\alpha_i & x_i^0 \cos\alpha_i + y_i^0 \sin\alpha_i & 1 \end{bmatrix}$$

in which x_i^0, y_i^0 represent the global coordinates of the origin of the local coordinates of member i, α_i represents angle from $\bar{o}_i \bar{x}_i$ to ox.

Eqs. (8) are the equilibrium equations in terms of global displacements; they are flexure–torsion coupled and coupled between substructures. The boundary conditions and connecting conditions are given as follows:

The bottom ends of each substructure are clamped:

$$For \quad z = 0, \quad \{U\}_{l1} = \{0\} \quad and \quad \{U'\}_{l1} = \{0\}, \quad l = 1, \cdots, s \tag{9}$$

The top ends of each substructure are free (or subjected to horizontal concentrated loads and torque $\{P\}_l = [P_x \quad P_y \quad M_z]_l^T$):

$$\left.\begin{array}{l} For \quad z=H, \quad \left\{U''\right\}_{l2}=\left\{0\right\}, \quad l=1, \cdots, s \\ [A]_{l2}\left\{U'''\right\}_{l2}-[B]_{l2}\left\{U'\right\}_{l2}=\left\{0\right\}\left(or -\left\{P\right\}_{l}\right) \end{array}\right\} \tag{10}$$

Connecting conditions between the upper and lower parts: For $z=H_1$

$$\left\{U\right\}_{l1}=\left\{U\right\}_{l2}, \quad \left\{U'\right\}_{l1}=\left\{U'\right\}_{l2} \tag{11}$$

$$\left.\begin{array}{l} [A]_{l1}\left\{U''\right\}_{l1}=[A]_{l2}\left\{U''\right\}_{l2} \\ [A]_{l1}\left\{U'''\right\}_{l1}-[B]_{l1}\left\{U'\right\}_{l1}=[A]_{l2}\left\{U'''\right\}_{l2}-[B]_{l2}\left\{U'\right\}_{l2} \end{array}\right\} \tag{12}$$

5. Horizontal vibrations of frame–shear wall–thin walled core skewly combined system with floor slab deformation considered.

The 'space series–parallel lumped mass model' can be used to analyze the space vibrations of this system[8], but the computational work is very large.

In this paper, the continuous techniques are used and a flexure–torsion coupled series–parallel continuous distributed mass model (Fig.2) is established. The vibrational differential equations for each substructure are obtained as below:[13]

$$\begin{aligned} &[A]_{ln}\left\{u''''\right\}_{ln}-[B]_{ln}\left\{u''\right\}_{ln}+[C_1]_{ln}\left\{u\right\}_{ln}+[C_2]_{l-1,n}\left\{u\right\}_{l-1,n} \\ &+[C_3]_{l+1,n}\left\{u\right\}_{l+1,n}=-[M]_{ln}\left\{\ddot{u}\right\}_{ln}, \quad l=1,\cdots,s; \quad n=1,2 \end{aligned} \tag{13}$$

where

$$[M]_{ln}=\begin{bmatrix} m & 0 & -my^{\circ} \\ 0 & m & mx^{\circ} \\ -my^{\circ} & mx^{\circ} & J \end{bmatrix}_{ln}$$

in which m and J represent distributed mass and moment of inertia of the mass about z axis; u, v and θ represent the displacements along x,y directions and rotation angle about z axis for the mass centroid (x° ,y°); other symbols are the same as before.

Assume $\left\{u(z,t)\right\}_{ln}=\left\{U(z)\right\}_{ln} \sin\omega t$

Substituting into Eqs. (13), the following ordinary differential equations are obtained.

$$[A]_{ln}\left\{U''''\right\}_{ln} - [B]_{ln}\left\{U''\right\}_{ln} + [C_1]_{ln}\left\{U\right\}_{ln} + [C_2]_{l-1,n}\left\{U\right\}_{l-1,n}$$

$$+ [C_3]_{l+1,n}\left\{U\right\}_{l+1,n} = \lambda[M]_{ln}\left\{U\right\}_{ln}, \quad l=1,\cdots,s; \quad n=1,2 \tag{14}$$

in which $\lambda=\omega^2$, ω represents natural frequency.

Eqs. (14) are the horizontal vibrational equations of the system, the boundary conditions and connecting conditions are the same as in Eqs. (9)~(12).

Ⅲ. SOLUTIONS BY ODE SOLVER AND EXAMPLES

The differential equations obtained above are divided into two categories: for static analysis, there are boundary value problems in ordinary differential equations; for dynamic and stability analyses, there are eigenvalue problems in ordinary differential equations. Both of them can be solved by ODE Solvers.

ODE Solvers are researched and developed by ODE specialists for general purpose ODE problems and are powerful tools for computational mechanics. In this paper, we use ODE code COLSYS which is capable of solving mixed–order systems of boundary value problems in ODEs. The method used in COLSYS is spline collocation at Gaussian points. The damped and modified Newton methods are employed for nonlinear iteration. In COLSYS, approximate solutions are computed on a sequence of automatically selected meshes until a user–specified set of tolerances is satisfied. Here we give a brief description of the features of the adopted ODE code. For more details, the reader is referred to Ref. [16]~[18].

For static analysis, the solutions can be obtained directly from COLSYS. For dynamic and stability analyses, since COLSYS are not equipped with eigensolution features, transformation techniques are used to reform eigenproblems to standard nonlinear ODEs for accurate and efficient solution by COLSYS. Various computational examples are given in Ref. [9]~[15], two of them are shown below.

Example 1 A six–storey reinforced concrete structure with two substructures is shown in Fig.3. Each substructure consists of four bracings. The geometric dimensions of the four bracings are shown in Fig.3. The storey height h = 300cm. The elastic modulus of the material $E=2.5\times 10^6 N/cm^2$, $G=0.4E$. The thickness of the floor slabs is 15cm. Uniformly distributed load 500N / cm subjected to substructure 2.

Partly obtained displacements are shown in table 1.

TABLE 1 Displacements (1×10^{-3}cm or rad.)

coordinates z(m)	substructure 1						substructure 2					
	bracing 1			bracing 2			bracing 1			bracing 2		
	$\bar{u}$	$\bar{v}$	$\bar{\theta}$	$\bar{u}$	$\bar{v}$	$\bar{\theta}$	$\bar{u}$	$\bar{v}$	$\bar{\theta}$	$\bar{u}$	$\bar{v}$	$\bar{\theta}$
18	110.29	79.52	0.17	25.12	50.96	0.17	−88.46	401.54	0.21	352.61	258.93	0.21
12	64.94	42.13	0.10	10.14	27.57	0.10	−56.09	233.03	0.13	207.12	149.73	0.13
6	21.42	12.04	0.03	1.91	7.87	0.03	−20.54	77.34	0.05	69.95	49.34	0.05

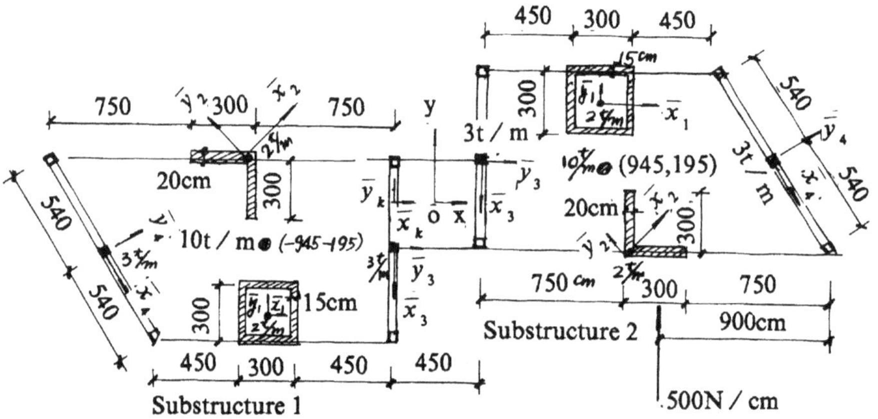

Figure 3 Structural plan of examples

Example 2 The structure is the same as in example 1. Building plan and the distributed masses of the four bracings of each substructure are shown in Fig.3. The total distributed masses of substructures 1 and 2 are $m_1 = m_2 = 100$kg / cm, the moments of inertia of the masses about mass centroids are $J_1 = J_2 = 3.987 \times 10^7$ kg • cm^2 / cm.

Two cases are computed: 1. The whole structure is uniform in the height direction. 2. The whole structure is divided into two parts, the lower three storeys plan and data still shown in Fig.3; the upper three storeys plan is the same as in Fig.3, but the stiffness and the mass are r times of that in the lower part.

The obtained natural frequencies are given below; the relevant vibrational modes can be seen in Ref. [13]

For $r = 1$, $\omega_1 = 3.60$, $\omega_2 = 4.01$, $\omega_3 = 4.06$, $\omega_4 = 11.23$, $\omega_5 = 19.42$

$\omega_6 = 20.33$, $\omega_7 = 22.70$, $\omega_8 = 23.48$, $\omega_9 = 41.46$, $\omega_{10} = 41.50$

For $r = 0.1$, $\omega_1 = 8.18$, $\omega_2 = 8.31$, $\omega_3 = 8.74$, $\omega_4 = 13.10$, $\omega_5 = 17.87$

$\omega_6 = 18.40$, $\omega_7 = 20.57$, $\omega_8 = 21.33$, $\omega_9 = 42.22$, $\omega_{10} = 45.94$

IV. FURTHER RESEARCH AND COOPERATION OUTLINE

A brief introduction to author's research work concerning the analyses of tall building structures by the method of analytic ODE Solver is given in the above of this paper.

Another research works concerning the analyses of tall building structures by the method of semi–analytic ODE Solver are being done by the author of this paper and its research group. Another title of the method of semi–analytic ODE Solver is called finite element method of lines (FEMOL) that has been proposed by associate professor Yuan, member of the research group, during his post-Ph. D. in London. FEMOL is a new semi–analytical method based on ODE Solvers. In FEMOL firstly by using the finite element method techniques and the variational principle of energy functional, the governing partial differential equations are semi–discretized to ordinary differential equations expressed in terms of nodal line functions; secondly solutions are directly obtained by ODE Solvers. The applications of two normal and key techniques (Finite Element and ODE Solver) make FEMOL attractive and competitive as a new method. FEMOL has been used in general computational mechanics and good results have been obtained. The author of this paper expand the FEMOL to the static, dynamic and stability analyses of tall building tubular structure and preliminary results have been obtained.

The author of this paper prepares to use these two methods to the analyses of the varieties of tall building structures and establish the analysis series of tall building structures by the methods of analytic and semi–analytic ODE Solvers. Firstly research works are putting on the normally used tall building structures, and then the advanced problems, such as: static dynamic and stability analyses of complicated structural systems, the combined analysis of the soil–foundation– structure interaction problem, nonlinear analysis problem, analysis of earthquake response, new structural systems (megastructure and megaframe, suspended system, cellular structure, composite system and bridged structure). The computational works of these problems by completely discrete method are all extremely large. Since the numbers of ODEs are not large in the methods of analytic and semi–analytic ODE Solvers, more advantages would be seen.

In the development of tall building structural analyses, both discrete methods which are suitable for large-scale computer and analytical, semi–analytical methods which are suitable for small computer should be developed. The methods of analytic and semi–analytic ODE Solvers proposed by the author of this paper belong to the latter. The uses of these methods in the analyses of tall building structures are for the first time both within and outside the country, and would have good prospects.

The author of this paper and his research group would like to cooperate with the other side in these aspects: to exchange research problems and academic viewpoints; to do cooperating research work for problems interested by both sides; to participate special topical discussion, lecture and conference; co–writing of book; to apply for the U.S. funds in cooperating research project and so on.

ACKNOWLEDGEMENT

The work in this paper is supported by the National Natural Science Foundation of China and the Structural and Vibrational Engineering Research Laboratory of Tsinghua University.

REFERENCES

1. Bao Shihua and Fang Ehua, **Design of Tall Building Structures,** Tsinghua University Press, 1st Edition, 1985; 2nd Edition, 1990.
2. China Academy of Building Research, **Design of Tall Building structures,** Scientific Press (China), 1982.
3. Bao Shihua, **Analyses of Tall Building Structures,** High Educational Press, 1990.
4. G.Q. He and S.F.Luo, The State of Art of Tall Building Structures in China, **Proc. 3rd Inter. Conf. on Tall Building,** Dec. 1984.
5. Zhao Xi'an, Development on the Methods of Analysis of Tall Building Structures in China. **Proc. 3rd Inter. Conf. on Tall Building,** Dec. 1984.
6. Zhao Xi'an, An Analytic Method for Tall Building Structures Considering Floor Deformation, **China Civil Engineering Journal** vol. 16, No.4, 1983.
7. Liang, Q.Z. and Xie, L., The Second–Order Analysis of Frame–Shear wall Structure, **J. of Building Structures,** No.5, 1986.
8. Zhao Xi'an, Direct Dynamic Analysis of Tall Building Structures Considering Floor Deformation, **Computational Structural Mechanics and Application,** vol.7, No.4, 1990.
9. Bao Shihua and Yuan Si, Analysis of Tall Building Structures with Floor Slab Deformations Considered by the Continuous Method and ODE Solvers, to be published in **Building Structures.**
10. Bao Shihua and Yuan Si, Dynamic Analysis of Tall Building Structures with Floor Slab Deformation Considered by the Continuous Method and ODE Solvers, to be published in **J. of Building Structures.**
11. Bao Shihua and Zhang Yiguo, Overall Stability Analysis of Tall Building Structures with Stepped Parameters and with Floor Slab Deformations Considered, to be published in **China Civil eng. J.**
12. Bao Shihua and Zhang Yiguo, Analysis of Frame–Shear Wall– Thin Walled Core with Stepped Cross–Sections, Skewly cooperated and with Floor Slab Deformations Considered, to be published in **J. of Eng. Mech.**
13. Bao Shihua and Zhang Yiguo, Dynamic Analysis of Frame–Shear Wall–Thin Walled Core with Stepped Cross–Sections, Skewly Cooperated and with Floor Slab Deformations Considered, to be published in **J. of Earthquake Eng. and**

Eng. Vibr.

14. Bao Shihua and Yuan Si, Coupled Vibrations for Frame–Shear Wall–Thin Walled Core Cooperating System with Stepped Cross–Section by ODE Solvers, to be published in **Computational Structural Mechanics and Applications.**
15. Yuan Si and Bao Shihua, Horizontal Vibrations for Shear Wall, Shear Wall on Flexible Support and Framed Wall Cooperating System with Stepped Cross–Section by ODE Solvers, to be published in **Engineering Mechanics.**
16. U.Ascher, J. Christiansen and R.D.Russell, Collocation Software for Boundary–value ODEs, **ACM Trans. Math. Software,** vol.7, No.2, 1981.
17. U. Ascher, J. Christiansen and R.D.Russell, Algorithm 569, COLSYS: Collocation Software for Boundary–Value ODEs [D_2], **ACM Trans. Math. Software,** vol.7, No.2, 1981.
18. S.Yuan, An Introduction to the general purpose program for Boundary–value ODEs, **Computational Structural Mechanics and Applications,** No.2, 1990.
19. U. Ascher and R.D.Russell, Reformulation of Boundary Value Problems into 'Standard' Form, **SIAM, Rev. 23, April 1981.**
20. S. Yuan and Y.Zhang, A Solution Strategy for Structural Eigenproblems in ODEs by Standard ODE Solver, to be published.
21. S. Yuan, Finite Element Method of Lines in Computational Mechanics, to be published.

NONLINEAR AND POSTBUCKLING ANALYSIS OF ELASTIC FRAMES

Wang, Pao-Hsii* and Chen, Yung-Ti**
Department of Civil Engineering
Chung-Yung Christian University
Chung-Li, Taiwan, China

ABSTRACT

A nonlinear Finite Element computation procedure using the corotational formulation for nonlinear and postbuckling analysis of elastic frames is presented in the study. The beam-column approach is ultilized herein and the convected (corotational) coordinate system is used in each beam element for seperating rigid-body motions from local deformations. The plane and space beam elements and their coordinate transformation coefficients are introduced here. The linearized system equations are solved in increment-iterationwise. In order to overcome the numerical difficulties at the occurrence of multiple critical points, the generalized displacement control method is adopted to trace the complete load-deflection curve of frames. With the guide of a general stiffness parameter (GSP), the load step size will be adjusted and the loading direction will be changed during the computation and the equilibrium path can be traced fully automatically. The snap-through, snap-back and looping behaviors of a right-angle frame, a circular arch and a spatial frame are investigated numerically in details.

Keywords: Postbuckling Analysis, Large Deflection Analysis, Elastic Frames

INTRODUCTION

Large deflection analysis has received a lot of attention in the structural mechanics over the last two decades. Considerable effort has been devoted recently to the development of finite element formulations and of improved numerical techniques for nonlinear structural analysis. Different finite element formulation stratagies has been presented. One of the most widely used formulations for large deflection analysis is the Corotational (CR) formulation[11,12,14]. The CR-formulation has mainly been applied to large displacement and large rotation problems, but with small strain. In the CR-formulation a Cartesian coordinate system is used and embedded within each finite element so that it translates and rotates with the element during deformations, but it does not deform with the element. Such a local Cartesian coordinate system attached to each element is called the corotational (or convected) coordinate system. Based on the convected coordinate system

* Professor ** Graduate Student

a set of element coordinates is chosen , which completely determine the deformation, but do not deal with rigid body motions, i.e. the deformation of element can be separated from its rigid body motion and both of them will be treated separately. Within the CR-formulation , either a Total Lagrangian (TL) or a Updated Lagrangian (UL) approach can be employed. In the TL-formulation all static and kinematic variables are referred to the initial configuration, while in the UL-formulation the current one (last known) is chosen as the reference configuration. Bathe & Bolourchi [2] concluded that same results are obtained in both TL-and UL-formulations, but the latter is more computationally effective in large deflection, small strain analysis. In the analysis of large deflection but small strain problems the CR-formulation works very efficiently and accurately. Recently, a series of papers concerning with large deflection problems handled with CR-formulation has been presented.

The nonlinear structure often exhibits a complex behavior. For the structures having a monotonously increasing or decreasing load-deflection curve, the large deflection analysis can be easily performed, in which an increment-iteration procedure based on the Newton-Raphson method can be successfully applied. When the structure has multiple critical points, such as limit points (snap-through), snap-back points and bifurcation points, the nonlinear analysis may become extremely complicated. Because of the occurrence of critical points, the numerical computation may fail at the neighborhood of such critical points. More sophisticated solution schemes for tracing complex load-deflection curves in nonlinear and postbuckling analysis have been developed in the past and are called as the path-following -type methods, such as displacement control method [1,29], arc–length method [12-14,19-21,26] and work control method [27] etc. Most recently, a new method named the generalized displacement control method is proposed by Yang & Shieh [28]. This method is believed to be superior to most existing methods and has the following advantages: (1) The numerical stability can always be achieved in regions near the critical points, (2) the load increment can be adjusted with the variation in nonlinearity of the structure, and (3) a sign change in a so-called generalized stiffness parameter (GSP) at limit points serves as a good indicator for reversing the loading direction [28].

The aim of the paper is to present a nonlinear computation procedure using the corotational formulation for nonlinear and postbuckling analysis of elastic frames. A solution algorithm adopting the generalized displacement control method will also be built up and used to trace the pre-and post-buckling load deflection curves of the frames. Two plane frames and one space frame are selected and analyzed in detail for demonstrating the efficiency of the computation procedure and examining their nonlinear postbuckling behavior.

NONLINEAR FINITE ELEMENT FORMULATION

Nonlinear system equations

The nonlinear system equations of a finite element model of structures in nonlinear statics may be written as follows [23,24]

$$K^j \cdot b^j_\alpha - \sum_{el} S_j a_{j\alpha} = 0 \qquad \alpha = 1,2,\ldots N \quad \text{D.O.F.} \tag{1}$$

where $b^j_\alpha = \frac{\partial W^j}{\partial q_\alpha}$ = basic vectors, $a_{j\alpha} = \frac{\partial u_j}{\partial q_\alpha}$ = transformation coefficients,

$P_\alpha = K^j \cdot b^j_\alpha$ = generalized external forces, $T_\alpha = \sum S_j a_{j\alpha}$ = generalized internal forces,

$\mathbf{K}^j$= external nodal load vectors, $\mathbf{W}^j$= displacement vectors corresponding to $\mathbf{K}^j$, $S_j = KE_{jk} u_k + S_j$ = generalized element forces, S_j = initial generalized element forces KE_{jk} = element stiffness matrix, $\sum_{el}$= Summation over all elements, and N = number of degrees of freedom. The superscript j denotes the nodal number. The subscripts $\alpha, \beta, \gamma ...$ denote the number of system coordinate and j,k,l,...the number of element coordinate. The index summation convention is used here for the super-and subscripts. The letters printed in bold type e.g. $\mathbf{K}^j, \mathbf{b}^j_\alpha$ represent vectors. The dot notation between vectors means scalar product. In the large deflection analysis of structures, the material behaves linearly elastic and the deflection is large, but with small strain. In Eqs.(1) $\mathbf{b}^j_\alpha$, $a_{j\alpha}$, u_j may be nonlinear functions of the system coordinates q_α when large deflections occur and the nodal load vectors may also become function of q_α, if they are not displacement independent. Eqs.(1) represents a set of nonlinear algebraic equations and can be solved by load increment methods or iteration methods.

Linearized System Equations

For the sake of solving large deflection problem in increment-iterationwise the linearized system equations have to be derived.By taking the first order of Taylors' expansion of the nonlinear system equations, the linearized system equations for a small load interval is obtained as follows:

$$\Delta P^n_\alpha + {}_uP^n_\alpha = {}^2K^n_{\alpha\beta}\, \Delta q^n_\beta \qquad \text{for } P^n_\alpha \leq P_\alpha \leq P^{n+1}_\alpha \tag{2}$$

where $\Delta P^n_\alpha = P^{n+1}_\alpha - P^n_\alpha$ = load increments, $\Delta q^n_\beta = q^{n+1}_\beta - q^n_\beta$ = displacement increments, ${}_uP^n_\alpha = P^n_\alpha - \sum_{el} S^n_j a^n_{j\alpha}$ = unbalanced forces in statics , ${}^2K^n_{\alpha\beta} = \sum_{el} KE_{jk} a^n_{j\alpha} a^n_{k\beta} + \sum_{el} S^n_j a^n_{j\alpha,\beta} - {}^n\mathbf{K}^j \cdot {}^n\mathbf{b}^j_{\alpha,\beta} - {}^n\mathbf{K}^j \cdot {}^n\mathbf{b}^j_\alpha$ = tangent system stiffness matrix, $a_{j\alpha,\beta} = \frac{\partial a_{j\alpha}}{\partial q_\beta}$= transformation coefficients of second order and $\mathbf{b}^j_{\alpha,\beta} = \frac{\partial \mathbf{b}^j_\alpha}{\partial q_\beta}$. The superscript "n" denotes the number of load step , and the "2" means iteration matrix of second order. The tangent system stiffness matrix ${}^2K^n_{\alpha\beta}$ consists of four terms. The first term is the elastic stiffness matrix, the second and the third terms are the so-called geometric stiffness matrix induced by large deflections. And the last term represents also a geometric stiffness matrix, but is induced by displacement dependent forces, which will not be taken into consideration in the study. The coefficients $a_{j\alpha}$ and $a_{j\alpha,\beta}$ are called coordinate transformation coefficients of first and second order respectively and they are determined for each element, based on deformed position.

Beam-Column Elements

In the study some assumptions are made for beam elements as follows. The material is homogeneous and isotropic and behaves as linearly elastic. The member is straight and it has constant cross section with double symmetry; thus the torsional stiffness is uncoupled to bending and axial stiffness. Shear deformations and warping effects are neglected. So the coupling between axial forces and twist is neglected thus excluding torsional stability problems. The external loads are applied at the nodes as well as being displacement independent. Large displacements and large rotations are allowed, but the strain is small.

For deriving the beam-column element the so-called corotational (or convected coordinate) formulation is ultilized in the study, in which the relative deformations of a member are referred to a local, Cartesian convected coordinate system attached to the member itself. The local convected coordinate system, rotates and translates with the member but it does not deform with the member. The corotational formulation has mainly been applied to large displacement and large rotation problems, but the strain is small. Based on the local convected coordinate system the element coordinates are so chosen that they can completely determine the deformations of the element but they do not deal with rigid body motions. These element coordinates are also called as the relative element coordinates by Schrader (1969) or as the natural (mode) coordinates in [1]. Since the rigid body motion and local deformations of an element are separated by using the corotational formulation, the derivation of the nonlinear element model of frame structures is divided into two major parts: one is the derivation of element force-displacement relations and the other is the derivation of the transformations that relate the element coordinates to the system coordinates.

(1) Plane Beam-Column Element

Based on the local, convected coordinate system, the relative element coordinates u_j of the plane beam-column element are chosen as shown in Fig.2.1. The three coordinates u_j, $j=1,2,3$ completely determine the flexural and axial deformation of the element, where u_1 and u_2 represent the relative end angles to member chord, and u_1 the relative axial displacement between two end sections. Corresponding to the element coordinates u_j, the element forces S_1,S_2,and S_3 denote bending moments at end sections and axial force in element respectively. The linear element stiffness matrix of the beam-column element can be easily found in literatures. It is not difficult to describe the exact kinematics of a beam element moving in plane when large deflection occur. The exact nonlinear transformation coefficients of a beam element can be derived according to the actual large deflected position of the element and can be found in [24]. Since the exact transformation coefficients exist, the nonlinear analysis of plane frames are carried out in a manner of the Total Lagrangian approach in the study.

(2) Space Beam-Column Element

Similarly, based on the local, convected coordinate system,the relative element coordinates u_j of a space beam-column element are chosen as shown in Fig 2.2. The six coordinates u_j, $j=1,2,...6$ completely determine the deformation of the element, where u_1,u_2,u_3,u_4 represent the relative end angles to the member chord, u_5 the relative twist angle (relative rotation of member end sections about the x-axis), and u_6 the relative axial displacement. Corresponding to the element coordinates u_j, the element forces S_1,S_2,S_3 and S_4 denote bending moments and S_5,S_6 are the torsion and the axial force respectively. The linear element stiffness matrix of the space beam–column element can also be found in literatures. Large displacements and large rotations may occur in the large deflection problems of frames. It is well known that finite rotations about

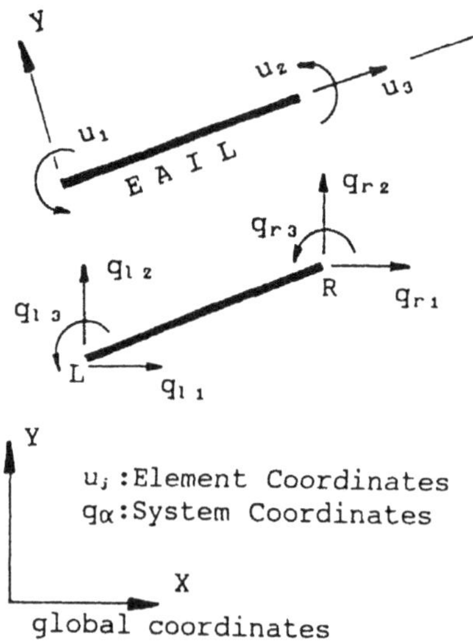

Fig.2.1 Plane Beam Element

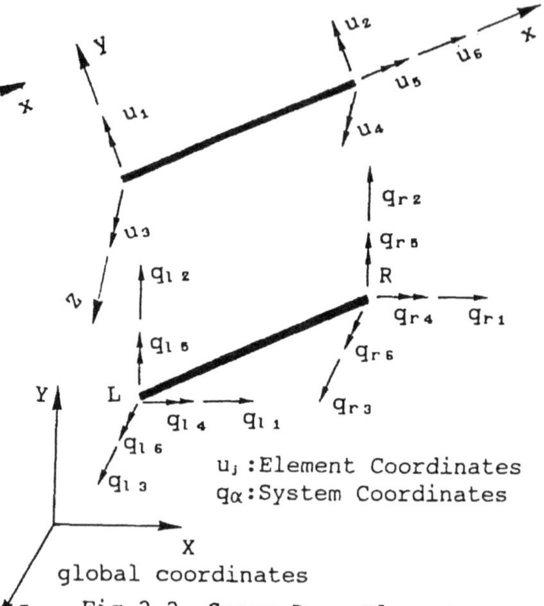

Fig.2.2 Space Beam Element

differently oriented axes in space are not commutative and the results will depend on the order in which rotations take place. So finite rotations can not be treated as vector quantities. Therefore, it is difficult to describe the exact kinematics of a beam element moving in three-dimensional space when large deflection occur and the exact nonlinear transformation coefficients of a space beam element are impossible to find. Therefore the large deflection problem of space frames can only be solved in incrementwise and treated as a series of piecewise linearized problems. In each load increment, the deflection is assumed to be small and the rotations are so small that it can be treated as vector quantities. Hence, based on slightly deformed position, only the linearized coordinate transformation coefficients of first and second order can be found for a space beam element. They have been presented in [25] and expressed explicitly by the member orientation matrix proposed by Oran [15]. In each load step, the linearized transformation coefficients are determined according to the last equilibrium position. Therefore, the nonlinear computation of space frames is performed in such a manner of the Updated Lagrangian approach.

Much more accuracy of frame analysis can be achieved, when the bowing effect of the element, i.e. the axial shortening due to element bowing is taken into account. When the element bowing effect is considered, the element bending moments will be dependent on axial force and the element axial force depends quadratically on element displacements. By differentiating the nonlinear element force-displacement equations, a tangent element stiffness matrix relating the increment element forces and increment element displacements can be found and is referred to [25].

SOLUTION ALGORITHM FOR NONLINEAR ANALYSIS

The linearized system equations (2) are ultilized and solved step-by-step incrementally in the nonlinear analysis. It is well known that a pure increment computation will bring large numerical errors. An increment-iteration solution procedure is recommended in order to get more accurate solution. In the increment-iteration procedure the nonlinear problem is stepwise linearized and the solution is corrected by additional equilibrium iterations. The most often used method is the Newton-Raphson method. When a structure exhibits a simple monotonously increasing or decreasing load-deflection curve, the nonlinear computation can be easily and accurately performed by using the Newton-Raphson method. For the structures having critical points, such as limit points (snap-through), snap-back points and bifurcation points in the load-deflection curves, the Newton-Raphson method fails to converge in the neighborhood of critical points and it becomes very difficult to trace the complete load-deflection curve.

In the past, several more sophisticated solution schemes for tracing the complex load -deflection curves in nonlinear and post-buckling analysis have been developed and are called as the path-following-type methods, in which the iteration computation is forced to follow a specified iteration path. An ideal solution scheme should be able to trace the entire pre-and post- buckling path of a structure, which may have multiple critical points and the bifurcation path. The popularly used methods are the displacement control method proposed by Argyris 1965 [1] and Zienkiewicz 1971 [29], the constant arc-length method introduced independently by Riks 1972 [20] and Wempner 1971 [26], the modified arc-length method by Crisfield 1981 [7], and the equal work control method by Yang & McGuire 1984 [27]. Most recently, the generalized displacement control method is proposed by Yang & Shieh [28] and is believed to be superior to most of the existent methods. In this study the Yang's generalized displacement control method is adopted for the nonlinear and postbuckling analysis. In the following a brief summary of the Yang's method is given and the details of the method are referred to [28].

In Yang's method, the load increments will be adjusted automatically in each iteration and can be represented by

$$ {}^{(i)}\Delta P_{\alpha}^{n} = {}^{(i)}\lambda^{n} \; {}^{ref}\Delta P_{\alpha} $$

where ${}^{ref}\Delta P_{\alpha}$ is a given reference load increment and λ is an unknown load parameter, which has to be determined in each iteration. The superscript n denotes n-th load increment step and i the i-th iteration.

According to Batoz and Dhatt [3], the linearized system equations (2) can be replaced by the following two set linear equations:

$$ {}^{ref}\Delta P_{\alpha} = {}_{2}^{(i)}K_{\alpha\beta}^{n} \cdot {}^{(i)}\Delta \bar{q}_{\beta}^{n} $$

$$ {}^{(i)}_{u}p_{\alpha}^{n} = {}_{2}^{(i)}K_{\alpha\beta}^{n} \cdot {}^{(i)}\Delta \bar{\bar{q}}_{\beta}^{n} $$

where ${}^{(i)}\Delta \bar{q}_{\beta}^{n}$, ${}^{(i)}\Delta \bar{\bar{q}}_{\beta}^{n}$ are displacement increments induced by the reference load increment ${}^{ref}\Delta P_{\alpha}$ and the unbalanced force ${}^{(i)}_{u}p_{\alpha}^{n}$, and the total displacement increment can be obtained by superposition ,i.e. ${}^{(i)}\Delta q_{\beta}^{n} = {}^{(i)}\lambda^{n} \; {}^{(i)}\Delta \bar{q}_{\beta}^{n} + {}^{(i)}\Delta \bar{\bar{q}}_{\beta}^{n}$.

Since the load parameter λ has to be determined in each iterations, there are N+1 unknowns (N displacement increments plus one load parameter) in the analysis. An extra constraint equation has been introduced by Yang [28] as follows:

$$ C_{\beta} \; {}^{(i)}\Delta q_{\beta}^{n} + k \; {}^{(i)}\lambda^{n} = {}^{(i)}H^{n} $$

where C_{β}, k and H are the selected constant column matrix and constants. And the load parameter is given by

$$ {}^{(i)}\lambda^{n} = \frac{ {}^{(i)}H^{n} - C_{\beta} \cdot {}^{(i)}\Delta \bar{\bar{q}}_{\beta}^{\,n} }{ C_{\beta} \cdot {}^{(i)}\Delta \bar{q}_{\beta}^{\,n} + k } $$

For the generalized displacement control method it is suggested to take

$$ C_{\beta} = {}^{(1)}\lambda^{n} \; {}^{(1)}\Delta \bar{q}^{\,n-1} , \quad k = 0 \text{ and } {}^{(i)}H^{n} = 0 \text{ for } i \geq 2. $$

Hence the load parameter becomes

$$ {}^{(i)}\lambda^{n} = \frac{ {}^{(i)}H^{n} - {}^{(1)}\lambda^{n} \; {}^{(1)}\Delta \bar{q}_{\beta}^{\,n-1} \cdot {}^{(i)}\Delta \bar{\bar{q}}_{\beta}^{\,n} }{ {}^{(1)}\lambda^{n} \; {}^{(1)}\Delta \bar{q}_{\beta}^{\,n-1} \cdot {}^{(i)}\Delta \bar{q}_{\beta}^{\,n} } $$

Since ${}^{(i)}\Delta \bar{\bar{q}}_{\beta}^{\,n} = 0$ for $i = 1$ and ${}^{(i)}H^{n} = 0$ for $i = 2$, we have

$$ {}^{(1)}\lambda^{n} = \left(\frac{ {}^{(1)}H^{n} }{ {}^{(1)}\Delta \bar{q}_{\beta}^{\,n-1} \cdot {}^{(1)}\Delta \bar{q}_{\beta}^{\,n} } \right)^{1/2} \quad \text{for the increment step } i = 1 $$

and

$$ {}^{(i)}\lambda^{n} = - \frac{ {}^{(1)}\Delta \bar{q}_{\beta}^{\,n-1} \cdot {}^{(i)}\Delta \bar{\bar{q}}_{\beta}^{\,n} }{ {}^{(1)}\Delta \bar{q}_{\beta}^{\,n-1} \; {}^{(i)}\Delta \bar{q}_{\beta}^{\,n} } \quad \text{for the iteration step } i \geq 2 $$

For $n = 1$, $i = 1$ and let ${}^{(1)}\Delta \bar{q}_{\beta}^{\,0} = {}^{(1)}\Delta \bar{q}_{\beta}^{\,1}$ we have

$$ {}^{(1)}\lambda^{1} = \left(\frac{ {}^{(1)}H^{1} }{ {}^{(1)}\Delta \bar{q}_{\beta}^{\,1} \cdot {}^{(1)}\Delta \bar{q}_{\beta}^{\,1} } \right)^{1/2} $$

hence

$$ {}^{(1)}H^{1} = \left({}^{(1)}\lambda^{1} \right)^{2} \; {}^{(1)}\Delta \bar{q}_{\beta}^{\,1} \cdot {}^{(1)}\Delta \bar{q}_{\beta}^{\,1} $$

and

$$ {}^{(1)}\lambda^{n} = - {}^{(1)}\lambda^{1} \left(\frac{ {}^{(1)}\Delta \bar{q}_{\beta}^{\,1} \cdot {}^{(1)}\Delta \bar{q}_{\beta}^{\,1} }{ {}^{(1)}\Delta \bar{q}_{\beta}^{\,n-1} \; {}^{(1)}\Delta \bar{q}_{\beta}^{\,n} } \right)^{1/2} = {}^{(1)}\lambda^{1} \cdot (G\,S\,P)^{1/2} $$

where $G\,S\,P = \dfrac{ {}^{(1)}\Delta \bar{q}_{\beta}^{\,n} \cdot {}^{(1)}\Delta \bar{q}_{\beta}^{\,1} }{ {}^{(1)}\Delta \bar{q}_{\beta}^{\,n-1} \; {}^{(1)}\Delta \bar{q}_{\beta}^{\,n} }$ and is defined as the generalized stiffness parameters (G S P).

NUMERICAL EXAMPLES

Three examples are presented in the study to demonstrate the solution algorithms described in the previous sections. All the load- displacement curves of the examples are traced automatically by using the generalized displacement control method and the iteration convergence tolerance $\epsilon = 10^{-4}$ is used. In the examples E,I,and A denote the Young's modulus, the moment inertia of cross-section and the cross-section area of members.

Example 1 : Right-Angle Frames

The right-angle frame shown in Fig.4.1a was first studied by Lee [13]. Its geometric and physical properties are given in Fig.4.1a. For its finite element model, 10 simple plane beam elements per each member are used and the system has totally 59 degree of freedoms. A vertical concentrated load P is applied at point D having a distance f from the rigid joint B counterclockwise. For $f > 0$, the load P is on the member BC and when $f < 0$, P locals at the left hand side of joint B and will be replaced by a single load P plus a moment $P \cdot f$ applied at joint B for the analysis. In the nonlinear analysis the cases with $f = 0$, $f > 0$, and $f < 0$ are examined in details.

When $f = 0$, the load P is applied directly at the joint B and the problem becomes a pure stability problem. The buckling load of the frame can be found by the linear stability analysis and has the value of $P_{cr} = 8332.0$ N. The postbuckling behavior of the frame is also examined and shown in Fig.4.1b. For examining its postbuckling behavior the nonlinear large deflection analysis is carried out. At the beginning of the computation in this case, a very small disturbance (initial displacement or initial force) is given and the load–displacement curve is determined. Fig.4.1b exhibits a typical asymmetric postbuckling behavior, where the load required to maintain the equilibrium may increase or decrease with increasing deflection depending on the direction in which the frame deflects after buckling.

When $f > 0$ or $f < 0$, the problem belongs to deformation analysis and is often used to study the imperfection effect with changing the value f. For the case of $f > 0$, different values of f are taken in the study. The nonlinear load–displacement curves are first determined and shown with the deformed shapes of the frame in Fig.4.1c. No numerical difficulties appear in the computation when the described solution procedure is used. For the cases of $f < 0$, only $f = -20$ is taken and the load-displacement curve and deformed shapes of the frame are found and shown in Fig.4.1d. It is obviously to see in Fig.4.1d that the load-displacement curve becomes much more complicated than that in the case of $f > 0$. The curve in Fig.4.1d is very irregular and contains multiple critical points and loops. The deflected shapes of the frame exhibits higher buckling mode shapes as the load P increases.

Example 2 : Circular Arch

The hinged circular arch is taken from Ref.[5], and is shown in Fig.4.2a with its geometric and physical data. A vertical concentrated load P is applied at the apex of the arch. In the study the arch is modeled by 26 simple straight beam element of equal length and the whole system has 77 degree of freedoms. Two cases are examined herein; one is a symmetric case and the other is unsymmetric case. In the symmetric case, no imperfection is assumed, the arch has only symmetric deformation modes. And in the unsymmetric case , imperfection is assumed and induced by giving a very small disturbance (e.g. a horizontal force 0.01P) at the beginning of the computation. The complete load displacement curves of the both cases are traced fully automatically by the generalized displacement control method and shown in Fig.4.2b to 4.2d respectively, where the deformed shapes of the arch are also plotted. In the symmetric case since automatic reversing in the loading direction at the limit points takes place when the generalized displacement control method is used, the path after snap-through, curve f–b'–g–h–i–j–k in Fig4.2c, does not appear in the result in Fig.4.2b. The figure Fig.4.2c is taken from Fig.4.2b and enlarged, where the path f–b'–g–h–i–j–k after snap–through, determined with the Newton-Raphson method is added. The load increment is adjusted by hand when the Newton-Raphson Method is used in this case. In unsymmetric case, the arch deflects in both symmetric and unsymmetric deformation modes, and the critical loads at the limit points have evidently smaller values than that in symmetric case, see Fig.4.2d. Combining part of Fig.4.2d with the Fig.4.2c, we have the Fig.4.2e, from which it is obviously to see that the unsymmetric paths k–l–m and o–p–q–r, curves with circled mark are the bifurcation path of the arch. In Fig.4.2e the bifurcation points of the arch locate at k,m,o and r.

Example 3 : 24–member Hexagonal Star-Shaped Dome

The dome shown in Fig.4.3a is taken from [14]. Its supports are assumed to be pinned and restrained against translation motion. The interior joints are rigid joints. In the study the dome is idealized by using 24 space beam elements, one element per member, where the axial shortening due to element bowing is taken into account. The system has 39 degree of freedoms. Two loading cases are considered: in case 1, a vertical concentrated load P is applied at the apex (node 1) of the dome (symmetric case) and in case 2, except P at node 1, an extra load P/2 is added at node 2 (unsymmetric case). Two different cross-sections of member are also considered herein. The results of the dome determined by using the generalized displacement control method are plotted in Fig.4.3b and 4.3c. No snap-back points appear in the example. In the figures the load-displacement curves with the circled mark is determined by using space truss element instead of beam element. No numerical difficulties appear in the entire computation process. The results obtained in the study agree with that of Meek & Tan [14] very well.

CONCLUSIONS

A nonlinear finite element computation procedure using the corotational formulation combined the generalized displacement control method is presented for nonlinear and postbuckling analysis of elastic frames. The entire pre- and post-buckling load-deflection curves of elastic frames having multiple critical points can be traced successfully and automatically by this procedure. During the computation the load increment is adjusted automatically and the loading direction is reversed automatically when GSP changes sign at limit points. No numerical instability occurs in the computation of the selected examples. The numerical experiments in the study show that the herein proposed procedure works very accurately and efficiently for nonlinear and postbuckling analysis of elastic frames having complex behaviors.

REFERENCES

1. Argyris, J.H. , "Continua and Discontinua," Proceedings of First Conference on matrix Method in Structural Mechanics, Wright-Patterson Air Force Base, Ohio, pp. 11-89, (1965)
2. Bathe, K.J. and Bolourchi, S. "Large Displacement Analysis of Three-Dimensional Beam Structures",Int . J. for Num. Meth. in Engineering ,Vol.9, pp.961–986, (1979)
3. Batoz, J.L., and Dhatt, G. "Incremental Displacement Algorithms for Nonlinear problems" ,Int. J. for Num. Meth. in Engineering , Vol. 14, pp. 1262–1266 (1977)
4. Bergan, P.G. "Solution Techniques for Nonlinear Finite Element Problems", Int. J. for Num. Meth. in Engineering, Vol.12, pp. 1677–1696. (1978)
5. Brink, K. "Theories and Berechnung in einer Ebene Gekrummter Stabe unter grosser Verformungen ",Technisch–wissenschaftliche Mittelungen Nr. 79–1 , Institut fuer KIB, Ruhr–Universitaet Bochum , Deutschland, (1979)

6. Clarke, M.J. and Hancock , G.J. "A Study of Incremental Iterative Strategies for Non-linear Analysis ", Int. J.for Num. Meth. in Engineering, Vol. 29. pp. 1365–1391, (1990)
7. Crisfield, M.A. "A Fast Incremental/Iterative Solution Procedure that Handles Snap–Through", Computers and Structures, Vol.13, pp. 55–62 (1981)
8. Crisfield M.A."An Arc–Length Method Include Line Search and Accelerations" , Int. J. for Num. Meth. in Engineering, Vol.19, pp. 1269–1289, (1983)
9. Forde B. W.R and Stiemer S.F. "Improved Arc length Orthognality Method for Non-linear Finite element Analysis", Computers & Structures, Vol. 27 pp. 625–630 (1987)
10. Fujii F. "Scheme for Elasticas With Snap–Back and Looping", J. of Engineering Mechanics, ASCE, Vol.115, pp. 2166–2181 (1989)
11. Hsiao K.M., Horng H.J. and Chen Y.R., "A corotational Procedure that Handles Large Rotations of Spatial Structures ", Computers & Structures, Vol. 27, pp 769–781, (1988)
12. Honecker A. , Knothe K. , and Karamanlis D., "Large Deflection Finite Element Analysis of pre-and postcritical Response ot thin Elastic Frames", Proceedings of the 2nd International Symposium Weitgespannte Flaechentragwerken. Vol. 2. Stuttgart, Deutschland (1979)
13. Lee, S.L. and Manuel, F.S. "Large Deflections and Stability of Elastic Frames ", J. of Eng. Mech. Div. , ASCE pp. 521–546, (1968)
14. Meek, J.L. and Tan, H.S. ,"Geometrically nonlinear Analysis of Space Frames by an Incremental Iterative Technique", Computer Methods in Applied Mechanics and Engineering, pp. 261–282, (1984)
15. Oran, C. "Tangent Stiffness in Space Frames ",J. of the Str. Div. ,ASCE , Vol. 99, pp. 987–1001 (1973)
16. Powell , G. and Simons , J. "Improved Iteration Strategy for Non–linear Structures", Int. J. for Num. Meth. in Engineering, Vol. 17, pp. 1455–1467, (1981)
17. Ramm, E. "Strategies for Tracing the Nonlinear Response near Limit Points", in Nonlinear Finite Element Analysis in Structural Mechanics, edited by Wunderlich, W., Stein,E. and Bathe ,K.J. , Springer–Verlag, Berlin, pp. 63–69 (1981)
18. Remseth, S. N. "Nonlinear Static and Dynamic Analysis of Framed Structures ", Computers & Structures, Vol. 10, pp. 879–897, (1979)
19. Riks, E. "An Incremental Approach to the Solution of Snapping and Buckling Problems", Int. J. of Solids & Structures, Vol. 15, pp.529–551 , (1979)
20. Riks, E. "The Application of Newton's Method to the Problem of Elastic Stability" , J. of Applied Mechanics, Vol. 39, pp. 1060–1066.(1972)
21. Riks, E. "Some Computational Aspects of the Stability Analysis of Nonlinear structures", Computer Methods in Applied Mechanics and Engineering, pp. 219–259 (1984)
22. Schweizerhof K.H. and Wriggers P. "Consistent Linearization for Path Following Methods in Nonlinear F.E. Analysis ", Computer Methods in Applied Mechanics and Engineering pp. 261–279, (1986)
23. Schrader , K.–H. "Die Defurmationsmethode als Grundlage einer Programmierten Sprache", Hochschule B. I. 830, Bibliographisches Institute, Mannheim/Zuerich, (1969)
24. Schrader, K.–H. : MESY3 Benutzerhandbuch, Version CDC CYBR 175, FORTRAN 77, Institute fuer KIB, Ruhr–Universitaet Bochum, Dentschland, (1983)
25. Wang, P.H. , Lai, S.J. and Shen , H.Y. "Large Deflection Analysis of 3–D Beam Structures", Proceedings of The 3rd Summer Colloquium on Shell and Spatial Structures, Taegu, Korea Aug. pp 169–179, (1990)
26. Wempner, G.A., "Discrete Approximation Related to Nonlinear Theories of Solids," International Journal of Solids and Structures , Vol.7, pp. 1581–1599 (1971)
27. Yang, Y.B. and McGuire, W. "A Work Control Method for Geometrically Nonlinear Analysis ", Proceedings of The Int. Conf. on Num. Meth in Engg. - Theory and Applications, University College Swansea , Wales, pp.913–921 (1985)
28. Yang, Y.B. and Shieh, M.S. "Solution Method for Nonlinear Problems with Multiple Critical Points", AIAA Journal, Vol.28, No. 12, pp.2110–2116, Dec. (1990)
29. Zienkiewicz, O.C. , "Incremental Displacement in Non–Linear Analysis, " International Journal for Numerical Methods in Engineering, Vol.3, pp. 587–588 (1971)

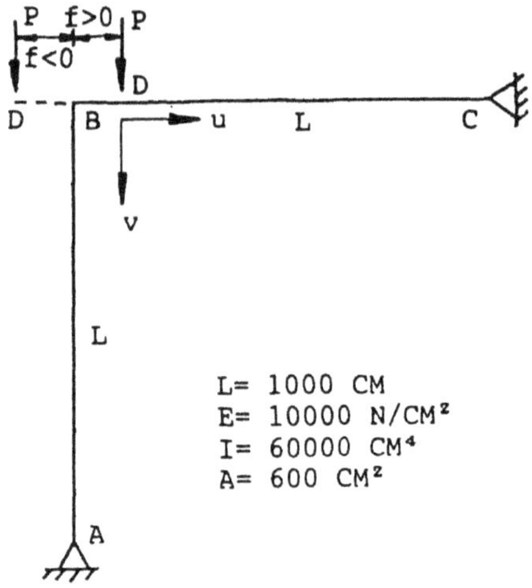

Fig.4.1a Right Angle Frame

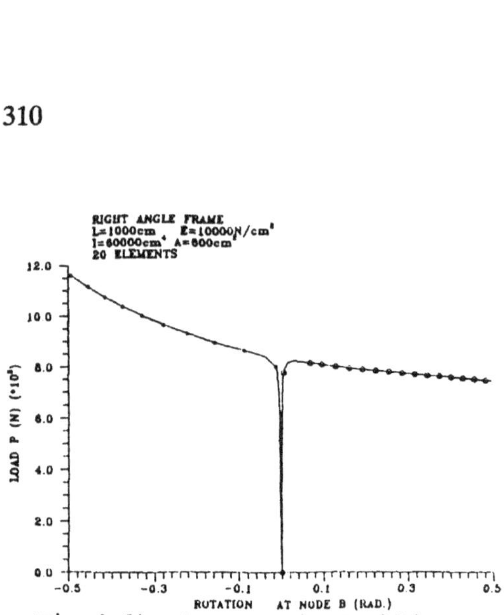

Fig.4.1b Asymmetric Buckling of Right Angle Frame

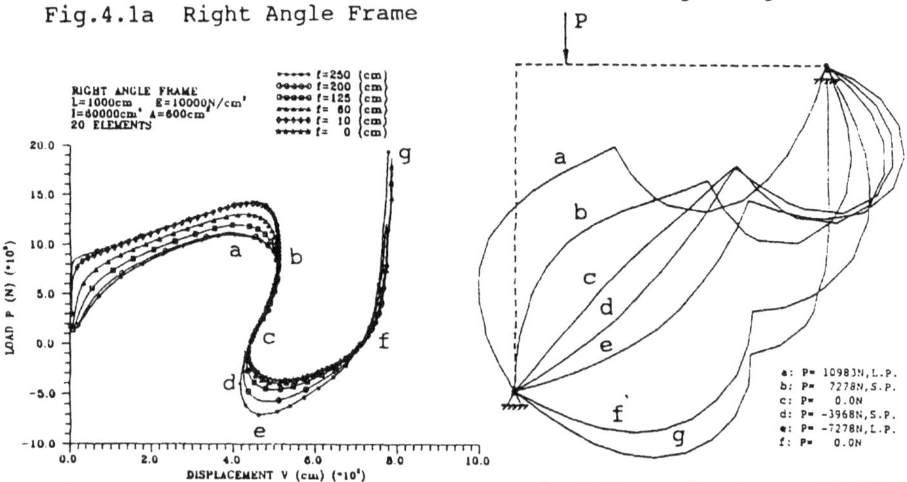

Fig.4.1c Load-Displacement Curve of Right Angle Frame (f>0)

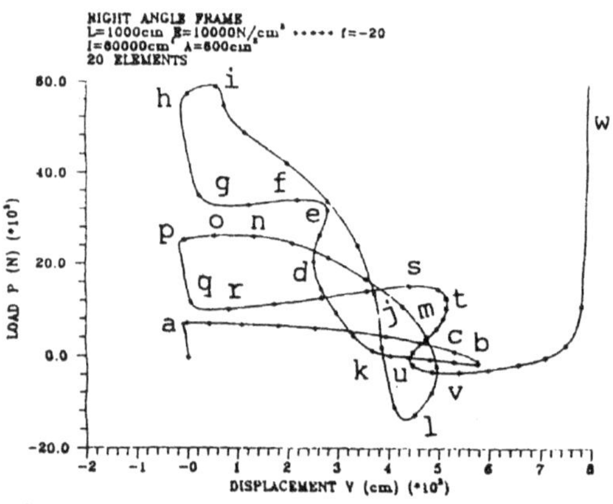

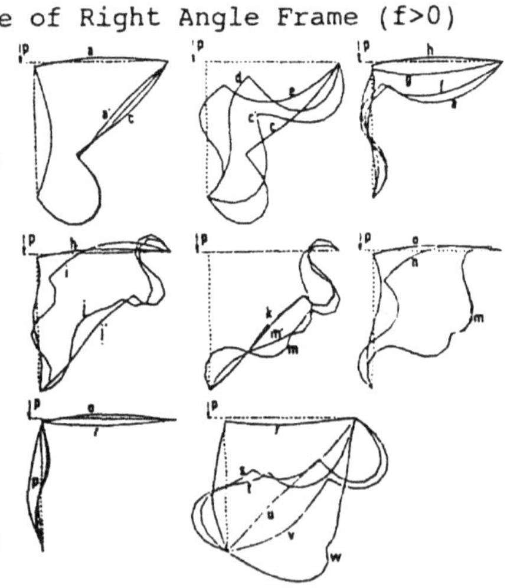

Fig.4.1d Load-Displacement Curve of Right Angle Frame (f<0)

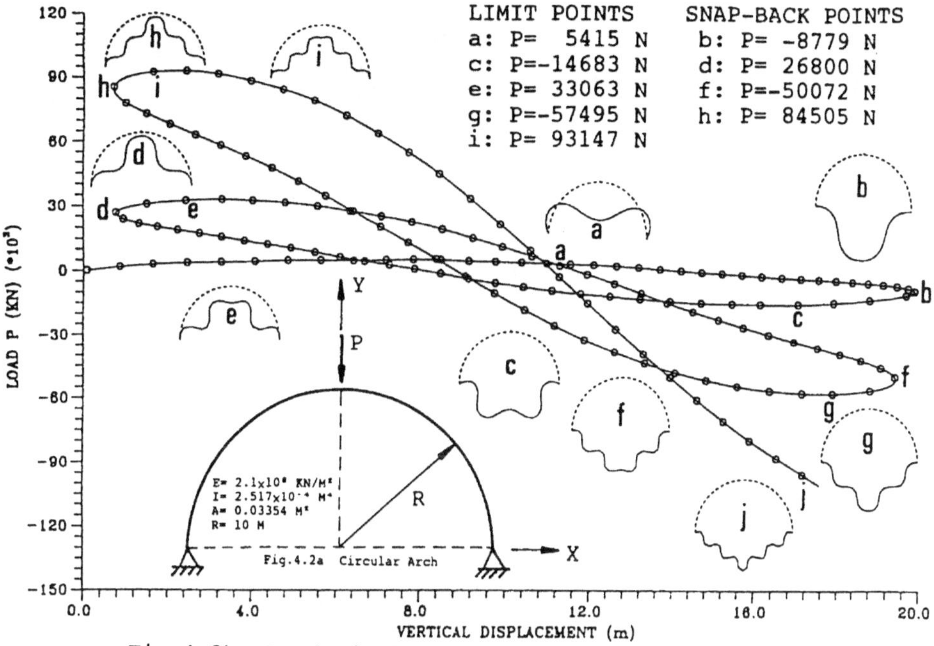

Fig.4.2b Load-Displacement Curve of Circular Arch Under Symmetric Loading

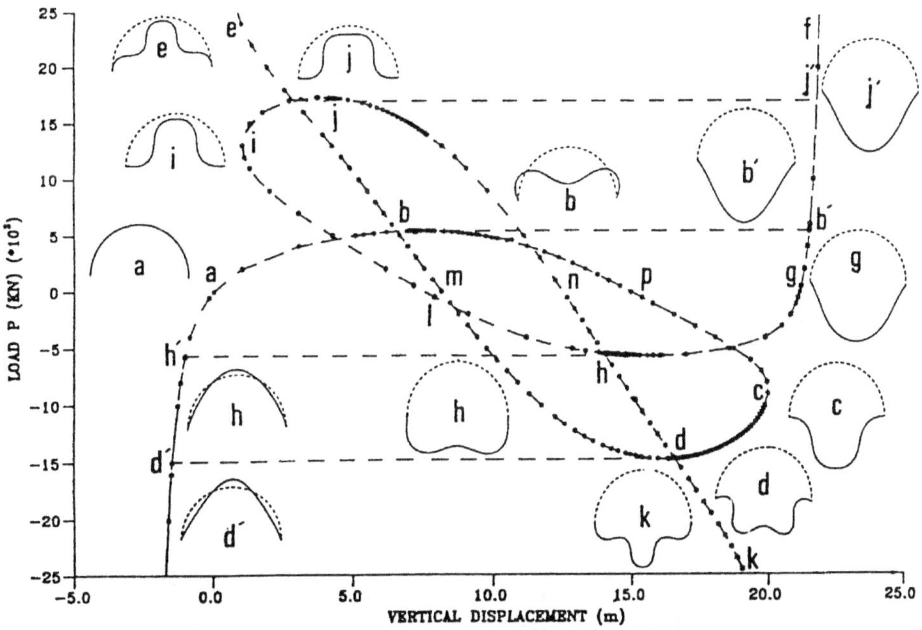

Fig.4.2c Load-Displacement Curve of Circular Arch

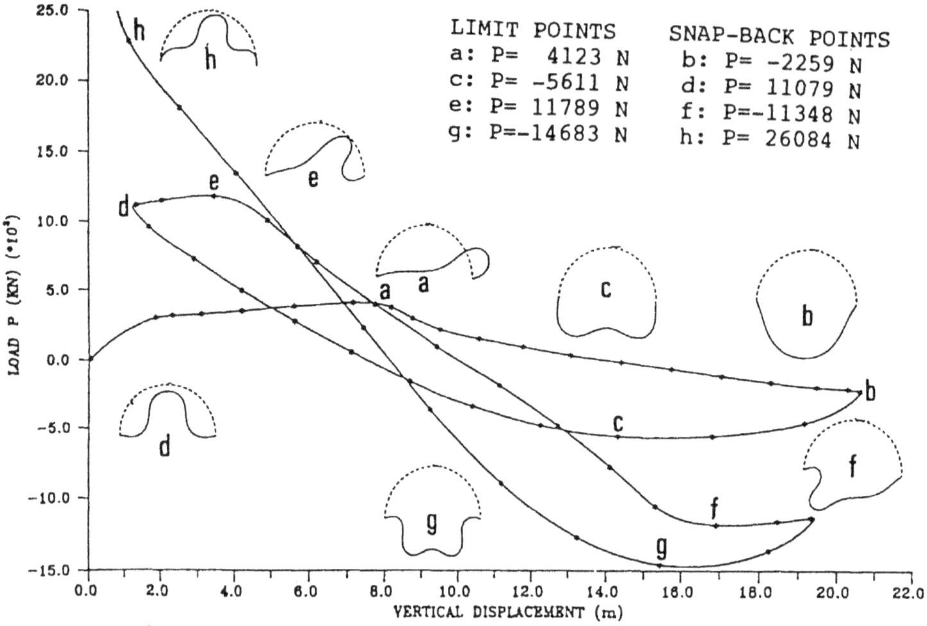

Fig.4.2d Load-Displacement Curve of Circular Arch With Slight Imperfection

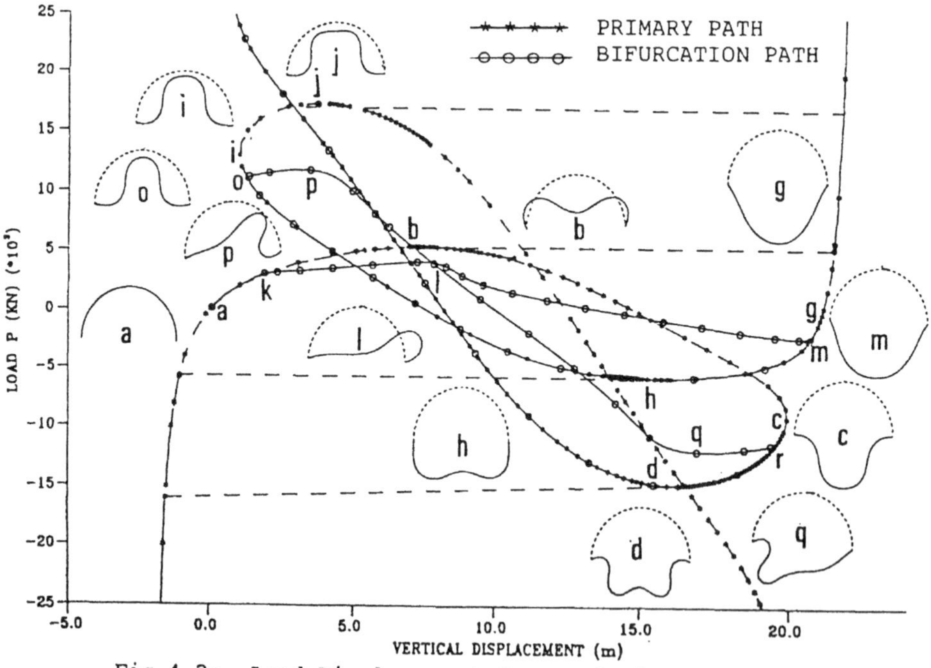

Fig.4.2e Load-Displacement Curve of Circular Arch With Bifurcation Path

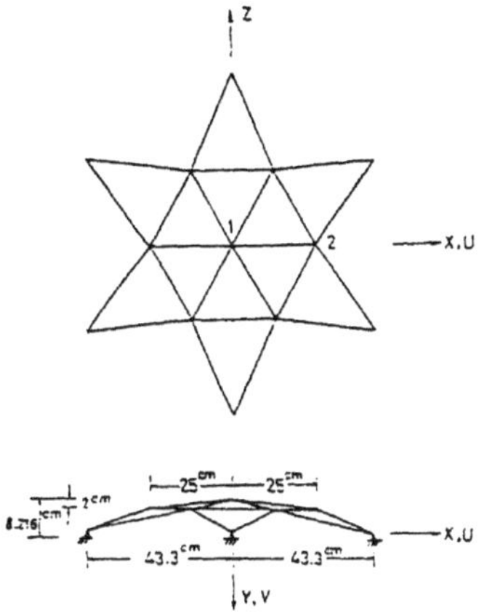

Fig.4.3a 24-Member Hexagonal Star-Shaped Dome

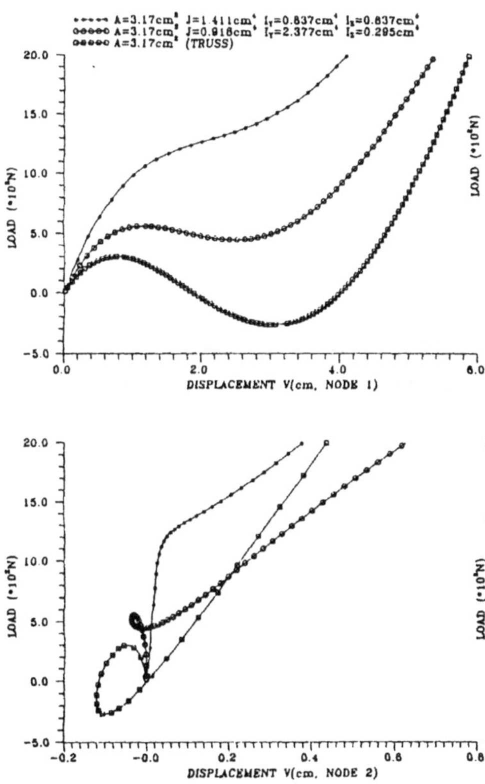

Fig.4.3b Load-Displacement Curve of Star-Shaped Dome Under Symmetric Loading

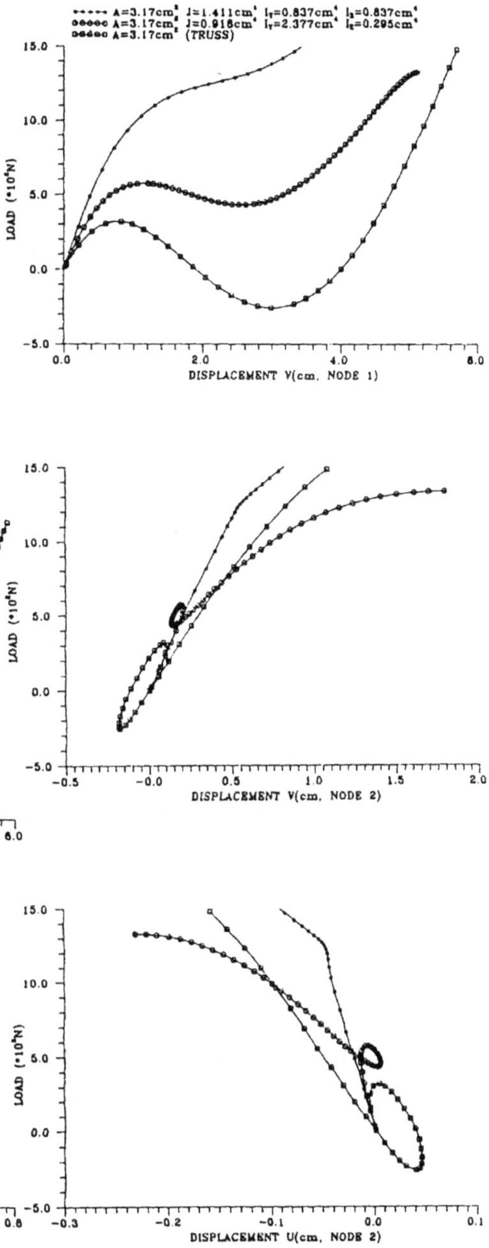

Fig.4.3c Load-Displacement Curve of Star-Shaped Dome Under Unsymmetric Loading

OPTIMUM STRUCTURAL DESIGN UNDER UNCERTAINTIES AND IMPRECISIONS

LI XINGSI QIAN LINGXI
Research Institute of Engineering Mechanics,
Dalian University of Technology,
Dalian 116024, CHINA

ABSTRACT

This paper is primally concerned with the incorporation of different forms of uncertainties and imprecisions into the formulation and solution of optimum structural design problems. Constraint boundaries are treated as "soft" boundary regions and the statistical uncertainties inherent in material properties and structural loadings are modelled by a first–order second–moment reliability analysis approach. Non–statistical uncertainties (imprecisions) arising from model idealizations, approximate structural analyses and quality aspects of design and construction stages are handled by fuzzy set operations.

INTRODUCTION

An optimum structural design problem is usually formulated as a mathematical programming problem in the form:

$$(P0) \qquad \text{Minimize} \quad W(\mathbf{x}) \qquad (1)$$

$$\text{Subject to} \quad g_i(\mathbf{x}) \leqslant b_i, \qquad i = 1,\ldots,m \qquad (2)$$

where W is an objective function representing some efficiency criterion of the design (weight, cost, etc.); **x** is a vector of design variables (dimensions, sizes); constraint functions $g_i(\mathbf{x})$ represent limitations upon the design such as are specified by structural mechanics, codes of practice, etc.; b_i are known constraint limits.

The optimum structural design problem as posed by (1) and (2) can be solved by very many different mathematical methods to yield an optimum design **x**, but all these methods are essentially *deterministic* in nature. They require W and g_i, $i = 1,\ldots,m$, to be deterministic functions of **x**, and b_i, $i = 1,\ldots,m$, to be known constants. This poses no difficulty when the optimum structural design is carried out deterministically as was the case many years ago and is still dominant today. In recent years, however, the deterministic context of structural design has partly yielded to a non–deterministic view which attempts to handle better many uncertainties and imprecisions that are endemic in the design process.

In many codes of practice for the design of reinforced concrete structures, the limit state design philosophy is adopted, and both the non–deterministic nature of structural design and the need for deterministic methods suitable for practical design use are recognized. For example, CP110 [1] recognizes two classes of uncertainties in the design process:

(1) uncertainties associated with structural loadings and material strengths, whose random variations can be represented *statistically* as probability distributions of load and strength.

(2) imprecisions associated with specific aspects of a design such as the importance and value of the structure, the quality of the construction, the skill of the design team and the appropriateness and accuracy of the used analysis methods, which cannot be modelled

statistically.

While accepting the non–deterministic nature of these two classes of uncertainties, CP110 actually represents the first class by using specified characteristic values of load and strength (the 95 percentile of an assumed normal distribution), and the second class by specifying partial safety factors which modify the load and strength characteristic values. In this paper, Problem (P0) is examined for the optimum structural design and the constraints are modelled in such a way that both classes of uncertainties noted above are incorporated as fully as possible, yet a deterministic constraint form is set up which is entirely suitable for solution by (deterministic) mathematical programming methods.

CLASS I (STATISTICAL) UNCERTAINTIES

These are associated with random variations in load and strength parameters. Using a deterministic limit state philosophy, a typical constraint in (P0) has the form:

$$S_i(\mathbf{x}) - R_i(\mathbf{x}) \leqslant 0 \tag{3}$$

where $S_i(\mathbf{x})$ and $R_i(\mathbf{x})$ are the applied load and resistive strength in the i–th limit state, respectively. This constraint simply requires the design to be such that the resistive strength at least exceeds the applied load for the i–th limit state. Using this form for all limit states $i = 1,...,m$ would result a feasible design. If S_i and R_i are now assumed to be *probabilistic*, the constraint form for the i–th limit state becomes

$$p_i(\mathbf{x}) \equiv p(R_i(\mathbf{x}) - S_i(\mathbf{x}) < 0) \leqslant \bar{p}_i \tag{4}$$

where p_i is the probability of failure in the i–th limit state and is defined as the probability that the resistive strength R_i does not exceed the applied load S_i in State i. Constraint (4) requires the failure probability in the i–th limit state to be smaller than some prescribed permissible bound $\bar{p}_i$. However, it cannot be used directly in (P0) because its left side is not a deterministic function of the design vector **x**. To set up a suitable form of Constraint (4), a ″level 2″ (first–order second–moment) reliability analysis [2] can be used.

Define M_i to be the safety margin, $R_i - S_i$, in the i–th limit state and, for notational convenience only, drop the functional dependence upon **x**. The probability distribution of M_i is clearly related to those of R_i and S_i which are not always known accurately; the best information available is often limited to knowledge only of mean values, $\bar{R}_i$ and $\bar{S}_i$, and estimates of variances of the strength and load, $Var(R_i)$ and $Var(S_i)$. With this information, it is clear that the distribution of M_i must have the following mean and variance,

$$\bar{M}_i = \bar{R}_i - \bar{S}_i \tag{5a}$$

$$Var(M_i) = Var(R_i) + Var(S_i) \tag{5b}$$

The least–biased choice of distributions for a specified mean and variance is the normal (Gaussian) distribution which is therefore chosen for M_i. Accordingly, the failure probability, $p(M_i<0)$, can be shown to be

$$p_i = \Phi(-\bar{M}_i / \sqrt{Var(M_i)}) \tag{6}$$

where $\Phi(.)$ is the cumulative normal distribution function and the term

$\bar{M}_i / \sqrt{Var(M_i)}$ is known as the safety index, β_i. Thus, Eq.(6) can be written as

$$p_i = \Phi(-\beta_i) = 1 - \Phi(\beta_i) \tag{7}$$

which can be inverted to express the safety index in terms of the failure probability as

$$\beta_i = \Phi^{-1}(1 - p_i) \tag{8}$$

Replacing the failure probability in (4) by the corresponding safety index gives the following:

$$1 - \Phi(\beta_i) \leqslant 1 - \Phi(\bar{\beta}_i) \tag{9}$$

where $\bar{\beta}_i$ is the safety index corresponding to $\bar{p}_i$. Inequality (9) is further simplified to give

$$\beta_i \geqslant \bar{\beta}_i$$

i.e.,

$$\bar{M}_i / \sqrt{Var(M_i)} \geqslant \bar{\beta}_i \tag{10}$$

which is the constraint form used in Problem (P0). On substituting Eq.(4) for $\bar{M}_i$ and $Var(M_i)$, the left–hand side of the constraint (10) becomes a function of R_i and S_i, and hence of the design vector **x**. Some elements of R_i and S_i may be deterministic while others may be probabilistic parameters. For example, the design vector **x** must be deterministic but loads, yield stresses and elastic moduli may be random. Collecting the random parameters together as **u** and **v** gives

$$R_i = R_i(\mathbf{x}, u_1, \ldots, u_r) \tag{11a}$$

$$S_i = S_i(\mathbf{x}, v_1, \ldots, v_s) \tag{11b}$$

By carrying out first–order expansion of the random variates, the quantities $\bar{R}_i$, $\bar{S}_i$, $Var(R_i)$ and $Var(S_i)$ are then given as

$$\bar{R}_i = R_i(\mathbf{x}, \bar{u}_1, \ldots, \bar{u}_r) \tag{12a}$$

$$\bar{S}_i = S_i(\mathbf{x},\bar{v}_1,...,\bar{v}_s) \tag{12b}$$

$$Var(R_i) = \sum_{k=1}^{r} \sum_{l=1}^{r} \left(\frac{\partial R_i}{\partial u_k}\right)_0 \left(\frac{\partial R_i}{\partial u_l}\right)_0 C^u_{kl}\sigma(u_k)\sigma(u_l) \tag{12c}$$

$$Var(S_i) = \sum_{k=1}^{s} \sum_{l=1}^{s} \left(\frac{\partial S_i}{\partial v_k}\right)_0 \left(\frac{\partial S_i}{\partial v_l}\right)_0 C^v_{kl}\sigma(v_k)\sigma(v_l) \tag{12d}$$

where $\bar{u}_1,...,\bar{u}_r,\bar{v}_1,...,\bar{v}_s$ are mean values of the random parameters and the subscript 0 denotes derivatives evaluated at the mean values; $\sigma(.)$ denotes the standard deviation of the random variable in parentheses; C^u_{kl} and C^v_{kl} are correlation coefficients between pairs of the corresponding random variables. In the case where the elements of **u** and **v** are statistically independent, Eqs. (12c) and (12d) reduce to

$$Var(R_i) = \sum_{k=1}^{r} \left(\frac{\partial R_i}{\partial u_k}\right)_0^2 Var(u_k) \tag{13a}$$

$$Var(S_i) = \sum_{k=1}^{s} \left(\frac{\partial S_i}{\partial v_k}\right)_0^2 Var(v_k) \tag{13b}$$

CLASS II (NON–STATISTICAL) UNCERTAINTIES

The foregoing has established the optimum structural design formulation as that of minimizing W(x) subject to constraints of the form (10) which incorporate statistical variations of load and strength (Class I uncertainties). The formulation must now be modified to include Class II uncertainties. This will be done by fuzzifying the right–hand sides of (10).

Munro [3], Wang and Wang [4] and Qian [5] examined ways in which constraint limits or boundaries may be "softened" by some fuzzy operations so as to incorporate non–statistical uncertainties (imprecisions). Some of these ideas are used here. For a constraint of the form (10), the "hard" (crisp) boundary defined by the deterministic right–hand side, $\bar{\beta}_i$, is replaced by a softened region lying between $\beta_i^1 = \bar{\beta}_i$ and β_i^2 (and consequently corresponding to permissible probabilities $p_i^1 = 1-\Phi^{-1}(\beta_i^1)$and $p_i^2 = 1-\Phi^{-1}(\beta_i^2)$). Defining $d\beta_i = \beta_i^2 - \beta_i^1$ to be the transitive interval (boundary width), any value of safety indices within the interval can be expressed as

$$\beta_i(\alpha) = \beta_i^2 - \alpha d\beta_i \tag{14}$$

in which the fuzzifying parameter α takes values on the interval (0,1).

In this paper, α is treated as an unknown parameter whose value is calculated from available "subjective" information on Class II uncertainties in the design. This calculation is carried out by fuzzy set operations. The value $\alpha = 0$ corresponds to the outer boundary and represents a situation in which very little subjective confidence can be placed in the design – hence a tight boundary with an associated high value of the safety index (low failure probability) is specified. As the subjective confidence in the design process increases, α increases towards unity, the boundary becomes less tight and moves towards β_i^1, a relatively lower value of the safety index and higher probability of failure.

The interval (0,1) of α may be discretized into n subintervals $0 = \alpha_0, \alpha_1,..,\alpha_k,..,\alpha_n = 1$. Treating α as a fuzzy quantity with the support η_k(membership) for each discretized α_k, the fuzzy mean $\bar{\alpha}$ of α is evaluated by

$$\bar{\alpha} = \sum_{k=0}^{n} \eta_k \alpha_k \Big/ \sum_{k=0}^{n} \eta_k \tag{15}$$

Substituting (15) into (14) and then replacing the right–hand side of

(10) by (14), we have the final form of the optimum structural design problem as:

$$(P) \quad \textit{Minimize} \quad W(\mathbf{x}) \qquad (1)$$

$$\textit{Subject to} \quad \overline{M}_i / \sqrt{Var(M_i)} \geqslant \beta_i^2 - \bar{\alpha}\, d\beta_i \quad i = 1,...,m \qquad (16)$$

In order to calculate $\bar{\alpha}$ by Eq.(15) for use in Problem (P), it is necessary to know the supports η_k, k = 0,1,...,n. These will be evaluated by fuzzy set operations incorporating subjective measures of uncertainties as follows. First, suppose that n = 10 giving 11 discretized points in the (0,1) interval of α. Assume that statements such as "small", "medium" and "large" can be defined in terms of supports for discrete values as shown in Table 1. Suppose then that there are a number of factors leading to Class II uncertainties in a specific design and that for each factor, the significance level can be expressed subjectively by one of these fuzzy categories (or by appropriately defined concentrations such as "very small" and "medium large" as suggested by Zadeh [6]). For example, suppose there are three factors influencing a designer: 1) the accuracy of the structural analysis model; 2) the quality of site workmanship, and 3) the importance of the structure. Furthermore, suppose that qualitative statements about these three factors are that: 1) the analysis model is accurate, corresponding to the fuzzy category "large"; 2) the workmanship can be described only as "medium" and 3) the structure is not important. In this last case, an unimportant structure would warrant a low safety index, thus the fuzzy descriptor chosen should be "large" since this concentrates supports near unity which corresponds to a low safety index.

Table 1

Fuzzy sets	Supports (Memberships)										
	0	0.1	0.2	0.3	0.4	0.5	0.6	0.7	0.8	0.9	1
small	1	0.9	0.6	0.3	0	0	0	0	0	0	0
medium	0	0	0	0.3	0.6	1	0.6	0.3	0	0	0
large	0	0	0	0	0	0	0	0.3	0.6	0.9	1

From this information, a fuzzy relation matrix can be constructed by extracting from Table 1 the supports appropriate to each of the three factors. This matrix reads

$$T = \begin{bmatrix} 0 & 0 & 0 & 0 & 0 & 0 & 0 & 0.3 & 0.6 & 0.9 & 1 \\ 0 & 0 & 0 & 0.3 & 0.6 & 1 & 0.6 & 0.3 & 0 & 0 & 0 \\ 0 & 0 & 0 & 0 & 0 & 0 & 0 & 0.3 & 0.6 & 0.9 & 1 \end{bmatrix} \tag{17}$$

Each row in the above matrix corresponds to the fuzzy sets of supports for each of the three factors. In order to extract the supports η_k for α_k from the matrix (17), it must be premultiplied by another vector in which the normalized relative weights associated with each of the three factors are specified. Suppose this vector E is (0.5,0.3,0.2), i.e., the analysis model is deemed more significant (0.5) than the site workmanship (0.3) or structure importance (0.2). The supports η_k, i= 1,...,10, for α are then found by calculating the product of the matrices E and T. Thus we have

$$\begin{aligned} \alpha = & \ 0/0 + 0/0.1 + 0/0.2 + 0.09/0.3 + 0.18/0.4 + 0.3/0.5 \\ & + 0.18/0.6 + 0.3/0.7 + 0.42/0.8 + 0.63/0.9 + 0.7/1 \end{aligned} \tag{18}$$

where the notation η_k/α_k denotes the support η_k for the value α_k, k=0,1,...,n. From (18), the fuzzy mean $\bar{\alpha}$ can be calculated by Eq.(15) to be $\bar{\alpha} = 0.78$.

If we now suppose that for a particular limit state i, failure probabilities of 10^{-5} and 10^{-8} are appropriate for p_i^1 and p_i^2, respectively corresponding to high degrees of certainty and uncertainty, these values give $\beta_i^1 = 4.27$ and $\beta_i^2 = 5.61$ by Eq.(8). The boundary width thus becomes $d\beta_i = 1.34$. The right–hand side of Eq.(14) is then

$$\beta_i^2 - \bar{\alpha} d\beta_i = 5.61 - 0.78x1.34 = 4.57$$

A similar procedure to that described above establishes the right–hand sides of all constraints in Problem (P) which can then be solved by (deterministic) mathematical programming methods.

DISCUSSION

To establish a value of α for use in Eq.(14), the mean, as given by (15), is used. Zadeh[6] proposed the "rule of the preponderant alternative" to make such a choice. By this rule, α would simply be chosen as that α_k in Eq.(18) which has the highest support value, i.e., $\alpha = 1$. This highlights one aspect of subjective uncertainties and their fuzzy manipulations: because the information used is subjective the concept of an accurate or exact solution is non–existent; any "answer" resulting from subjective calculations can be "correct" so long as it can be justified in light of the subjective factors used to derive it.

In the above example, the designer was supposed to be satisfied that his analysis model did not give cause for major concern and that the structure itself was not of great importance. He was, however, concerned that fabricational workmanship was only likely to be of medium quality. With this subjective information, it is reasonable to suppose that the designer might wish to include in his design some safety factor to account for the workmanship but would not especially need extra factors for analysis accuracy or structural importance. The value of $\alpha = 0.78$ represents this balanced safety factor and is justifiable on subjective grounds since the closer α approaches to unity, the smaller the safety factor becomes. Using $\alpha = 1$ by Zadeh's "rule" implies no factoring for possible poor workmanship and hence is less justifiable. In the context of the present application, the use of Eq.(15) seems to be more appropriate than Zadeh's rule and is therefore adopted.

Clearly, the calculation of a value for α depends upon the nature of the subjective information used and the ways in which it is quantified. Changes in the subjective information or the ways in which it is translated into numerical information will change the value of α. The present method is neither correct nor incorrect, accurate nor inaccurate. The results of many test calculations, however, show that it gives sensible, justifiable results and permits the calculation of safety factors appropriate to particular design situations.

CONCLUSIONS

Limit state codes of practice, such as CP110, recognize the existence of both statistical and non–statistical uncertainties in the processes of structural design. The use of characteristic values of load and strength, and partial safety factors enables these uncertainties to be very crudely modelled in a deterministic design approach which is suitable for both present design practice and computer-aided optimum design.

This paper has addressed the problem of how to improve upon the modelling of design uncertainties by existing codes of practice without losing a deterministic final model which is suitable for both design practice and optimum design. It has shown that the use of characteristic values of load and strength may be replaced by a "level 2" reliability analysis which allows information on the mean and variance of random quantities to be incorporated in the calculations. Also it has shown that code–specified partial safety factors accounting for non–statistical uncertainties may be replaced by a method of calculating safety factors appropriate to different design situations from available subjective information through fuzzy set operations. The resulting model retains its deterministic nature and is therefore appropriate for solution by the many existing deterministic optimization methods.

The combination of a limit state approach, reliability analysis and fuzzy operations appears to offer possible lines of development for design methods and codes of practice which are not incompatible with parallel developments in computer - aided design and optimization. It is, of course, clear that work remains to be done in the future.

REFERENCES

1. British Standards Institution, CP110: The structural use of concrete, 1972.

2. Ang, A.H.S., Structural risk analysis and reliability–based design, ASCE J. Structural Division, Vol.99, No.ST9, pp. 1891–1910, 1973.

3. Munro, J., Fuzzy programming and imprecise data, Civil Engineering Systems, Vol.1, No.5,pp. 255–260, 1984.

4. Wang Guangyuan and Wang Wenquan, Fuzzy optimum design of structures, Engineering Optimization, Vol.8, No.4, 291–300, 1985.

5. Qian Lingxi, Some aspects of applied mechanics in China, Proceedings of 10th Canadian Congress of Applied Mechanics, London, Ontario,pp. 25–34, 1985.

6. Zadeh, L.A., Outline of a new approach to the analysis of complex systems and decision processes, IEEE Trans., Systems, Man and Cybernetics, SMC3, pp. 29–44, 1973.

OPTIMUM EARTHQUAKE RESISTANT DESIGN OF PLANE FRAME STRUCTURES WITH ELASTO-PLASTIC BEAM

TEIZO FUJIWARA
Disaster Prevention Research Institute, Kyoto University,
Gokasho Uji, Kyoto 611, JAPAN
and
MASAHITO AKIMOTO
Nuclear Facilities Division, Ohbayashi Corporation,
Sansei-Kajicho Bldg., 3-5 Kajicho, Kanda, Chiyoda-ku,
Tokyo 101, JAPAN

ABSTRACT

In order to obtain the optimum base shear coefficient of plane frame structures and the stiffness/strength distribution of members subjected to the filtered random excitation, a method of optimization of structures is presented. An equivalent linearization technique for the elasto-plastic beams of multi-story structures is introduced . The performance index is selected so as to minimize the sum of the second moment of the cross-sections of columns and beams taking into account the following constraints: the ductility ratio of beams and the fiber stress of column members. Root mean square responses of the optimum design are theoretically obtained and compared with the results of the deterministic responses obtained under simulated earthquake conditions.

INTRODUCTION

Many approaches to optimum dynamic aseismic design were proposed because it was deemed the most economical approach to designing building structures that would uniformly and closely distribute the ratio of the responses under severe earthquake conditions without exceeding the specified safety factor.

One approach is to analyze the deterministic response of a spring-mass system by parametric study. Hisatoku and Kobori [1] presented the optimum stiffness/strength distribution of a multi-mass-spring system when the ductility ratio was uniformly distributed to accord with recorded earthquakes, cited in Japan's new building code. Another is to analyze stochastic response of a spring-mass system by using the equivalent linearization technique. Kobori, Minai and Kawano [2] obtained the optimum stiffness distribution of a shear model under white noise input by parametric study, and Soda and Tani [3] extended the responses under nonstationary white noise. Kapoor and Rao [4] presented a similar problem by introducing total weight performance index.

In this paper, a method of optimum design of frame structures with beams that are weak under conditions of filtered random excitation, based on the paper by Kapoor and Rao, is presented and the results are compared with the responses for simulated excitation. Equivalent linearization technique [5] of elasto-plastic beams and complex modal superposition are used in this study. Allowable ductility ratio and frequency characteristics of input motion on the responses of optimum design are discussed.

THE CONSTITUTIVE EQUATION OF A FRAME STRUCTURE WITH WEAK BEAMS

In Japan it is common philosophy to design a building structure that behaves elastically under the moderate earthquakes occurring once or twice during the lifetime of the structure and that, without destructive damage, resists inelastically the biggest earthquake predicted. In order to realize this design philosophy, we must keep column members that are subjected to gravity forces in the elastic range and expect hysteresis damping in beam members.

The multi-story plane frame structures considered here are assumed to be constructed of elastic columns, and elasto-plastic beams as shown in Fig. 1. The relationships between the bending moment and the related rotation angle of the jth-story column and beam are shown in Eqs. (1) and (2) :

$$M_{j,j-1} = K_j (2\theta_j + \theta_{j-1} - 3R_j) \tag{1}$$

$$M_{Bj} = S_j F_j (\theta_j, t) \tag{2}$$

From the equilibrium equation of each nodal point, the rotation angle can be expressed by horizontal drift u.

$$[A]\{\theta\} + 2\{S\ F(\theta, t)\} = [B]\{u\} \tag{3}$$

in which, respectively, M and M_B: bending moments of column and beam, θ and R : node rotation and chord angle of a story, $K_j = 2EI_{Cj}/h_j$ and $S_j = 6EI_{Bj}/l$: initial stiffness of column and beam, {u} : horizontal displacement vector, [A] and [B] : constant matrices related to the shape of the structure and $F(\theta, t)$: hysteresis function of an elasto-plastic beam, are expressed as shown in Fig. 2.

The steady state response θ_i of the structure subjected to stationary white noise is assumed to be a sinusoidal vibration of amplitude Θ_i and frequency ω_i. Hysteresis function $F_i(\theta_i, t)$ can be replaced as in Eq. (4) by introducing the equivalent linearization technique.

$$F_i(\theta_i,t) = \kappa_i\theta_i + d_i\dot{\theta}_i + \varepsilon(\theta_i, t) \tag{4}$$

By minimizing the error term of the above equation during one cycle, the equivalent stiffness and damping coefficient are finally obtained as follows by introducing the ergodic process and the Rayleigh distribution of the amplitude:

$$\kappa_i = \frac{\langle \theta_i F_i(\theta_i, t)\rangle}{\langle \theta_i^2\rangle} = \frac{\left\langle \dfrac{1}{2\pi}\displaystyle\int_0^{2\pi} (-\Theta_i \cos \Phi_i)\, F(\Theta_i \cos \Psi_i, t)\, d\Psi_i \right\rangle}{0.5\langle\Theta_i^2\rangle}$$

$$= 1 - \frac{8\mu}{\pi}\int_0^{\infty}\left[\{(z+1)\lambda_i^2\}^{-1} z^{1/2} + (z+1)^{-3} z^{1/2}\right]\exp\left(-\frac{(z+1)^2}{\lambda_i^2}\right)dz \tag{5}$$

$$d_i = \frac{\langle \dot{\theta}_i F_i(\theta_i, t)\rangle}{\langle \dot{\theta}_i^2\rangle} = \frac{\left\langle \dfrac{1}{2\pi}\displaystyle\int_0^{2\pi} (-\omega_i\Theta_i \sin\Psi_i)\, F(\Theta_i \cos \Psi_i, t)\, d\Psi_i \right\rangle}{\frac{1}{2}\omega_i^2\langle\Theta_i^2\rangle}$$

$$= \frac{2\gamma}{\sqrt{\pi}}\frac{1}{\lambda_i\omega_i}\,\mathrm{erfc}\left(\frac{1}{\lambda_i}\right) \tag{6}$$

$$\lambda_i = \frac{\sqrt{2}\sigma_{\theta_i}}{\theta_{yi}}, \quad \theta_i = \Theta_i\cos(\omega_i t), \quad z = \frac{\Theta_i}{\theta_{yi}}, \quad \Psi_i = \omega_i t$$

Substituting Eqs. (5) and (6) into Eqs. (4) and (3), the equilibrium equation can be written as

$$\left([A] + [K_B^{eq}]\right)\{\theta\} + [C_B^{eq}]\{\dot{\theta}\} = [B]\{u\} \tag{7}$$

in which $[K_B^{eq}]$ = diag [2 S·κ], $[C_B^{eq}]$ = diag [2S·d] and γ : stiffness ratio of the 2nd branch to the initial stiffness.

The equation of motion can be expressed by using the column shear force which is a function of θ and u

$$[M]\{\ddot{u}\} + [C]\{\dot{u}\} - [B]^T\{\theta\} + [E]\{u\} = -[M]\{1\}\ddot{u}_g \tag{8}$$

and finally expressed by substituting Eq. (7) and its derivatives into Eq. (8),

$$\begin{aligned} &[M]\{\ddot{u}\} + [C]\{\dot{u}\} + [E]\{u\} \\ &+ \sum_{n=0}^{\infty} (-1)^{n+1}[B]^T\left[\left([A] + \left[K_B^{eq}\right]\right)^{-1}\left[C_B^{eq}\right]\right]^n\left([A] + \left[K_B^{eq}\right]\right)^{-1}[B]\{u^{(n)}\} \\ &= -[M]\{1\}\ddot{u}_g \end{aligned} \tag{9}$$

in which $\{u^{(n)}\}$ means the n-th derivatives of horizontal displacement related to time.

Assuming the term $([A] + [K_B^{eq}])^{-1}[C_B^{eq}]$ to be sufficiently smaller than unity and neglecting $n > 2$, the following approximate equation is obtained.

$$[M]\{\ddot{u}\} + [r]\{\dot{u}\} + [k]\{u\} = -[M]\{1\}\ddot{u}_g \tag{10}$$

$$[r] = [C] + [B]^T\left([A] + \left[K_B^{eq}\right]\right)^{-1}\left[C_B^{eq}\right]\left([A] + \left[K_B^{eq}\right]\right)^{-1}[B]$$

$$[k] = -[B]^T\left([A] + \left[K_B^{eq}\right]\right)^{-1}[B] + [E]$$

$$[C] = \frac{2h_1}{\omega_1}[k]$$

in which h_1: critical damping ratio of the 1st mode of vibration, ω_1 : fundamental circular frequency, [r], [k] : symmetrical matrix.

METHOD OF ANALYSIS

Complex modal analysis is used for solving the equation of motion, which is reduced to a first order differential equation [4].

$$[R]\{\dot{z}\} + [K]\{z\} = \{F(t)\} \tag{11}$$

$$[R] = \begin{bmatrix} [0] & [M] \\ [M] & [r] \end{bmatrix}, \quad [K] = \begin{bmatrix} -[M] & [0] \\ [0] & [k] \end{bmatrix}, \quad \{F(t)\} = \begin{Bmatrix} \{0\} \\ -[M]\{1\}\ddot{u}_g \end{Bmatrix}, \quad \{z\} = \begin{Bmatrix} \{\dot{u}\} \\ \{u\} \end{Bmatrix}$$

By replacing $\{z\} = e^{\alpha t}\{\Phi\}$, the homogeneous solution of Eq. (11) will yield 2f eigenvalues α_n and eigenvectors $\{\phi^{(n)}\}$.

$$\left\{\Phi^{(n)}\right\} = \left\{ \begin{matrix} \alpha_n\left\{\phi^{(n)}\right\} \\ \left\{\phi^{(n)}\right\} \end{matrix} \right\} \quad n = 1, 2, ---, 2f \tag{12}$$

where f is the number of degrees of freedom.

The nonhomogeneous solution of Eqs. (11) and (8) can be obtained by expanding into modal series

$$\{z\} = \sum_{n=1}^{2f} \left\{\Phi^{(n)}\right\}\xi_n(t) \tag{13}$$

$$\{u\} = \sum_{n=1}^{2f} \left\{\phi^{(n)}\right\}\xi_n(t) \tag{14}$$

in which

$$\xi_n(t) = \frac{1}{R_n}\int_0^t \exp(\alpha_n(t-\tau))F_n(\tau)d\tau = -\Gamma_n\int_0^t h_n(t-\tau)\ddot{u}_g(\tau)d\tau$$

$$R_n = 2\alpha_n\left\{\phi^{(n)}\right\}^T[M]\left\{\phi^{(n)}\right\} + \left\{\phi^{(n)}\right\}^T[r]\left\{\phi^{(n)}\right\} \qquad F_n(t) = -\left\{\phi^{(n)}\right\}^T[M]\{1\}\ddot{u}_g$$

$$h_n = \exp(\alpha_n t)H(t) \qquad H(t) \text{ :Heaviside unit function}$$

$$\Gamma_n = \frac{1}{R_n}\left\{\phi^{(n)}\right\}^T[M]\{1\}$$

Using the above modal solution of {u}, the mean square responses of displacement can finally be obtained under the condition of stationary random ground motion.

$$E\left[u_i u_j^*\right] = R_{u_i u_j^*}(0) = \frac{1}{2\pi}\sum_{l=1}^{2f}\sum_{m=1}^{2f} \phi_i^{(l)}\phi_j^{(m)^*}\Gamma_l\Gamma_m^* I_{lm} \tag{15}$$

$$I_{lm} = 2\pi S_0 B_{lm} = \int_{-\infty}^{\infty} H_{hl}(-i\omega)\, H_{hm}^*(i\omega)\, \varphi_{\alpha\alpha}(\omega)\, d\omega$$

in which $\phi_{\alpha\alpha}(\omega)$ is the power spectrum of the input ground motion and is shown in Eq. (16) for white noise and in Eq. (17) for white noise modified by the Kanai-Tajimi filter, respectively.

$$\phi_{\alpha\alpha}(\omega) = S_0 \tag{16}$$

$$\varphi_{\alpha\alpha}(\omega) = \frac{\omega_g^4 + 4h_g^2\omega_g^2\omega^2}{(\omega_g^2 - \omega^2)^2 + 4h_g^2\omega_g^2\omega^2} S_0 \tag{17}$$

In the above equations, S_0 represents the spectral density of white noise and * represents the conjugate complex. The mean square responses of velocity and acceleration are also obtained with similar equations by differentiating Eq. (15).

Mean square responses of node rotation related to the local ductility ratio are also derived as follows. From Eq. (7) and its derivatives, node rotation vector $\{\theta\}$ is expressed :

$$\{\theta\} = \sum_{n=0}^{\infty} (-1)^n \left[\left([A] + [K_B^{eq}]\right)^{-1} [C_B^{eq}]\right]^n \left([A] + [K_B^{eq}]\right)^{-1} [B] \{u^{(n)}\} \tag{18}$$

Assuming that the equivalent damping coefficient is rather small, approximate relationships are obtained by neglecting orders higher than the second order.

$$\begin{aligned} \{\theta\} &= [a]\{u\} + [b]\{\dot{u}\} \\ \{\dot{\theta}\} &= [a]\{\dot{u}\} + [b]\{\ddot{u}\} \\ [a] &= \left([A] + [K_B^{eq}]\right)^{-1} [B] \\ [b] &= \left([A] + [K_B^{eq}]\right)^{-1} [C_B^{eq}] \left([A] + [K_B^{eq}]\right)^{-1} [B] \end{aligned} \tag{19}$$

Mean square responses of rotation and of its derivative are expessed by the mean square responses of displacement and of its derivative designated in Eqs. (20) and (21).

$$\begin{aligned} E[\theta_i\theta_i^*] = \sum_{j=1}^{f}\sum_{k=1}^{f} & a_{ij}a_{ik}E[u_j u_k^*] - a_{ij}b_{ik}E[u_j\dot{u}_k^*] \\ & - b_{ij}a_{ik}E[\dot{u}_j u_k^*] + b_{ij}b_{ik}E[\dot{u}_j\dot{u}_k^*] \end{aligned} \tag{20}$$

$$\begin{aligned} E[\dot{\theta}_i\dot{\theta}_i^*] = \sum_{j=1}^{f}\sum_{k=1}^{f} & a_{ij}a_{ik}E[\dot{u}_j\dot{u}_k^*] - a_{ij}b_{ik}E[\dot{u}_j\ddot{u}_k^*] \\ & - b_{ij}a_{ik}E[\ddot{u}_j\dot{u}_k^*] + b_{ij}b_{ik}E[\ddot{u}_j\ddot{u}_k^*] \end{aligned} \tag{21}$$

Mean square values of fiber stress σ_s^2 of the column end can also be obtained using a similar method [6] by calculating $\sigma_s^2 = E[M\,M^*]/(2EI/h)^2$.

OBJECTIVE FUNCTION AND CONSTRAINTS

Many concepts of what constitutes optimum design, such as minimum cost and minimum weight and so on, have been proposed by researchers. In this paper, minimizing the sum of the second moments of the cross-section of each member is adopted for the performance index

$$\min\left\{\sum_i (EI_{Ci} + EI_{Bi})\right\} \tag{22}$$

under the following constraint conditions.

1. Maximum fiber stress of each column member is equal to or less than the allowable elastic limit stress S_y of the material.
2. Maximum ductility ratio of the rotation of each beam member is equal to or less than the allowable limit ductility μ.

The above concepts can be written in Eqs. (23) and (24).

$$P_s\left[\left|\frac{S(EI_C, EI_B, t)}{S_y}\right| > 1, 0 \le t \le T\right] < P_{fa} \tag{23}$$

$$P_\theta\left[\left|\frac{\theta(EI_C, EI_B, t)}{\theta_y}\right| > \mu, 0 \le t \le T\right] < P_{fa} \tag{24}$$

in which, respectively, P_s, P_θ : the probability of maximum fiber stress and rotation, S, S_y : fiber stress and allowable limit stress of the column member, θ, θ_y : rotation and elastic limit rotation angle of the beam member and P_{fa} : failure probability.

Assuming the Poisson Process, the probability that the maximum response of fiber stress exceeds its limit value at least once during the time interval is

$$P_s(T) = 1 - \exp\left(-\frac{1}{\pi}\frac{\sigma_{\dot{s}}}{\sigma_s} T \exp\left(-\frac{s^2}{2\sigma_s^2}\right)\right) \tag{25}$$

in which σ_s, $\sigma_{\dot{s}}$ are, respectively, the standard deviation of stress and stress rate.

Transforming an optimum problem with constraints into a problem without constraints by using the Lagrange multiplier theory, the optimum design parameters were calculated by Fortran minimization subroutines using quasi-Newton methods at the Computer Center of Kyoto University.

In this study, the allowable failure probability P_{fa} was set at 0.05. Attention must be paid to larger values of P_{fa} to prevent unstable computation conditions.

ANALYTICAL RESULTS AND DISCUSSION

Size and material of a specified structure are generally determined by the designer. A structural engineer must determine the base shear coefficient and stiffness/strength distribution considering the dynamic property of soil strata. Table 1 shows the dimensions of the frame structure considered here. Three parameters are important in this study:

1. Stiffness ratio of the 2nd branch to the initial stiffness of elasto-plastic beams
   $r = 1.0$ (linear), 0.5 and 0.25 $(r = 1 - \gamma)$
2. Allowable ductility ratio of beams
   $\mu = 1.0$ (linear), 1.5, 2.0 and 2.5
3. Soil predominant frequency
   $\omega_g = 1, 2, 3, 4, 6$ and white noise

Damping ratio of the Kanai-Tajimi Filter was assumed to be constant : $h_g = 30\%$.

TABLE Dimension of model structure

Span Length	$l = 750$ cm
Number of Story	$N = 3$ and 5
Column Length	$h = 350$ cm
Mass	$m = 25.5$ kgf $\sec^2/\text{cm}$
Column Depth	$h_c = 40$ cm and 50 cm
Beam Depth	$h_b = 50$ cm and 60 cm
Critical Damping Ratio	$h = 0.02$
Stiffness Ratio	$r = 1.0$, 0.5 and 0.25 $(r = 1 - \gamma)$
Allowable Ductility Ratio	$\mu = 1.0$, 1.5, 2.0 and 2.5
Young Modulus	$E = 2100$ tonf/cm^2
Power Spectral Density	$S_0 = 500$ cm^2/sec^3
Duration Time	$T = 20$ sec
Predominant Frequency	$f_g = 1, 2, 3, 4, 6$ and white noise
Failure Probability	$P_{fa} = 5\%$

In order to verify the above formulation, linear and nonlinear time history responses under artificial earthquake conditions were obtained and compared with the theoretical results. The 20th of the simulated filtered earthquakes was calculated from Eq. (26)

$$\ddot{u}_g(t) = \sum_{i=1}^{N} \sqrt{2S(\omega_i)\Delta\omega/\pi}\,\cos(\omega t + \phi_i) \tag{26}$$

$$S(\omega_i) = \varphi_{\alpha\alpha}(\omega_i),\quad S_0 = 500\text{cm}^2/\text{sec}^3,\quad T = 20\text{ sec}$$

in which $\Delta\omega = (\omega_U - \omega_L)/N$, $\omega_i = \omega_L + (i - 1/2)\Delta\omega$ and ϕ_i shows random numbers between 0 and 2π.

Figs. 3 and 4 show the distribution of displacement (a), stress of columns (b), ductility ratio of beams (c) obtained from optimum design of 5-story structures in broken lines. The solid line shows the deterministic response of the design structure subjected to simulated earthquakes. MR, MP and PP represent mean value of root mean square responses, mean value of peak response and peak value of peak response under 20th of responses subjected to simulated ground motion. Circles and dots show response under constraint ductility $\mu = 1.0$ ($r = 1.0$) and $\mu = 2.0$ ($r = 0.25$). The distribution of design value is quite similar to the deterministic average response. Fiber stresses of columns are uniformly distributed and peak stresses are within the allowable limit stress 2.4 t/cm^2, except for the first story where the effect of fixed base is remarkable. Ductility ratios of beam members are also in the elastic limit when $r=1.0$, while peak ductility responses are distributed close to the constraint ductility ($\mu = 2$) when the stiffness ratio = 0.25, which is approximately 3 times the average value of root mean square responses. Displacement response subjected to white noise is larger than the response under filtered white noise because, given differences in design base shear, the optimum stiffness/strength under white noise is smaller than that under filtered input. The difference between white noise input and filtered white noise is not clear in Figs. 3 and 4 and is shown in the following figures.

Distribution of the stiffness ratio and the fundamental frequency of Model A (depth of beam = 50 cm, depth of column = 40 cm) based on the column stiffness at the first story for optimum design is compared with Model B (depth of beam = 60 cm, depth of column = 50 cm) in Fig. 5. In this case, white noise is used for input ground motion. As shown on the right in Fig. 5, the fundamental frequency of Model A is quite similar to that of Model B. We can arbitrarily choose the depth of members without considering the base shear coefficient or fundamental frequency.

Frequency characteristics of ground motion input to building structures are affected by frequency characteristics of soil foundation. In order to make clear the effects of soil condition on optimum solution of the structures, the predominant frequency of the Kanai-Tajimi filter is focussed on the following figures. The effect of input motion frequency characteristics on the stiffness/strength distribution and fundamental frequency are shown in Fig. 6. Stiffness distribution of optimum design does not depend on the predominant frequency of input motion except for beam stiffness at the second floor in the case of $\mu = 2$. Optimum stiffness distribution of column members is approximatly 1.0 : 0.5 : 0.3, which is quite different from results obtained with spring mass system [2, 7]. We must make the base column stiffer than that of a structure with strong beams (shear model) . One reason is that the base column tends to be rigidly connected to the foundation.

Fig. 7 shows the effect of the constraint ductility ratio of beam members on stiffness distribution. The larger the allowable ductility ratio, the larger the stiffness/strength ratio of lower-story column members. Stiffness of the lower beams may be made considerably less than in the elastic design.

Fundamental circular frequency of optimum design is shown in Fig. 8, in which the horizontal axis represents the predominant frequency of the input motion and w represents white noise input. In the case of elasto-plastic behavior ($\mu = 2$), fundamental frequency decreases by about one half that of elastic behavior ($\mu = 1$) when the predominant frequency is higher than 20 rad/sec.

Optimum base shear coeffecient for a design subjected to different predominant ground motions is shown in Fig. 9. There are two points of optimum design that correspond to the predominant frequency of the soil. If the frequency of a structure is less than the predominant frequency, the optimum design value will be determined under white noise input. On the other hand, if the predominant frequency is low, i.e. soft soil is present, the amplification factor decreases at frequencies higher than the predominant frequency, and the optimum value will be obtained in a range close to but higher than the predominant frequency. As shown in Fig. 8, the effect of soil condition on the optimum base shear coefficient is similar to the effect on the fundamental frequency. The solid line without circle in Fig. 9 shows the coefficient of a 3-story elastic structure when the amplitude of input motion is one half of the case when $\mu = 1.0$. As mentioned above, engineers must design to satisfy two kinds of criteria such as inelastic case ($\mu = 2$) and elastic case ($\mu = 0.5$), that is, to clear the heavy line in Fig. 9.

CONCLUDING REMARKS

The optimum solution of structural members of frame structures with weak beams subjected to filtered random excitation was derived and some effects of soil characteristics, constraint ductility ratio and member depth on the optimum design values were examined. Results obtained here were restricted but the following conclusions were reached:

1. The method presented here demonstrates its superiority by the simulation analysis of average and peak responses of 3- and 5-story structures.

2. Stiffness distribution of optimum design depends upon the constraint condition of the allowable ductility ratio of beam rotation. The large allowable ductility ratio suggests a design with greater column stiffness and a lower beam value in lower stories.

3. Engineers must consider two aseismic design criteria: moderate excitation and catastrophic earthquake. The design value of the base shear coefficient must clear both of those criteria.

REFERENCES

1. Hisatoku, T. and Kobori, T., On the optimum aseismic design data for multi-story structures based on elasto-plastic earthquake responses. Proc. 3rd European Symposium on Earthq. Eng., 1970, 435-445.
2. Kobori, T., Minai, R. and Kawano, M., A method of optimizing earthquake responses of building structures. Annuals of the Disast. Prev. Res. Inst., Kyoto Univ., **14-A**, 1980, 315-332.
3. Soda, S. and Tani, S., Basic study on statistical aseismic design of tall buildings. Transactions of the Architectural Institute of Japan, No. 288, 1980, 97-106 (in Japanese).
4. Kapoor, M. A. and Rao, J. V. N., Optimum aseismic design of multi-story steel frames. J. Engineering Optimization, 11, 1987, 339-353.
5. Caughey, T. K. , Random excitation of a system with bilinear hysteresis. J. Applied Mech., ASME, Dec., 1960, 649-652.
6. Fujiwara, T. and Akimoto, M., Optimum design of plane frame structures with elasto-plastic beams considering soil effects. Annual of the Disaster Prevention Research Institute of Kyoto University, **33-B1**, 1991, 89-101 (in Japanese).
7. Kobori, T., Minai, R. and Fujiwara, T., Optimum design of the structural members due to ground motion, Proc. 6th World Conference on Earthquake Engineering, **2**, 1977, 5.79 - 5.84.

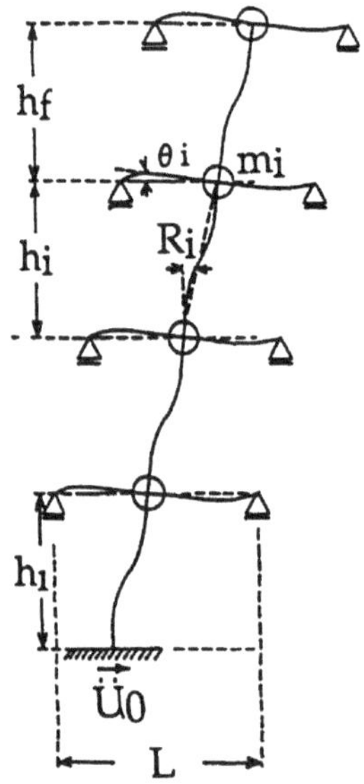

Figure 1. Model structure

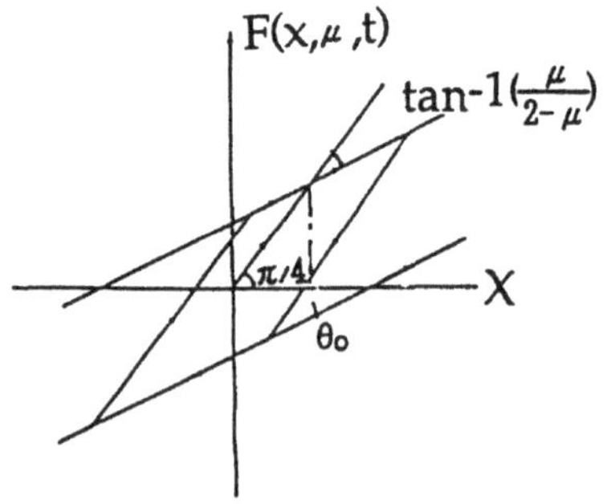

Figure 2. Bi-linear hysteresis

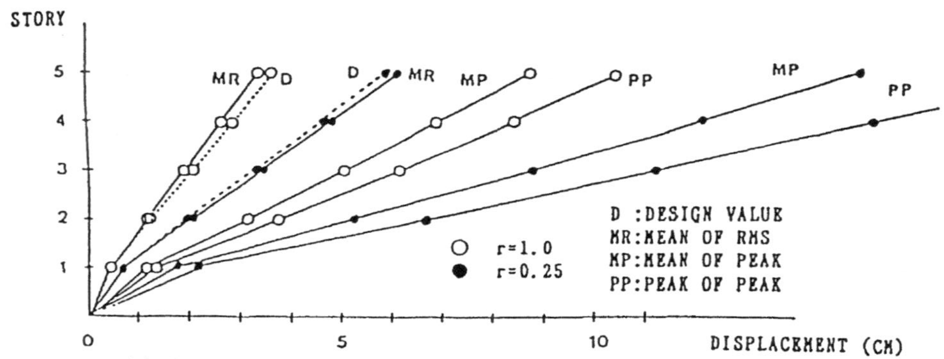

(a) Design and simulated value of displacement

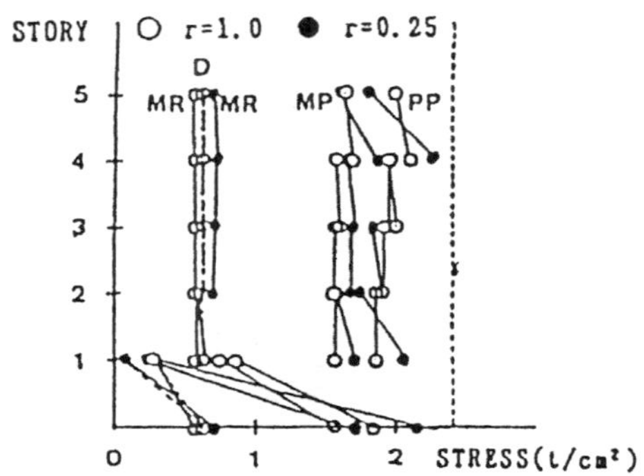

(b) Fiber stress of column members

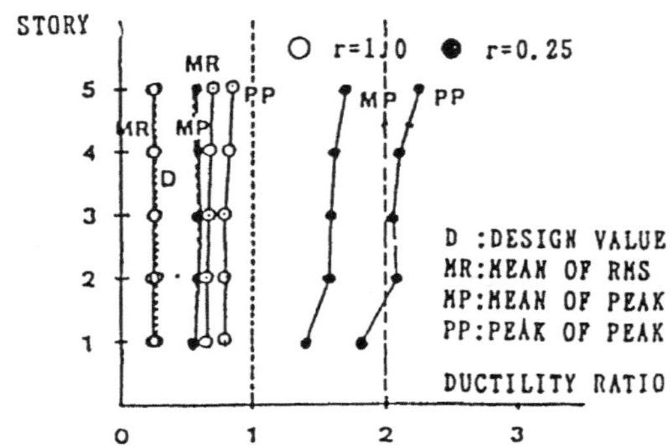

(c) Ductility ratio of beam members

Figure 3. Design and simulated response due to white noise

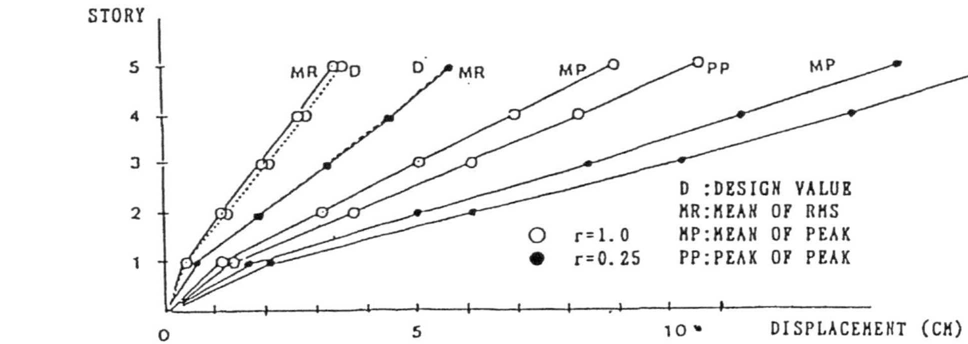

(a) Design and simulated value of displacement

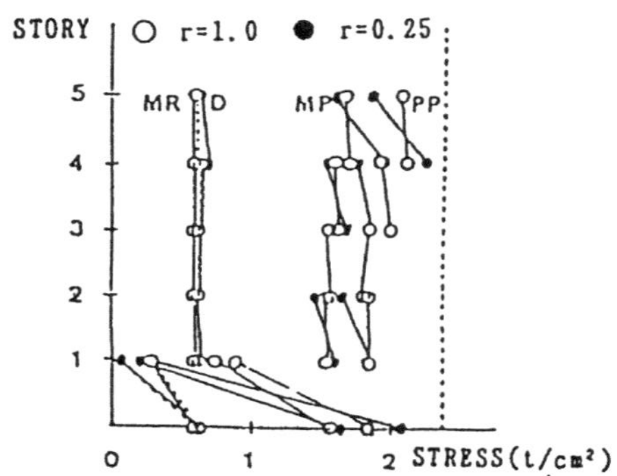

(b) Fiber stress of column members

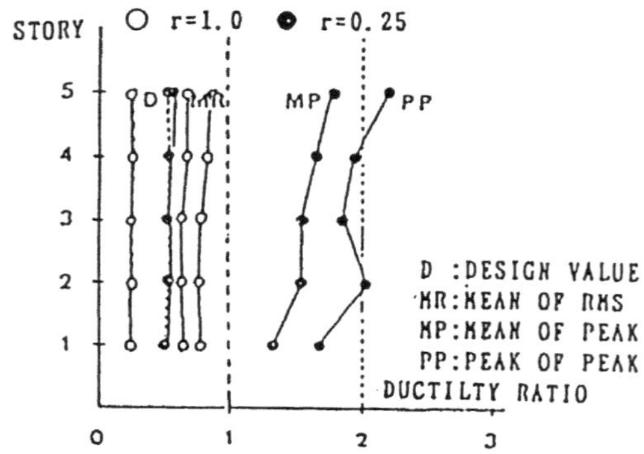

(c) Ductility ratio of beam members

Figure 4. Design and simulated response due to filtered white noise (Fg = 4Hz)

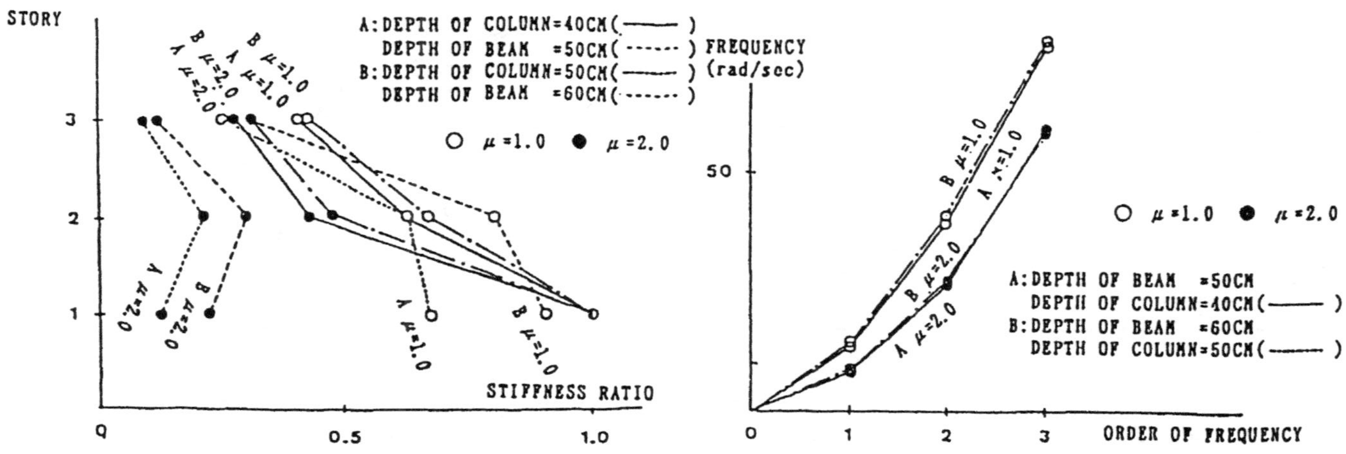

Stiffness ratio Frequency

Figure 5. Effect of the member depth on the stiffness distribution and frequency

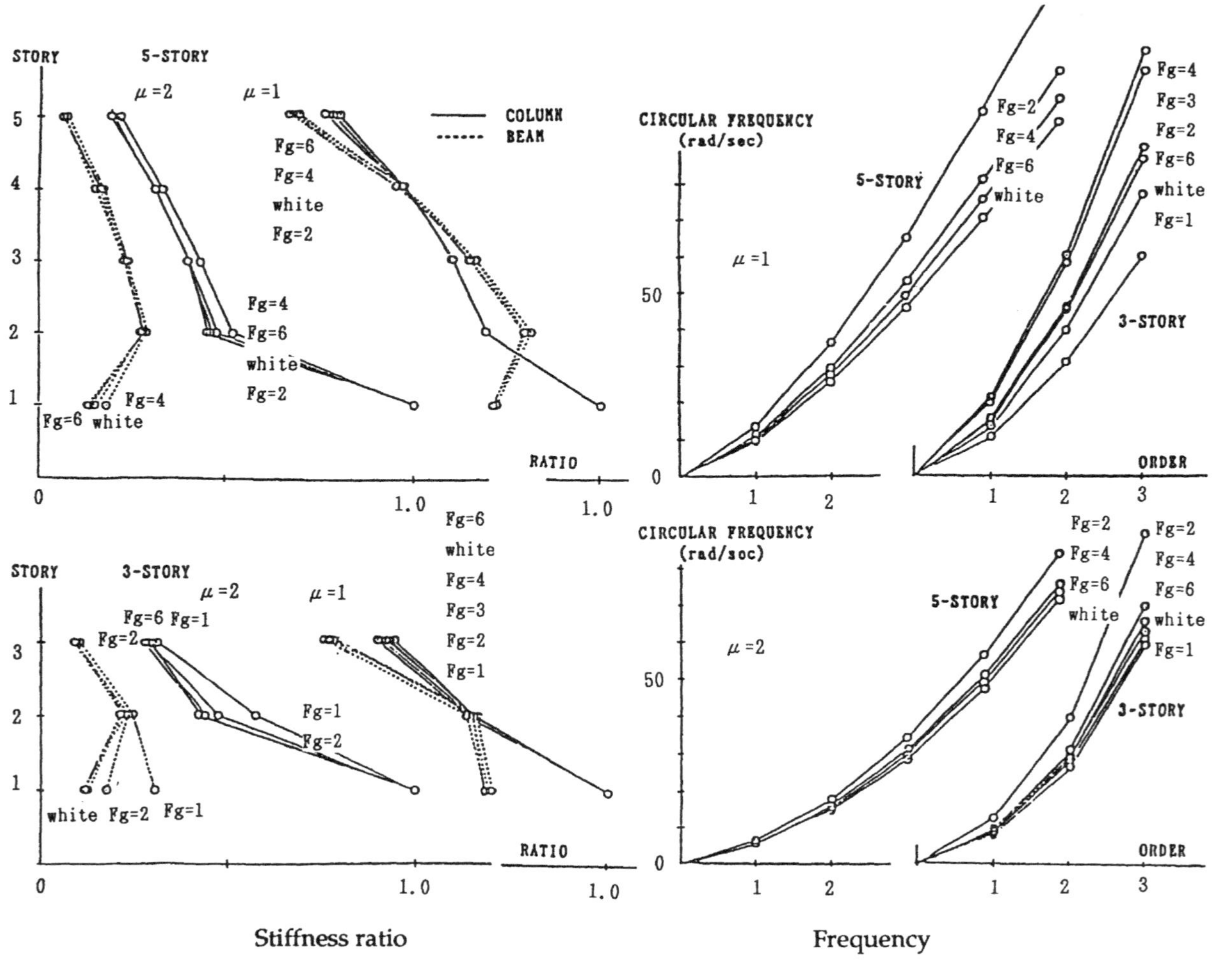

Figure 6. Effect of the input filter on the stiffness distribution and frequency

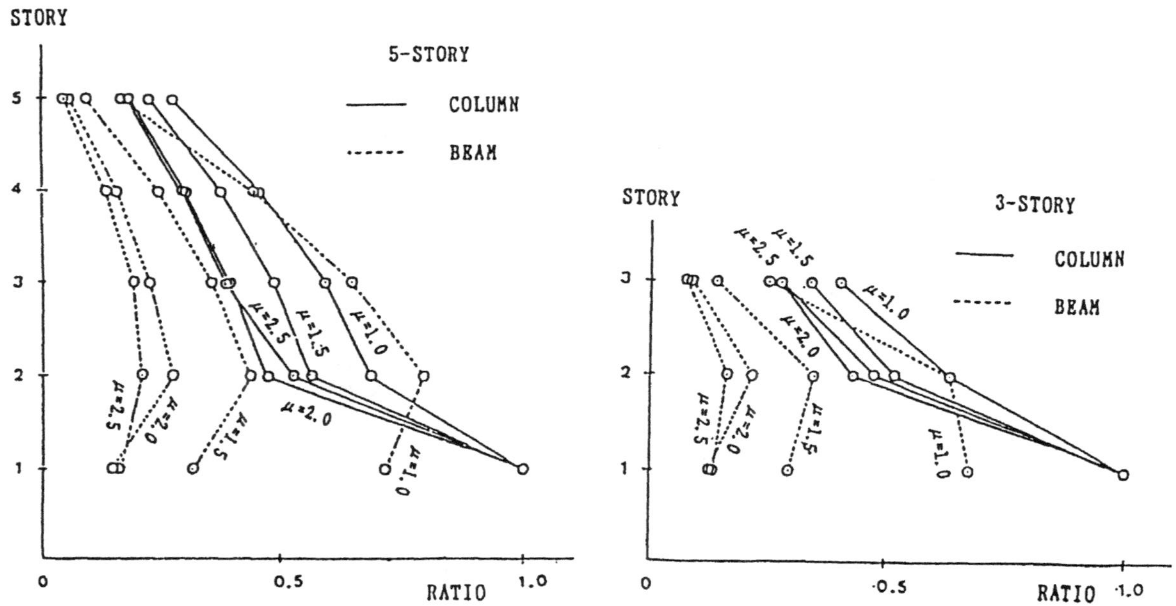

Figure 7. Effect of the constraint ductility on the stiffness distribution

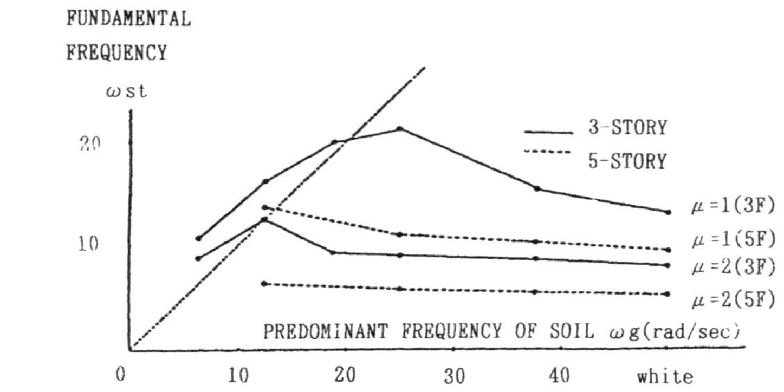

Figure 8. Optimum circular frequency versus predominant frequency

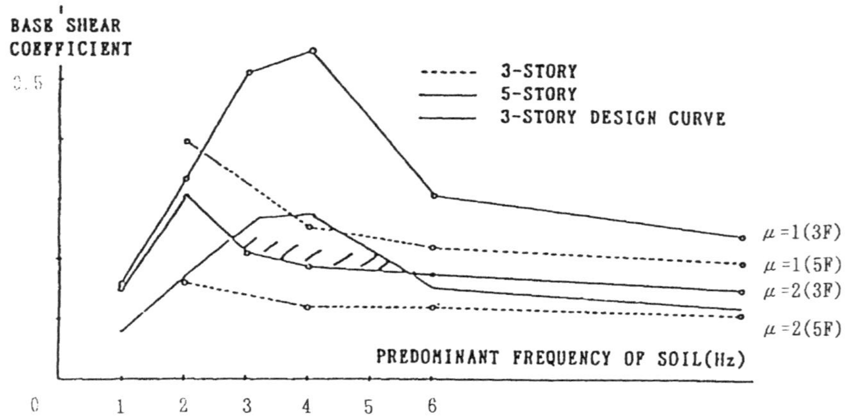

Figure 9. Optimum base shear coefficient versus predominant frequency

NEW SYSTEM OF EARTHQUAKE RESISTANT STRUCTURES IN SEISMIC ZONE

ZHOU Fu Lin
Guangzhou Urban Construction Institute
Guangzhou, 510405. China

ABSTRACT

This paper describes summarily the gradual development、present status and existent problems in the field of earthquake resistant structures in China, suggests a new system of earthquake resistant structures —— Seismic Response Control System of Structures, which is more safe、effective、simple and economic, is able to be used in all seismic zones with different Intensities of Earthquakes. The author also introduces the main contents、characters、design method、engineering application、calculation theory and the results of updated researching work about this system. Finally, the author discusses the developing tendency in the future about the field of earthquake resistant structures.

THE GRADUAL DEVELOPMENT AND PRESENT STATUS OF EARTHQUAKE RESISTANT STRUCTURES IN CHINA

During earthquake attacks, the building structure which is fixed on the ground will respond gradually increasing from the building bottom (ground) to the building top, liking an "Amplifier" (Fig. 1. a). This will result in damage or collapse of structure or results in damage of nonstructural components、decoration and equipment facilities in building due to the large response of structure. In order to reduce the response and avoid the damage of structure, some controlling methods have been developed:

1. Increasing the structural stiffness to be much more, form a "Rigid Structure System" (Fig. 1. b) whose structural response may be nearly as same as the ground motion. But this kind of structure system is very expensive and very difficult to realize in some cases.

2. Decreasing the structural stiffness to be very small, form a "Flexible Structure System" (Fig. 1. c) whose structural response should be also very small. But this kind of structure system is not suitable for normal use because it is too flexible under wind load or minor earthquakes.

3. Making the first story columns very soft and allowing it to deform into the inelastic range to dissipate the energy thus reducing the response of the upper structure, form a "Soft First Story Structure System" (Fig. 1. d). This system can reduce the upper structure response very effectively, but the building may collapse in severe earthquake because of the large inelastic deformation of first story columns.

4. Increasing the structural ductility and allowing the structural elements or joints to work in inelastic range to dissipate the energy of structure under-earthquakes, thereby reducing the structural response, by forming an "Inelastic Structure System" (Fig. 1. e). This is the general structural system for earthquake resistance in many countries at present. But its usefulness is limited or it is not very safe in some cases. First, it is difficult to control the structural damage level due to the inelastic deformation, and it may be quite dangerous in severe earthquake unpredicted before. Secondly, it is not able to be used in some important structures whose element is not allowed to work in inelastic range, such as some buildings whose decoration is very expensive, nuclear power plant, museum building, and so on. Thirdly, it is not possible to be used for buildings with precise instruments in them.

The structural systems described above for control of seismic response can not match the requirements today. So the structural system of base isolation, energy dissipation and seismic controlling is suggested to be used for earthquake resistant structures.

The Base Isolation System (Fig. 1. f) can reduce the structure response very effectively by keeping the advantages of "Soft First Story Structure System", but it changes from the soft first story columns into an isolation sliding Layer only.

The Energy Dissipation System (Fig. 1. g) can also reduce the structural response by keeping the advantages of "Inelastic Structure system", but it changes from allowing the structural elements or joints to work in the inelastic range become to allow only nonstructural components to work in inelastic range for dissipating energy and protecting the structure from damage in earthquake.

The Structural Controlling System (Fig. 1. h, i) keeps the advantages of "Rigid Structure System" and "Flexible Structure System" by changing the dynamic characteristics (Mass, stiffness or damping) of the structure.

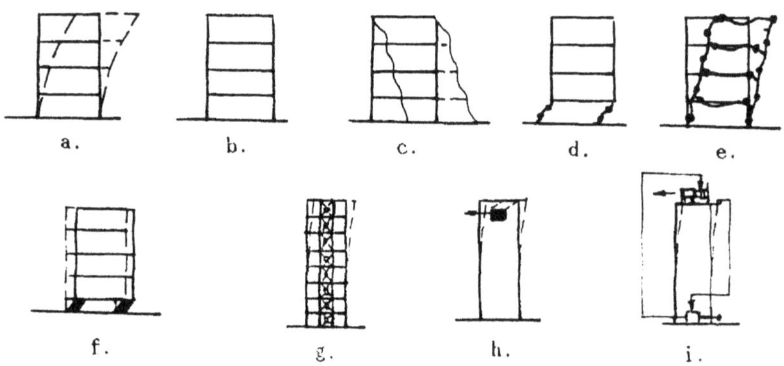

Figure 1. Development of earthquake resistant structure

Many countries in the world have paid more attention to find out a new way [1] [2] for earthquake resistant structure——Base Isolation, Energy Dissipation and Structural Controlling System. Some available results of research have been obtained, and the new system has been used successfully in engineering application in some countries including China.

BASE ISOLATION SYSTEM

The Characteristics of Base Isolation System

The base isolation system consists of structure and isolation device fixed between the ground and the bottom of the structure. In general case, the base isolation device requires to possess three basic characters: (Fig. 2)

1. Soft sliding: The structure can softly slide on the base in severe earthquake. This character can isolate the horizontal vibration from ground motion to structure, make the natural period of structure very long in order to reduce the acceleration response of structure effectively (Fig. 3).

2. Suitable horizontal stiffness K, it will provide the primary stiffness against wind load or minor earthquake while $P < Py$ (Fig. 4).

3. Certain amount of damping C, it will dissipate the energy input to the structure then attenuate the response of structure in earthquake (Fig. 5).

The "Base Isolation System" supplies a new method to control the seismic response of building structure and it is very effective, safe, simple and can be used in very wide range. This new method possesses the following advantages:

1. It can reduce the structure response very effectively [3] by keeping the advantages of "Soft First Story Structure System", but it changes from the soft first story columns into only an isolation sliding layer. So it is more effective and safe.

2. It can keep the structure working in elastic range in earthquake [4] [5], so it can be used for different kinds of important buildings in different seismic zones.

3. It changes from relying upon the whole structure to resist the earthquake into only relying upon the isolating device to isolate the earthquake. So it makes the design and construction work very simple and the restoring work is very easy.

4. It can be used in new buildings, as well as in existing buildings to improve its earthquake resistant ability.

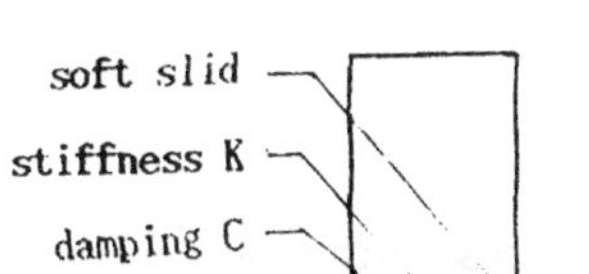

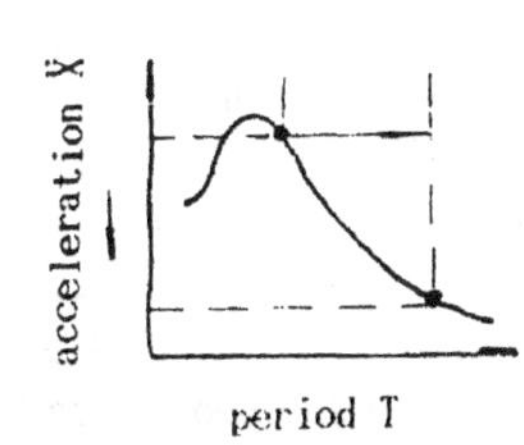

Figure 2. Characteristics of base isolation Figure 3. Acceleration reduce

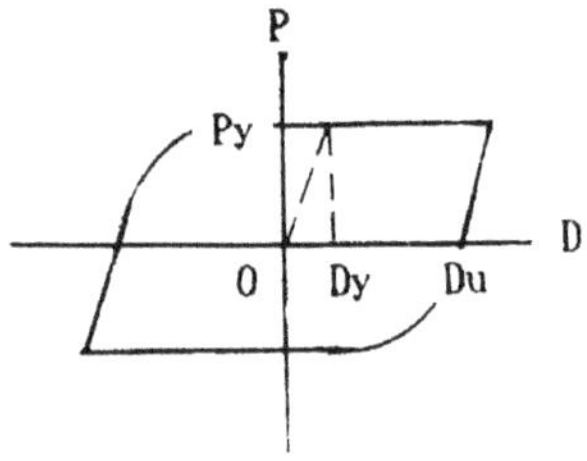

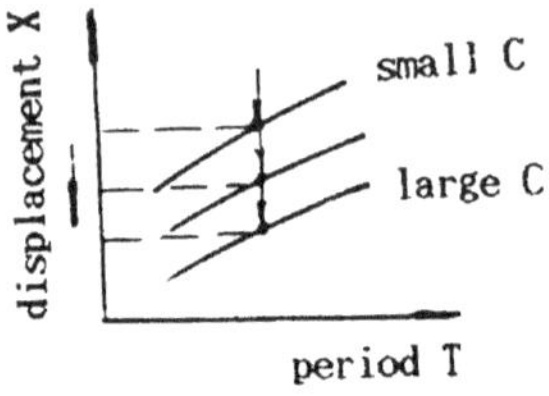

Figure 4. Load—displacement curve Figure 5. Displacement reduce

Design and Engineering Application of Base Isolation System

There are four kinds of combination of base isolation and energy dissipation system:

1. Rubber pad with steel plates reinforced as isolator (Fig. 5).

2. Roller as isolator combines with steel elements as energy dissipator [6] [7] [8] [9] (Fig. 7).

3. Dry friction layer or sand (or other material) sliding layer as isolator also energy dissipator [10] [11].

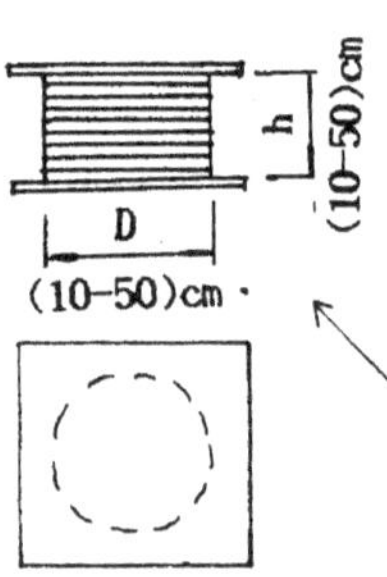

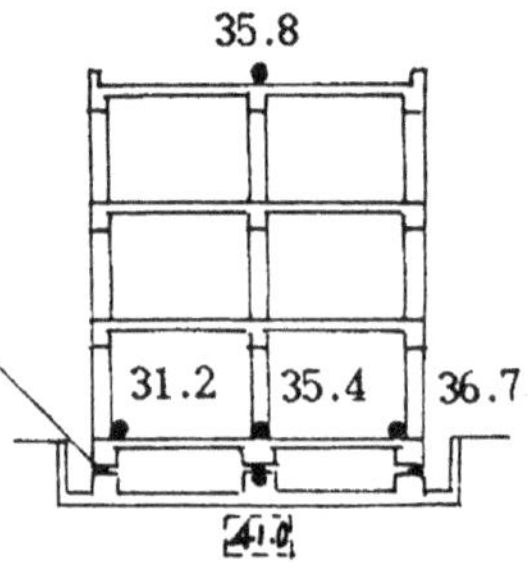

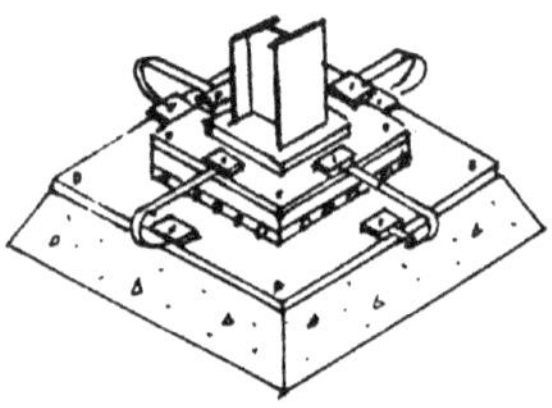

Figure 6. Rubber pad isolation Figure 7. Roller and steel plate isolation

Tests and Results of Base Isolation System

The tests includes two kinds of work:

(1) Tests of isolators and energy dissipators: Static tests of full scale elements, pseudo—static tests, low cycle fatigue failure tests.

(2) Shaking table tests for large scale structural model: The tests were carried out on an earthquake shaking table which has a floor dimension of 10ft × 10ft in THE UNIVERSITY OF BRITISH COLUMBIA. The overall dimension of four story steel frame model is 10ft × 4.6ft in plan and 12.8ft high. The total mass of structure model and loading concrete block is 16 kips (Fig. 8). The roller and mild steel curved plates were used as isolating and energy dissipating device.

A series of pseudo tests were finished for curved plates [8] before shaking table tests. Four sine—waveforms (Freq. w=1.0, 2.0, 3.0, 4.0 HZ) and three simulated earthquake records (El Centro, San Fernando, Parkfield) were input to the shaking table, whose predominant Freq. w was found from the Fourier spectra of acceleration record. The testing results show that: The acceleration responses on each story of structure model are nearly the same. It means that the elements and joints of structure with base isolation nearly work within elastic range only. The acceleration response on structure with base isolation is only (1/6—1/10) response on structure fixed on shaking table. It means the base isolation is more effective to attenuate the structural response in earthquake than any other methods (Fig. 9).

Figure 8. Shaking table tests

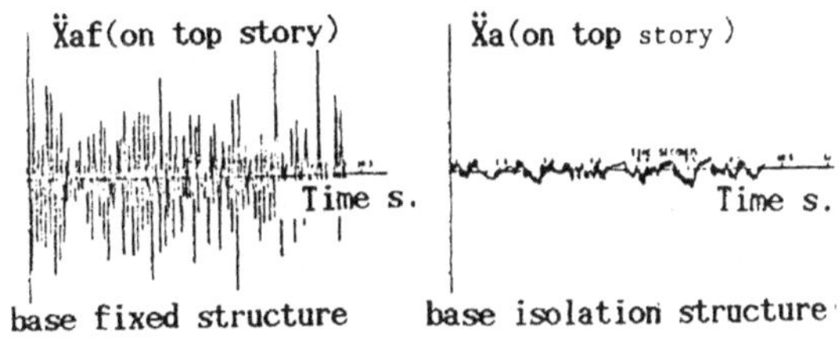

Figure 9. Acceleration response

In order to investigate the effect of base isolating system for highrise building, a 1/15 scale model of 9 story frame building structure was tested on a shaking table. Fig. 10 shows the testing records which relate to the shaking table vibration with Freq. w=15.5 HZ and the maximum acceleration of shaking table $\ddot{X}$ =0.49 g. Analyzing these results and know that:

First, the acceleration response at top story of structure model with base isolation is only 0.005 g, while the acceleration response at the same story with fixed base reaches 1.13 g. The ratio of $\ddot{X}$ (isolation) / $\ddot{X}$ (fixed base) =0.005 g/ 1.13g=1/226. It means that the base isolation is very significantly effective to attenuate the structure response for highrise building during severe earthquake.

Secondly, the acceleration responses at each story of structure model with base isolation are nearly the same. It means that the elements and joints of structure with base isolation will work within elastic range during earthquake.

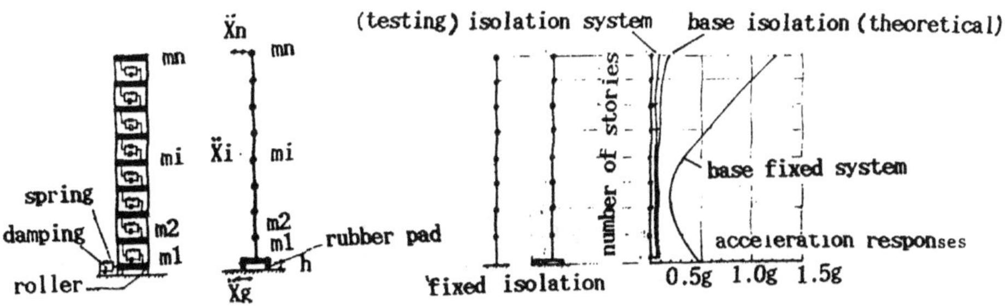

Figure 10. Testing results of shaking table with isolating and fixed base

Calculation Theory of Base Isolation System

From mathematic model, the basic differential equation of motion is given as:

$$M \ddot{X}a + Ce \dot{X}a + K Xa = Ce \dot{X}g + K Xg \quad (1)$$

Where M is the structural mass, Ce is the equivalent viscous damping of isolating and energy dissipating system, K is the elastic stiffness of isolating and energy dissipating system, Xg $\dot{X}g$ $\ddot{X}g$ are the ground response of displacment, velocity and acceleration respectively in earthquake. Xa $\dot{X}a$ $\ddot{X}a$ are the structure response of displacement, velocity and acceleration respectively in earthquake.

Define Ee = Ce / 2 M Wn as EQUIVALENT DAMPING RATIO,

Wn = K / M as the NATURAL FREQUENCY of system.

Solve Eq. (1) with finding transfer function and get:

$$Ee = \frac{1}{2\ (W/Wn)} \sqrt{\frac{1-AR^2\ [1-\ (W/Wn)^2]^2}{AR^2-1}} \quad (2)$$

Where AR = $\ddot{X}a / \ddot{X}g$ is called ACCELERATION ATTENUATION RATIO of system.

The basic differential equation of motion can also be written:

$$M \ddot{D}u + Ce \dot{D}u + K Du = -M \ddot{X}g$$

Where Du $\dot{D}u$ $\ddot{D}u$ are maximum relative displacement, velocity and acceleration respectively between structure and ground.

Solve this equation with finding transfer function and get:

$$Du = \frac{\ddot{X}g}{W^2} \sqrt{\frac{(1-AR^2)\ (W/Wn)^2}{(W/Wn)^2-2}} = Xg * r \quad (3)$$

Define $r = \sqrt{\frac{(1-AR^2)\ (W/Wn)^2}{(W/Wn)^2-2}}$ as DISPLACEMENT FACTOR

Comparing the theoretical value Du with measuring value [Du] from tests show that the Du/ [Du] = 0.98 — 1.67, Du is always greater than [Du]. It means Eq. (3) is suitable and conservative in designing.

The damping ratio Ee is related to the area enclosed by the hysteresis loop and find:

$$Ee = \frac{Ce}{2\ M\ Wn} = \frac{2\ (1-1/U)}{U\ \pi} \frac{1}{(W/Wn)} = B \frac{1}{(W/Wn)} \quad (4)$$

Define $B = \frac{2\ (1-1/U)}{U\ \pi}$ as ENERGY DISSIPATION DAMPING RATIO

U = Du / Dy as DUCTILITY FACTOR

Where Du and Dy are the relative displacements at ultimate point and yield point shown in Fig. 4. The value of B represents the basic of Ee, and depends on the value of U.

Substitute Eq. (4) into Eq. (2) and finally get:

$$AR = \frac{\ddot{X}a}{\ddot{X}g} = \sqrt{\frac{1 + 4\ B^2}{4B^2 + [1-\ (W/Wn)^2]^2}} \sqrt{\frac{W}{Wn}} \quad (5)$$

This is the final expression of ACCELERATION ATTENUATION RATIO

AR. Now, comparing the theoretical values AR from Eq. (5) with the measured values [AR], shown in Fig. 17, 18. and know: The ratio AR/ [AR] approaches 1. 0 ; it means that Eq. (5) gives reasonable estimation. The theoretical values AR are always larger than measured values [AR], It means that Eq. (5) gives conservative results for practical design. The author has also compiled a program EBIS—1 for design in engineering application.

ENERGY DISSIPATION SYSTEM

The Characteristics of Energy Dissipation System

Research indicates that the earthquake resistance of structure mainly depends on the energy dissipating ability of structure. During earthquake attack, certain amounts of energy are input to structure from ground motion. In general case, the energy is transformed and dissipated in four ways: (1) Potential energy. (2) Kinetic energy. (3) Energy dissipated by viscous damping of structure. (4) Energy dissipated by inelastic deformation in structure, components, joints or absorbers mounted in structures. The items (1) and (2) only transform one to the other but do not dissipate energy. The value of dissipating energy in item (3) is very small The item (4) is the main way to dissipate energy input to structure from ground motion.

During earthquake attack, some structural components are cracked, some structural joints are broken, or whole structure collapses, that just are the ways chosen by structure itself to dissipate the energy. A reasonable design of earthquake resistant structure means arranging a reasonable way chosen by designer to dissipate the greater part of energy input to structure and prevent the structure from damage during earthquake. This is the basic concept of energy dissipation design method.

Design and Engineering Application

There are two kinds of energy dissipation system to be used:

(1) Energy Dissipation Components:

(a) Energy Dissipating Bracing (Fig. 11): Such as square frame bracing, cycle bracing, K shape bracing. These bracings almost are made of mild steel which posseses large value of ductility, especially in bending state.

(b) Energy Dissipating Shear Wall (Fig. 12): Such as, the shear wall with vertical cracks, the shear wall with horizontal cracks, the shear wall with round cracks. During earthquake attack, the "crack" will be developed and dissipate the energy preventing the structure from damage.

(2) Energy Dissipating Damper (Fig. 13) : These dampers may be fixed on certain position (such as on bracing, shear wall, joints) and dissipate the energy while the deformation of structure is large during earthquake attack.

Author has finished many tests and got some results [12] [13], and has built an industrial structure with energy dissipating bracing.

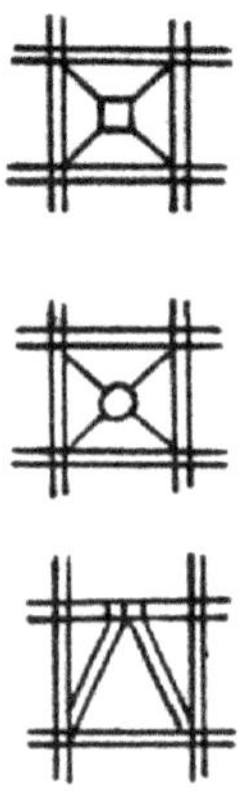

Figure ll. Bracing

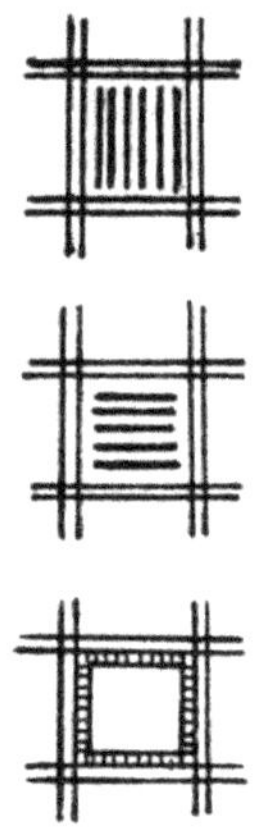

Figure 12. Shear wall

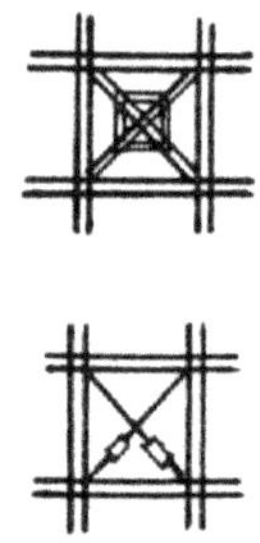

Figure 13. Damper on bracing

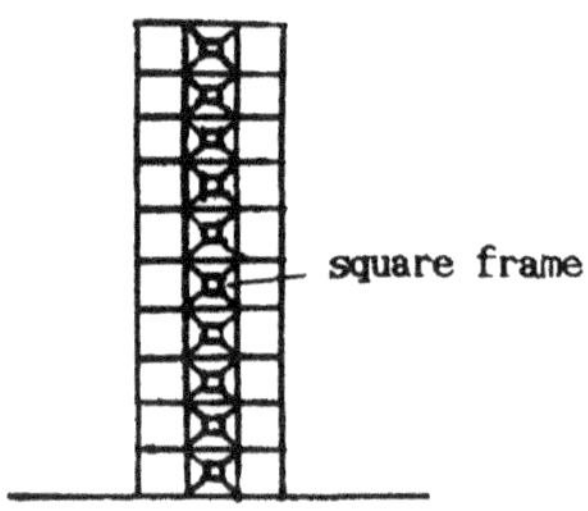

Figure 14. Building with bracing

Figure 15. Pseudo test for bracing

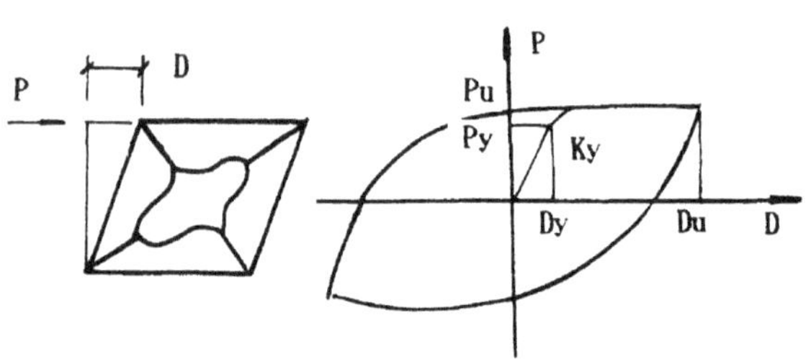

Figure 16. Testing results of bracing

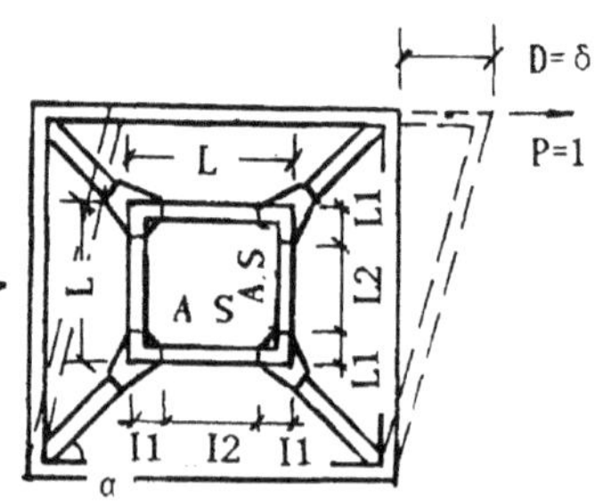

Figure 17. Calculating model

Tests and Research of Energy Dissipation System

Based on this concept, author suggests an energy dissipating bracing system (Fig. 14) for using in earthquake resistant high rise building structures. In order to prove the behaviour of this new bracing system, 5 sets of full scale models of this bracing have been tested. To bracing plane (Fig. 15), pseudo—static tests were controlled by displacement, the max. displacement Du. The lateral load P and displacement D were recorded by X—Y recorder. The P—D loop curves were shown in Fig. 16.

The main parameters for designing bracing system are lateral yielding force Py and elastic stiffness Ky (Fig. 16). Based on the analysis and testing results, authors suggest two sets of calculating formula for designing. [14]

(1) Yielding force Py of bracing system: From Fig. 17, write the static equilibrium equations and after solving, the final result can be written:

$$Py = 4\,A\,S\,Fy \,/\, (2\,S + A\,L\tan\alpha) \qquad (6)$$

Where Py——lateral yielding force acting at the top joint of one set of bracing.

A——cross steel section area of one component in "square frame".

S——elastic section modulus of one component in "square frame".

L——length of depth of "square frame".

α——angle between diagonal bar and horizontal level.

Fy——specified minimum yield strength of mild steel in the "square frame".

Comparing the theoretical values Py from Eq. (6) with the measuring values (Py) from tests. It shows that both values are very close, and Py is always smaller than (Py). It means that conservative results can be obtained from Eq. (6).

(2) Stiffness Ky of bracing system: From Fig.. 17, write the static equilibrium and solve, the final result can be written as:

$$\sigma = \tan\alpha\ (\tan\alpha+1)\ (L_1/EI_1+L_2/6EI_2)\ L_2^2/4$$

$$K = 0.6/\delta \qquad (7)$$

Where δ——horizontal displacement at top joint of one set of bracing under unit horizontal load P=1; Ky——lateral elastic stiffness of one set of bracing; L, I, —— length and moment of inertia of component at joint zone in "square frame"; L2, I2——length and moment of inertia of component between adjacent zones in "square frame"; E——elastic modulus of mild steel,the components made of a——angle between diagonal bar and horizontal level.

Comparing the theoretical values Ky from Eq. (7) with the measuring values (Ky) from testing, it shows that both values are very close, and Ky is always larger than (Ky). It may be induced by accounting the effect if compression in diagonal bar in Eq. (7), but its effect is very small in fact.

Because the steel elements were used as energy dissipater, these may deform into plastic regions withstanding large number of cyclic loads, and may fail due to the low cyclic fatigue in earthquake. The permanence of resisting low cycle fatigue failure can be represented by a parameter——Number of loading cycles to fail N. Many works indicate that N mainly depends on the absolute maximum strain value ε on the surface of mild steel elements in cyclic loads. So, the relationship between ε and N is the key to predict the low cycle fatigue failure for mild steel elements.

In order to get the accurate results, a series of X — shaped plates made of mild steel were chosen as main specimens to be tested, because the vertical strain of section is nearly uniformly distributed during horizontal bending under horizontal loads at top or bottom of specimen (Fig. 18, 19).

There are 36 pieces of X—shaped mild steel plates with different thickness t and length L cyclically loaded to fail under certain strain ε on surface. The tests were carried out on the "MTS Model 904. 55 Structure Testing System". Tests were displacement controlled. The loading cyclic frequency is 0. 1 HZ. The number of loading cycles to fail N was recorded by recorders automatically. The strain ε was measured by 4 pieces of strain gauges placed on each side of the X—shaped plate (Fig. 18). The testing results of values ε and N were plotted in Fig. 20.

From statistic analysis [15], a theoretical curve (Fig. 20) to represent the relationship between ε and N was expressed by equation:

$$\varepsilon = 0.22/N \quad (8)$$

Where ε——The maximum strain on surface of steel elements,

N——The number of loading cycles up to fail.

The results of tests and analysis indicate that the theoretical curve is very close to the testing records, especially in the strain range $\varepsilon = (0.5-3.0)\%$ which covers the general cases. The theoretical N values from theoretical curve are always smaller than the testing values. It means the appropriate conservative results can be obtained from Eq. (8) for design. The ability of resisting low cycle fatigue failure for mild steel elements under bending is very great. If yielding strain for mild steel is $\varepsilon = 0.15\%$, while maximum strain ε on surface of elements in earthquake is about $\varepsilon = 1.5\%$ (equivalently ductile factor $U = 10$ about), then the allowed number of cyclic loading N reaches more 200 from Eq. (8). It means this energy dissipator can reliably withstand tons of severe earthquake attack continuously. The designer can use Eq. (8) to preaict the permanence of resisting low cycle fatigue failure by simply controlling the strain value ε on the surface of steel elements.

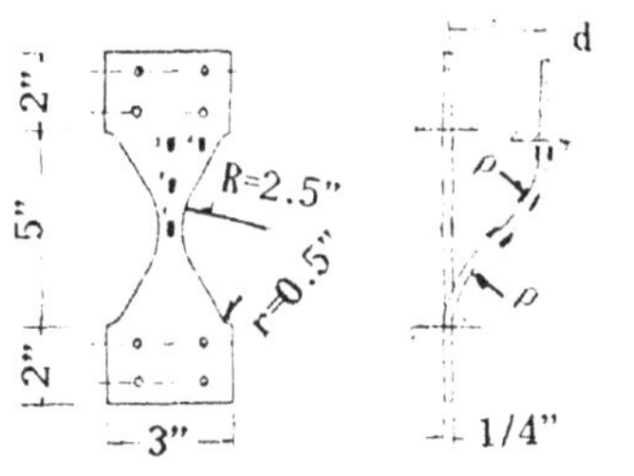

Figure 18. Strain gauges on X shaped plate

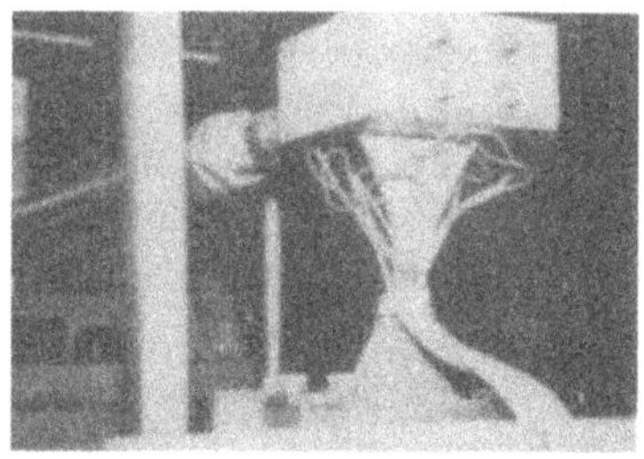

Figure 19. Test of cycle loads

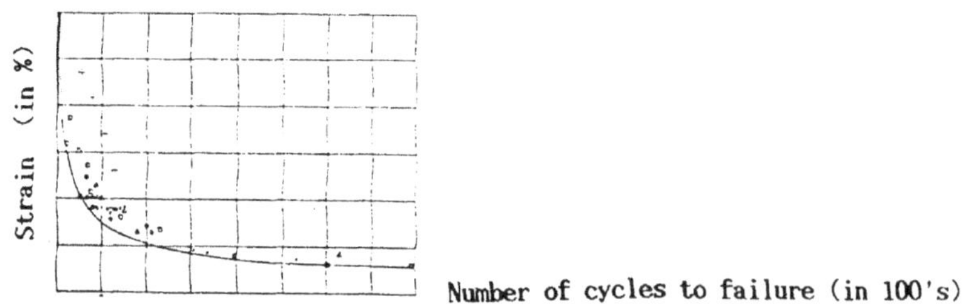

Figure 20. Relation of strain ε and loading cycles up to fail N

Design and Calculation of Energy Dissipation System

For high rise building structure with energy dissipating bracing, the energy balance equation at any instant of time during earthquake is:

$$Ein = Ep + Ek + Ep + Eb \quad (9)$$

Where Ein—— energy input to the structure; Ep—— potential energy in Structural vibration; Ek——Kinetic energy in structural vibration; Ed——energy dissipated by viscous damping of structure; Eb——energy dissipated by bracing system (energy dissipator).

Research indicates that the energy dissipator can dissipate about 90% of the total energy input at the end of earthquake. So, some items whose effect is relatively small can be neglected in Eq. (9). Then the energy dissipating design for earthquake resistance need to be satisfied with:

$$Ein < Eb \quad (10)$$

For calculating energy input Ein, the system can be considered as multi-degree of freedom system. If the number of stories (lumped mass) i=1, 2,..., n, the number of vibration mode j=1, 2,..., m, the mode shape Xj (i). The mass Mi, the mode participation factor rj, the velocity spectra value Sj, because Sj approximately is constant in a very wide range of frequency, Sj=S, then the energy input ot structure during earthquake can be got as below:

$$Ein = 0.5 \sum_{i=1}^{n} Mi \left[\sum_{j=1}^{m} Rj\, Sj\, Xj\ (i)\right]^2 = 0.5\, S^2 \sum_{i=1}^{n} Mi \left[\sum_{j=1}^{m} Rj\, Xj\ (i)\right]^2$$

$$\sum_{j=1}^{m} Rj\, Xj\ (i) = 1, \qquad \sum_{i=1}^{n} Mi = M$$

$$Ein = 0.5\, M\, S^2 \quad (11)$$

Where M is the sum mass of structure, S is the velocity spectra of structure which can be got from many previous works.

The energy dissipated by bracing system depends on the area enclosed by P—D loop curve, written finally as:

$$Eb = 1.26\, N \left(\sum_{i=1}^{n} Pyi\, Dui \right) F \quad (12)$$

Where 1. 26 is the factor for calculating the area of ellipse enclosed by P—D loop curve; N——number of loading cycles which depends on the structural natural period and the duration of earthquake. In general case, N = 30 — 60. Pyi——yielding force of i th bracing set calculated. Dui——max. horizontal displacement at the top joint of i th bracing set which needs to be satisfied the enough stiffness Ky and the capacity of resisting low cycle fatigue failure; i=1, 2..., n, the number of bracing sets; F——the factor considered the bracing sets as not fully working in whole structure, about 0. 5.

PASSIVE CONTROLLING STRUCTURE SYSTEM

The system , on certain position (such as on roof) in structure, places an additional Filial Structure (possesses mass M, stiffness K and damping C) to change the dynamic characteristics of original structure. During the earthquake attack, the Filial Structure will move against the direction of original structural vibration and reduce the response of original structure. This Filial Structure is called TMD (Tuned Mass Damper). The system operates without any outside energy sources, so it is also called "Passive Controlling Structure System".

This system can reduce the response of structure very effectively, is safe and very economic. It has been used in earthquake (or wind) resistant highrise buildings, tower structure and some large span structures.

There are three kinds of TMD used now:

(1) Supporting TMD (Fig. 21, a): The Filial Structure is supported on the roof or other places in original structure. The Mass can move in bilateral directions.

(2) Hanging TMD (Fig. 21, b): The Filial Structure is hung at top of the original structure. The Mass can sway in any direction . In some cases, the Water Tank on the top of the building may be designed as a Filial Structure hung at the top of building.

(3) Impacting TMD (Fig. 21. c): The Filial Structure is a Weight hung at the top place of structure. During earthquake, the hanging Weight will impact the original structure and reduce the response of structure.

At present, TMD has not been used in engineering application in China.

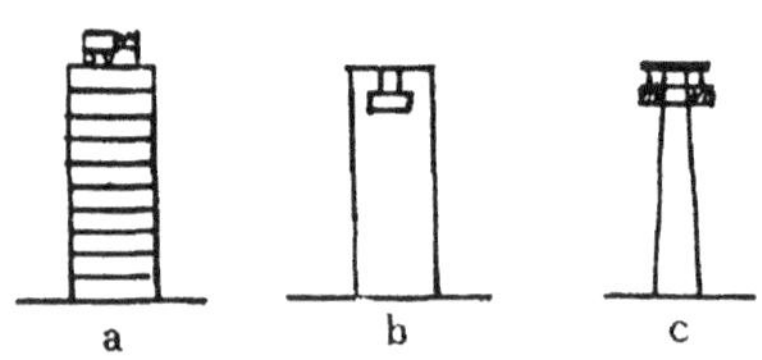

Figure 21. TMD System

ACTIVE CONTROLLING STRUCTURE SYSTEM

The system , on certain position in structure, places an additional Filial Structure which operates by computer control and immediately changes the dynamic characters of the structure to reduce the response of structure during earthquake. This system is controlled by outside energy sources, so is called "Active Controlling Structure System".

Two kinds of the system: one , the Filial Structure possesses Mass, Stiffness and Damping, is called AMD (Active Mass Damper). Another , the Filial Structure is a Bracing, is called Active Controlling Bracing.

The Active Controlling Structure System is just begun to be researched in China and other countries.

LIQUID MASS DAMPER SYSTEM

A damper unit consists of a pair of axially extensible cylindrical vessels, to which a conduit tube is connected. The cavity of the vessels and the tube is filled with liquid. When the unit is so incorporated in a structure that the structural vibration induces flow of the liquid in the tube, acceleration of the liquid causes large inertial as well as viscous resistance against the movement of the structural mass, thus acting as a powerful mass damper.

There are three different schemes of their incorporation in the structure, i. e. "direct damper", "gapped damper" and "open circulation damper".

The inertial and viscous resistance of the liquid flow oscillating in a circular tube is evaluated on the basis of laminar flow theory. The theoretical models of vibrating one—mass systems with these dampers are presented and steady responses to harmonic excitation are analyzed.

Also, small portal frame models having respective type of the dampers are subjected to harmonic excitation tests using vibrators and a shaking table.

The analyses and the tests reveal the different types of the dampers.

The gapped damper provides the structure with strong dependence of natural characteristic effects of the mass, resulting wide suppression band of frequency between two peaks of resonance.

The gapped damper provides the structure with strong dependence of natural frequency on displacement amplitude, which results truncated peaks of response in the resonance curves.

The open circulation damper exerts powerful damping action without creating any higher mode resonance.

THE FUTURE DEVELOPING TENDENCY ABOUT EARTHQUAKE RESISTANT STRUCTURE

The traditional earthquake resistant structure system and its design method have resulted in many problems in application. Since the Intensity of Earthquake is always uncertain, the ground motion and structural response are too complicated, and the requirements of structures with modern contents are more and more strict. The traditional system and its design method for earthquake resistant structures are not satisfied for usage. The new system of earthquake resistant structure will give a new way to resist earthquake for structures in seismic zone. The new system is more safe, effective, economic and simple. Many earthquake engineers and scientists predict: "The coming decade will be a fast-developing decade in Base Isolation, Energy Dissipation and Controlling Structure System!"

Some suggestions for future work:

(1) Cooperation research projects.

(2) Tests and research for some key problems in new system.

(3) Engineering application.

(4) Design code or design standard for new system.

(5) Establish more perfect style of new system in China and other countries.

REFERENCES

1. Zhou Fu Lin: Development and application for Technique of Isolation and Energy Dissipation and Control of Structural Response (lst Part). International Earthquake Engineering. Vol. 4, 1989 (in Chinese).
2. Zhou Fu Lin: Development and Application for Technique of Isolation and Energy Dissipation and Control of Structural Response (2nd Part). International Earthquake Engineering. Vol. 4, 1989 (in Chinese).
3. Chow Foo Lin: Experimental Investigations of Solid State Steel Energy Absorbers for Earthquake Resistant Structures. Publication of The University of British Columbia. Vancouver, Canada. Nov. 1983.
4. Zhou Fu Lin, Stiemer S. F. and Cherry S.: A New Isolation and Energy Dissipating System for Earthquake Resistant Structures. Proc. of 9th European Conference on Eanthquake Engineering. Moscow, Sept. 1990.
5. Zhou Fu Lin: Application of Base Isolation for Earthquake Resistant High Rise Buildings. Proc. of Ninth Symposium on Earthquake Engineering, India. Vol. 1. Roorkee, Dec. 1990.
6. Bao Zhen: The Research of Seismic Isolation Device Mounted on the Top of Flexible Support Frame. Proc. of UX/PRC Workshop on BIPEA, May 1990. San Francisco.
7. Zhou Fu Lin, S. F. Stiemer, S. Cherry: Tests, Analysis and Calculation for Base Isolation System. Proc. of US/PRC Workshop on Base Isolation and Passive Energy Absorption. San Francisco, USA. July 1990.
8. S. F. Stiemer, Chow Foo Lin: Curved Plate Energy Absorbers for Earthquake Resistant Structures. Proc. of 8th World Conference on Earthquake Engineering, San Francisco, USA, July 1984.
9. Zhou Fu Lin, S. F. Stiemer, S. Cherry: Design Method of Isolating An Energy Dissipating System for Earthquake Resistant Structures. Proc. of 9th World Conference on Earthquake Engineering. Tokyo—Kyoto. Aug. 1988.
10. Li Li: Base Isolation Measure for Aseismic Building in China. Prof. of 8WCEE, Vol. 6, p. 791.
11. Liu De Xin, Zen Guo Lin, Liu Guang Jun: Base Isolation with Friction and Its Deformation Control. Proc. 3rd National Conference on Earthquake Engineering, China, 1990, (in Chinese).
12. Zhou Fu Lin, Yu Gong Hua: Testing and Design for Energy Dissipation Bracing. Report on Earthquake Engineering, Vol. 1. 1982. (in Chinese)
13. Chen Zong Min, Zhang Qing: Investigation for Multistory Energy Dissipation Bracing with Friction Shear Hinge. Earthquake Resistant Engineering, Vol. 1 1991, (in Chinese)
14. Zhou Fu Lin, S. F. Stiemer, S. Cherry: Energy Dissipating Design Methed for Earthquake Resistant High Rise Building Structures. Proc. of 4th International Conference on Tall Building. HongKong & Shanghai, May 1987.
15. Zhou Fu Lin, S. F. Stiemer, S. Cherry: Prediction of low Cycle Fatigue Failure for Mild Steel Structural Components. Proc. of International Conference on Structural Failure. Singapore, April 1987.

OBSERVATIONAL CONSTRUCTION CONTROL SYSTEM FOR SOIL STRUCTURES BY COMPUTER NETWORK

TAKASHI MATSUMOTO [1] and MIKIO SHOJI[2]
[1] Manager of R & D Section, [2] General Manager
Information Processing Center/Kajima Corporation
1-2-7 Motoakasaka Minatoku, Tokyo 107, JAPAN

ABSTRACT

This paper presents the observational construction control system for soil structures having been developed by the authors in these three years, mainly an efficient use of finite element back analysis of two-dimensional elastic consolidation. We also discuss an applicability of sensitivity analysis so as to select the points where tell-tale instruments should be installed. Furthermore, how to estimate elastic plastic soil parameters in the constitutive model proposed by Sekiguchi and Ohta, based on the back analyzed elastic parameters, is also presented. As one of the typical examples, an application of our system to the project of retaining wall, which is now under construction, is demonstrated.

INTRODUCTION

To build safely, effort is always made at the designing stage to predict the behavior of structures both under construction and in service. However, due to various uncertainties and unknowns, the actual behavior of structures, especially soil structures such as embankments and excavations, sometimes differs substantially from the predicted behavior. This unpredictability may cause collapse during and after construction, and may lead to deterioration of surrounding structures, especially in urban areas.

The observational procedure is a method to close the gap between predicted and actual behavior, which was first proposed by Terzaghi and Peck(1948). This procedure consists of making appropriate observations soon enough during construction to detect any signs of departure of the real conditions from those assumed by the designs and of modifying either the design or the method of construction in accordance with the findings.

In Japan, the philosophy of the method was first applied to an embankment on soft ground for the Meishin Express Highway project by Japan Highway Public Corporation in 1965. By the late 1970's the use of the method had been adopted by many others as well. In these ten years field engineers are now able to examine the existing condition of the site and accurately judge the situation more quickly by combining the use of personal computers programed to use field measurements. Concurrently the development of large-scale digital computers enablestaking advantage of the finite element method with some sophisticated constitutive laws to predict deformations and stresses of soil structures as well as surrounding soil *at the design stage.*

Under these circumstances back analysis using a combination of the finite element method and the mathematical programming technique attracts much attention as the method of predicting soil behavior in the subsequent construction stage *during construction.* The analysis makes it possible to estimate the value of material properties that lead to results as close as possible to the corresponding in situ measurements.

In this paper the observational construction control system for soil structures developed by the authors in these three years, in which a finite element back analysis of two-dimensional elastic consolidation plays a principal role, is presented. Since it must be computed by a large-sized computer in headquarters, it is necessary to set up a computer network system using personal computers at the sites as terminals so as to allow real-time handling and processing of data sent from project sites. As one of the typical examples, an application of our system to the project of retaining wall, which is now under construction, is demonstrated.

DEVELOPED CONSTRUCTION CONTROL SYSTEM

The concept of a new control system proposed by the authors is shown in Fig. 1, which includes four sophisticated programs working in the large-sized computer and some easy-to-operate I/O programs in small computer at project site. The four programs were developed in accordance with the following purposes; ①the finite element back analysis of two-dimensional elastic consolidation called BACCHUS, ②the optimum location of measuring points (DEMETER), ③the estimation of elastic plastic soil parameters (SOMBAC), and ④the finite element analysis of elastic plastic two-dimensional consolidation called LIMDAC. The satellite computer in branch offices is suitable to store the data sent from project sites and to maintain the security of network. The total system is called TOPCON.

Finite Element Back Analysis of Two-Dimensional Elastic Consolidation

A governing equation, utilizing the two-dimensional finite element consolidation analysis, has been formulated in various forms by many researchers [e.g., Shoji and Matsumoto(1976), Akai and Tamura(1978) etc.]. In this study, the same

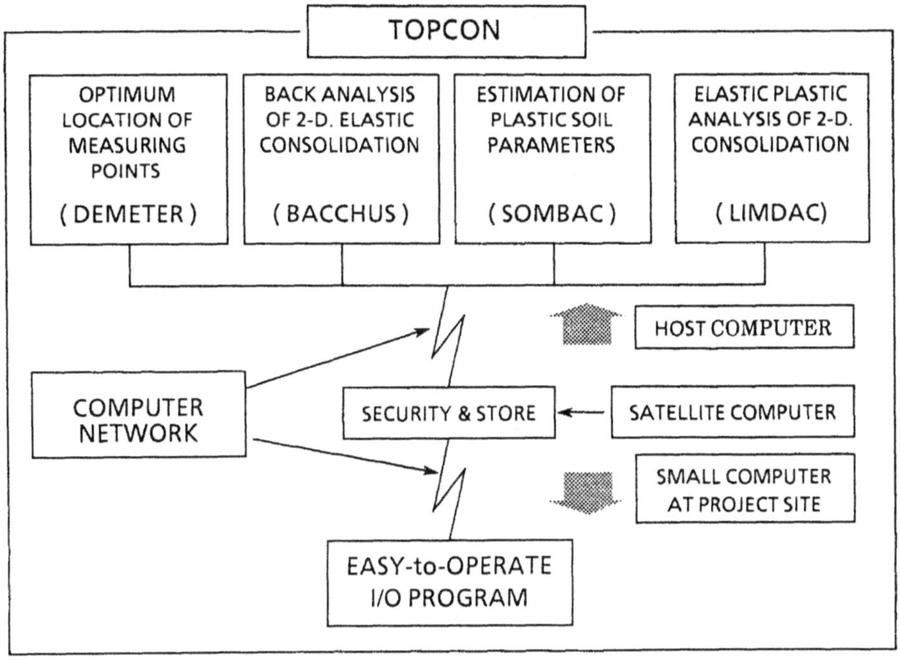

Figure 1. Concept of a new control system

form as that proposed by Akai and Tamura(1978) is used and expressed as follows:

$$\begin{aligned}&[K_e]\{u^{n+1}\}+\{L\}\,p^{n+1}=\{f^{n+1}\}\\&\{L\}^T\{u^{n+1}\}-\alpha\cdot p^{n+1}+\sum_i(\alpha_i\cdot p_i^{n+1})=V^n\end{aligned}\tag{1}$$

where $[K_e]$: element stiffness matrix, $\{u^n\}$: nodal displacements, p^n : element pore water pressure, p_i^n : pore water pressure of neighboring element, $\{L\}$: vector to calculate the element volume change from nodal displacements, $\{fn\}$: nodal forces, V^n : volume change of the element, and n denotes the discretized time step. In addition, α and α_i are expressed as,

$$\alpha=\frac{k\Delta t}{\gamma_w}\sum_i\frac{b_i}{s_i}\quad;\quad\alpha_i=\frac{k\Delta t}{\gamma_w}\cdot\frac{b_i}{s_i}\tag{2}$$

where k : permeability, Δt : increment of time step, s_i : length between the centers of an element and its surrounding i-th element, b_i : width of drainage surface of an element adjacent to its surrounding i-th element, and γ_w : unit weight of water.

To get the values of soil parameters (Young's modulus E, Poisson's ratio ν and permeability coefficient k) in this FEM model that lead to results as close as possible to the corresponding in-situ measurements, the objective function in the back-analysis is formulated as follows:

$$\text{objective function}:J=\frac{1}{2}\sum_{j=1}^{M}\sum_{i=1}^{Md}(U_i^j-V_i^j)^2\rightarrow min \tag{3}$$

$$\text{constraints}: \quad 0<E^n\ ;\ 0<v^n<0.5\ ;\ 0<k^n\quad (n=1,\cdots\cdots,N) \tag{4}$$

where J is the objective function, M is the number of time steps, M_d is the number of observational displacements, V_i^j is the measured displacement of the concerned nodal point at time j, U_i^j is the computed displacement corresponding to V_i^j, and N is the number of materials. This formulation is proposed by Arai, Ohta and Kojima(1984) and Shoji, Ohta, Arai, Matsumoto and Takahashi(1990).

Most of the iteration procedures in the numerical method are used by following the flow chart shown in Fig.2, as it is not easy to solve the formulated optimization problem analytically. This is due to the high nonlinearity in the

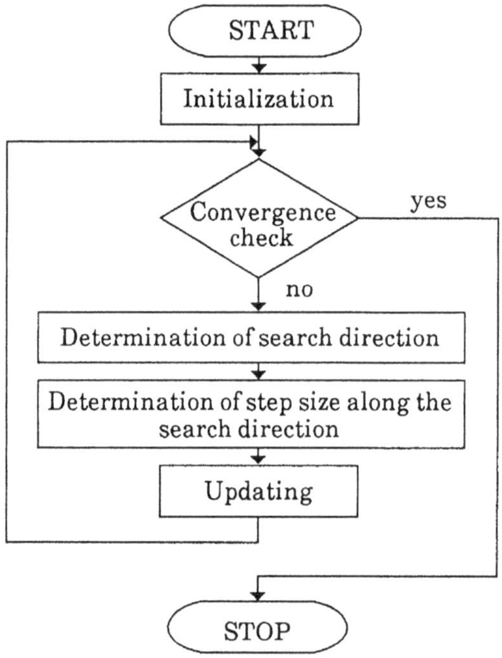

Figure 2. Flow chart for iterative procedure

objective function with respect to E^n, v^n and k^n, which are soil parameters to be estimated and are expressed as follows:

$$x=\begin{Bmatrix} E^n \\ v^n \\ k^n \end{Bmatrix} \tag{5}$$

Soil parameters estimated at the $(l+1)$-th iteration procedure can be written as follows:

$$x^{l+1} = x^l + a_l d^l \tag{6}$$

in which d^l : the search direction at the current point, a_l : the step size along the search direction. The most important implementation in the flow chart is "Determination of Search Direction". In this study, the quasi-Newton method with the BFGS formula which is assessed to be one of the best procedures for the unconstrained minimization problems [e.g., see Konno and Yamashita(1978)] is adopted. In the quasi-Newton method, the search direction d^l can be computed by

$$d^l = -H_l g^l \tag{7}$$

in which g^l : the gradient vector of J evaluated at the current point, and H_l : the quasi-Newton updating matrix which is updated at each step to approximate the inverted Hessian matrix at the current point. The BFGS formula is given as follows:

$$H_{l+1} = H_l + \left(1 + \frac{(y^l)^T H_l y^l}{(s^l)^T y^l}\right)\frac{s^l (s^l)^T}{(s^l)^T y^l} - \frac{H_l y^l (s^l)^T + s^l (y^l)^T H_l}{(s^l)^T y^l} \tag{8}$$

in which $y^l = g^{l+1} - g^l \quad ; \quad s^l = x^{l+1} - x^l$ (9)

Optimum Location of Measuring Points

It is important to select the points where tell-tale instruments should be installed within the limits of the project budget, as much of the reliability of the estimated soil parameters depends on the information from the in-situ measurements. So far, installation points have been chosen according to the results of prediction analysis and/or precedents. However, in the back-analysis of elastic consolidation, soil parameters are estimated by minimizing the differences between calculated and measured quantities, so that estimated parameters vary as construction progresses. Hence, it is difficult to conduct highly qualified observational construction unless the instruments are set up at points where the value of the computed displacement drastically changes in proportion to the variation of soil parameters estimated by the back-analysis.

This study employs a type of sensitivity analysis to consult the decision of instrument installation points. The sensitivity is obtained from differentiation of the computed displacements or pore water pressures with respect to soil parameters. This gives the magnitude of the variation of the computed values due to the fluctuation of the soil parameters estimated by back-analyses. It might be noticed that the sensitivity changes with the elapsed time. For the sake of simplicity, the following expression is substituted in place of Eq.(1).

$$[A]\{U^n\} = \{F^n\} \tag{10}$$

where [A] is a governing matrix composed of elastic constants , coordinates of nodal points and time, $\{U^n\}$ is a vector composed of nodal displacements and pore water pressures and $\{F^n\}$ is a force vector in the total system at time step n. Thus a sensitivity vector is given by differentiating $\{U^n\}$ with respect to each soil parameter and written as follows:

$$\frac{\partial\{U^n\}}{\partial x}=\frac{\partial([A]^{-1}\{F^n\})}{\partial x}=-[A]^{-1}\cdot\frac{\partial[A]}{\partial x}\cdot[A]^{-1}\{F^n\}+[A]^{-1}\cdot\frac{\partial\{F^n\}}{\partial x}$$

$$=-[A]^{-1}\left(\frac{\partial[A]}{\partial x}\cdot\{U^n\}-\frac{\partial\{F^n\}}{\partial x}\right) \qquad (11)$$

where x is the soil parameter to be noticed: either effective Young's modulus, effective Poisson's ratio or permeability coefficient. The computer program is called DEMETER.

In order to demonstrate the applicability of this idea, an example is examined to an actual test embankment which was built to test the consolidation properties of the foundation ground composed of an organic soil layer (layer 1) and a silty clay layer (layer 2). As one of typical results Fig.3 shows the distribution of the normalized sensitivity coefficients of horizontal and vertical displacements with respect to the Young's modulus of layer 2 (E_2). The actual measurement

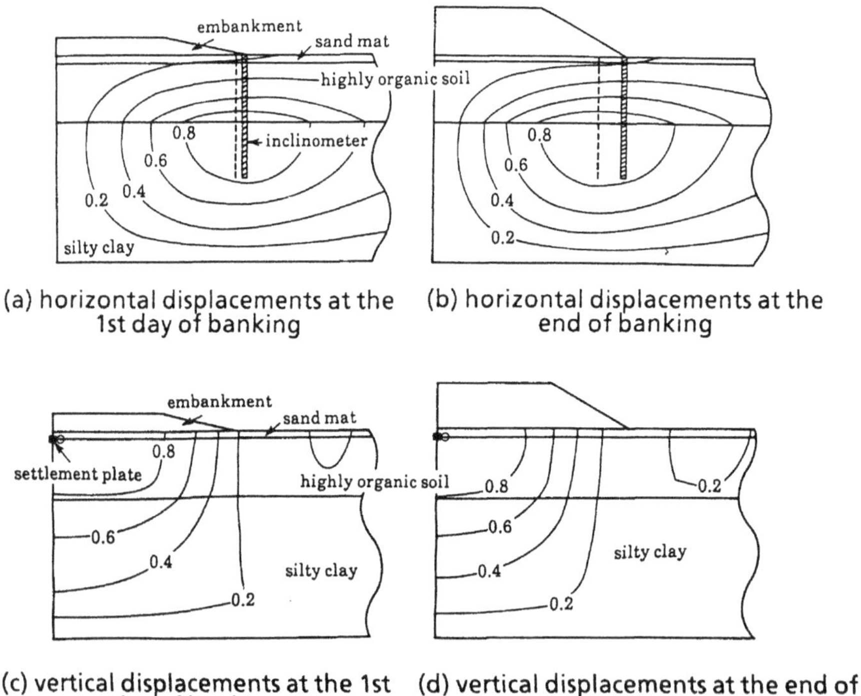

(a) horizontal displacements at the 1st day of banking (b) horizontal displacements at the end of banking

(c) vertical displacements at the 1st day of banking (d) vertical displacements at the end of banking

Figure 3. Sensitivity distribution of displacements with respect to E_2

points chosen according to the precedents are also represented in this figure: a solid line for the inclinometer and a solid square for the settlement gauge. Furthermore judging from the results of the sensitivity analysis, this figure illustrates the points where the instruments should have been installed too: the broken line for the inclinometer, the open circle for the settlement gauge. The results suggest that the best position of the inclinometer would be in a position of a few meters inside the toe of the slope.

Estimation of Elastic Plastic Soil Parameters and Elastic Plastic Two Dimensional Consolidation Analysis

A linear elastic consolidation analysis has some difficulties in predicting the stability of soil structures accurately since a large reduction in stiffness can occur when the critical state is reached. An elastic plastic two-dimensional consolidation finite element analysis is one of the most powerful and useful ones to overcome this shortcoming. A sophisticated and established elastic plastic constitutive relation can be easily employed and fabricated into the computer program. However, it had been difficult for the behavior of soil elements which come to the perfect plastic region such as the failure state because of its incremental strain becoming infinity. The authors(1988) improved the conventional methods by introducing an imaginary viscosity method first proposed by Zienkiewicz and Cormeau (1974) to analyze the behavior of soil from the initial stress state to the failure state accurately. The finite element formulation is basically same as Eq. (1) though the first equation in Eq.(1) has the additional external nodal force equivalent to the imaginary visco-plastic strain. The computer program is called LIMDAC.

The elastic plastic constitutive model proposed by Sekiguchi and Ohta (1977) is adopted in LIMDAC to take the anisotropic strength of K_0-consolidated clays into account. The input parameters required for this model are the following nine constants; (i) irreversibility ratio $\Lambda(=1-\kappa/\lambda)$, where λ and κ are the compression and swelling indices, respectively, (ii) critical state parameter M, (iii) dilatancy coefficient D, (iv) Poisson's ratio in terms of effective stress ν, (v) pre-consolidation vertical pressure σ'_{vo}, (vi) coefficient of earth pressure at rest K_O, (vii) effective current overburden pressure σ'_{vi}, (viii) coefficient of in-situ earth pressure at rest K_i, and (ix) coefficient of permeability k.

It is difficult to estimate the elastic plastic soil parameters by directly applying a mathematical programming technique to an objective function when the constitutive model has a strong non-linearity depending on the stress level. Therefore, there is no other way but to search the parameters by repetition of trial and error. However, it is not prudent, in the engineering sense, to estimate these nine parameters as independent to each other, as enormous iterations may be required. Thus, the following two engineering terms are utilized, so as to reduce the computation time: (i) The experimental or theoretical relationship between the effective internal friction angle ϕ' and the nine parameters, and (ii) the values of Young's moduli E, the Poisson's ratios ν, the permeability coefficient k of each soil layer and the effective stress and strains obtained from "BACCHUS". Under these assumptions the principle idea is to detect ϕ' such as will make the difference minimum between the equivalent elastic secant and bulk moduli in each finite element to the computed nonlinear stress-strain relation and those obtained from the back analysis of two-dimensional elastic consolidation. The range of change in value of ϕ' is very limited, since each type of soil has an inherent effective internal friction angle. Therefore, only a few repetitions are required for convergence. The nine parameters are determined once the optimal effective internal friction angle is calculated. This procedure is processed in the computer program SOMBAC and its conceptual flow chart is shown in Fig.4. The detailed explanation might be better understood in the paper by Shoji, Ohta, Matsumoto and Morikawa(1989).

A housing lot embankment constructed in Kochi Prefecture in Japan is employed as an example to describe how the procedure is applied. Fig.5 illustrates the computation results of the ultimate height at the time of 90 days from the beginning of banking (the banking height is 2.6 m) by using LIMDAC and SOMBAC. In the process to calculate the ultimate height of embankment, fictitious embankment loads are increased until reaching its ultimate limit under undrained conditions. The relationship between settlement and embankment height is presented in Fig. 5(a). In Fig. 5(b), a failure region under the banking

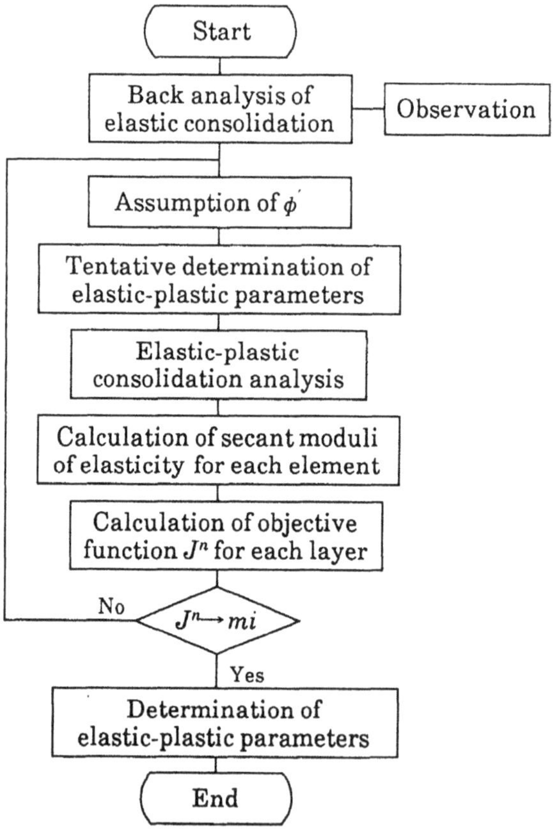

Figure 4. Procedure of estimating elastic-plastic soil parameters

height of $H=8.6$m, 9.6m and 10.6m, of which the safety factors correspond to 1.2, 1.1, and 1.0, respectively, is illustrated. The failure region spreads to the upper section as the height increases. Upon reaching $H=10.6$m, a sliding surface is formed in a way surrounding the non-failure region under the slope of embankment. Thus, the safety factor F is calculated to be 4.1 under the current banking height. The results during the other monitoring periods represent the same tendency. This fact demonstrates that the embankment was built with complete safety ($F>4.0$) during the construction. From the view-point of just the mechanical stability the banking speed could be increased and the project completed quicker. Of course many other factors should be considered such as the schedule for earth-moving machines etc. when we change the working schedule.

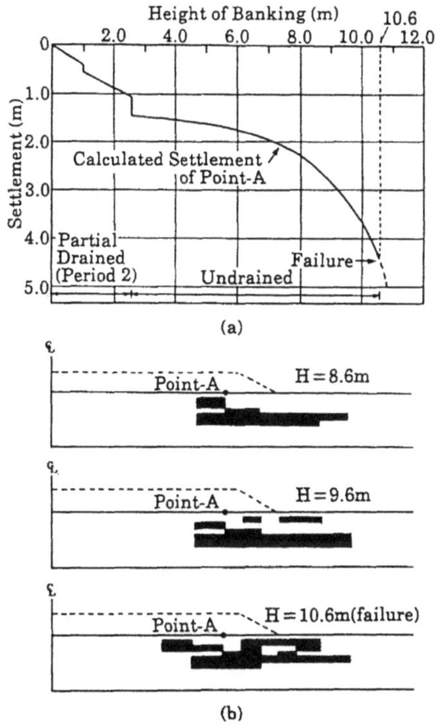

Figure 5. Ultimate banking height and development of failure region

Computer Network

The network system shown in Fig.6 has been constructed in Kajima. This system enables computer communication among all offices and allows real-time handling of data sent from gauges at project sites.

APPLICATION TO ACTUAL LARGE-SCALE EXCAVATION

Our observational construction control system has been applied to many actual construction projects in order to maintain their safety. As one of typical examples an excavation to construct a foundation for a multipurpose building located in

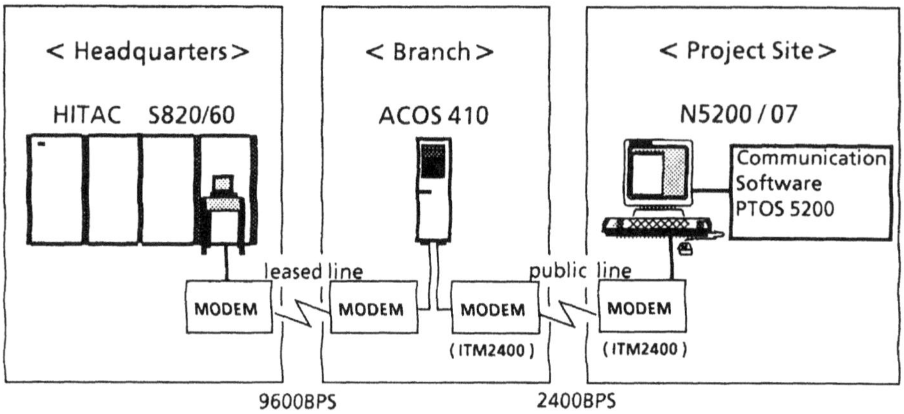

Figure 6. Computer network system

Nagoya, which contains a high-voltage substation, is presented here. Especially an application of the computer code BACCHUS is discussed.

Summary of Construction Site

Since the substation with high capacity is installed under the building, a huge underground soil structure of 33m depth ,though such a structure has never been constructed before in Nagoya, should be prepared. Because of its depth, the inverted concrete shooting construction technique, which places concrete just after the excavation step by step, must be employed, and a continuous wall in ground, whose thickness and depth are 1.2m and 62m respectively, is set up as retaining wall as well as cut-off wall. Soil profile in the site seems relatively proper, which mainly consists of sand and gravel layers, though clay layer can be partially detected. However, since there are many important structures such as viaduct for the expressway, high buildings and

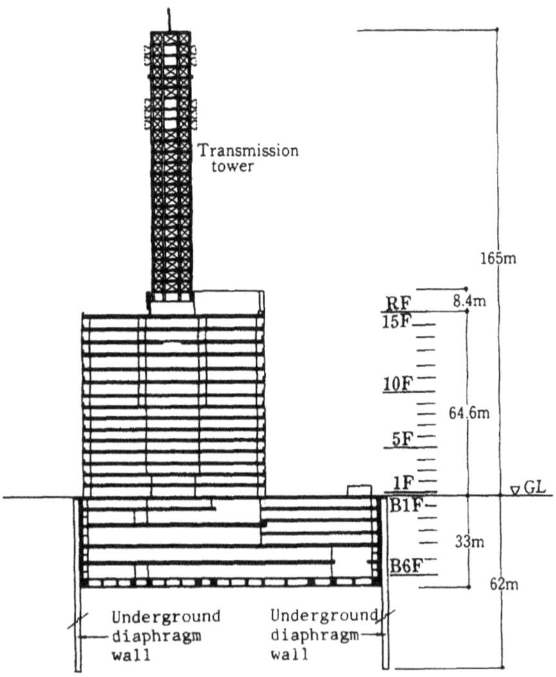

Figure 7. Cross-section of the building

private houses in the vicinity of the construction site, sophisticated construction management should be carried out. To clarify the construction site more visually, a cross-section of the planned building is shown in Fig.7, and a cross-section of the retaining wall with soil properties is shown in Fig.9, respectively.

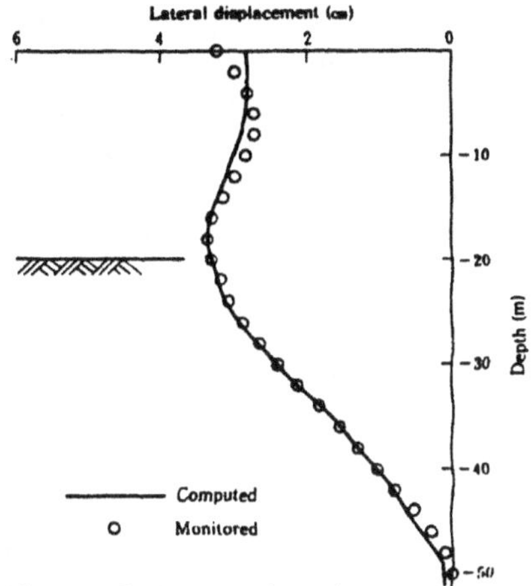

Figure 8. Comparison between monitored and computed displacements

Discussion

This construction project is now under construction. As of July, 1991, it has already finished Stage-4 of excavation, which corresponds to 20m depth (see Fig.9). Back-analysis has been carried out by using all the monitored displacements obtained during the period up to Stage 4. Fig.8 demonstrates comparison between monitored and computed displacements, where the latter can be obtained by using back analyzed soil parameters. This figure shows fairly good agreement between them, though slight difference can be observed near the ground surface. Fig.9 shows the prediction of the future lateral displacements of retaining wall after Stage 5 of excavation, which are much smaller than those predicted at the design stage. This figure also shows the cross-section of the excavation with soil properties. If it can be concluded that the strut to be set up after Stage 8 of excavation is not necessary, it is expected that the period of construction could be reasonably reduced, which also leads to cutback on construction expenses. The broken line in Fig.9 indicates the predicted lateral displacement just after the completion of excavation without this strut. In fact, as shown in Fig.10, the prediction of the bending moment for retaining wall will not exceed the prescribed allowable value, even when the strut after Stage 8 is not installed.

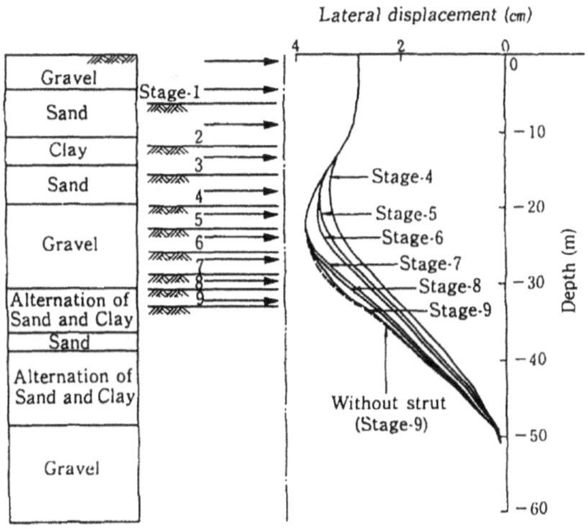

Figure 9. Cross-section of the excavation and prediction of future displacements

Finally, Fig.11 shows the predicted deformations in the whole area just after the final stage, Stage 9. It clearly indicates that this construction project gives little effect on the surrounding structures. Based on these results the project manager is now under consideration of cancelling this strut, though further observation during excavation must be continued. The developed control system including SOMBAC and LIMDAC, which have not been used because of excavating with complete safety until Stage 4, is supposed to be continuously applied to this construction project in order to increase the degree of its safety and rationality.

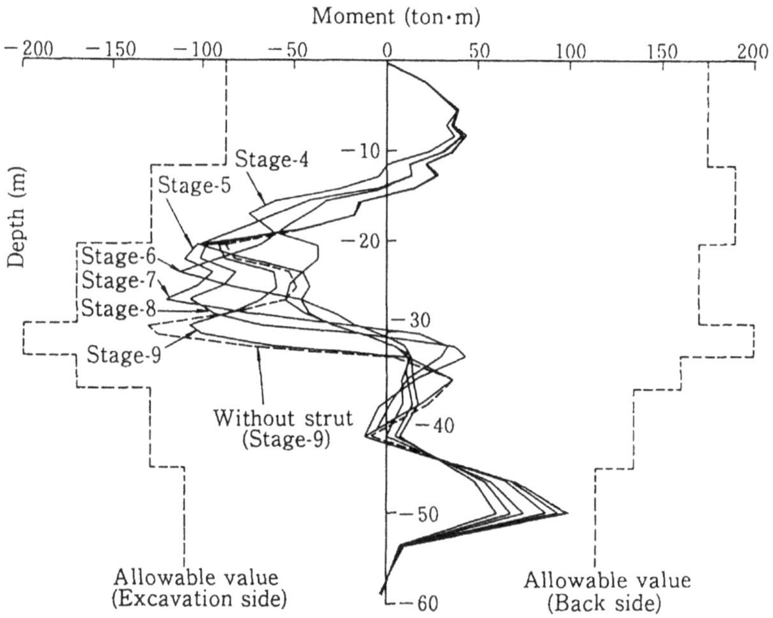

Figure 10. Prediction of the bending moments for retaining wall

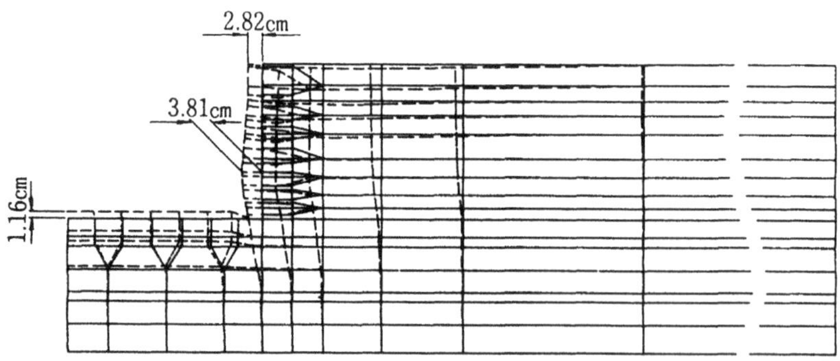

Figure. 11. Predicted deformation in the whole area just after the final stage

CONCLUSION

It is the civil engineers' responsibility to avoid the collapse of totally or partially constructed structures and to conduct construction projects safely, speedily, and economically. An observational method or information-oriented-design and construction would be one of the most promising tools to achieve that responsibility.

The observational construction control system, presented in this paper, has been developed to respond the above needs, and practically applied to more than 10 actual construction sites. The results obtained through its application are summarized as follows:

① Improvement in safety during construction:
Since the future behavior can be predicted by making use of back-analysis with monitored displacements, it is expected that construction projects will have been completed under high reliability. In fact, if symptom of dangerous behavior can be detected at an early stage, appropriate counter-measures can be taken to compensate for the undesirable possibility that may cause serious problems.

② Reduction in construction expenses:
To make use of back-analysis enables us to grasp the safety of the whole area in a construction site. Therefore, when the initial design can be diagnosed as too excessively safe, it may be possible to relax that design to some extent, so that the total construction period and expenses can be reduced in fairly reasonable way.

③ Reduction in expenses on monitoring instruments:
To make use of sensitivity analysis enables us to efficiently determine rational installation positions of soil-monitoring instruments, which also leads to reduction of the expenses on monitoring instruments.

④ Improvement in efficiency of processing data:
Since the time required to transfer the monitored quantities and the numerical results through our computer network may be at most a few minutes, construction management has been readily carried out with smooth communication among the computers at the headquarters and project sites.

Finally, the authors would like to express sincere appreciation to Professor Hideki Ohta, Kanazawa University, for his valuable comments on our system, especially the fundamental idea and the analytic approaches contained in the programs. The authors also thank Mr. Seiji Morikawa, Mr. Toshihiko Takahashi, and many other researchers in Information Processing Center, Kajima Corporation, for their collaboration in developing our system.

REFERENCES

1 Akai, K. and Tamura, T., Numerical analysis of multidimensional consolidation accompanied elasto-plastic constitutive equation. Proc. of JSCE, 1978, No.269, pp.95-104 (in Japanese).

2 Arai, K., Ohta, H. and Kojima, K., Estimation of soil parameters based on monitored movement of subsoil under consolidation. Soils and Foundations, 1984,Vol.24, No.4, pp.95- 108.

3 Konno, H. and Yamashita, H., Nonlinear programming, Union of Japanese Scientists and Engineers, Tokyo, 1978 (in Japanese).

4 Sekiguchi, H. and Ohta, H., Induced anisotropy and time dependency in clays. Proc. 9th ICSMFE, Tokyo, Specialty Session 9, 1977, pp.229- 237.

5 Shoji M.and Matsumoto, T., Consolidation of embankment foundation. Soils and Foundations, 1976, Vol.16, No.1, pp.59-74.

6 Shoji, M., Matsumoto, T., Morikawa, S., Ohta, H. and Iizuka, A., Coupled elasto-plastic deformation-flow finite element analysis using imaginary viscosity procedure. Proc. 6th Int. Conf. Numerical Methods in Geomechanics, Innsbruck, 1988, Vol. 1, pp.299-304.

7 Shoji, M., Ohta, H., Matsumoto, T. and Morikawa, S., Safety control of embankment foundation based on elastic-plastic back analysis. Soils and Foundations, 1989, Vol.29, No.2, pp.112-126.

8 Shoji, M., Ohta, H., Arai, K., Matsumoto, T. and Takahashi, T., Two-dimensional consolidation back analysis. Soils and Foundations, 1990, Vol.30, No.2, pp.60-78.

9 Terzaghi, K. and Peck, R.B., Soil mechanics in engineering practice, John Wiley & Sons, Inc., 1948, Introduction.

10 Zienkiewicz, O. C. and Cormeau, I. C., Visco-plasticity–Plasticity and creep in elastic solids- A unified numerical solution approach. International Journal for Numerical Methods in Engineering, 1974, Vol.8, No.4, pp.821~845.

MINIMUM SEARCH FOR LAGRANGE MULTIPLIER METHOD IN CONSTRAINED OPTIMIZATION PROBLEMS

TSUNG-WU LIN, JIN-TEN HUANG AND MING-HUEI JOUNG
Department of Civil Engineering
National Taiwan University
Taipei, Taiwan 10764, China

ABSTRACT

This paper uses a method proposed by the authors for modifying the original Lagrangian with indefinite Hessian matrix into a new form with positive definite Hessian matrix. The original Lagrangian is obtained by using Lagrange multipliers to transform a constrained optimization problem into an unconstrained optimization problem. The modified Lagrangian is obtained by adding to the original one a quadratic term $\frac{1}{2}(\mathbf{x}-\mathbf{x}_N)^T D_k(\mathbf{x}-\mathbf{x}_N)$, where D_k is a diagonal matrix to make the modified Hessian matrix $H_k + D_k$ positive definite, H_k is the original Hessian matrix at the k-th iterating point $\mathbf{x}_k$, and $\mathbf{x}_N$ is the Newton point of both the original and the modified Lagrangian. Therefore, we can use stable and efficient search methods to locate the minimum point of the modified Lagrangian and get the stationary point of the original Lagrangian. The modification is made whenever the new Newton point is found in each iteration. In this paper a procedure is proposed to update the modified Hessian matrix by using variable metric method such as the PSB update. Several numerical examples are given to indicate the applicability of the proposed procedure.

INTRODUCTION

Constrained optimization problems can be solved by methods for unconstrained optimization problems. Constrained problems can be converted to unconstrained problems by the penalty functions or the Lagrange multipliers. A number of penalty functions have been proposed and investigated [1,2,3,4,5,6]. The unconstrained problem converted by the penalty functions is usually only an approximate one. Therefore, it is required to solve a sequence of unconstrained problems with different penalty parameters.

The unconstrained problem converted by the Lagrange multipliers is exact and differentiable, but its solution corresponding to the original constrained minimization is always a saddle point of the Lagrangian function. Therefore, the globally convergent one-dimensional search methods cannot be applied directly to the converted unconstrained problem. The method of multipliers (MOM) [7,8] is a well-known method based on the modification of the Lagrangian [7]:

$$Min.\ L(\mathbf{x};\lambda,c) = f(\mathbf{x}) + \sum \lambda_i g_i(\mathbf{x}) + c\sum g_i^2(\mathbf{x}) \tag{1}$$

where $c > 0$ is a penalty coefficient and λ_i is an approximation of the Lagrange multiplier for the constraint $g_i(\mathbf{x}) = 0$.

The method in reality is also a penalty type method. The Lagrange multipliers remain constant in each iteration as the control parameters and are updated from iteration to iteration by certain rules. With this method, it is also required to solve an infinite sequence of unconstrained problems with different multipliers as the parameters.

To overcome this difficulty, Fletcher [9,10] introduced in the augmented Lagrangian a multiplier vector continuously dependent on $\mathbf{x}$, which requires only a single minimization. This method, however, requires a matrix inversion at each function evaluation, which tends to limit its applicability.

Pillo and Grippo [11,12] also introduced a new class of augmented Lagrangian obtained by adding to the augmented Lagrangian of the MOM a penalty term relating to the first order necessary condition $\nabla f(\mathbf{x}) + \sum \lambda_i \nabla g_i(\mathbf{x}) = 0$:

$$Min.\ S(\mathbf{x},\lambda;c) = f(\mathbf{x}) + \sum \lambda_i g_i(\mathbf{x}) + c\sum g_i^2(\mathbf{x}) + \left\| M(\mathbf{x})\left(\nabla f(\mathbf{x}) + \sum \lambda_i \nabla g_i(\mathbf{x})\right)\right\|^2 \tag{2}$$

where $M(\mathbf{x})$ is an appropriate weighting matrix.

For a finite value of the penalty coefficient c, this method requires only a single unconstrained minimization of $S(\mathbf{x},\lambda;c)$ with respect to both $\mathbf{x}$ and λ. The disadvantages with this method are the presence of first order derivatives of the function $f(\mathbf{x})$ and $g_i(\mathbf{x})$ in the augmented Lagrangian and the fact that $S(\mathbf{x},\lambda;c)$ may be unbounded with respect to λ. This method has been further investigated in [13,14].

The authors proposed a method to locate the stationary (saddle) point of the original Lagrangian [15]

$$L(\mathbf{x},\lambda) = f(\mathbf{x}) + \sum \lambda_i g_i(\mathbf{x}) \tag{3}$$

by searching the minimum point of the modified Lagrangian

$$Min.\ \bar{L}(\mathbf{x},\lambda) = f(\mathbf{x}) + \sum \lambda_i g_i(\mathbf{x}) + \frac{1}{2}(\mathbf{x}-\mathbf{x}_N)^T D(\mathbf{x}-\mathbf{x}_N) + \frac{1}{2}\sum d_i(\lambda_i - \lambda_{N\,i})^2 \tag{4}$$

where the subscript N denotes the Newton point, and D is a diagonal matrix.

The addition of the quadratic terms has two main purposes: (1) to make the modified Lagrangian have the same Newton point as the original Lagrangian, (2) to make the modified Lagrangian have a positive definite Hessian matrix. Consequently, one can use one-dimensional search method to locate the minimum point of the modified Lagrangian and get the stationary point of the original Lagrangian.

This paper proposes a procedure for updating the modified Hessian matrix by using variable metric methods such as the PSB update [16]. The applicability of the variable metric methods is important for it will eliminate the time-consuming computation of the Hessian matrix. Several numerical examples are tested, of which the results are shown to be satisfactory both in effectiveness and in simplicity.

THE ADAPTIVE MODIFICATION METHOD

To simplify the notation, the Lagrange mutipliers λ_i will be included in the original design variables, i.e., the vector $\mathbf{x}$ contains both $\mathbf{x}$ and λ_i. By denoting the Lagrangian $L(\mathbf{x}, \lambda)$ by $f(\mathbf{x})$, equation (4) can be simplified as

$$\bar{f}(\mathbf{x}) = f(\mathbf{x}) + \frac{1}{2}(\mathbf{x} - \mathbf{x}_N)^T D(\mathbf{x} - \mathbf{x}_N) \tag{5}$$

These two functions, $f(\mathbf{x})$ and $\bar{f}(\mathbf{x})$, have the same Newton point $\mathbf{x}_N$, which can be expresssed as equations (6) or (7).

$$\mathbf{x}_N = \mathbf{x} - H^{-1}\nabla f \tag{6}$$

$$\mathbf{x}_N = \mathbf{x} - (H + D)^{-1}(\nabla f + D(\mathbf{x} - \mathbf{x}_N)) \tag{7}$$

Note that the Hessian matrix of the modified function is $H + D$ and it can be made positive definite by adding the appropriate diagonal matrix D even if the Hessian matrix of the original function, H, is indefinite. Therefore we can use the globally convergent search method to locate the minimizer of the modified function instead of just using the Newton point which may not converge at all from a poor start.

Following are the procedures for finding the stationary (saddle) point of the function $f(\mathbf{x})$:

0. Compute ∇f_o and H_o of the original function $f(\mathbf{x})$ at the initial point $\mathbf{x}_o$.

For each iteration $k \geq 0$:

1. Find the diagonal matrix D_k such that $H_k + D_k$ is positive definite. The diagonal matrix D_k can be found by the perturbed Cholesky decomposition [17].
2. Compute the Newton point $\mathbf{x}_N$ by the following equation:

$$\mathbf{x}_N = \mathbf{x}_k - H_k^{-1}\nabla f_k \tag{6'}$$

3. Search the minimizer of the following modified function $\bar{f}(\mathbf{x})$

$$\bar{f}(\mathbf{x}) = f(\mathbf{x}_k) + \frac{1}{2}(\mathbf{x} - \mathbf{x}_N)^T D_k(\mathbf{x} - \mathbf{x}_N) \tag{5'}$$

The search may be along a Bezier curve defined by the three points $(\mathbf{x}_k, \mathbf{x}_C, \mathbf{x}_N)$ as follows [18]:

$$\mathbf{x}(u) = u^2\mathbf{x}_N + 2u(1-u)\mathbf{x}_C + (1-u)^2\mathbf{x}_C \tag{8}$$

where $\mathbf{x}_C$ is the Cauchy point of the modified function $\bar{f}(\mathbf{x})$, and u is the parameter of the curve. The Cauchy point is

$$\mathbf{x}_C = \mathbf{x}_k - \lambda_k \nabla \bar{f}_k \tag{9}$$

where

$$\nabla \bar{f}_k = \nabla f_k + D_k(\mathbf{x}_k - \mathbf{x}_N) \tag{10}$$

$$\lambda_k = \frac{\nabla \bar{f}_k^T \nabla \bar{f}_k}{\nabla \bar{f}_k^T (H_k + D_k) \nabla \bar{f}_k} \tag{11}$$

4. Set the minimizer as the next point $\mathbf{x}_{k+1}$ for the next iteration.
5. Stop if the stop criteria are met.
6. Compute ∇f_{k+1} at the next point $\mathbf{x}_{k+1}$.
7. Compute $\mathbf{s}_k = \mathbf{x}_{k+1} - \mathbf{x}_k$, $\quad \mathbf{y}_k = \nabla f(\mathbf{x}_{k+1}) - \nabla f(\mathbf{x}_k)$
8. Compute H_{k+1} from H_k by using the following PSB update [16]:

$$H_{k+1} = H_k + \frac{(\mathbf{y}_k - H_k \mathbf{s}_k)\mathbf{s}_k^T + \mathbf{s}_k(\mathbf{y}_k - H_k \mathbf{s}_k)^T}{\mathbf{s}_k^T \mathbf{s}_k} - \frac{< \mathbf{y}_k - H_k \mathbf{s}_k, \mathbf{s}_k > \mathbf{s}_k \mathbf{s}_k^T}{(\mathbf{s}_k^T \mathbf{s}_k)^2} \tag{12}$$

Note that the added functions are modified whenever the Newton point is recalculated, i.e., the modifications of the objective function are adaptable in each iteration. Therefore, the proposed method is more effective than the MOM, which updates the Lagrange multipliers after the final minimizer has been found for a fixed set of Lagrange multipliers.

A method proposed by Lin, Shiau and Huang [19] has been used to find the Newton point $\mathbf{x}_N$ from the perturbed Cholesky decomposition matrices obtained in step 1. By adding an extra column and row to the original Hessian matrix for each modified diagonal element, equation (6′), rewritten as $H_k \Delta\mathbf{x} = \mathbf{P}$, where $\Delta\mathbf{x} = \mathbf{x}_N - \mathbf{x}_k$ and $\mathbf{P} = -\nabla f_k$, can be expanded to the following form:

$$\begin{pmatrix} h_{11} & \cdots & h_{1i} & \cdots & h_{1j} & \cdots & h_{1n} & & \\ \vdots & \ddots & \vdots & \ddots & \vdots & \ddots & \vdots & & \\ h_{i1} & \cdots & h_{ii}+d_1 & \cdots & h_{ij} & \cdots & h_{in} & -d_1 & \\ \vdots & \ddots & \vdots & \ddots & \vdots & \ddots & \vdots & & \\ h_{j1} & \cdots & h_{ji} & \cdots & h_{jj}+d_2 & \cdots & h_{jn} & & -d_2 \\ \vdots & \ddots & \vdots & \ddots & \vdots & \ddots & \vdots & & \\ h_{n1} & \cdots & h_{ni} & \cdots & h_{nj} & \cdots & h_{nn} & & \\ & & -d_1 & & & & & d_1 & \\ & & & & -d_2 & & & & d_2 \end{pmatrix} \begin{Bmatrix} \delta x_1 \\ \vdots \\ \delta x_i \\ \vdots \\ \delta x_j \\ \vdots \\ \delta x_n \\ \delta x_{n+1} \\ \delta x_{n+2} \end{Bmatrix} = \begin{Bmatrix} p_1 \\ \vdots \\ p_i \\ \vdots \\ p_j \\ \vdots \\ p_n \\ 0 \\ 0 \end{Bmatrix} \tag{13}$$

where n is the order of the matrix H_k. By noting that $\delta x_{n+1} = \delta x_i$ and $\delta x_{n+2} = \delta x_j$, it is clear that the expanded equation has the same solution as $H_k \Delta\mathbf{x} = \mathbf{P}$ for the original n-variables. Note that no interchange of rows or columns is required, thereby preserving the sparseness of matrix H_k. Such a scheme is very suitable for large systems.

NUMERICAL EXAMPLES

To illustrate the applicabilit y of the PSB update, three examples are solved by two methods of the determination of the Hessian matrix H. The method 1 uses the Hessian matrix computed by $\nabla^2 f$. The method 2 uses the Hessian matrix computed by PSB update.

The number of iterations and the number of function calls for these examples using the proposed methods are shown in Tables 1, 3 and 5. The function values at each iteration for these examples are shown both in Tables 2, 4 and 6 and in Figures 1-3.

Example 1: (Source: [20])

$$\begin{aligned}
\text{Minimize}: \quad & f(x_1,x_2) = (x_1-2)^2 + (x_2-1)^2 \\
\text{Subject to}: \quad & x_1^2 - x_2 \leq 0 \\
& x_1 + x_2 - 2 \leq 0 \\
\text{Lagrangian}: \quad & L(x_1,x_2,s_1,s_2,\lambda_1,\lambda_2) = (x_1-2)^2 + (x_2-1)^2 \\
& \qquad + \lambda_1(x_1^2 - x_2 + s_1^2) + \lambda_2(x_1 + x_2 - 2 + s_2^2) \\
\text{Starting point}: \quad & (2,2,0.5,0.5,1,1) \\
\text{Solution}: \quad & L(1,1,0,0,.471404,1.37437) = 1.00000
\end{aligned}$$

TABLE 1
Example 1 – Number of iterations and function calls

Method	No. of Iterations	No. of Function Calls		Function Values
		For H & ∇f	Linesearch	
1	5	230	26	1.000000
2	12	216	66	1.000000

P.S.: Starting Point (2,2,.5,.5,1,1)

TABLE 2
Example 1 – Function values

Iteration No.	Method 1	Method 2
0	5.50	5.50
1	0.951863	0.951725
2	0.983471	0.937508
3	0.999961	1.016810
4	1.000000	1.000580
5	1.000000	0.999045
6		1.000170
7		0.999993
8 – 12		1.000000

P.S.: Starting Point (2,2,.5,.5,1,1)

Example 2: (Source: [20])

$$\begin{aligned}
\text{Minimize:} \quad & f(x_1, x_2) = (x_1 - 2)^2 + (x_2 - 1)^2 \\
\text{Subject to:} \quad & x_1 - 2x_2 + 1 = 0 \\
& x_1^2/4 + x_2^2 - 1 \leq 0 \\
\text{Lagrangian:} \quad & L(x_1, x_2, s_1, \lambda_1, \lambda_2) = (x_1 - 2)^2 + (x_2 - 1)^2 \\
& \quad + \lambda_1(x_1 - 2x_2 + 1) + \lambda_2(x_1^2/4 + x_2^2 - 1 + s_1^2) \\
\text{Starting point:} \quad & (2, 2, 0.5, 1, 1) \\
\text{Solution:} \quad & L(.822876, .911438, 0, 1.78270, 3.80684) = 1.39346
\end{aligned}$$

TABLE 3
Example 2 – Number of iterations and function calls

Method	No. of Iterations	No. of Function Calls		Function Values
		For H & ∇f	Linesearch	
1	6	216	31	1.39346
2	20	291	128	1.39346

P.S.: Starting Point (2,2,.5,1,1)

TABLE 4
Example 2 – Function values

Iteration No.	Method 1	Method 2
0	4.250	4.250
1	0.787921	0.787921
2	1.231570	0.930800
3	1.380150	1.390830
4	1.393310	1.363740
5	1.393470	1.353920
6	1.393460	1.370370
7		1.388440
8		1.391440
9		1.392330
10		1.392660
11		1.393470
12 – 20		1.393460

P.S.: Starting Point (2,2,.5,1,1)

Example 3: (Source: [11])

$$\begin{aligned}
\text{Minimize}: \quad & f(\mathbf{x}) = (x_1-1)^2 + (x_1-x_2)^2 + (x_2-x_3)^2 \\
& \qquad + (x_3-x_4)^4 + (x_4-x_5)^4 \\
\text{Subject to}: \quad & x_1 + x_2^2 + x_3^3 - 2 - 3\sqrt{2} = 0 \\
& x_2 - x_3^2 + x_4 + 2 - 2\sqrt{2} = 0 \\
& x_1 x_5 - 2 = 0 \\
\text{Lagrangian}: \quad & L(\mathbf{x}, \lambda_1, \lambda_2, \lambda_3) = f(\mathbf{x}) + \lambda_1(x_1 + x_2^2 + x_3^3 - 2 - 3\sqrt{2}) \\
& \qquad + \lambda_2(x_2 - x_3^2 + x_4 + 2 - 2\sqrt{2}) + \lambda_3(x_1 x_5 - 2) \\
\text{Starting point}: \quad & (2,2,2,2,2,1,1,1) \\
\text{Solution}: \quad & L(1.1911, 1.3626, 1.4728, 1.6350, 1.6790, \\
& \qquad -.24629, -.035482, -.00040847) = 0.0787768
\end{aligned}$$

TABLE 5
Example 3 – Number of iterations and function calls

Method	No. of Iterations	No. of Function Calls		Function Values
		For H & ∇f	Linesearch	
1	6	487	35	0.0787768
2	16	372	109	0.0787768

P.S.: Starting Point (2,2,2,2,2,1,1,1)

TABLE 6
Example 3 – Function values

Iteration No.	Method 1	Method 2
0	1.000	1.000
1	0.2100870	0.2005900
2	0.0928118	0.2076200
3	0.0791606	0.0882852
4	0.0787768	0.0786638
5	0.0787768	0.0793041
6	0.0787768	0.0789924
7		0.0788255
8		0.0787820
9		0.0787769
10		0.0787770
11 - 16		0.0787768

P.S.: Starting Point (2,2,2,2,2,1,1,1)

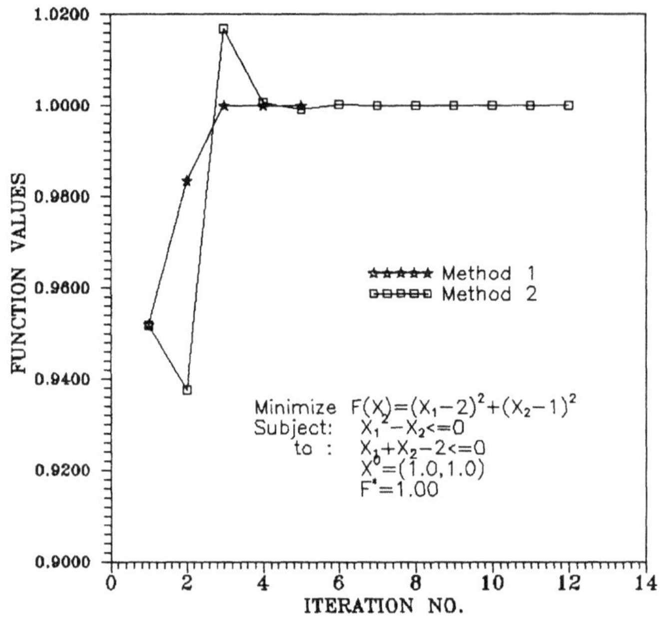

Fig.1 Function values of numerical example 1

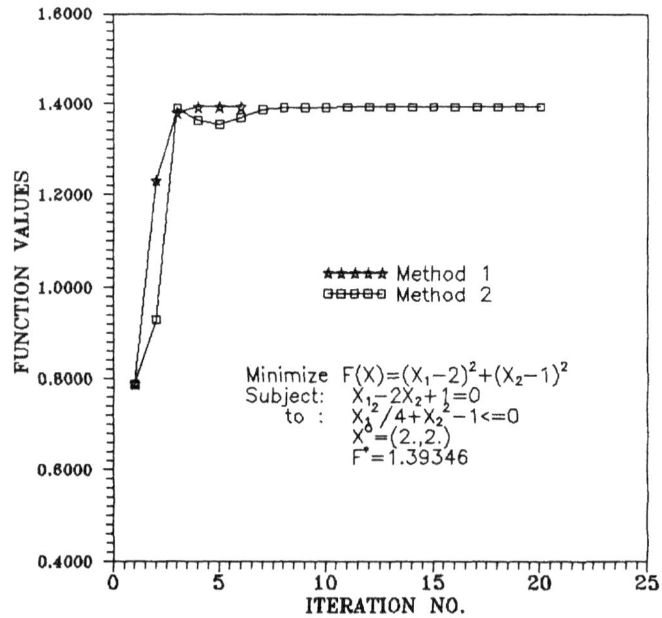

Fig.2 Function values of numerical example 2

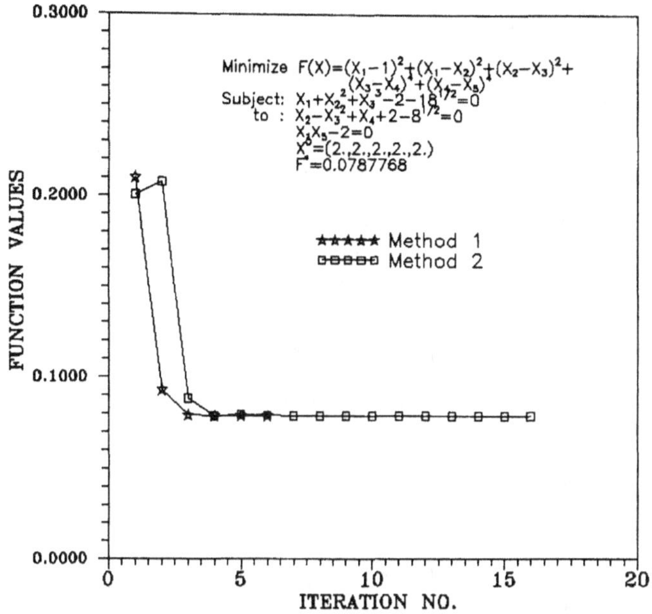

Fig.3 Function values of numerical example 3

DISCUSSION

From the numerical results listed in Tables 1-3, we can observe that the proposed procedure based on the PSB update can find the optimal solutions of all three constrained optimization problems very efficiently. The PSB update requires a few more iterations to get the same accuracy as the standard Newton method. But the saving on the direct computation of the Hessian matrix is very obvious.

By using the one-dimensional search along a curve defined by a Bezier polynomial, the method is globally convergent and locally q-quadratically convergent. From Figures 1-3 we can see that the proposed method approaches near-minima in just three or four iterations.

Note also that the proposed method can find the exact optimal solution in just one iteration for a quadratic objective function subject to linear equality constraints.

REFERENCES

1. Box, M. J., A new method of constrained optimization and a comparison with other methods, Computer J., Vol.8, 1965, pp.42-52.

2. Box, M. J., A comparison of several current optimization methods and the use of transformation in constrained problems, Computer J., Vol.9, 1966, pp.67-77.

3. Cassis, J. H. and Schmit, L. A., On Implementation of the Extended Interior Penalty Function, Int. J. Num. Meth. Eng., Vol.10, 1976, pp.3-23.

4. Fiacco, A. V. and McCormick, G. P., The sequential unconstrained minimization technique for nonlinear programming, a primal dual method, Management Sci. J.,

Vol.10, 1964, pp.360-366.

5. Haltka, R. T. and Starnes, J. H., Application of a Quadratic Extended Interior Penalty Function for Structural Optimization, AIAA J., Vol.14, 1976, pp.718-724.

6. Lin, T.W., Well Behaved Penalty Functions for Constrained Optimization, Journal of the Chinese Institute of Engineers, Vol.13, No.2, 1990, pp.157-166.

7. Hestenes, M.R., Multipliers and Gradient Methods, in Computing Methods in Optimization Problems, Vol.2, eds. L.A. Zadeh, L.W. Neustadt, A.V. Balakrishnan, Academic Press, New York, 1969, pp.143-163.

8. Powell, M.J.D., A Method for Nonlinear Constraints in Minimization Problems, Opti- mization, ed. Fletcher, R., Academic Press, New York, 1969, pp.283-298.

9. Fletcher, R., An Exact Penalty Function for Nonlinear Programming with Inequalities, Math. Programming, Vol.5, 1973, pp.129-150.

10. Fletcher, R., Methods Related to Lagrangian Functions, Numerical Methods for Constrained Optimization, eds. P.E. Gill and W. Murray, Academic Press, New York, 1974, pp.219-239.

11. Pillo, G. Di and L. Grippo, A New Class of Augmented Lagrangians in Nonlinear Programming, SIAM J. Control Optim., Vol.17, 1979, pp.618-628.

12. Pillo, G. Di and L. Grippo, An Augmented Lagrangian for Inequality Constraints in Nonlinear Programming Problems, J. Optim. Theory Appl., Vol.36, 1982, pp.495-519.

13. Lucidi, S., New Results on a Class of Exact Augmented Lagrangians, J. Optim. Theory Appl., Vol.58, 1988, pp.259-282.

14. Vinante, C. and S. Pintos, On Differentiable Exact Penalty Functions, J. Optim. Theory Appl., Vol.50, 1986, pp.479-493.

15. Lin, T.W., Huang, J.T. and Joung, M.H., Adaptive Modification of Objective Function for Lagrange Multiplier Method in Constrained Optimization Problems, Submitted to Journal of the Chinese Institute of Engineers.

16. Powell, M.J.D., A New Algorithm for Unconstrained Optimization, in Nonlinear Programming, eds. J.B. Rosen, O.L. Mangasarian and K. Ritter, Academic Press, New York, 1970, pp.31-65.

17. Gill, P.E., Murray, W. and Wright, M.H., Practical Optimization, Academic Press, London, 1981.

18. Lin, T.W., Joung, M.H. and Huang, J.T., Solving Optimization Problems by Searching Along a Curve, Accepted for publication in Journal of the Chinese Institute of Engineers.

19. Lin, T.W., Shiau, H.T. and Huang, J.T., Decomposition of Singular Large Sparse Matrix by Adding Dummy Links and Dummy Degrees, Submitted to Journal of the Chinese Institute of Engineers.

20. Brocken, J. and McCormick, G.P., Selected Applications of Nonlinear Programming, John Wiley & Sons, Inc., New York, 1968.

COMPUTATIONAL DAMAGE ANALYSIS AND RELIABILITY ASSESSMENT

A. H-S. ANG
University of California
Irvine, California, U.S.A.

ABSTRACT

Computational methods for structural damage analysis and collapse reliability assessment are summarized. A seismic damage model for reinforced concrete buildings is formulated as a function of the maximum structural deformation and dissipated hysteretic energy. Such a model has been calibrated with observed damages from past earthquakes. Reliability against ultimate collapse can be evaluated effectively with the stable configuration approach. Numerical calculation of the required probabilities can be accomplished through an adaptive Kernel method of importance sampling.

INTRODUCTION

In linear and nonlinear structural analysis, the properties of the elements or components of a structure are the basic building blocks from which the prediction of the behavior or response of the structure is determined through theoretical analysis. Following the same premise, the damage and collapse analyses of structures may also be determined on the basis of damage and failure information of the respective components. Indeed, a theory of structural damage analysis may be formulated on this basis. The concepts for this latter purpose are reviewed and developed in terms of probability and discussed with regard to seismic damage assessment.

In general, the ultimate collapse of a structure may be modeled as a brittle mode of failure; i.e. upon collapse the load-carrying capacity is reduced to zero. On this basis, the system reliability against collapse of a structure can be effectively assessed through the stable configuration approach, which is the dual formulation of the failure mode approach. An efficient importance sampling algorithm is also presented for the numerical calculation of the required probabilities.

DAMAGE MODELING AND ANALYSIS

One of the purposes of modeling and systems analysis is to estimate the potential damage of a structure to an applied load or disturbance and thus assess its safety against such a load. In this regard, the ductility ratio may be a measure of structural damage; indeed, this has been used widely for this purpose. The ductility ratio, which is a measure of the maximum deformation of a structure, is a major contributor to its total damage; however, damage can be induced also through the oscillatory response of a structure under repeated load applications. Damage models that include the effect of repeated cyclic loadings have been proposed (e.g. Banon and Veneziano, 1982; Stephens and Yao, 1987; and Park and Ang, 1985).

In Park and Ang (1985), the total damage is defined in terms of an index

$$D = \frac{\delta_{max}}{\delta_u} + \frac{\beta}{Q_y \delta_u} \int dE \tag{1}$$

where: δ_{max} = the maximum response deformation;
$\int dE$ = the dissipated hysteretic energy;
δ_u = the ultimate static deformation;
Q_y = the static yield strength; and
β = a normalizing constant, dependent on the material.

In Eq. 1, the first term is the damage caused by maximum deformation (i.e. synonymous with the ductility ratio), whereas the second term represents the damage caused by the absorbed hysteretic energy. Increasing values of D, of course, represent increasing degrees of damage; $D = 0$ means no damage, and $D \geq 1.0$ means total collapse.

For structural damage assessment, the nonlinear-inelastic response of a structure (including the range approaching collapse) is obviously necessary. For this purpose, and in order to accommodate the oscillatory or repeated loadings, a nonlinear-hysteretic restoring force model of the structure is required. Several such models have been suggested; e.g. Clough and Johnston (1966) for reinforced concrete, and Kato and Akiyama (1973) for steel structures. The total restoring force may be represented as

$$g = \alpha x + (1 - \alpha) z \tag{2}$$

where: x = the nonhysteretic component of response,
z = the hysteretic component of response, and
α = ratio of post-yield to pre-yield stiffnesses.

Differential equation forms for z have also been suggested (e.g. as discussed in Suzuki and Minai, 1987). A large class of smooth hysteresis can be represented by the following differential equation model (Wen, 1976),

$$\dot{z} = \frac{1}{\eta}[A\dot{x} - \nu(\beta|\dot{x}||z|^{n-1}z - \gamma\dot{x}|z|^{n})] \tag{3}$$

where the parameters A, β, γ, η, ν, and n define the amplitude, hysteretic loop shape, and inelastic properties. Both softening and hardening inelasticities can be described with Eq. 3. Also, degradation in strength and/or stiffness can be included through the parameter A; e.g.

$$A(t) = A_o - \delta_A \epsilon(t) \tag{4}$$

where: δ_A = rate of degradation, and

$\epsilon(t)$ = $(1-\alpha)k \int_0^t z(\tau)x(\tau)d\tau$, the dissipated hysteretic energy.

Degradation may also be a function of the response amplitude (Sues et al. 1985).

The above smooth hysteretic restoring force model has been generalized for two directions. To include the interaction in the two directions, the hysteretic components, z_x and z_y, are described by the following coupled differential equations:

$$\dot{z}_x = A\dot{u}_x - \beta|\dot{u}_x z_x|z_x - \tau\dot{u}_x z_x^2 - \beta|\dot{u}_y z_y|z_x - \tau\dot{u}_y z_x z_y \tag{5}$$

$$\dot{z}_y = A\dot{u}_y - \beta|\dot{u}_y z_y|z_y - \tau\dot{u}_y z_y^2 - \beta|\dot{u}_x z_x|z_y - \tau\dot{u}_x z_y z_x \tag{6}$$

where u_x and u_y are the displacements in the x and y directions, respectively. In this case, the hysteretic energy is defined by

$$\epsilon(t) = \int_0^t [z_x(\tau)\dot{u}_x(\tau) + z_y(\tau)\dot{u}_y(\tau)]d\tau \tag{7}$$

The versatility and capability of the above restoring force models have been demonstrated, including for complex force-displacement relationships of reinforced concrete structures. For example, in Fig. 1 are shown the experimental results for a reinforced concrete frame and corresponding analytical representation with Eq. 3, whereas similar comparison of results for biaxial loadings of a reinforced concrete column are shown in Fig. 2.

Response Statistics

With a differential equation form of the restoring force, and modeling the earthquake excitation with a filtered shot noise, the nonlinear stochastic response analysis necessary for damage analysis may be formulated on the basis of the Markov vector process, in which the transition probability is governed by the Fokker-Planck equation. Theoretical solutions to the resulting partial differential equation for the transition probabilities, however, are difficult to obtain and are limited to single-degree-of-freedom systems (e.g. Wen, 1976).

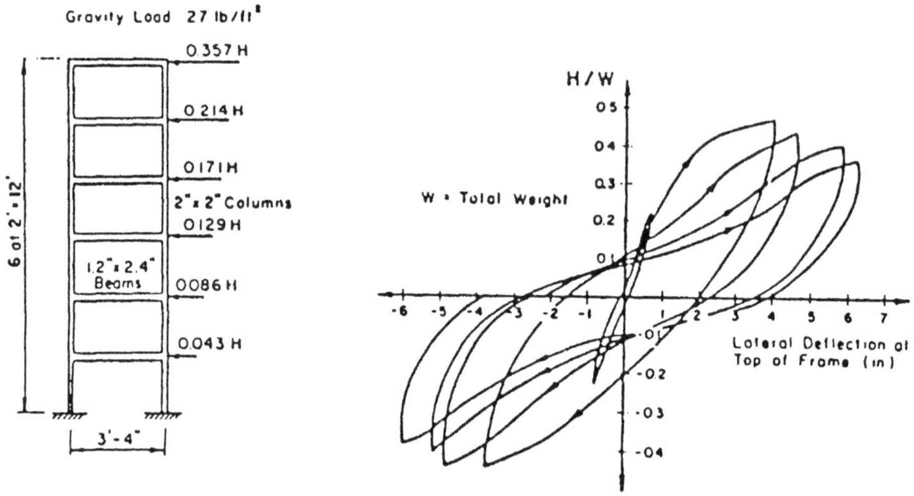

a) Experimental Load-Deflection Curves for Reinforced Concrete Building Model

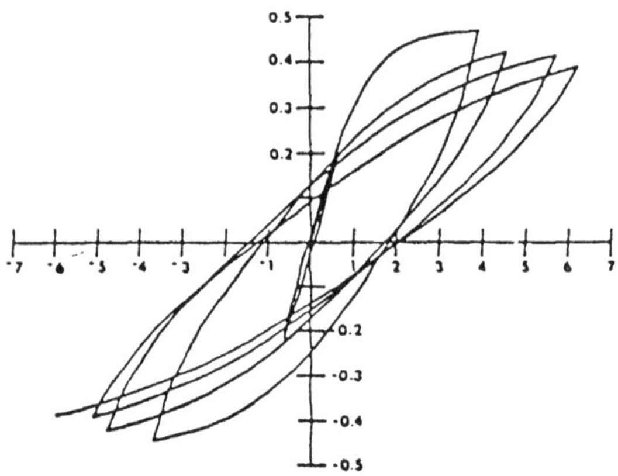

(b) Hysteretic Curves from Displacement-Based Stiffness Degrading Model

Fig. 1 Comparison of Test (Park and Paulay, 1975) and Analytical Hysteretic Curves

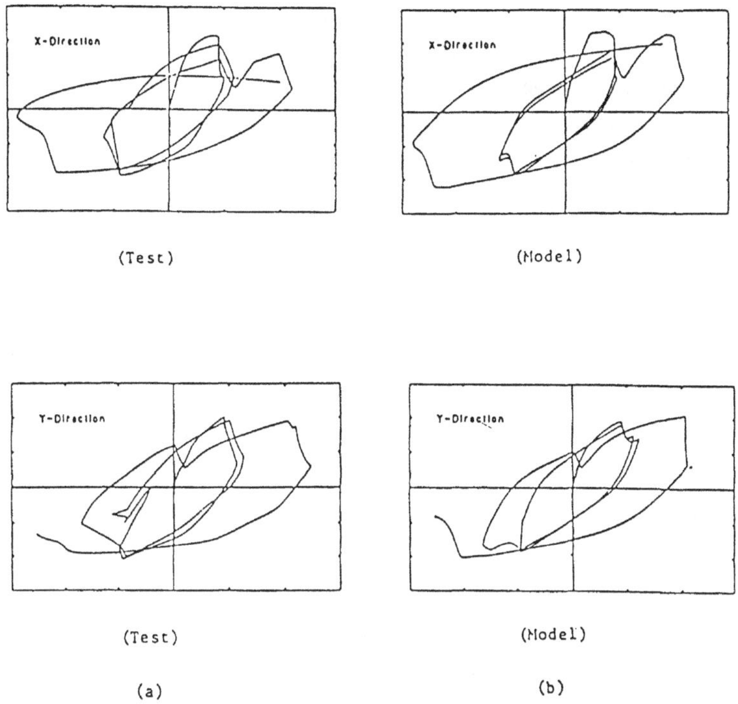

Fig. 2 Comparison of Test (Takizawa and Aoyama, 1976) and Analytical Bi-axial Hysteretic Curves

With the smooth hysteresis model of Eq. 3, approximate solutions can be obtained through stochastic linearization, including solutions for degrading multi-degree-of-freedom systems which are necessary for damage assessment. Briefly, the restoring force of Eq. 3 can be linearized as

$$\dot{z} + C_1 \dot{x} + C_2 z = 0 \tag{8}$$

where the coefficients C_1 and C_2 are evaluated on the basis of minimum mean squared error; for example, for Gaussian excitation and response,

$$C_1 = \sqrt{\frac{2}{\pi}} \left[\tau \frac{E(\dot{x}z)}{\sigma_{\dot{x}}} + \beta \sigma_z \right] - A \tag{9}$$

$$C_2 = \sqrt{\frac{2}{\pi}} \left[\tau \sigma_{\dot{x}} + \beta \frac{E(x\dot{z})}{\sigma_z} \right] \tag{10}$$

where E(xz) represents the covariance of x and z. Eqs. 9 and 10 and the equation of motion then describe the equivalent linearized system of governing equations for a SDF system. The response covariance matrix, S, of the system response variables (x, $\dot{x}$, and z) is governed by the following matrix ordinary differential equation:

$$\frac{dS}{dt} = GS + SG^t + B \tag{11}$$

where t stands for transpose, and

G = matrix describing the structural system, including the linearization coefficients and excitation filter parameters; and

B = matrix of the response vector and excitation.

Because the linearization coefficients are functions of the response statistics, iterative numerical solution for the covariances is generally required.

The accuracy and practical effectiveness of the above linearization technique have been demonstrated extensively (e.g. Baber and Wen, 1981; Asano, 1987).

According to Eq. 1 the seismic response statistics needed for damage assessment are the mean and variances of the maximum deformation of a structure, δ_{max}, and of the dissipated or absorbed hysteretic energy, $\int dE$. These statistics can be evaluated from the covariances obtained as solutions to Eq. 11.

Damage Assessment of Reinforced Concrete Buildings

The above damage model has been applied to reinforced concrete structures, particularly buildings. In this regard, the damage indices of over 400 reinforced concrete members (consisting of beams and beam-columns) that were tested to final failure were evaluated; the specimens were subjected to monotonic or repeated cyclic loadings, and under static and dynamic conditions. The combined results show that at failure, the damage index can be described by a lognormal distribution with a median of 1.0 and a c.o.v. of approximately 50%; these results are shown graphically in Fig. 3. On this basis, the damage index may be assumed to have a lognormal distribution.

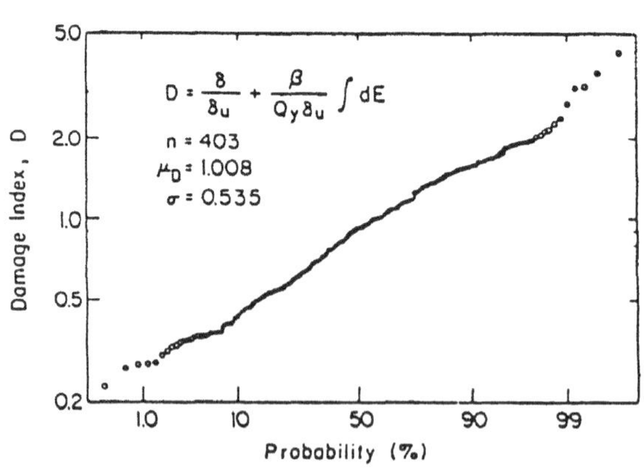

Fig. 3 Damage Index, D, Plotted on Lognormal Probability Paper

Therefore, the cumulative distribution function (CDF) of the damage at the collapse stage is

$$F_{D_i}(d_u) = \Phi\left(\frac{ln\, d_u - ln\, 1.0}{0.50}\right) = \Phi\left(2\, ln\, d_u\right) \tag{12}$$

For a building with n stories, the damage of the building may be defined as

$$(D \le d) = \prod_i (D_i \le d) \tag{13}$$

i.e. for the building to have damage $D \le d$, all stories must have $D_i \le d$. Then, the CDF of the building damage is,

$$F_D(d) = P[\prod_i (D_i \le d)] \tag{14}$$

where D_i is evaluated with the procedure described above.

At collapse, if the story damages are uniformly distributed and statistically independent,

$$F_D(d_u) = [\,\Phi(2\, ln\, d_u)]^n \tag{15}$$

This distribution will tend to the Type II distribution of the largest value for large n.

Calibration of Calculated Damage

In order to specify or define a tolerable degree of damage, the damage index must be related to actual observed damages. For this purpose, the damage index of Eq. 1 is calibrated with respect to the damages of buildings observed from past earthquakes; in particular, the 1971 San Fernando and 1978 Miyagiken-Oki earthquakes. The buildings are listed in Table 1 with the observed seismic damage descriptions. The building details and results of extensive damage inspection of these buildings are available in U.S. Department of Commerce (1973) for building A, and Architectural Institute of Japan (1980) for buildings B through I. Table 1 also gives the calculated overall damage indices D_T and the corresponding ground motion intensity in terms of I_c (Park, Ang, and Wen, 1985). The same results are also summarized in Fig. 4, which are used as the basis for calibrating the damage index D.

Based on the calibration results below and the subsequent disposition of the buildings that were damaged during these two earthquakes, it appears that R. C. structures with $D_T < 0.4$ can be repaired, whereas those with $D_T \ge 0.4$ must be demolished. Therefore, a limiting (or tolerable) seismic structural damage may be specified at $D_T = 0.4$.

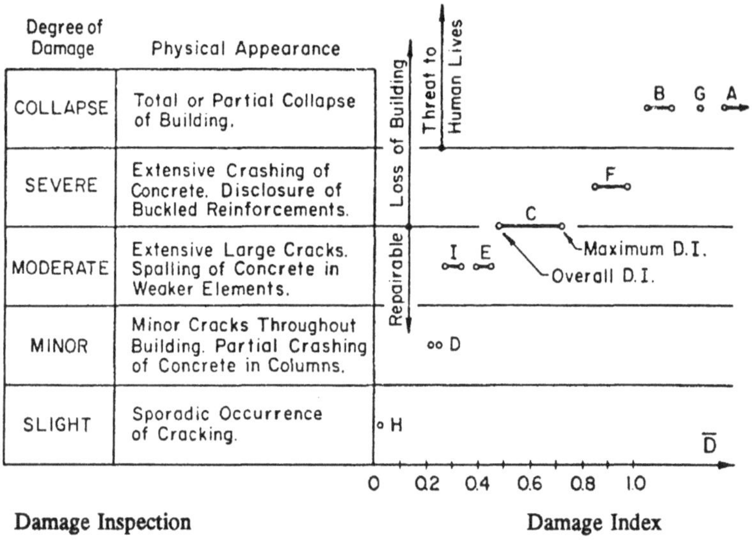

Figure 4. Comparison of Calculated Damage Index with Damage Inspection

TABLE 1
List of Damaged Buildings

Name of Buildings	Number of Stories	Observed Damage	Ground Motion I_c	Damage Index D_T
Olive View Hospital	6	Collapse*	0.31	1.47
Taiyo Gyogyo Building	3	Collapse*	0.23	1.05
Tohoku Togyo University	4	Moderate*	0.12	0.48
Saigo School	2	Minor	0.19	0.22
Tonan High School	3	Moderate*	0.19	0.39
Kinoshita Menko Building	3	Severe*	0.23	0.85
Obisan Office Building	3	Collapse*	0.23	1.25
Fukushi Kaikan Building	2	Very Minor	0.12	0.02
Izumi High School	3	Minor	0.23	0.27

*Subsequently demolished

COLLAPSE ANALYSIS

Collapse of structures may occur gradually such as of ductile systems or suddenly as represented by brittle fracture. Collapse of perfectly plastic systems through plastic hinge mechanisms is a very idealized type of behavior that can seldom be realized in real structures. More generally, ultimate collapse of a structure or structural component is characterized by the loss of its load-carrying capacity, and therefore may be modeled as <u>brittle</u> in the sense that upon failure the structural capacity is reduced to zero.

The analysis of the reliability against collapse of a brittle system can be accomplished most effectively with the stable configuration approach, as presented below.

The Stable Configuration Approach

The stable configuration approach (SCA) has been shown to be particularly effective for brittle systems (Quek and Ang, 1990); in contrast to the failure mode approach (FMA) which depends on the failure sequences of the components, resulting in a very large number of potential failure modes for brittle systems; the SCA is independent of such failure sequences. Moreover, as illustrated below, only the few dominant configurations are essential with the SCA to obtain useful results for practical systems. Also, it is of significance to observe that the FMA will invariably underestimate the failure probability (i.e. unconservative), whereas with the SCA any approximation in the estimated failure probability will invariably be conservative.

The Failure Graph

In general, a stable configuration is the practical (approximate) equivalence of a cut-set in a theoretical graph network. In certain cases, however, the equivalence is exact; for example, for the three parallel bar system of Fig. 5, the failure graph is represented in Fig. 6 with the following notations:

Node -- represents a damaged state of the structure;

Branch -- represents failure of a component.

The initial node, O, therefore represents the initial (undamaged) structure, whereas the terminal node represents the collapse state of the system. An intermediate node represents a partially damaged state of the system.

Any path, consisting of branches connecting the initial node to the terminal node, is a potential failure mode of the structure. In Fig. 6, the path $\pi_1 = B_1 \cap B_{2|1} \cap B_{3|1,2}$ represents the collapse of the system in Fig. 5 in which the three bars fail sequentially in the order $1 \rightarrow 2 \rightarrow 3$. Similarly, the path $\pi_2 = B_2 \cap B_{1|2} \cap B_{3|1,2}$ represents the collapse in which the same three bars fail sequentially in the order $2 \rightarrow 1 \rightarrow 3$. In fact, the six different sequences in which the three bars can fail constitute six possible paths $\pi_1, \pi_2, \ldots, \pi_6$. The system will collapse if any of the potential paths is realized; thus the collapse of the system is $E = \pi_1 \cup \pi_2 \cup \pi_3 \cup \pi_4 \cup \pi_5 \cup \pi_6$.

A cut-set contains one branch from each path of the network, and represents the damage (i.e. failure of one or more components) to the corresponding configuration. For example, in Fig. 6, the cut-set $\bar{C}_0 = B_1 \cup B_2 \cup B_3$ is the damage to the original system, whereas $\bar{C}_1 = B_2 \cup B_3 \cup B_{2|1} \cup B_{3|1} = B_{2|1} \cup B_{3|1}$, is equivalent to the damage of the sub-configuration with bar #1 removed.

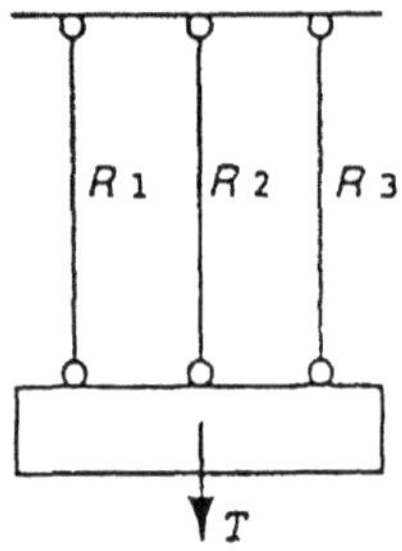

Figure 5. Three Parallel Bar Structure

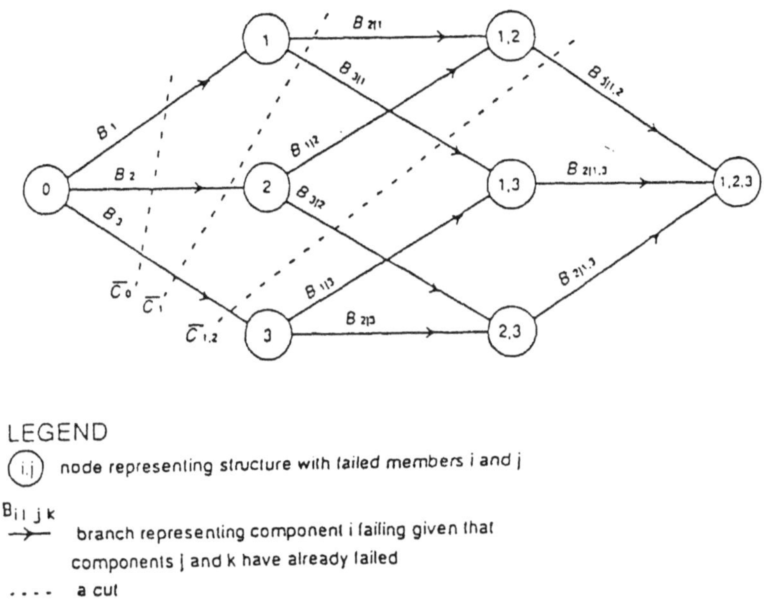

Figure 6. Failure Graph for Three Bar Structure (from Quek and Ang, 1990)

Finally, in this example, $\bar{C}_{1,2} = B_3 \cup B_{3|2} \cup B_{3|1} \cup B_{3|1,2} = B_{3|1,2}$, which is the damage to the sub-configuration with the single bar #3 remaining. For this system, all the possible configurations are given in Table 2. The system will collapse if all the possible configurations (i.e. sub-systems) are damaged; therefore, in this case,

$$E = \bar{C}_0 \cap \bar{C}_1 \cap \bar{C}_2 \cap \bar{C}_3 \cap \bar{C}_{1,2} \cap \bar{C}_{1,3} \cap \bar{C}_{2,3}$$

In general, the collapse of a system is $E = \cap \bar{C}_i$, which is the intersection of all cut-sets or damages to all possible configurations of a system.

TABLE 2
Damage of Configurations

Configurations	Performance Functions	Component Failures
$\bar{C}_0$	$g_i(\underline{X}) = R_i - T/3$	$B_1 \cup B_2 \cup B_3$
$\bar{C}_1$		$(B_2 \cup B_3) \cup B_{2\|1} \cup B_{3\|1}) = B_{2\|1} \cup B_{3\|1}$
$\bar{C}_2$	$g_i(\underline{X}) = R_i - T/2$	$B_{1\|2} \cup B_{3\|2}$
$\bar{C}_3$		$B_{1\|3} \cup B_{2\|3}$
$\bar{C}_{1,2}$		$B_3 \cup B_{3\|1} \cup B_{3\|2} \cup B_{3\|1,2} = B_{3\|1,2}$
$\bar{C}_{2,3}$	$g_i(\underline{X}) = R_i - T$	$B_{1\|2,3}$
$\bar{C}_{1,3}$		$B_{2\|1,3}$

Each cut-set involves the union of branches (component failures), and thus its probability is

$$P(\bar{C}_i) = P(B_{1i} \cup B_{2i} \cup ... \cup B_{ki}) \tag{16}$$

For structural purposes (with low component failure probabilities), the geometric mean of the second-order bounds (Ang and Tang, 1984) of $P(\bar{C}_i)$ may be used.

The collapse probability can be expanded as

$$\begin{aligned} P(E) &= P(\cap \bar{C}_i) \\ &= \sum_{i=1}^{n} P(\bar{C}_i) - \sum_{i=1}^{n-1} \sum_{j>i}^{n} P(\bar{C}_i \cup \bar{C}_j) \\ &\quad + \sum_{i=1}^{n-2} \sum_{j>1}^{n-1} \sum_{k>j}^{n} P(\bar{C}_i \cup \bar{C}_j \cup \bar{C}_k) - .... \end{aligned} \tag{17}$$

where n is the number of cut-sets. Quek and Ang (1990) has shown that for structural problems, the initial configuration plus those with one failed component give sufficiently accurate results -- this is also demonstrated below for the 3-bar system. On this basis, Eq. 17 is reduced to

$$P(\cap \bar{C}_i) = \sum_{i=0}^{k} P(\bar{C}_i) - \sum_{j=1}^{k} P(\bar{C}_0 \cup \bar{C}_j) \tag{18}$$

where: $\bar{C}_0$ = damage to the original structure;

$\bar{C}_i$ = further damage to the configuration with one failed component.

Numerical Illustration

Consider the brittle system of three parallel bars of Fig. 5 subjected to the load T; the statistics of the load and fracture resistances of the individual bars are shown in Table 3.

TABLE 3
Load and Resistances for the Three Bar System

Load T		Resistances R_1, R_2, R_3	
Mean(kips)	COV	Mean(kips)	COV
120	0.25	75	0.15

The performance functions and component failures involved in the respective configuration damages are summarized in Table 2. For this example, the damage probabilities of the pertinent configurations are as follows:

$$P(\bar{C}_0) = 0.0275$$
$$P(\bar{C}_i) = 0.3109; \quad i = 1,2,3$$
$$P(\bar{C}_0 \cup \bar{C}_j) = 0.3121; \quad j = 1,2,3$$

Progressively, Eq. 18 would then yield the following collapse probabilities:

Considering only the damage to the initial configuration,

$$P(\bar{C}_0) = 0.0275$$

whereas, by including the configurations with one failed component, Eq. 3 yields

$$P(\bar{C}_0\ \bar{C}_1\ \bar{C}_2\ \bar{C}_3) = 0.0239$$

Finally, by including all possible configurations,

$$P(\bar{C}_0\ \bar{C}_1\ \bar{C}_2\ \bar{C}_3\ \bar{C}_{1,2}\ \bar{C}_{1,3}\ \bar{C}_{2,3}) = 0.0239$$

The corresponding failure surfaces for the above events are shown in Fig. 7. Therefore, including configurations beyond those with one failed component does not improve the results. Hence, the collapse probability for this system is P(E) = 0.0239.

PROBABILITY COMPUTATIONS

In many reliability problems, the calculation of the pertinent probabilities often require numerical sampling or Monte Carlo methods. This includes the evaluation of the probability of intersection of cut-sets or stable configurations in the collapse analysis of structural systems. The basic Monte Carlo method, however, is often inefficient and expensive, particularly when very small failure probabilities are involved such that very large samples are required for sufficiently precise determination of the desired probability. To overcome this weakness, various variance reduction techniques have been proposed. One such technique is the importance sampling method, which is of special significance in structural reliability.

For time-invariant problems, the probability of failure of a structure can be given as

$$P_F = \int_{D_f} f_{\underline{X}}(\underline{x})d\underline{x} \tag{19}$$

which is the volume integral in the $\underline{X}$ space over the failure domain D_f, and $f_{\underline{X}}(\underline{x})$ is the joint probability density function (PDF) of the random variables $\underline{X}$.

Eq. 19 can be rewritten as

$$P_F = \int_{D_f} I[g(\underline{x})] \cdot f_{\underline{X}}(\underline{x})d\underline{x} \tag{20}$$

where:

$$I[g(\underline{x})] = \begin{cases} 1 & \textit{if } g(\underline{x}) \le 0 \\ 0 & \textit{if } g(\underline{x}) > 0 \end{cases}$$

in which $g(\underline{x})$ is the performance function. Then, by Monte Carlo sampling, the probability of failure is estimated as

$$P_F = \frac{1}{N}\sum_{i=1}^{N} I[g(\underline{x}_i)] \tag{21}$$

where N is the number of Monte Carlo samples.

If P_F is small, N has to be very large to obtain sufficient number of failure samples satisfying $g(\underline{x}_i) \le 0$. To increase the number of failure samples for given N, the sampling can be biased toward the failure domain as follows:

$$P_F = \int_{D_f} I[g(\underline{x})] \cdot \frac{f_{\underline{X}}(\underline{x})}{h_{\underline{X}}(\underline{x})} h_{\underline{X}}(\underline{x})d\underline{x} \tag{22}$$

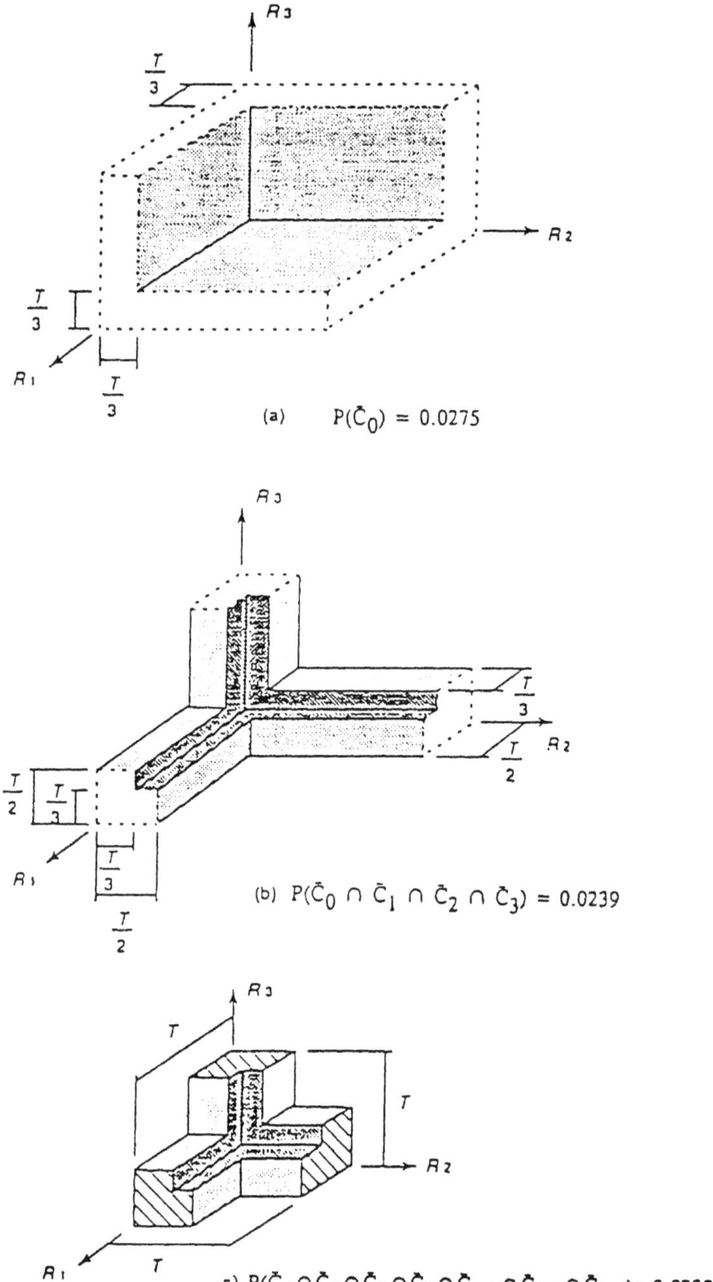

Figure 7. Limit State Surfaces for Three Bar Structure with Brittle Behavior

where $h_{\underline{X}}(\underline{x})$ = importance sampling PDF. Accordingly, the Monte Carlo estimate of the failure probability becomes

$$P_F = \frac{1}{N}\sum_{i=1}^{N} I[g(\underline{x}_i)]\,\frac{f_{\underline{X}}(\underline{x}_i)}{h_{\underline{X}}(\underline{x}_i)} \tag{23}$$

Clearly, the appropriate choice of the importance sampling PDF, $h_{\underline{X}}(\underline{x})$, is crucial for the proper Monte Carlo estimate of P_F with Eq. 23.

With the Kernel method, Ang (1990) suggested the following for constructing the required importance sampling PDF:

$$\hat{h}_{\underline{X}}(\underline{x}) = \frac{1}{M}\sum_{i=1}^{M} \frac{1}{(\lambda_i w)^d}\, K\left(\frac{\underline{x} - \underline{x}_i}{w x_i}\right) \tag{24}$$

where $\underline{x}_i$ = failure samples generated from the original joint PDF, $f_{\underline{X}}(\underline{x})$; $K(\cdot)$ = Kernel estimator satisfying $\int_{\infty}^{\infty} K(y)dy = 1$; w = window width; and λ_i = a scale parameter. Criteria for selecting w and λ_i are suggested in Ang (1990) including a Gaussian PDF for $K(\cdot)$.

The above Kernel method suffers from having to generate the M failure samples from $f_{\underline{X}}(\underline{x})$ through basic Monte Carlo. To overcome this weakness, the original Kernel method can be modified by introducing an adaptive sampling scheme for generating the M failure samples.

Observe that the statistical error of Eq. 23 reduces to zero if $h_{\underline{X}}(\underline{x})$ is the original PDF, $f_{\underline{X}}(\underline{x})$, conditional on the failure domain D_f; i.e.

$$h_{\underline{X}}(\underline{x}) = f_{\underline{X}}(\underline{x} \,|\, \underline{x} \in D_f) \tag{25}$$

It is, however, difficult to find this function. As an alternative, the condition may be imposed on the first and second moments only; namely,

$$E_h(\underline{X}) = E_f(\underline{X} \,|\, \underline{x} \in D_f) \tag{26}$$

$$E_h(\underline{X}\underline{X}^T) = E_f(\underline{X}\underline{X}^T \,|\, \underline{x} \in D_f) \tag{27}$$

where the subscripts h and f refer to the expected values with respect to the respective PDF's.

The essence of the above proposed "Adaptive Kernel Method" is to select an appropriate point $\bar{\underline{x}}_h$ from which the Kernel importance sampling density can be constructed efficiently. Instead of the mean of each variate as the starting point, it is proposed that the starting point be moved to a point which is close to the failure domain.

The algorithm for the above purpose may be described as follows:

(i) Choose a point close to the failure domain, i.e., $\mu_{\underline{x}}' = (\mu_{x1}', \mu_{x2}', \ldots, \mu_{xd}')^T$.

(ii) With $x_j = \mu_{xj}'$, where $j = 1,2,\ldots,d$, and with $x_i = a$ for $i \neq j$ determine the value of a from the limit-state equation $g(a, \mu_{xj}') = 0$, and the conditional failure probability $P_{Fi} = P(X_i < a)$ or $P(X_i > a)$. If X_i is a normal variate, i.e. $N(\mu_{xi}, \sigma_{xi})$, one of the following two equations can be used to obtain the required conditional mean:

(28)

$$\bar{x}_{h_{ci}} = E[x_i | x_i > a] = \mu_{x_i} + \sigma_{x_i} \cdot \frac{e^{-\frac{1}{2}\left(\frac{a-\mu_{x_i}}{\sigma_{x_i}}\right)^2}}{\sqrt{2\pi}\ \Phi\left(-\frac{a-\mu_{x_i}}{\sigma_{x_i}}\right)} \tag{29}$$

where μ_{xi} and σ_{xi} are the original mean and standard deviation of X_i.

(iii) Repeat step (ii) for all i=1,2,...,d to obtain $\underline{\bar{x}}_{hc}=(\bar{x}_{h_{c1}}, \bar{x}_{h_{c2}}, \ldots, \bar{x}_{h_{cd}})$

(iv) The point $\underline{\bar{x}}_h=(\bar{x}_{h1},\bar{x}_{h2},\ldots,\bar{x}_{hd})^T$, is then obtained through

$$\bar{x}_{h_i} = (P_{F_i}\bar{x}_{h_{ci}} + \mu'_{x_i}\sum_{\substack{j=1\\ j\neq i}}^{d} P_{Fj}) \,/ \sum_{j=1}^{d} P_{F_j} \tag{30}$$

Once the point $\underline{\bar{x}}_h$ is determined, the conditional standard deviation, $\underline{\sigma}_{xh}$, can be obtained using the following two steps:

(i) With $x_j=\bar{x}_{hj}$, where j=1,2,...,d, and with $X_i=\beta$ for $i\neq j$, determine the value of β from $g(\beta,\underline{\bar{x}}_h)=0$. Then $\underline{\sigma}_{xh}$ is calculated using the following equation:

$$\sigma_{x_{h_i}} = \sigma_{x_i}\sqrt{1 + \left(\frac{\beta - \mu_{x_i}}{\sigma_{x_i}}\right)\left(\frac{\bar{x}_{h_i} - \mu_{x_i}}{\sigma_{x_i}}\right) - \left(\frac{\bar{x}_{h_i} - \mu_{x_i}}{\sigma_{x_i}}\right)^2} \tag{31}$$

(ii) Repeat step (i) for all i=1,2,...,d, to obtain $\underline{\sigma}_{xh}=(\sigma_{xh1},\sigma_{xh2},\ldots,\sigma_{xhd})$.

With $\underline{\bar{x}}_h$ and $\underline{\sigma}_{xh}$, the M failure points $y_1,y_2,\ldots,y_M$ can be generated efficiently. The Kernel importance sampling density, $\hat{\underline{h}}_x(\underline{x})$, is then constructed from Eq. (24).

N samples can then be generated from the Kernel density function, $\hat{\underline{h}}_x$, and the failure probability of the system is calculated with Eq. (23).

ACKNOWLEDGMENTS

The concepts and material presented here were obtained under the support of research grants from the National Science Foundation, including the current grant BCS 8903080 of the Earthquake Hazard Mitigation Program, monitored by Dr. S. C. Liu. These supports are gratefully acknowledged.

The results for the adaptive Kernel sampling method were obtained by Grace Wang, graduate student currently working on her doctoral dissertation at the University of California, Irvine.

REFERENCES

1. Ang, G., Ang, A.H-S. and Tang, W-H., Multi-dimensional kernel method in importance sampling Proc. 6th Conf. on Appl. of Stats. and Prob. in Civil Eng., Mexico City, June 1991, 289-296.
2. Architectural Institute of Japan, 1978 Miyagiken-Oki Earthquake Damage Report, Tokyo, 1980.
3. Asano, K., Application of probabilistic approach to aseismic safety analysis of soil-building structure systems Proc. US-Japan Joint Sem. on Stoch. Approaches in Earthq. Eng., Florida, May 1987.
4. Baber, T.T. and Wen, Y.K., Random vibration of hysteretic degrading systems J. of Eng. Mech., ASCE, 107, 1981, 1065.
5. Banon, H. and Veneziano, D., Seismic safety of reinforced concrete members and structures Earthq. Eng. and Struc. Dyn., J. Wiley & Sons, 1982, 10, 179-193.
6. Clough, R. W. and Johnston, S.B., Effect of stiffness degradation on earthquake ductility requirements Proc. 2nd Japan Earthq. Eng. Symp., Tokyo, October 1966, 227-232.
7. Kato, B. and Akiyama, H., Theoretical prediction of the load-deflection relationship of steel members and frames Symp. on Resistance and Ultimate Deformality of Struc., IASBSE Prelim. Publ., Lisbon, 1973.
8. Park, R. and Paulay, T., Reinforced Concrete Structures, J. Wiley & Sons, New York, 1975.
9. Park, Y-J. and Ang, A.H-S., Mechanistic seismic damage model for reinforced concrete J. of Struc. Eng., ASCE, 111, 4, 1985, 722-739.
10. Park, Y-J., Ang, A.H-S. and Wen, Y.K., Seismic damage analysis of reinforced concrete buildings J. of Struc. Eng., ASCE, 111, 4, 1985, 740-757.
11. Quek, S-T. and Ang, A.H-S., Reliability analysis of structural systems by stable configurations J. of Struc. Eng., ASCE, 116, 1990, 2656-2670.
12. Stephens, J.E. and Yao, J.T.P., Damage assessment using response measurements J. of Struc. Eng., ASCE, 113, 1987, 787-801.
13. Suzuki, Y. and Minai, R., Application of stochastic differential equations to seismic reliability analysis of hysteretic structures Proc. US-Japan Joint Sem. on Stoch. Approaches in Earthq. Eng., Florida, May 1987.
14. Takizawa, H. and Aoyama, H., Biaxial effect in modeling earthquake response of R/C structures J. Earthq. Eng. and Struc. Dyn., 4, 1976, 523-552.
15. U.S. Department of Commerce, San Fernando, California Earthquake of February 9, 1971, Washington D.C., 1973.
16. Wen, Y.K., Method for random vibration of hysteretic systems J. of Eng. Mech., ASCE, 102, 1976, 249-263.

DYNAMIC BEHAVIOR OF BRIDGES UNDER RANDOM LOADING AND DYNAMIC RELIABILITY PROBLEM *

Ying Jun Chen, Daotang Wang
(Civil Engineering Department, Northern Jiaotong University
Beijing 100044, China)

Keywords: Structural reliability; Bayesian analysis; Fatigue / Fracture reliability; Fuzzy reliability; Random vibration; Dynamic reliability; Vehicle–Girder–Pier system; Cable–stayed bridge; Curved plate bridge; Stiffened truss bridge; Earthquake response; Stochastic process; First passage time;Random vehicle flow; Research train; Vortex–induced vibration; Deteriorated structure.

I. Introduction

The last two decades have seen healthy developments in bridge vibration and dynamic reliability theory in China. The structural reliability committee, an affiliation of the bridge and structural engineering institute, China Civil Engineering Society, has formally been established in 1986. The Chinese Conference on Structural Safety and Reliability is held every two years. As for ICOSSAR (the International Conference On Structural Safety And Reliability), Chinese scholars began to attend from 1985, the 4th ICOSSAR. They have also attended some other conferences concerning structural vibration and reliability. Some main research achievements may be summarized as follows:

(1) About 20 cable–stayed bridges have been built since 1978, the AVS(ambient vibration survey) method has been used to determine the dynamic characteristics of the bridge [1], the wind design problems have been studied [2][3], the dynamic reliablity of

* This research was supported by Science Foundation of China State Education Commission.

the tower of a floating type cable–stayed bridge considering the horizontal seismic force has also been studied [4][5].

(2) Application of Bayesian analysis. Considering the random properties of the strength of concrete, the non–destructive field test of structural concrete has been regarded as an important approach. Xing Chen [6] suggested a method to analyze the test result, expert knowledge and the subjective judgement of inspector on the basis of Bayesian inference, and proposed a normal model to adjust the regression equation.

Ying Jun Chen [7][8] studied the problem of estimation of seismic risk on the basis of Bayes′ theorem by using the mine of historical earthquake information in China.

(3) Fatigue / Fracture reliability. The probability density function of fatigue strength at a prescribed life was derived from that of fatigue life at a given stress level and a dynamic stress–strength interference model is established by Z. Gao, H. Fu, and W. Li [9][10].

(4) Fuzzy reliability. The problem of reliability analysis for the structural system in which the states of basis events are continuous random state variables has been researched (Mingzhu Duan, Guangyuan Wang [11]). A fuzzy random model of earthquake ground motion is established and a method to analyze the fuzzy random response of the aseismic structure is proposed (Guangyuan Wang). The fuzziness of the damage grades is further taken into account and the fuzzy safe criterion is presented (Jinping Ou, Guangyuan Wang [12]).

(5) Random vibration. China started research in the area of nonlinear seismic response analysis in 1977, e.g. Lian Wei [13], Deyuan Zhou, Bolong Zhu [14]. The random vibration and dynamic reliability analyses of various buildings under earthquake excitation as well as wind action are studied by Guiqing Li, Weilian Qu, Hong Cao, Daotang Wang et al. [15][16][17][18].

It can be seen that many research groups are actively engaged in various areas of structural vibration and reliability in China. In recent years, some practical approaches for code drafting and design practices had been proposed, e.g. Guofan Zhao [19], M. C. Yao [7].

In this paper, some problems of bridge vibration and dynamic reliability, in which the authors have participated, are discussed in more detail , and a calculation method of dynamic reliability of deteriorated structures has also been studied. Lastly, the problems to be further studied are also mentioned.

II. Random vibration and approximate reliability analysis of highway bridge

Based on Chinese traffic reality, a method for the dynamic response analysis for stiffened curved plate bridge, Fig. 2–1, which has been widely used in interchange bridges in China, under the action of random traffic flow and roughness of road surface, has been proposed [22][23].

1. Basic Assumption

(1) Only vertical vibration of vehicles is considered; the analytical model is as shown in Fig. 2–2

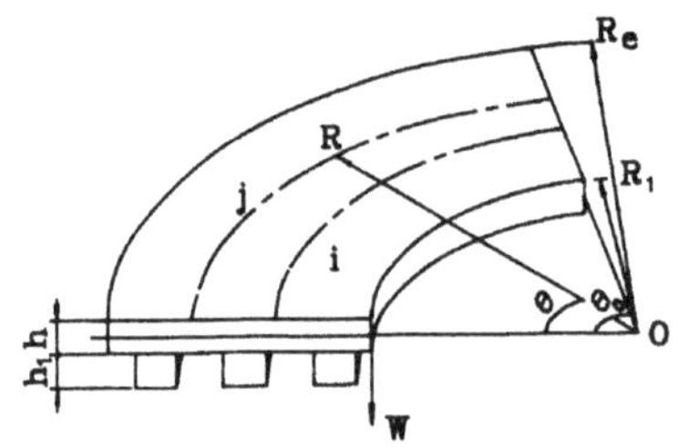

Fig. 2–1 Sectoral plate with stiffened ribs

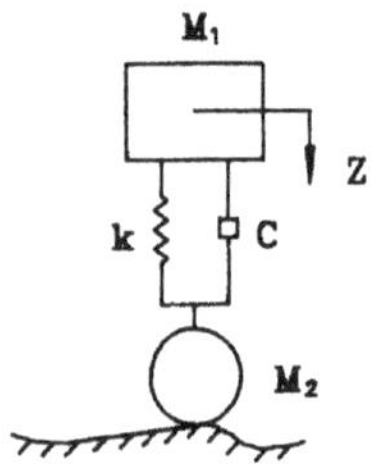

Fig. 2–2 Vehicle model

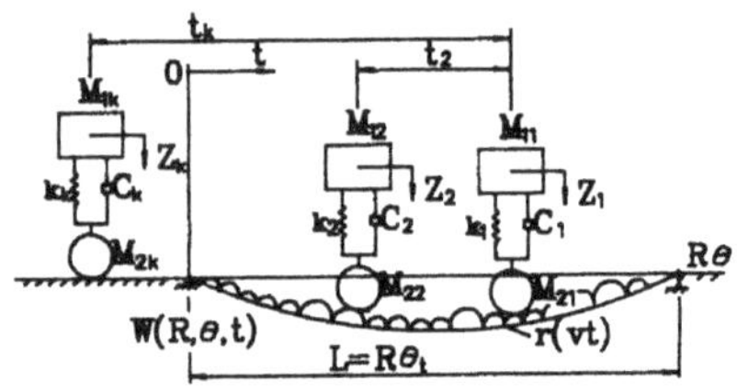

Fig. 2–3 Vehicle flow

(2) Spectral density function of road surface roughness r(vt) is given by Eq. (2.1).

$$S(\Omega)=a/(\Omega^2+\beta^2) \quad \text{or} \quad S(\omega)=2\pi Va/(\omega^2+\alpha^2) \tag{2.1}$$

in which Ω is the number of waves, a is parameter representing road surface evenness, β is a small constant, V,ω are velocity and cycle frequency respectively, and $\alpha=2\pi V\beta$. The autocorrelation function is given by Eq. (2.2).

$$R_r(\tau)=\int_{-\infty}^{+\infty}S(\omega)\exp(i\omega\tau)d\omega=\pi a\beta^{-1}exp(-\alpha|\tau|) \tag{2.2}$$

(3) Dynamic characteristics of bridge is determined by some lower vibration modes.

2. Basic Equations and Solving Method

Let L, $T=L/V$ denote span of bridge and passing time respectively. Refer to Fig. 2–3. According to FSE method, the deflection of bridge is [20]

$$W(R,\theta,t)=[C(R)][\bar{Z}]\{p(t)\}Y(\theta) \tag{2.3}$$

in which $[\bar{Z}]$ is mode matrix, and for simply supported bridge,

$$Y(\theta)=\sin(\pi\theta/\theta_0), \quad \theta=V(t-t_l)/R \tag{2.4}$$

The control equations of vehicle–bridge system are given by Eq. (2.5–2.7)

$$\ddot{Z}_l(t-t_l)+2\xi_l\omega_l\dot{Z}_l(t-t_l)+\omega_l^2 Z_l(t-t_l)$$
$$=\omega_l^2\{U(t,t_l)\{[C(R\ )][\overline{\Phi}]\{p(t)\}Y(V(t-t_l))\}+\gamma(V(t-t_l))\} \quad (2.5)$$

$$\dot{Z}_l(0)=\dot{z}_{0,l};\quad Z_l(0)=z_{0,l};\quad l=1,2,\dots n(t)$$
$$[I][\ddot{p}(t)]+[\overline{C}][\dot{p}(t)]+[\Lambda]\{p(t)\}$$
$$=\sum_{l=1}^{n(t)}[\Phi]^T[C(R\ )]^T U(t,t_l)M_l\, g[\eta_l-g^{-1}\ddot{Z}_l(t-t_l)]Y(V(t-t_l)) \quad (2.6)$$
$$\{\dot{p}(0)\}=\{0\},\qquad \{p(0)\}=\{0\}$$
$$U(t,t_l)=1 \qquad 0\leqslant t-t_l\leqslant T$$
$$U(t,t_l)=0 \qquad T<t-t_l<0 \quad (2.7)$$

in which $n(t)$ is a Poisson process with parameter λ, M_l is a random variable with logarithmic normal distribution, while $r(vt)$ is a stationary Gaussian process [20].

Let deflection of bridge be $\quad W=W_s+W_d \quad$ (2.8)
in which W_s, W_d represent static and dynamic deflection respectively.

$$W_s=[C][\overline{Z}][\wedge]^{-1}Y(\theta)\sum_{l=1}^{n(t)}M_l' g[\Phi]^T[C]^T Y(V(t-t_l))$$
$$=[a]\sum_{l=1}^{n(t)}M_l'[b]Y(V(t-t_l)) \quad (2.9)$$

as $M_1', M_2', \dots M_l', \dots$ are independent variables with identic distribution, W_s is a generalized filter Poisson process. Hence the mean, mean square value, and variance of W_s can be obtained.

The dynamic deflection W_d is given by

$$W_d=[C][\overline{Z}]Y(\theta)\{q(t)\} \quad (2.10)$$

in which $\quad \{q(t)\}=\{q_1(t),q_2(t),\dots q_m(t)\}^T \quad$ (2.11)

$$q_i(t)=-a_l\omega_i^{-1}\sum_{l=1}^{n(t)}M_l U(t,t_l)\ \cdot$$
$$\int_0^{t-t_l}\ddot{Z}(\tau)Y(V_\tau)\exp\{-\xi_i\omega_i(t-t_l-\tau)\}\sin(\omega_i'(t-t_l-\tau)d\tau$$
$$=\sum_{l=1}^{n(t)}M_l\ \overline{q}_i(t-t_l) \quad (2.12)$$

hence $W_d=[\overline{a}]\sum_{l=1}^{n(t)}M_l\{\overline{q}(t-t_l)\} \quad$ (2.10′)

For approximate analysis, taking $N=\lambda$, the mean value of traffic flow process, so that

$$E[W_d^2]=[\overline{a}]^T[\overline{a}]\sum_{l=1}^{N}\sum_{j=1}^{N}E[M_l M_j]E[\{\overline{q}(t-t_l)\}\{\overline{q}(t-t_j)\}^T] \quad (2.13)$$

as $\quad E[M_l,M_j]=E[M_l^2] \quad$ *when* $l=j$
and $\quad E[M_l,M_j]=0 \quad$ *when* $l\neq j \quad$ (2.14)

hence $E[W_d^2]=[\bar{d}]\sum_{l=1}^{N}E[M_l^2]E[\{\bar{q}(t-t_l)\}\{\bar{q}(t-t_l)\}^T]$ (2–13′)

while $\sigma_{Wd}^2=E[W_d^2]-E[W_d]^2=E[W_d^2]$

from Eq. (2.8) $E[W]=E[W_s];\ \sigma_W^2=\sigma_{Ws}^2+E[W_d^2]$ (2.15)

$E[\{\bar{q}(t-t_l)\}\{\bar{q}(t-t_l)\}^T]$ can be obtained by using the method suggested in [21]. In bridge design, the extreme deflection is more attractive. As W is a stochastic process in time range $[0,T_0]$, we define $W_n=E[W_s]+\gamma\sigma_W$ as nominal deflection, and

$$W_{ne}=\max W_n \tag{2.16}$$

in $[0,T_0]$ as the extreme value of nominal deflection, which is a random variable.

If we consider the difference of traffic flow in different areas, and also the difference for traffic loads and road condition, λ ,$E[M_l]$ and parameter a in Eq. (2.1) must be treated as random variables. For simplicity, assume λ, $E[M_l]$ and a to have uniform distribution. $E[W]$ and σ_W are only a member in stochastic process sets.

Within $[0,T_0]$, we can get a series of extremes $\overline{W}_{ne}=\{W_{ne1},W_{ne2},\ldots,W_{nem},\ldots\}$. Obviously, $\overline{W}_{ne}$ is a random series. As its numbers are the extreme value of W_n, one may assume that it obeys Gumbel distribution. The validity of the assumption can be proved.

According to the assumption, the CDF and PDF of W_{ne} are

$$F_{Wne}(w)=\exp\{-\exp[-c(w-u)]\};$$
$$f_{Wne}(w)=c\exp\{-c(w-u)-\exp[-c(w-u)]\} \tag{2.17}$$

If the data used in obtaining W_{ne} is based on yearly statistics, F_{Wne} will represent yearly extreme deflection. Based on the fact of independent and identic distribution of yearly extremes, CDF of W_{ne} for life time of bridge can be obtained.

$$F_l(w)=[F_{Wne}(w)]^n=\exp\{-n\exp[-c(w-u)]\} \tag{2.18}$$

The probability of no deflection W_{ne} exceeding x_0 is $P\{W_{ne}<x_0\}=F_{Wne}(x_0)$, and the mean first passage time $T_n(x_0)$ is given by Eq. (2.19)

$$T_n=[1-F_{Wne}(x_0)]^{-1}=n[-\ln(F_l(x_0))]^{-1} \tag{2.19}$$

For given T_n and F_{Wne} or F_l , the threshold x_0 can be obtained. Some numerical examples are given in [20]–[23].

Ⅲ. Theoretical calculation for the lateral vibration of the V–G–P (Vehicle–Girder–Pier) system

The dynamic analytical model of the V–G–P system is established by the following substructures of vehicle and bridge.

1. Vehicle model

The vehicle model used here is a research train which consists of two diesel locomotives, two ballast hopper cars, a research car and a sleeping coach. Based on some assumptions, Fig. 3–1 can be taken as the vehicle model. Thus the dynamic equilibrium equations of the ith vehicle are as follows:

$$\begin{bmatrix} M_i & 0 & 0 \\ 0 & J_{\theta i} & 0 \\ 0 & 0 & J_{\psi i} \end{bmatrix} \begin{bmatrix} \ddot{Y}_i \\ \ddot{\theta}_i \\ \ddot{\psi}_i \end{bmatrix} + \begin{bmatrix} C_i & -h_i C_i & 0 \\ -h_i C_i & C_{\theta i} & 0 \\ 0 & 0 & C_{\psi i} \end{bmatrix} \begin{bmatrix} \dot{Y}_i \\ \dot{\theta}_i \\ \dot{\psi}_i \end{bmatrix} + \begin{bmatrix} K_i & -h_i K_i & 0 \\ -h_i K_i & K_{\theta i} & 0 \\ 0 & 0 & K_{\psi i} \end{bmatrix} \begin{bmatrix} Y_i \\ \theta_i \\ \psi_i \end{bmatrix}$$

$$= \sum_{j=1}^{2N_{wi}} \frac{1}{2N_{wi}} \begin{Bmatrix} K_i y_{wij} + C_i \dot{y}_{wij} \\ -h_i (K_i y_{wij} + C_i \dot{y}_{wij}) + 2(K_{\theta wi} \theta_{wij} + C_{\theta ij} \dot{\theta}_{wij}) \\ SN_j S_{ij} (K_i y_{wij} + C_i \dot{y}_{wij}) / 2 \end{Bmatrix} \tag{3.1}$$

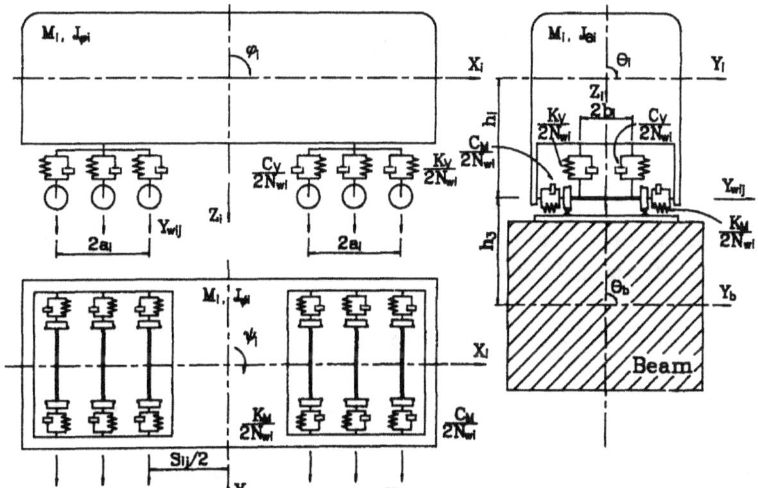

Fig. 3–1 Vehicle model

where SN_j = the symbol function of the jth wheel–set,
$SN_j = 1$, when wheel j belongs to front bogie;
$SN_j = -1$, when wheel j belongs to rear bogie.

and $K_{\theta wi} = K_{vi} b_i^2$; $K_{\theta i} = 2(h_i^2 K_{Hi} + K_{\theta wi})$;
$C_{\theta wi} = C_{vi} b_i^2$; $C_{\theta i} = 2(h_i^2 C_{Hi} + C_{\theta wi})$;

$C_{\psi i}=S_i^2 C_{Hi}/2;\qquad K_{\psi i}=S_i^2 K_{Hi}/2$;

K_{Hi} = total lateral spring stiffness of the ith bogie;

K_{vi} = total vertical spring stiffness of the ith bogie;

C_{Hi} = total lateral damping coefficient of the ith bogie;

C_{vi} = total vertical damping coefficient of the ith bogie;

N_{wi} = the number of the wheel–sets on the ith bogie.

2. Dynamic equations of bridge

Firstly, the modes and the frequencies of the free vibration of the bridge as a whole are solved. According to the orthogonality of the modes, and to the factor that the dynamic response of the structure is dominantly influenced by several lowest modes, the dynamical responses can often be obtained by considering only a few modes of vibration; the modes and natural frequencies of the structure are solved by means of the general structure analysis program SAP5.

Two more assumptions are made as follows: a) There is no relative displacement between the girders and the track, i.e., the elasticity of the track on bridges is not taken into account; b) The shape of the cross sections of the girders keeps constant in the course of vibration.

Thus, the lateral displacement of any cross section of a bridge can be determined by lateral displacement $Y_b(x)$ and torsional angle $\theta_b(x)$. If the first NQ orders of modes are considered in calculation, we obtain

$$Y_b(x)=\sum_{n=1}^{NQ} A_n\varphi'(x),\quad \text{and}\quad \theta_b(x)=\sum_{n=1}^{NQ} A_n\varphi''(x) \tag{3.2}$$

in which $\varphi'(x)$ and $\varphi''(x)$ are the lateral displacement component and the torsional angular component of the nth mode function respectively. A_n is the generalized coefficient, namely the modal amplitude.

The modal equation related to the nth mode is

$$M_n\ddot{A}_n+C_n\dot{A}_n+K_nA_n=F_n \tag{3.3}$$

where F_n is the generalized force of the nth mode, and here follows the derivation of F_n.

See Fig.3–2. when the jth wheel–set of the ith vehicle is on the bridge, forces act on the bridge through the wheel–set include: a) inertial forces of the wheel–set: lateral force of inertia $P_{uij}=-m_{wij}\ddot{y}_{wij}$ and moment of the inertial force $P_{\theta ij}=-J_{wij}\ddot{\beta}_{wij}+h_3P_{uij}$; b) forces through springs and dampers: lateral force P_{usij} and moment $P_{\theta sij}$. Therefore, the resultants act on the bridge are $F_{uij}=P_{uij}+P_{usij}$; $F_{\theta ij}=P_{\theta ij}+P_{\theta sij}$. These forces of the wheels cannot be summed directly to form the generalized force F_n because these forces act at the bridge at different positions. Suppose there are N_v vehicles on the bridge and the ith vehicle has N_{bi} wheel–sets, the generalized force

$$F_n=\sum_{i=1}^{Nv}\sum_{j=1}^{Nbi}\varphi_n'(x_{ij})F_{uij}+\varphi_n''(x_{ij})F_{\theta ij}=\sum_{i=1}^{Nv}\sum_{j=1}^{Nbi}(f_{ij}^n\ F_{uij}+g_{\theta ij}^n\ F_{\theta ij}) \tag{3.4}$$

where x_{ij} shows the position of the jth wheel–set of the ith vehicle on bridge.

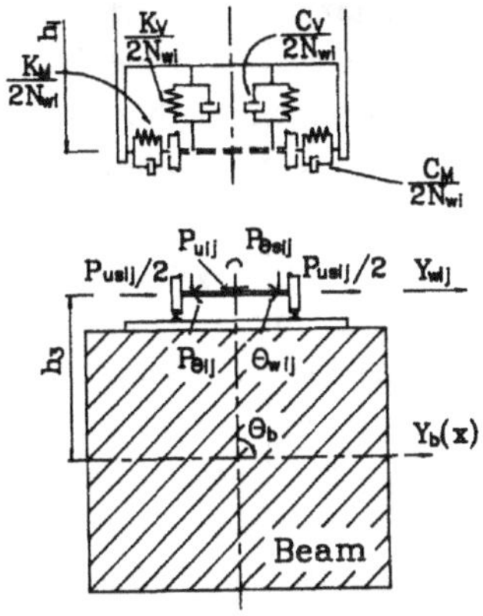

Fig. 3–2 Forces through wheel–set

Because that the vehicles move through the bridge continuatively while the mode–shape vector from the finite element analysis is discrete, when wheel j is not at any point, the mode function values f_{ij}^n and g_{ij}^n (namely $\varphi_n'(x_{ij})$ and $\varphi_n''(x_{ij})$) can be obtained by any kind of interpolation.

3. Displacements of wheels (linking equations)

When a train is moving on a bridge, the lateral displacement y_{wij} and the rotary angle θ_{wij} of the wheel–set, the hunting movements $Y_s(x_{ij})$ and $\theta_s(x_{ij})$, and the lateral bent–and–torsional deformations $Y_b(x_{ij})$ and $\theta_b(x_{ij})$ of the girder should satisfy the relations (cf. (3.2)):

$$y_{wij}=Y_b(x_{ij})+h_3\theta_b(x_{ij})+Y_s(x_{ij})\ ;\quad \text{and} \quad \theta_{wij}=\theta_b(x_{ij})+\theta_s(x_{ij}) \tag{3.5}$$

The hunting movements may be obtained by experiments and computer simulation. Substitute (3.2) into (3.5),

$$y_{wij}=\sum_{n=1}^{NQ}(f_{ij}^n+h_3g_{ij}^n)A_n+Y_s(x_{ij})\ ;\text{and}\quad \theta_{wij}=\sum_{n=1}^{NQ}g_{ij}^nA_n+\theta_s(x_{ij}) \tag{3.6}$$

Therefore, y_{wij} and θ_{wij} can be expressed by the linear composition of the generalized coefficients $A_n(n=1,2,3...NQ)$; hence the equations of wheel–sets are not independent and they need not be incorporated in the dynamical eqilibrium equations of the V–G–P system. The function of the equations of wheel–set is to link up the vehicle equations with the bridge equations.

4. Dynamic equilibrium equations of V–G–P system and its response

By substituting Eqs.(3.6) into the vehicle equations and the bridge equations, and combining them together, the dynamic eqilibrium equations of V–G–P system can be obtained (see [24]). The system equations are 2–order linear differential equations with variable coefficients as they are always varying while the train is moving on the bridge. The equations are solved by the Newmark β method with $\beta = 1/4$, and a program for the step–by–step integration procedure is worked out.

The dynamic model of the V–G–P system established on the principle of substructure method and the computer simulating solution of the dynamic responses of vehicles and high piers can reflect the main vibrational behaviors rather correctly. This analytical approach is proved to be practicable in studying the dynamic interactions between bridge and vehicles. The calculated results agree well with those obtained in the field tests. The influence of pier vibration on the running stability of vehicles is studied in [24]. This is an important problem in the serviceability reliability.

IV. Truss bridge stiffened by flexible arch

The truss bridge stiffened by a flexible arch at top has frequently been used in China especially for railroad construction in mountainous regions. On the basis of the method presented in the previous article, the dynamic analysis of the bridge under train–loads is studied [25]. The analysis on the earthquake response [26][27] and the problem of the aerostatic and aerodynamic characteristics of the H–shaped slender hanger have also been studied [28][29].

1. Earthquake response analysis

In order to simplify the calculation, a panel element method is proposed. The bridge is separated into N panel elements which are composed of two sets of elements, element I and J. The element I consists of ribs, rigid hangers and the truss; the element J consists of a rib brace, hangers and cross beam of the truss. Every panel element is of four nodes (two arch nodes and two truss nodes, the former with one degree–of–freedom, the latter with two).

In Fig. 4–1, W_r = lateral displacement of the rib, W_s = lateral displacement of the truss, θ = torsional angle of the truss, $E_j(t)$ = lateral seismic excitation at jth support of the bridge $(j=1,2,...M)$, M = number of supports, $q_i(t)$ = generalized displacement of the panel–element node (i = 1,2,...,6), l = span of the bridge, L = length of the element.

In references [26][27], different earthquake responses in three directions are obtained by using microcomputer, and the results are compared with those obtained by program SAP5; another model considering the interaction of piers has also been used in the calculation [26]. A spatial analysis for the lateral bending–torsional earthquake response is derived from the probability view of earthquake response; the power spectrum approach is also discussed.

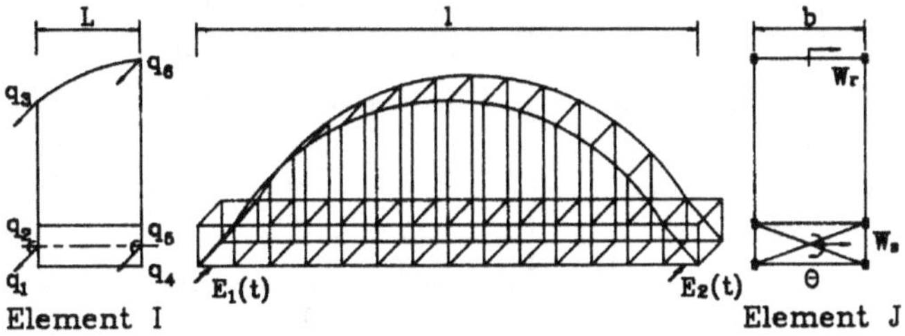

Fig. 4–1 model for calculation

For example, the analysis is applied to a bridge (span l = 112m, arch rise f = 16m) under the input of the displacement power spectrum of earthquake ElCentro N–S (1940), the scale factor of the acceleration response spectrum is 98.1 m / s^2, the lateral displacement power spectrum at mid–span of the bridge is given in Fig. 4–2, in which LDPSVOR = lateral displacement power spectrum value of the rib, LDPSVOT = lateral displacement power spectrum value of the truss, No. k = number of kth section of the bridge, k = 1,2,...15

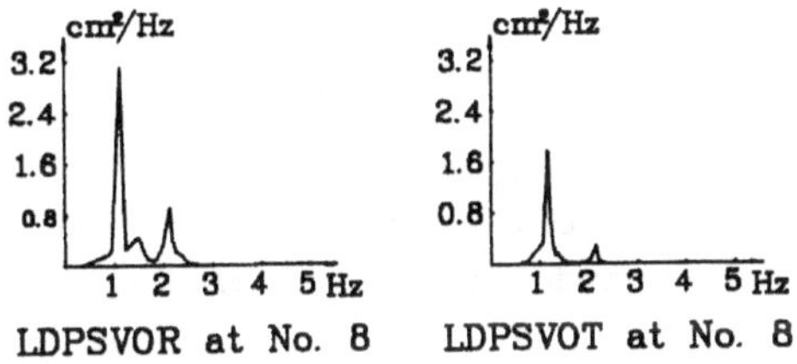

Fig. 4–2 power spectrum at mid–span section of the bridge

2. Effect of vortex–induced vibration on slender hangers

The aerostatic and aerodynamic characteristics of slender hangers (Fig. 4–3) are studied by wind tunnel tests [28][29].

The research result shows that the St curves of models change a little because of the effect of perforations in webs. For H–shaped models, the maximal values of St are 0.114–0.12, for box–shaped models, they are 0.109–0.115, in the directions of 0 ° and 10 ° . Models H2,H3,B2 and B3 may not vibrate by vortex–shedding excitation in the range of the angle of attack $\alpha=70°-90°$. It is an efficient solution for weakening the fatigue effects of vortex–induced vibration to set perforations in the webs.

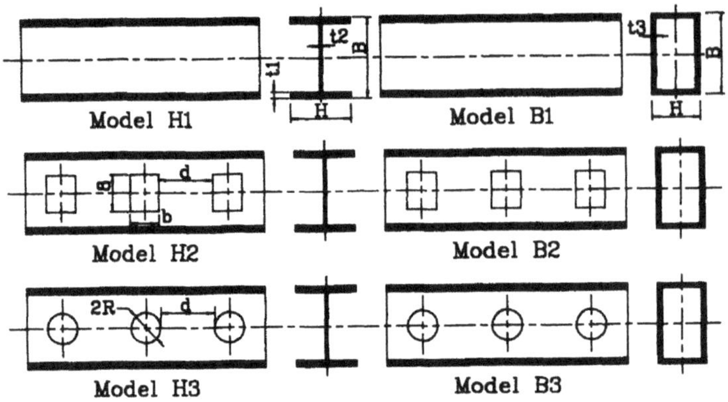

Fig. 4–3 Schematic diagram of cross section and perforations in the webs of sectional models

V. Dynamic reliability of deteriorated structures

The proposed method [30] is a combination of the reliability theory and dynamic experiment. Suppose the mass of the deteriorated structure such as aged bridge does not vary with time and the stiffness and damping are random processes slowly varying with time. Thus the motion equation of a linearly vibrational deteriorated system may be written as

$$M\ddot{x}(t)+C(T)\dot{x}(t)+K(T)x(t)=f(t) \tag{5.1}$$

where the unit of T is year, the unit of t is second. Assuming that the response $x(t)$ of a SDF structure is a stationary random process, the exciting force is a normal white noise whose power spectral density is Φ_0, the structure deterioration leads the natural frequency to decrease $\omega(T)=\omega_0\exp(-\alpha T)$, the damping to increase $\xi(T)=\xi_0\exp(\beta T)$, the threshold of the response to decrease $a(T)=a_0\exp(-\gamma T)$. The numbers of the response crossing the threshold in time $(0,t)$ can be taken as Poisson process. The crossing assumption is taken as the two–state Markov crossing. Then the dynamic reliability function $P_s(T)$ can be obtained

$$P_s(T)=\exp\left\{-\frac{\omega_0}{\pi}\int_0^T e^{-\alpha T}\frac{1-\exp\{4\xi_0\omega_0^{3/2}a_0e^{(\beta-3\alpha/2-\gamma)}/\sqrt{\pi\Phi_0}}{\exp\left[\dfrac{\xi_0\omega_0^3a_0^2\exp[(\beta-3\alpha-2\gamma)T]}{\pi\Phi_0}\right]-1}dT\right\} \tag{5.2}$$

Ⅵ. On the course of joint research

The following topics are of great interest; the above-mentioned studies can also be improved by taking the following into consideration. Future research should place more emphasis on the corresponding items.

(1) Dynamic reliability of deteriorated structures [30].Deterioration of resisting capacity for aged bridges, cumulative damage.

(2) Dynamic reliability of bridge by using the vehicle–girder–pier system model [24][27]. System reliability, effect of soil–structure interaction, stochastic finite element.

(3) Structural safety evaluation based on system identification approaches [31]. Distinction between randomness and uncertainty, fuzziness of parameters.

(4) Natural disaster reduction, aseismic reliability analysis [9][16].

(5) Wind–induced vibration and dynamic reliability of bridges [2][3][28].

(6) Reliability–based aseismic design, wind load, various limit states, load combination [32].

References

[1] Ying Jun Chen, Guobin Xu, "Application of ambient vibration survey to bridge vibration problems", Proc. of Sino–American symposium on bridge and structural engineering, Beijing, 1982

[2] Haifan Xiang, "Developments in flutter analysis of cable–stayed bridges", Recent advances in wind engineering vol. 2, Proc. 2nd APSOWE, Beijing, 1989

[3] Guohao Li, "On the postcritical torsional vibration of cable–stayed bridges", idem.

[4] Ruilin Li, Haifan Xiang, "A first–passage failure criterion based upon dynamic ultimate displacement" (in Chinese), Proc. of the 2nd conference on structural reliability, China, 1989

[5] Ruilin Li, Haifan Xiang, "Approximate analysis for aseismic reliability of floating type cable–stayed bridge" (in Chinese), Proc. of the 1st conference on structural reliability, China, 1987

[6] Xing Chen, "Application of Bayesian inference to non–destructive field test of concrete" (in Chinese), Proc. of the 1st conference on structural reliability, China, 1987

[7] Q. G. Cheng, M. C. Yao, X. X. Zhou and Y. J. Chen, "Research on the safety and reliability analysis of railway structures", Proc. of ICOSSAR' 85, May, 1985

[8] Ying Jun Chen, "On some aspects of bridge vibration problems" (in Chinese), Journal of Southwest Jiaotong University, 1986

[9] H. Fu, Z. Gao, "Probability density function of fatigue strength", Structural Safety and Reliability, vol. Ⅲ, Proc. of ICOSSAR' 85, Japan

[10] Weidong Li, "stochastic process of fatigue damage", Doctoral thesis in Chinese Beijing institute of aeronautics and astronautics, 1989

[11] Mingzhu Duan, Guangyuan Wang, "Fuzzy reliability analysis for structural systems with continuous states of failure", Structural safety and reliability Vol. I, Proc. of ICOSSAR' 89

[12] Jinping Ou, Guangyuan Wang, "Fuzzy random dynamic reliability analysis of aseismic structures", structural safety and reliability Vol. 3, Proc. of ICOSSAR' 89, USA, 1989

[13] Lian Wei, "Nonlinear response analysis for reinforced concrete structures in China", Proc. of the European mechanics colloquium, press, 1987

[14] Deyuan Zhou, Bolong Zhu, "Nonlinear earthquake response analysis of masonry multi–story building under bi–directional seismic excitation", Proc. of the international conference on structural engineering and computation, Beijing, 1990

[15] W. L. Qu, J. P. Ou, G. Q. Li, "Dynamic reliability analysis of shear–type tall buildings under earthquake excitation", Proc. of the 3rd international conference on tall buildings, 1984

[16] Qiusheng Li, Hong Cao, "Reliability analysis of earthquake–resistant structures considering fuzzy and random factors" (in Chinese), China Journal of Civil Engineering, Vol. 21, No. 4, 1988

[17] Guiqing Li, Hong Cao, "Dynamic reliability analysis of high– rise structures under wind loads" (in Chinese), Chinese Journal of Civil Engineering, Vol. 20, No. 1, 1987

[18] Hiroshi Tanaka, Daotang Wang, "The buffeting of tall building, Parametric study", Proc. of the international conference on structural engineering and computation, Beijing, 1990

[19] Guofan Zhao, "A practical FOSM method for structural reliablity analysis", Proc. of ICOSSAR' 85, May, 1985

[20] N. Wang and Ying Jun Chen, "Dynamic analysis of orthotropic sectorial plate and its stochastic response", Research Report of Northern Jiaotong University, 1985

[21] Ying Jun Chen and N. Wang, "A simple method for the coupled vibration problem in the stochastic vibration analysis of highway bridge structures and a discussion about first–passage problem", (in Japanese), Symposium JCOSSAR' 87, pp. 369–374, 1987

[22] Ying Jun Chen, Naisong Wang, "A simplified method for random vibration analysis of curved plate bridge" (in Chinese), Proc. of the 5th annual meeting of China Civil Engineering Society, 1990

[23] Ying Jun Chen, Naisong Wang, "Approximate reliability analysis of highway bridges under the action of random vehicles", Structural safety and reliability vol. 3, published by ASCE, Proc. of ICOSSAR' 89

[24] He Xia, Ying Jun Chen, "Dynamic response of high pier under train loads and its influense on running vehicle stability", Proc. of international symposium on

geomechanics, bridges and reliability analysis", Lanzhou, 1987

[25] Ying Jun Chen, He Xia, Ri Gau, "Dynamic analysis of truss bridge stiffened by flexible arch under train–loads", Proceedings EASEC–3, Shanghai, 1991

[26] Ri Gao, Guiping Yan, He Xia, Ying Jun Chen, "Analysis on the earthquake response to truss bridge stiffened by flexible arch", (in Chinese), Technical Report, No. SVL 88–004, Research Laboratory for Dynamics and Vibration of Structures, Northern Jiaotong University, 1988

[27] Ri Gao, Ying Jun Chen, "Lateral response of truss bridge stiffened by flexible arch to seismic excitation", Proc. of 2nd international conference on education, practice and promotion of computational method in engineering using small computers, Guangzhou, 1987

[28] Jianming Yin, Ying Jun Chen, Xizhe Yu, "Vortex–induced vibration of slender hanger member of a railway bridge", Proceedings of 2nd EASEC, Thailand, 1989

[29] Ying Jun Chen, Xizhe Yu, Jianming Yin, "A study on the aerostatic and aerodynamic characteristics of an H–shaped slender member and its modified shapes", Recent advances in wind engineering, Proc. 2nd APSOWE, vol. 1, Beijing, 1989

[30] Guo Qin Shi, Ying Jun Chen, "A calculation method of dynamic reliability of deteriorated structures", Research report, Northern Jiaotong University, 1990

[31] He Xia, Ying Jun Chen, "Application of AR model to modal parameter identification of bridges" (in Chinese), Research report, Northern Jiaotong University, 1991

[32] Ying Jun Chen, Li Wang, "On some problems of design wind load for Chinese railway bridges" (in Chinese), Proc. of the 2nd conference on structural reliability, China, 1989

FUZZY RANDOM ANALYSIS AND DESIGN OF ENGINEERING STRUCTURES

Wang Guangyuan   Ou Jinping   Tan Dongyao
(Harbin Architectural and Civil Engineering Institute,
Harbin 150006, China)

ABSTRACT

In structure itself and its working environments, there usually exist quite a number of intensely uncertain information, such as the randomness about the occurrence of future events, the fuzziness of the evaluation standard and the unascertained factors used by decision makers. A reasonable and advanced structural analysis and design should be able to consider these uncertain informations. This paper introduces one such kind of structural design theory proposed by us, which contains structural fuzzy random optimum design, structural generalized reliability, fuzzy random vibration analysis and design of aseismic structures, global optimization of large scale engineering systems.

FUZZY FACTORS IN STRUCTURAL ENGINEERING[2]

Fuzziness of Structural Design Objectives

Present structural design objective stresses too much on the reduction of structural initial construction cost, but overlooks its long term benefit. A more reasonable design objective should consider them both, i.e. a design scheme X should make following objective be minimum:

$$\mathbf{W}(X) = C(X) + \mathbf{E}(X) \rightarrow \min \tag{1}$$

where C is structural initial construction cost, **E** is the sum of structural maintenance fee and loss expectation deduced from its possible damages during its service period. It has both fuzziness and randomness since it is determined by the possible hazards that the structure would meet, whose grade and occurrence are both fuzzy and random. Also, the loss expectation reflects the importance of the structure, since the more important the structure is, the larger the loss it would bring about. The idea of considering the initial construction cost and future loss expectation together as one objective is so important that it will be often mentioned in fuzzy decision for optimum structural defence level, structural fuzzy optimum design, global optimization of engineering systems. In this paper, the bold symbols mean that the quantities they represent are fuzzy.

Fuzziness about the Allowable Interval for Structural Responses

The strength of structure or its element has relation with its allowable damage degree which has fuzzy sense. For example, structural damage degree can be divided into: None, Minor, Moderate, Heavy, etc. All of these grades are fuzzy. Even for the allowable stress $[\sigma]$ which is common to us, it is not appropriate to stipulate a deterministic interval for it. For instance, if the allowable stress of a kind of steel is required to satisfy $-1600 < [\sigma] < 2000(kg/cm^2)$, it means that $\sigma = 2000$ is allowed, while $\sigma = 2001$ is not allowed, but there is no essential difference between them. This is a contradiction caused by making an originally fuzzy set to a deterministic classical set. In automated structural optimum design, such a contradiction will cause some problems.

Similar to the allowable stress, the allowable displacement, size of element and structural natural frequency, etc. are essentially fuzzy set in the domain of discourse of real numbers, i.e. there should be a transient boundary between "absolutely allowable" and "absolutely not allowable" for these sets.

Fuzziness and Randomness of Earthquake Intensity

For historical reasons, discrete universe of discourse of earthquake intensity which contains twelve grades is widely used in the study for earthquakes and earthquake engineering. As earthquake intensity is the comprehensive measurement for the influence of the earthquakes, its domain of discourse in nature is continuous. If the grade has to be divided for application purpose, the boundaries between different grades should be transient and not clear-cut. It is suggested that the continuous universe of discourse [0,12] be used to replace the discrete universe of discourse of twelve grades. The intensity grades in [0,12] are fuzzy, called fuzzy earthquake intensity.

In structural fuzzy optimum design and fuzzy random analysis of structures subjected to earthquakes, the prospected earthquake intensity is described as a fuzzy set on a continuous domain of discourse.

Fuzziness in Site Soil Classification

In structural aseismic design, the predominate period of the ground motion is a very important parameter in determining structural earthquake actions. This parameter depends on the class of site soil. The evaluation for site soil classification should consider soil character, thickness of surface layer, earth layered structure, ground water level, mean shear wave velocity and bearing capacity of the soil. Fuzzy comprehensive evaluation procedure is suitable for carrying out the site soil classification. As the site is decided first, its classification has no randomness, but only fuzziness. After the fuzzy classification vector is evaluated, the predominate parameter used in structural aseismic design could be calculated by using weighted average method.[1]

STRUCTURAL FUZZY RANDOM OPTIMUM DESIGN

Mathematical Model

The mathematical model of structural fuzzy random optimum design which has quite a lot of generality can be expressed by following mathematical programming:

$$\left.\begin{array}{lll}\text{Find structural design scheme } X \text{ to make} & & \\ \mathbf{V-min} & \mathbf{f}_i(X,\ f_i,\ \zeta_i) & (i=1,2,\cdots,I) \\ \text{s.t.} & \mathbf{g}_j(X,\ g_j,\ \xi_j)\subset \mathbf{G}_j(g_j,\ \eta_j) & (j=1,2,\cdots,J)\end{array}\right\} \tag{2}$$

where **V−min** represents fuzzy minimum of multiple objectives; $\subset$ represents the fuzzy inclusion; $\mathbf{f}_i$ is the fuzzy random objective function; $\mathbf{g}_j$ is the fuzzy random constraint function, $\mathbf{G}_j$ is the fuzzy random allowable set of $\mathbf{g}_j$; f_i and g_j are basic variables of $\mathbf{f}_i$ and $\mathbf{g}_j$, respectively; ζ_i , ξ_j , and η_j are random variables in $\mathbf{f}_i$, $\mathbf{g}_j$ and $\mathbf{G}_j$ appeared as parameters.

Solving Method

The way of solving the above programming has been given in [3]. It is the development of the solving methods for fuzzy programming which were proposed in [4,5,6]. The key step in the procedure of finding solutions for this programming is to define the satisfaction degrees to fuzzy random constraints (β_j) and objectives (α_i). Because $\mathbf{f}_i$, $\mathbf{g}_j$ are functions of design scheme X, the event that a fuzzy random objective or constraint function fuzzily included in its satisfaction or allowable set will form a fuzzy sub–field in the design space. α_i and β_j concern the membership degrees of the design scheme to these fuzzy sub–fields, respectively.

The intersections of the above fuzzy sub–fields corresponding to objectives and constraints form fuzzy satisfactory field Θ and allowable field Ω , respectively. The membership degrees of the design scheme to Θ and Ω can be calculated according to the principle of intersection operation between fuzzy random sets as follows:

$$A(X) = \prod_{i=1}^{I} \alpha_i(X) \tag{3}$$

$$B(X) = \min_{j=1}^{J} \beta_j(X) \tag{4}$$

Here, A(X) and B(X) are called comprehensive satisfaction degrees to objective and constraint, respectively.

By using A(X) and B(X), the fuzzy random programming expressed by Eq.(2) can be changed into a series of deterministic programming:

$$\left.\begin{array}{ll}\text{Find } X \text{ to make} & \\ \text{s.t.} & A(X)\geqslant \lambda_A \\ & B(X)\geqslant \lambda_B\end{array}\right\} \tag{5}$$

where λ_A and λ_B are thresholds for A(X) and B(X), respectively. A series of satisfactory solutions can be obtained when λ_A and λ_B vary in interval [0, 1]. Therefore, a higher level selection for optimum solution can be made among the obtained satisfactory solutions.

Practical Structural Optimum Design in Uncertain Environment

Although the above solving method is generally in the theoretical sense, it is sometimes difficult to get applied. The principle difficulty is how to obtain the joint probability density function of the random variables. In addition, it has much too computational work. To simplify the problem for practical design purpose, some available research results should be used.[1]

First, the occurrence probability of hazard of different grades during structural service period is estimated by using hazard risk analysis method. Therefore, the randomness in structural design that comes from the probabilistic character of hazard load is treated. Second, the fuzzy resistance and ductility of conceptual structure (which refers to the structure not yet concretely designed) are obtained by using fuzzy comprehensive evaluation method. Then, the probability that the conceptual structure will damage in certain degrees can be calculated, so can also the loss expectation which results from these damages. By considering structural initial construction cost and its long-term benefit, an optimum defence level for conceptual structures can be decided. Take the aseismic structural design for example; its optimum defence level I_d is decided by following programming:

$$\left.\begin{array}{l}\text{Find } I_d \text{ to make}\\ W(I_d) = C(I_d) + E(I_d) \rightarrow \min\end{array}\right\} \tag{6}$$

where $C(I_d)$ and $E(I_d)$ are structural construction cost and loss expectation from hazard, respectively.

In this way, the originally fuzzy random loads and their caused structural responses will become deterministic if the defence level is decided first. Thus the structural fuzzy random optimum design is degenerated into a simple fuzzy optimum design:

$$\left.\begin{array}{ll}\text{Find X to make} & \\ & C(X) \rightarrow \min \\ \text{s.t.} & g_j(X) \subset G_j\end{array}\right\} \tag{7}$$

The satisfaction degrees to the constraints in this simple situation is:

$$\beta_j = \mu_{G_j}(g_j) \tag{8}$$

where $\mu_{G_j}(g_j)$ is the membership degree of g_j to its fuzzy allowable set G_j. Because structural long term benefit has already been considered in the decision for its optimum defence level, the objective in programming (7) only considers its construction cost. The λ-level cut method can be used to solve programming (7), and a series of satisfactory solutions will be obtained.

GENERALIZED RELIABILITY OF STRUCTURES

Structural reliability is defined as the probability that it can normally work under certain conditions during its service period. As some of the structural resistance and some loads acting on it are fuzzy and random, the event that the structure works normally will then be fuzzy and random. Hence, its reliability should be calculated by using fuzzy probabilistic method.

In structural reliability analysis, structural design scheme X is known. The event that the structure works normally can be described by following fuzzy random constraints which are similar to those given in structural fuzzy random optimum design in Section Ⅲ. They are

$$\omega_j = \{\mathbf{S}_j(X,\eta) \subset \mathbf{R}_j(\eta)\} \quad (j=1,2,\cdots,J) \tag{9}$$

where $\mathbf{S}_j$ and $\mathbf{R}_j$ are structural fuzzy random response and its fuzzy random allowable interval. η is a random vector composed by the random variables in structural design; its joint probabilistic density function is f_η. The satisfaction degree $\beta_j(X,y)$ (y is a realization of η) to fuzzy random constraint ω_j can be defined the same as that given in Section III.

The "minimum safe set" refers to a subset of the entire constraints in which only if all constraints in it were satisfied will the structure be safe. Denote E_l as the l–th minimum safe set, and the l–th fuzzy safe field is the intersection of the sub–fields formed by the fuzzy random constraints in E_l, i.e.

$$\Omega_l = \bigcap_{j \in E_l} \omega_j \quad (l=1,2,\cdots,L) \tag{10}$$

Since each safe state would guarantee safety of the structure, the entire fuzzy safe field is then the union of the fuzzy safe fields Ω_l, i.e.

$$\Omega = \bigcup_{l=1}^{L} \Omega_l = \bigcup_{l=1}^{L} (\bigcap_{j \in E_l} \omega_j) \tag{11}$$

Since $\beta_j(X, y)$ has the sense of membership degree that the design scheme X belongs to the fuzzy safe field ω_j when η has realization y according to the principle of the intersection and union operations between fuzzy sets, the membership degree that the design scheme X and realization y belong to fuzzy safe field Ω(or in other word, structural normal work state) can be calculated as:

$$\mu_\Omega(X,y) = \max_{l=1}^{L} [\min_{j \in E_l} \beta_j(X,y)] \tag{12}$$

The probability that fuzzy random event Ω occurs can be defined, according to fuzzy probabilistic theory, as the expectation of its membership function, i.e.:

$$\Psi(X) = P(\Omega) = \int f_\eta(y)\mu_\Omega(X,y)\, dy \tag{13}$$

The above reliability is called "fuzzy random reliability" which will degenerate into ordinary reliability when there is only randomness or into fuzzy reliability when there is only fuzziness. This concept can be expanded and conclusion can be obtained that any uncertainties could result the problem of reliability.

FUZZY RANDOM VIBRATION ANALYSIS AND DESIGN OF ASEISMIC STRUCTURES

Fuzzy Random Model of Earthquake Ground Motion

Generally, earthquake ground motion can be simulated as a random process. But the parameters in the random model of ground acceleration employed in the design of aseismic structures are often determined according to the aseismic intensity and site soil grade. As mentioned above, seismic intensity and site soil classification have strong fuzziness. By taking this fuzziness into account, a fuzzy random model of earthquake ground acceleration is established, which is a random process with a fuzzy parameter and can be expressed as:

$$A(\xi,t) = \sigma_A(\xi,t)F(\zeta_g,\omega_g,t) \tag{14}$$

in which σ_A^2 is the variance of A, F is a stationary random process, ξ is a fuzzy variable corresponding to the fuzzy intensity $\mathbf{I}_j$ or fuzzy predictive intensity **I**. ζ_g and ω_g are respectively F–expectations of the damping ratio and predominent frequency determined according to the site soil fuzzy grade.

Fuzzy Random Response Analysis of Aseismic Structures

The motion equation of a SDOF hysteretic structure subjected to earthquake expressed by the fuzzy random model (14) can be expressed as :

$$m\ddot{x} + c_0\dot{x} + f_s(x,\dot{x}) = -mA(\xi,t) \tag{15}$$

Equation (15) is a random differential equation with fuzzy parameter ξ. Noting that the fuzzy excitation $A(\xi,t)$ may be expressed by a series of F–sample process such as $A(u,t)$, $u\in U=\{u|\pi_\xi(u)>0\}$, we define that $x(\xi,t)$ is the a.f.m.s solution of Eq.(15), if for any $u\in U$, $x(u,t)$ is the m.s solution of Eq.(15). According to this definition, by means of the stochastic equivalent linearization technique, the analytical and numerical methods are presented for the analysis of fuzzy random responses of SDOF and MDOF aseismic structures[9].

Fuzzy Random Dynamic Reliability Analysis of Aseismic Structures

In the earthquake engineering, the damage grades of a structure are often divided into the following five grades:

$$\{\mathbf{B}_1,\mathbf{B}_2,\mathbf{B}_3,\mathbf{B}_4,\mathbf{B}_5\} = \{\text{Slight,Moderate,Severe,Destructive,Collapsed}\} \tag{16}$$

The definition of these damage grades evidently possess strong fuzziness and cannot use clear–cut values of structural response as the boundaries between these grades. Therefore, every damage grade $\mathbf{B}_i(i=1,2,3,4,5)$ should be a fuzzy subset on the value region of the response.

Let $\mathbf{B}_i^*$ represent the fuzzy safe region in which $\mathbf{B}_i$ or more severe damage will not occur for the structure, the corresponding dynamic reliability can be defined as:

$$P_s(\xi,\mathbf{B}_i^*,T) = P\{x(\xi,t)\subset\mathbf{B}_i^*,0\leqslant t\leqslant T\} \tag{17}$$

in which T is the duration of the earthquake. Since the response $x(\xi,t)$ is not only random but also fuzzy and the safe region is also fuzzy, the dynamic reliability defined by Eq.(17) is called fuzzy random one. Based on the definition (17) and by use of the analysis results of fuzzy random earthquake response, basic formula of fuzzy probability and analysis method of the dynamic reliability, the computational formulas are derived for the fuzzy random dynamic reliability of SDOF and MDOF aseismic structures.

The fuzzy random dynamic reliability mentioned above has comprehensively taken account of the fuzziness and randomness of the earthquake ground motion, structural hysteresis and fuzziness of damage grades. Thus, according to its definition and computational formulas, the reliability that a structure under moderate or strong earthquake will not suffer from the moderate or collapsed damage can be easily obtained. Therefore, the reliability analysis above provides a feasible and reasonable method for the safety estimation and probabilistic design of aseismic structures.

Fuzzy Damage Prediction and Loss Assessment of Aseismic Structures

The relationship between the earthquake intensity and duration time is further considered and the predictive fuzzy random model $\{\pi_{\xi}(u),A(u,t),T(u)\}$ is presented for the earthquake ground motion during the future Γ years. After the failure probability $P_{f\Gamma}(\mathbf{B}_i)$ $(i=1,2,3,4,5)$ that the structure will suffer from only $\mathbf{B}_i$ damage grade during the future Γ years is respectively obtained and the failure loss $C(\mathbf{B}_i)$ corresponding to the damage grade $\mathbf{B}_i$ which the structure will suffer from is estimated, the expected failure loss can be obtained as:

$$L_{\Gamma} = \sum_{i=1}^{5} L(\mathbf{B}_i)P_{f\Gamma}(\mathbf{B}_i) \tag{18}$$

Such a method to find the expected failure loss is the basis to realize the optimum design of aseismic structures based on the double objectives (initial construction cost and expected failure loss) and dynamic reliability constraints.

Design of Nonlinear Aseismic Structures Based on the Fuzzy Random Dynamic Reliability

A method is presented for the design of nonlinear aseismic structures (including steel, reinforced concrete and masonry structures) based on the analysis of fuzzy random dynamic reliability. First, a technique is presented for the determination of the parameters in the restoring force model corresponding to the design variables of structures by means of the structural analysis and experimental results. Then, a comprehensive damage index model taking account of the effects of deformation and cumulative hysteretic energy is established by using the low cycle fatigue experimental results. Since this index comprehensively takes account of the effects of deformation and cumulative hysteretic energy on the structural damage and its value region is the interval [0,1], the fuzzy damage grade $\mathbf{B}_i$ and fuzzy safe region $\mathbf{B}_i^*$ can be well determined by the index.

Let the comprehensive damage index be D_c, the probability that D_c falls in the fuzzy safe region $\mathbf{B}_i^*$:

$$P_s(\mathbf{B}_i^*) = P\{D_c \subset \mathbf{B}_i^*\} \tag{19}$$

is the reliability that the structure will not suffer from $\mathbf{B}_i$ or more severe damage. Then, the design of aseismic structures based on the fuzzy dynamical reliability can be formulated as:

$$\left.\begin{array}{ll} \text{Find:} & \text{Design Vector } X \\ \text{S.t.} & P_s(\mathbf{B}_i^*, X) \geqslant [P_i^*] \end{array}\right\} \tag{20}$$

This design formulation can well express the allowable damage grade $\mathbf{B}_i$ of a structure under earthquake with different intensity and the corresponding lowest allowable reliability $[P_i^*]$.

Design of Nonlinear Aseismic Structures Based on the Loss Assessment
The basic principle of loss assessment is presented and the loss function corresponding to the damage grades $\mathbf{B}_i$ is established for aseismic structures. Then, an optimum design method is presented and realized in the numerical examples for nonlinear steel, reinforced concrete and masonry structures based on the double objectives and dynamic reliability constraints. This optimum design can be formulated as:

$$\left.\begin{array}{ll}\text{Find:} & \text{Design Vector } X \\ \text{Min:} & W_T = C_I + L_f \\ \text{S.t.} & P_s(\mathbf{B}_i^*, X) \geq [P_i^*]\end{array}\right\} \tag{21}$$

in which C_I and L_f are respectively the initial construction cost and expected failure loss.

GLOBAL OPTIMIZATION OF LARGE-SCALE ENGINEERING SYSTEMS

Proposal of the Problem
Present engineering optimal design methods are for structures only. However, the engineering system combined with the structures optimized individually, which is similar to a structure combined with the elements optimized individually, will probably not be optimal. This is because, first, some local benefits should be sacrified in order to make an optimal global benefit; second, the individual optimization of local components cuts down their organic relations. In order to obtain an optimization for an engineering system, it must be designed globally. This is the idea of the global optimization of engineering system.

Large-Scale Engineering System Model of an Engineering Project
After the feasible analysis of the engineering project has been made and approved, a global allocation should be made for it, which includes how many subsystems the project will have and how many structures each subsystem will have, etc. The system obtained by this type of global allocation is refered to as "The Engineering System". By doing function analysis described as follows for the engineering project, a hierarchical system model can be established for it. First, according to its required function relations, the engineering project can be divided into a series of first-level subsystems in which if any of them loses efficiency, the entire system will lose efficiency. This will result in the subsystems having a series relationship. Then, the first level subsystems can be further decomposed into a series of second-level subsystems followed by third-level subsystems which are formed by several series or parallel structures.

Parameter Selection for the Global Optimization
"Low Cost and High Quality" are requirements for any production. In structural engineering, "High Quality" can be described by reliability since it is the probability that the structure or engineering system works normally under certain conditions during its service period, and "Low Cost" can be evaluated by considering both the present investment and future loss expectation because these two represent both present and long-term benefits of the systems. Therefore, the construction cost C and loss expectation L are selected as decoupling parameters, and the reliability ψ as the coordinating parameters.

Relation between the Construction Cost and Reliability of a Structure and its Loss Expectation

In general, structural construction cost C_j will increase with the increment of structural reliability ψ_j, so function $C_j(\psi_j)$ is an increasing function. Following equation is suggested to represent the relation between the construction cost and reliability of a structure:

$$C_j(\psi_j) = [1 - (1/\alpha_j)\ln(1 - \psi_j)]C_{0j} \tag{22}$$

where α_j and C_{0j} are empirical parameters.

If the loss caused by structural failure is D_j, the loss expectation L_j can be calculated as:

$$L_j = (1 - \psi_j)D_j \tag{23}$$

where the estimation for D_j has relation with structural failure criterion.

Relation between the Construction Cost and Reliability of an Engineering System and its Loss Expectation

Given a reliability ψ for an engineering system, its corresponding construction cost C is defined as the minimum one among the construction costs which can maintain this reliability. According to this definition, the relation between the construction cost and reliability of a series engineering system can be derived approximately as follows when subsystems are independent of each other.

$$C(\psi) = [1 - (1/\alpha)\ln(1 - \psi)]C_0 \tag{24}$$

where α and C_0 can be calculated from parameters α_j and C_{0j} of structures[10].

The loss expectation of an engineering system has relations both with the estimation for the loss of the structures it contains, and with the reliability distribution within the engineering system. After the possibility that two or more than two structures fail simultaneously is neglected, the loss expectation of a series engineering system can be calculated as[10]:

$$L = \psi \sum_{j=1}^{J} (1 - \psi_j)D_j / \psi_j \tag{25}$$

Thus, the equivalent loss value of the engineering system is

$$D = \frac{L}{1-\psi} = \frac{\psi}{1-\psi} \sum_{j=1}^{J} (1 - \psi_j)D_j / \psi_j \tag{26}$$

Optimization Criterion

Since construction cost and loss expectation are increasing and decreasing functions of reliability, respectively, there must exist a value of reliability which can optimally balance the construction cost and loss expectation, i.e. optimally balance the present and long-term benefits. This is the optimization criterion. It can be expressed as following programming:

$$\text{Find } \psi \text{ to minimum } C(\psi) + L(\psi) \tag{27}$$

Substitute Eq.(22) and Eq.(23) or Eq.(24) and Eq.(26) into the above equation; the optimal reliability can be derived:

$$\psi^{*} = 1 - C_0 / \alpha D \tag{28}$$

This equation is valid to both structures and engineering systems.

Preparation Stage of Global Optimization

First, the parameters which are necessary to global optimization procedure should be calculated with following steps(take three–level hierarchical engineering system as an example):

1).Estimate the parameters C_{0ij} and α_{ij} which are contained in the relation between structural construction cost and its reliability as well as the loss value D_{ij} caused by structural failure.
2).Calculate the corresponding parameters C_{0i} and α_i for each subsystem with the above structural parameters.
3).Calculate the corresponding parameters C_0 and α for entire system with the above parameters of subsystems.

Then, the initial values for optimization iteration are determined with following three steps:

1).Calculate the initial optimal reliability $\psi_{ij}^{(0)}$ for structures with their parameters C_{0ij}, α_{ij} and D_{ij}.
2).Calculate the reliability $\psi_i^{(0)}$ and equivalent loss value $D_i^{(0)}$ for each subsystem with structural $\psi_{ij}^{(0)}$ and D_{ij}.
3).Calculate the reliability $\psi^{(0)}$ and equivalent loss value $D^{(0)}$ for top system with $\psi_i^{(0)}$ and $D_i^{(0)}$ of the subsystems'. Then, determine the initially optimized reliability $\psi^{(1)}$ for the top system according to C_0, α, and $D^{(0)}$.

Iteration Stage of Global Optimization

There are six steps in this stage.

1).Calculate the optimized reliability ψ_i^{*} for each subsystem by making optimal apportionment for the initially optimized reliability $\psi^{(1)}$.
2).Calculate the optimized reliability ψ_{ij}^{*} for each structure by making optimal apportionment for the obtained ψ_i^{*}.
3).Calculate the loss expectation L_{ij}^{*} for each structure according to its obtained optimized reliability ψ_{ij}^{*}, and then report it to the subsystem to which the structure belongs.
4).Calculate the loss expectation L_i^{*} and equivalent loss value $D_i^{(1)}$ for each subsystem according to the reported L_{ij}^{*}, and then report it to the top system.
5).Calculate the equivalent loss value $D^{(1)}$ for the top system according to the reported L_i^{*}, and then a new optimized reliability for the top system $\psi^{(2)} = \psi^{*}$ can be determined.
6).The above five steps finish the first iteration of the global optimization. The second iteration starts by replacing the old $\psi^{(1)}$ with $\psi^{(2)}$ and returning to step 1). The iteration will be repeated untill the optimal reliabilities ψ_i^{*} and ψ_{ij}^{*} convergent.

ACKNOWLEDGEMENT Many thanks for the support of Chinese Natural Science Foundation and Ministry oi Construction of China in our research work.

REFERENCES

[1] Wang Guangyuan, 《Theory of Soft Design in Engineering》, Science Press, 1991(in chinese)

[2] Wang Guangyuan, "Methods of Treatment for Fuzzy Factors in Earthquake Engineering", *proc. of 9th World Conference on EarthquakeEngineering*, Tokyo–Kyoto, Japan, 1988

[3] Wang Guangyuan and Tan Dongyao, "General Programming in Random Fuzzy Environment", 《*Advancement ofFuzzy Theory and Systems in China and Japan*》, International Academic Publishers, 1990

[4] Wang Guangyuan and Wang Wenquan, "Fuzzy Optimum Design of Structures", 《*Engineering Mechanics in Civil Engineering*》 (Edited by A.P. Boresi and K.P. Chong), published by ASCE, 1984

[5] Wang Guangyuan and Wang Wenquan, "Fuzzy Optimum Design of Structures", *Engineering Optimization*, Vol.8, No.4, 1985

[6] Wang Guangyuan and Wang Wenquan, "Fuzzy Optimum Design of Aseismic Structures", *Earthquake Engineering and Structural Dynamics*, Vol.13,No.6, 1985

[7] Wang Guangyuan and Wang Wenquan, Fuzzy Reliability Analysis of Aseismic Structures", *proc. ofInternational Symposium on Fuzzy Mathematics in Earthquake Researches*, Beijing, 1985

[8] Wang Guangyuan, Wang Wenquan and Ou Jinping, "Generalized Reliability of Engineering Systems", *proc. of Second International Fuzzy Systems Association Congress*, Tokyo, Japan, 1987

[9] Wang Guangyuan and Ou Jinping, "Fuzzy Random Vibration of Multi–degree–of–freedom Hysterestic Systems Subjected to Earthquakes", *Earthquake Engineering and Structural Dynamics*, Vol.15, No.5, 1987

[10] Wang Guangyuan and Tan Dongyao, "Global Optimization of Large Scale Engineering Systems", *proc. of The International Conference on Structural Engineering and Computation*, Beijing, 1990

SUPERCOMPUTER APPLICATIONS

CHARLES R. FARRAR
CHARLES A. ANDERSON
Advanced Engineering Technology
Los Alamos National Laboratory
Los Alamos, NM 87545

ABSTRACT

Supercomputers can be used to solve structural problems characterized by complex material behavior and geometry. This paper presents three examples of supercomputer applications in structural mechanics: a seismic buckling analysis of a reactor containment; an analysis of the time-dependent behavior of a piping system; and an analysis of time-dependent deformation caused by a rupture along a fault in the Earth's crust.

INTRODUCTION

The development of numerical methods and associated computer software for the analysis of multidimensional structures when loaded by long-term time varying or constant loads has had an explosive growth in the last 20 years. Thus, today it is possible to analyze structural problems with enormous geometric complexity, with sophisticated inelastic constitutive models, and with inclusion of physical phenomena such as large deformations, hydrothermal diffusion, pore pressure, and chemical effects, which, until now, have been treated only in an *ad hoc* fashion. Indeed, with computer software available today, and with high-speed and expanded memory capacity of modern scientific computers, the primary limits on our structural analysis capabilities are the availability of experimental data to constrain constitutive models, experimental information on the behavior of joints, faults, and interfaces, and the ability to post-process the large amounts of data that are generated by these sophisticated analyses.

This paper discusses applications of supercomputers and the finite element method to solving complex problems in structural mechanics that have arisen at a large national laboratory. In this context supercomputers refer to machines such as a CRAY Y-MP that has 8 processors and 128×10^6 64 bit words of directly addressable memory. This machine can perform 2×10^9 floating point operation per second. Examples that will be illustrated include the analysis of time-dependent behavior of a nuclear power plant piping system, the analysis of deformations caused by cyclical fault rupture, and the prediction of the structural stability of containment structures under seismic loading.

The calculations described in this paper were carried out at the world's largest supercomputer facility located at the Los Alamos National Laboratory. The supercomputer facility currently possesses the computing power equivalent to over 70 CRAY-1 machines. The mainframe machines are networked to workstations with state-of-the-art graphics capabilities. Also included in this facility is an advanced computing laboratory featuring massively parallel computer architectures. Software for structural applications includes the well-known computer codes NASTRAN [1], ABAQUS [2], and PATRAN [3] (for graphics), as well as many special purpose computer codes for studying impact phenomena, structural acoustics, and thermomechanical diffusion.

SEISMIC BUCKLING OF A NUCLEAR REACTOR CONTAINMENT SHELL

This analysis is an example of a linear, elastic problem that is impractical to solve unless supercomputers are employed. The structure that is analyzed consists of a thin-walled (4.45-cm) steel cylindrical containment vessel with a hemispherical upper steel closure. The lower closure is ellipsoidal in shape and is cast in concrete. The radius-to-thickness ratio (R/t) of this containment is 645. To put this into perspective, a typical soft drink can has an R/t of approximately 260.

First, a 3 degree-of-freedom, lumped mass model that is based on simple beam theory was used to estimate the buckling capacity of the containment, as defined by ASME Boiler and Pressure Vessel Code Case N-284 [4]. Response spectrum analysis techniques were employed, and the containment was found to reach an incipient buckling state at 0.34 g's. A study of this analysis procedure revealed that simple beam theory does not capture the effects of cross-section ovalization, which can produce significant compressive hoop stresses that further reduce the buckling capacity of the structure. A finite element model was developed to accurately model the ovalization. An axisymmetric half model of the containment shell, as shown in Fig. 1, was generated using the eight-node quadratic shell element available in ABAQUS, a general purpose finite element structural analysis code.

Using ABAQUS, a conventional modal analysis was first performed to determine the frequencies of the various response modes of the containment shell. Results are shown in Fig. 2 for some of the lower shell modes. The fundamental shear-bending mode (n=1, m=1), shown in Fig. 3, occurs at approximately 7.1 Hz. Figure 2 shows that, even within a frequency band of 1 Hz about the fundamental shear-bending mode, a large number of shell modes are present. This excessive

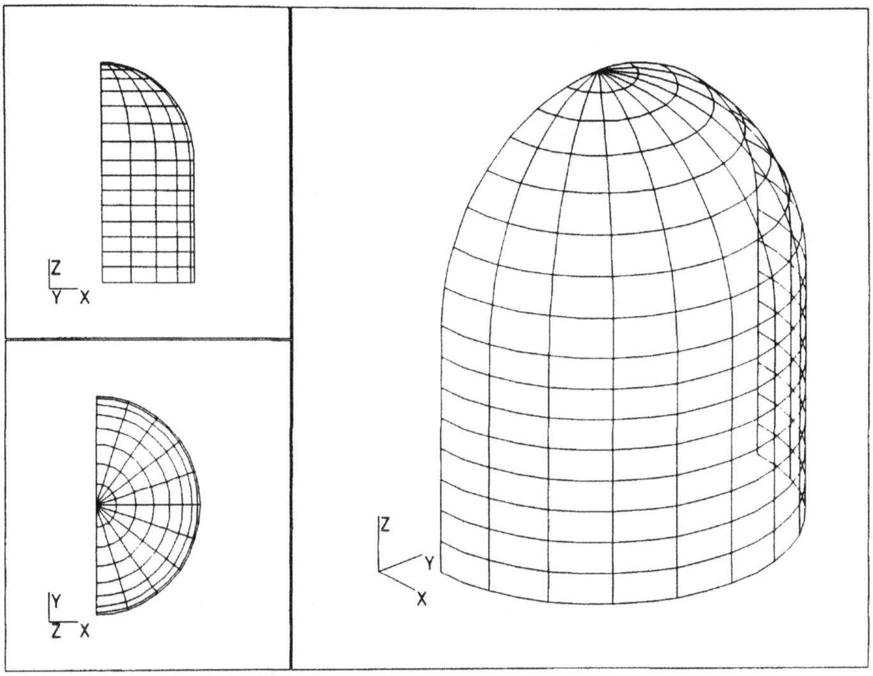

Figure 1. Finite element model of a steel containment structure.

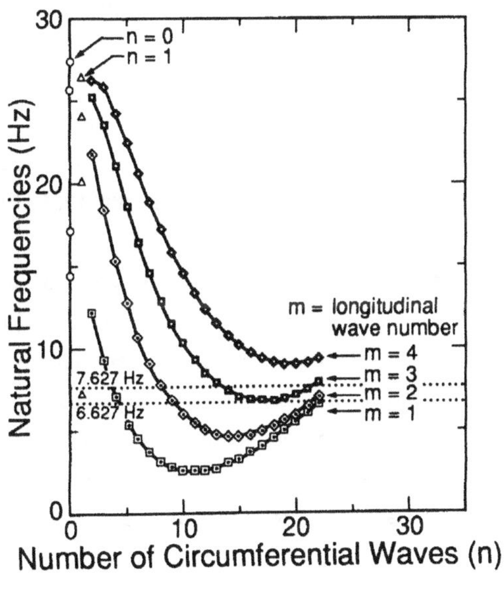

Figure 2. Resonant frequencies for the lower modes of the steel containment shell. M equals the longitudinal wave number.

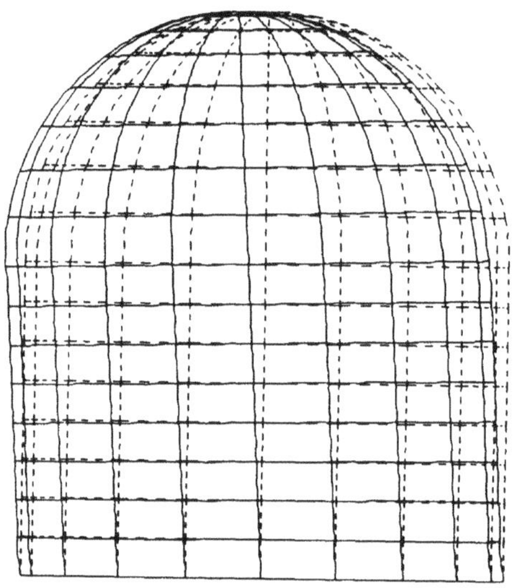

Figure 3. The shear-bending (n=1, m=1) mode of the steel containment structure.

number of low-frequency shell modes makes solution of the problem by the commonly used modal response spectrum approach impractical. Rather, transient-time integration was chosen, using earthquake acceleration-time histories as input to the base of the structure.

Actual recorded earthquake signals were used in the time history analysis. Various combinations of seismic excitation were applied to the base of the ABAQUS containment shell model in the two horizontal directions (global X and Y) and the vertical direction (global Z). Seismic acceleration-time histories used in this study are the strong motion portions (6-s duration) of the 1940 El Centro N-S component, the 1949 Olympia N86E component, and 1935 Helena west component scaled to 0.3 g's.

Runs were first made to determine the appropriate equilibrium tolerances that were necessary to obtain convergence. The model was then excited in the Y and Z directions, using the Olympia and Helena earthquake acceleration histories, respectively. A second axisymmetric model was run next, with the El Centro acceleration-time history applied in the y direction. Results of the time integration of the second run were then rotated 90-degrees (which simulated X-direction excitation) and were superposed with the earlier results of the Olympia-Helena run. This superposition of stress components results in the three-dimensional response of the containment shell from transient base excitation in the three orthogonal directions. Superposition of stress components by this method is valid when the response is small-deflection, elastic behavior. A total of approximately 10 hours of CPU time on a CRAY Y-MP was required for the combined analysis.

Results of combined stresses are shown on the ASME Code's longitudinal compressive/shear stress interaction curve in Fig. 4 for 1-percent damping. The ASME Code buckling interaction curve is also shown in Fig. 4, along with all the combined stress points calculated for each time step at critical integration points near the base of the containment. Figure 4 clearly shows that, for 1-percent damping, a substantial number of combined stress points lie outside the design interaction curve, implying that Code Case N-284 buckling criteria have been exceeded.

Because of the linear nature of this problem, these results can be used to determine the peak g-level to which the input signals should be scaled such that a condition of incipient buckling occurs. When this relation, which is quadratic in g, is solved, 0.18 g's is identified as the peak acceleration by which to scale the three inputs to produce an incipient buckling condition.

Results for 2-percent damping are shown in Fig. 5. As anticipated, considerably fewer stress combinations are outside the interaction curve. These results imply that the input transients would have to be scaled to a peak acceleration level of 0.22 g's to cause a state of incipient buckling.

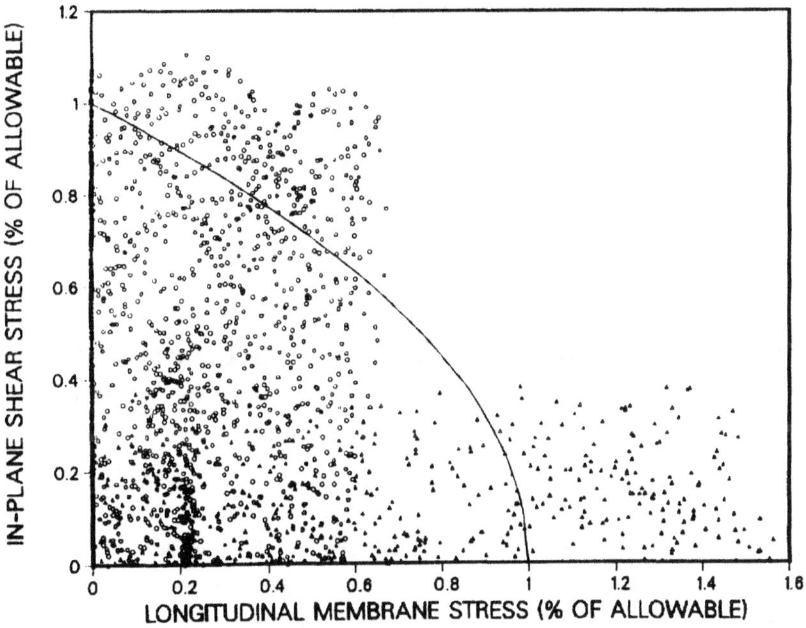

Figure 4. The normalized stress state for critical locations at the base of containment compared with the buckling interaction equation when 1% damping is assumed.

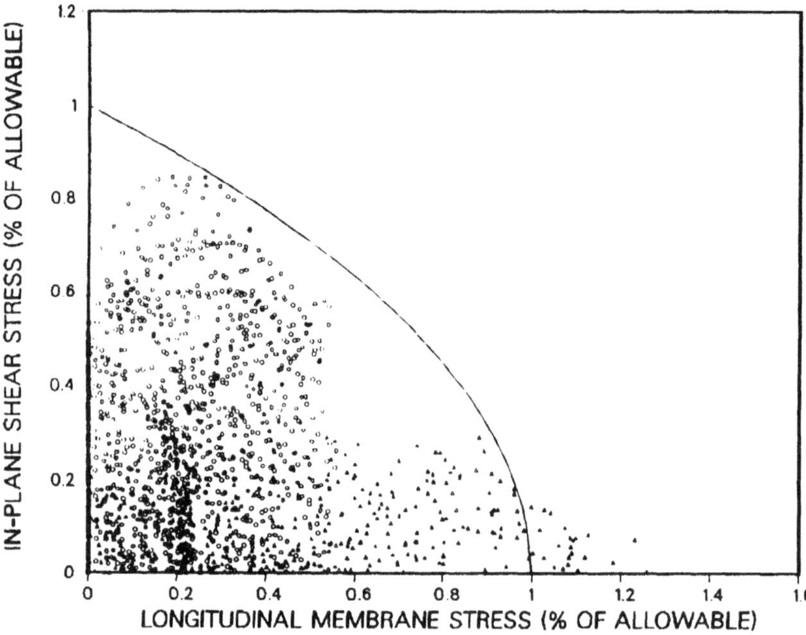

Figure 5. The normalized stress state for critical locations at the base of the containment compared with the buckling interaction equation when 2% damping is assumed.

For this type of analysis, supercomputers provide the means to accurately model the actual deformation that the structure undergoes and, hence, accurately predict the stresses that the structure experiences. In addition, computers allow us to vary unknown parameters, such as damping, and to study the sensitivity of the results to such parameters.

TIME-DEPENDENT INELASTIC ANALYSIS OF NUCLEAR REACTOR PIPING

Accurate assessment of creep in high-temperature piping systems, over long time periods, poses a problem that is best analyzed with supercomputers. In the problem discussed here, the ABAQUS finite element code was used to determine the nonlinear response of a typical hot leg in the primary heat transport system (PHTS) sodium loop of a liquid metal fast breeder reactor (LMFBR). The piping system response is caused by the loads specified in a histogram developed by the reactor designers, which envelops all the expected normal, upset, and emergency load conditions throughout the 30-year operating life of the system. The load histogram includes the effects of gravity, internal pressure, thermal transients including creep-hold time, and differential movement of the end boundaries during temperature variations.

The piping system analyzed is the portion of the PHTS piping between the primary sodium pump and the intermediate heat exchanger (IHX), as shown in Fig. 6. This portion of the PHTS consists of 0.61 m o.d., 316 stainless , insulated pipe, 12.7 mm thick. The pipe is rigidly fixed to the vessels at each end and supported at intermediate locations by five hanger supports. The operating temperature of the system is 546°C, and the operating pressure is 1.16 MPa.

The portion of the load histogram that was analyzed is summarized in Table I. The initial step of the sequence is used to calculate hanger reactions. The rest of the load histogram is intended to encompass 863 possible thermal transient events identified by the reactor-plant designers. Load steps 1-18 comprise the 36 most severe thermal downshock transients and represent the portion of the load histogram that was analyzed.

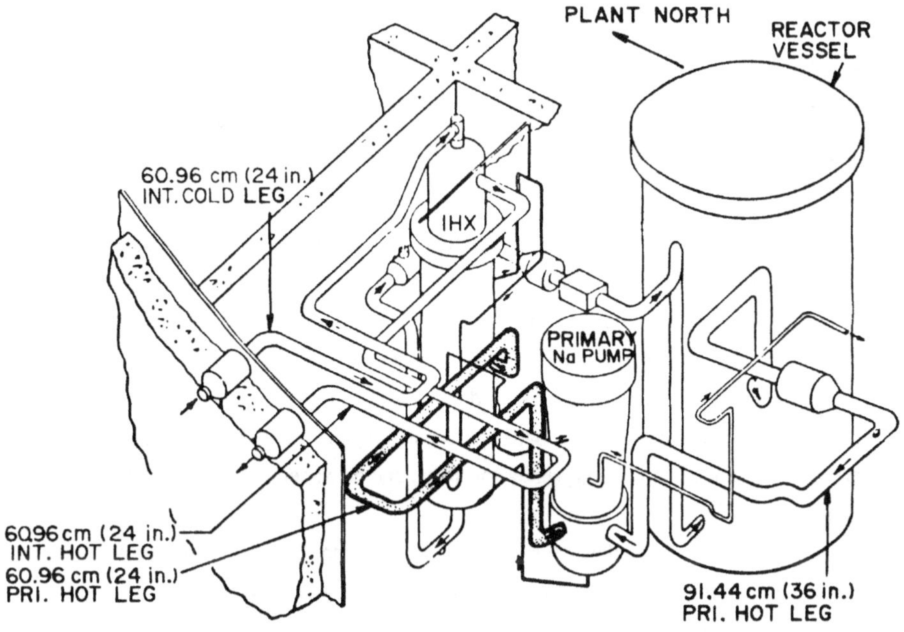

Figure 6. Typical in-containment Primary Heat Transport System (PHTS) piping; highlighted section is the portion of the PHTS being analyzed.

TABLE I
Load Histogram

Load Steps	Temperature Range[a]	Description
i	21.1	Set hanger forces: pressure (1.16 MPa), dead weight at room temperature.
1	21.1-.546.	First heat-up transient: increase pressure and dead weight to three times the value used in step i. Apply three times the constant hanger forces calculated in step i as concentrated forces at hanger locations. Increase temperature to system operating temperature
2	546.	Creep-hold time of 8258 h.
3	546.	Reduce pressure to 2.32 MPa.
4-18[b]	546.-204.	First downshock transient.

[a]Temperatures are given in degrees Celsius.

[b]Multiple load steps are required to incorporate the nonuniform radial temperature distribution in the pipe during the transient.

The pipe and pipebend elements of ABAQUS allow three temperatures to be specified through the thickness of the pipe. A previously performed heat transfer analysis provides temperature distributions through the pipe thickness during the thermal transient events. Temperatures are assumed to be constant in the axial direction for the entire system.

When thermal transients occur, the IHX and primary sodium pump experience unequal thermal expansions and contractions. The loads induced in the pipe system by this differential end movement are incorporated into the analysis by specifying end displacements as a function of temperature.

The pipe system is modeled with the elbow elements of the ABAQUS finite element code. To model both the predominantly shell behavior of the pipe bends and the predominantly beam behavior of the straight segments with a single type of element, the code considers the pipe as a beam with a deforming thin-walled cross-section and restricts the response to small displacements. This restriction allows the strains from shell and beam behavior to be analyzed separately and the results linearly superposed. The combination of shell and beam properties enables the elbow elements to accurately predict the response of the piping system to such effects as (1) increased stiffness from internal pressure, (2) increased flexibility from ovalization of an elbow, (3) the transition from shell behavior of an elbow to beam behavior of the the adjacent straight segment, and (4) warping from curved beam bending and torsion of the ovalized section.

The six elbows of the system being analyzed and, where geometrically possible, the adjacent 0.61 m of straight pipe are modeled with the elements that allow both warping and ovalization of the cross-section. The straight portions are included to provide the shell/beam behavior-transition range mentioned above. The remaining straight portions of the pipe are modeled with elements that allow only ovalization. The ovalization in the elements is constrained to uniform radial expansion on the straight segments of pipe away from the elbows to model the effects of internal pressure.

To maintain continuity, boundary conditions must be specified at the interface between locations where warping was allowed to occur and locations where warping is constrained. Both warping and ovalization must be constrained at this point, reducing all cross-sectional deformations to uniform radial expansion. The user, therefore, specifies the transition point from shell to beam behavior. The ovalization and warping are also constrained at the ends of the system where the pipe intersects the IHX and primary sodium pump. The pipe/vessel connections are both modeled as rigid, and the nozzle stiffness at these locations is neglected. The constant hanger supports are assumed to be axial members only, and any rotational restraints imposed by the hangers are neglected.

The temperature-dependent mechanical properties are used in the analysis along with a user-specified creep law. Environmental effects on material properties (such as irradiation, thermal aging, and corrosion, etc.) have been neglected. A kinematic hardening law is used, and the stress-strain curve is idealized as bilinear. The finite element discretization is shown in Fig. 7.

Postprocessing subroutines were developed to provide graphical displays of the results. The displays fall into two categories: profile plots that, for a particular load step, show the circumferential distributions of a variable at a cross-section in the piping system and history plots that describe the change in a particular variable over the entire load history.

The graph shown in Fig. 8 provides a three-dimensional time history of the maximum equivalent stress calculated for each element during the first two load steps. Load increments corresponding to creep-hold time (increments 7-99) representing 8258 h do not represent uniform time increments because ABAQUS automatically increases the time interval incrementally while integrating the creep-strain-rate equation.

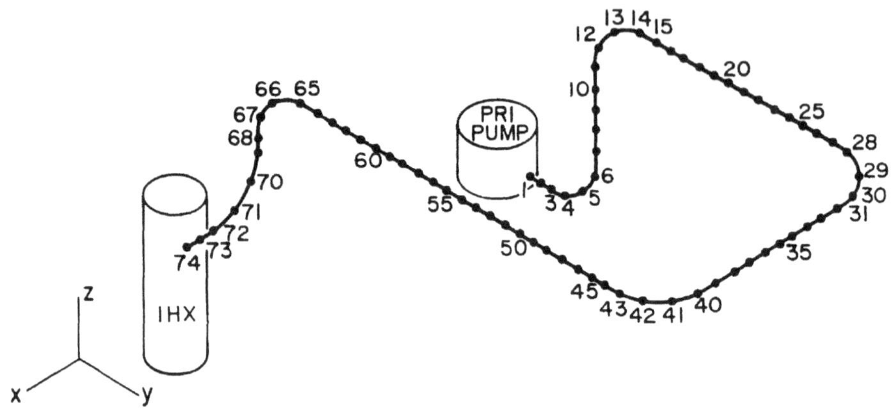

Figure 7. The finite element model of the primary heat transport system piping. An element is defined by two node points.

Because an elbow behaves like a shell, bending causes the cross-section to ovalize, which, in turn, reduces the flexural rigidity of the system. In the presence of bending caused by thermal loads, the system deformations will tend to be concentrated at the elbows; the concentration of deformation, which results from shell behavior, causes higher stresses and strains at these locations. Figure 8 depicts the equivalent-stress level in the piping system when the system is exposed to bending from thermal loads. As expected, six distinct peaks in the equivalent-stress plots occur at locations corresponding to the six elbows of the system. The peaks are maintained throughout the portion of the load histogram that has been analyzed. The location of the maximum equivalent stress varies from elbow 4 to elbow 5.

The effects of creep can be seen in Fig. 8. Most apparent in this figure is the strong dependence of the equation representing creep-strain rate on the equivalent-stress level. The rate of stress relaxation is a function of creep-strain rate which, in turn, is a function of equivalent stress and temperature. When the temperature is held constant in load step 2, the rate of relaxation is greatest at the locations where equivalent stress is greatest, that is, at the elbows. In the straight runs of pipe where equivalent stress is less, relaxation is not discernible.

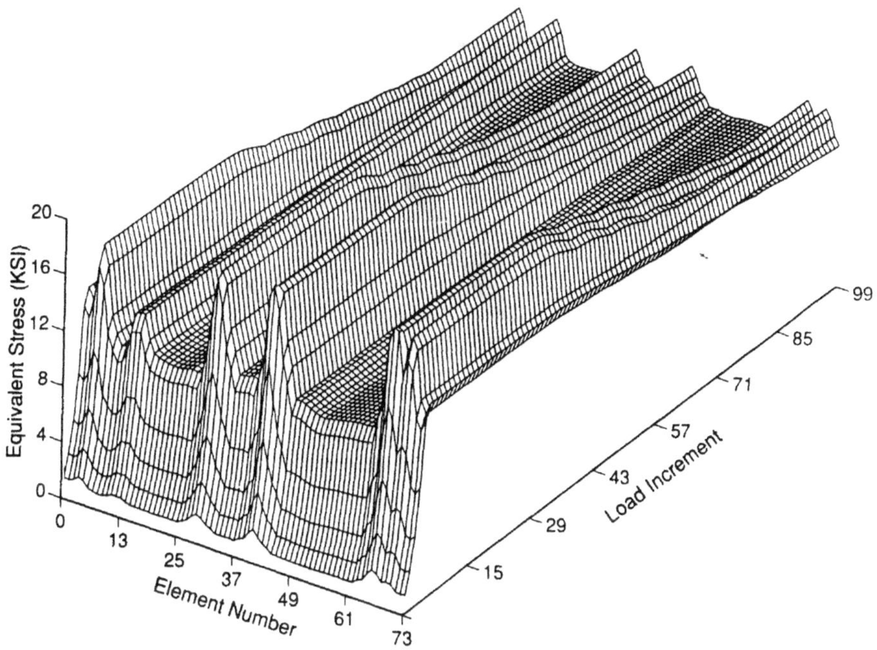

Figure 8. Maximum von Mises equivalent stress calculated for each element during the first heat-up transient and the creep-hold time.

In summary, the nonlinear analysis capabilities of the ABAQUS finite element code have been used to model the response of a typical LMFBR elevated-temperature piping system subjected to normal, upset, and emergency loads. The results obtained [5] show that radial thermal gradients caused by the first thermal downshock outlined in Table I produce the most severe stress levels experienced by the system. Repeated cycles of similar thermal transients will produce thermal ratcheting, leading to progressive deformation and strain accumulation. Analysis of this piping system required approximately 3 h of central processor time on a CRAY-1 computer.

MODELING DEFORMATION DURING GREAT CALIFORNIA EARTHQUAKE CYCLES

Li and Rice [6] have recently explored an analytic model for crustal deformation in great California earthquake cycles. As shown in Fig. 9, their model features an upper elastic layer and lower viscoelastic layer corresponding to a crustal asthenosphere. The top layer is divided by a transform fault whose upper seismogenic part alternately slips and locks and whose lower part (still in the elastic layer) freely slips consistent with the

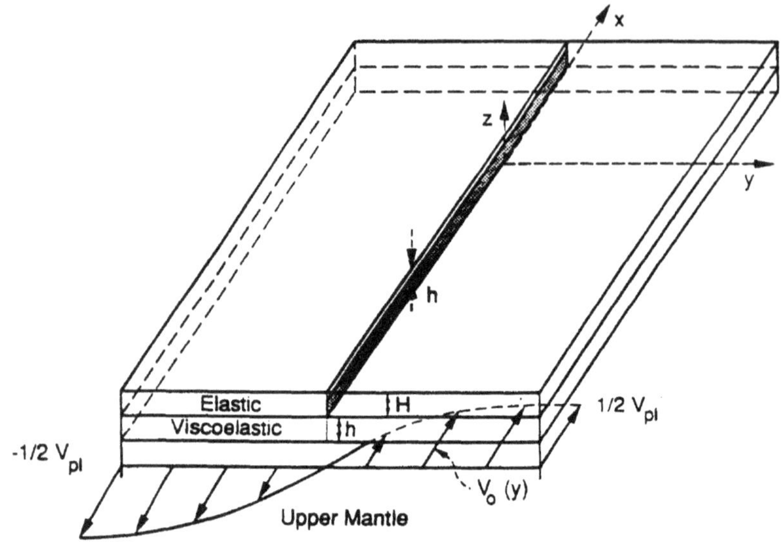

Figure 9. Elastic lithosphere coupled to a viscoelastic asthenosphere driven by deep mantle movement.

steady relative motion on the opposite sides of the fault. Following Elsasser [7], tectonic loading of the crust by the motions in the underlying mantle, are transmitted to the model crust through the viscoelastic crustal asthenosphere. The model is able to produce deformation associated with sudden earthquake slip in agreement with geodetic observations of deformation at various locations along the San Andreas fault. Matching of model results to observations constrains values of model parameters describing the subsurface structure in the vicinity of the fault. The model is, however, a relatively simplified one, and it cannot provide a variety of the information on subsurface structure and properties that would be useful to have. For example, only average stresses are considered within the two layers of the analytic model and only linear rheological behavior can be simulated. The analytic model cannot be used to treat variations in fault behavior parallel to the strike of the fault. In general, three-dimensional effects associated with multiple faults, curved fault planes, patchiness along faults, and the like, are beyond the capability of the analytic model. It is therefore appropriate and important to extend and generalize the Li and Rice model to a fully three-dimensional numerical finite element model that can bring more realism to fault behavior simulations and, in turn, give more information on subsurface fault structure and material properties after constraining model results with geodetic data. The Li and Rice model provides a reference against which to benchmark the more complicated and realistic finite element model.

The Li and Rice problem was solved exactly (to within numerical accuracy) with the finite element code ABAQUS. The mesh that was used is illustrated in Fig. 10. There is only one row of elements parallel to the fault, a simplification made possible by the anti-plane strain nature of the problem. The elastic layer is 22.5 km thick and the viscoelastic layer is 13.9 km thick. The mesh extends 312.5 km perpendicular to the fault and contains 990 nodes and 448 elements. The calculation is carried out on only one side of the fault to take advantage of the symmetry of the problem. The fault extends throughout the elastic layer. Other parameter values of the model problem are earthquake cycle time t_{cy} = 160 years, shear modulus in the elastic layer G = 1.05 x 10^5 MPa, shear modulus in the viscoelastic layer G_1 = 1.05 x 10^4 MPa, Poisson's ratio = 0.25, viscosity in the viscoelastic layer η = 2.1 x 10^5 MPa-y, velocity applied at the end of the model $V_p/2$ = 8.75 mm/year. The shear modulus G_1 in the viscoelastic layer is a factor of 10 smaller than G in the elastic layer because this is the circumstance in which the Elsasser model is most realistic. The corresponding parameter values for the generalized Elsasser model of Li and Rice are a relaxation time t_r = η/G_1 of 20 years and a coupling parameter b = 139 m.

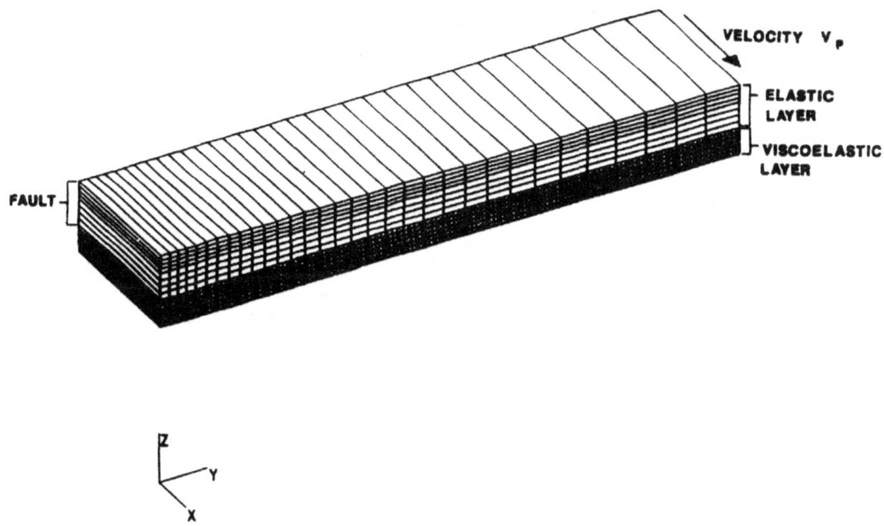

Figure 10. Mesh used in the finite element model of the anti-plane strain problem.

Boundary conditions for the numerical calculation are as follows. The end of the model far from and parallel to the fault is moved at the constant horizontal fault parallel velocity $V_p/2$ for a time of t_{cy}. Vertical velocity and fault perpendicular velocity are zero on this end. The top surface is stress free. The bottom surface and the extension of the fault plane downward into the viscoelastic layer are also moved with the fault parallel velocity $V_p/2$. The fault plane in the elastic layer initially is held fixed. It is pinned during t_{cy}. Finally, it is displaced parallel to itself and horizontally by an amount $t_{cy} \cdot V_p/2$ during a very short time interval at the end of a cycle to catch up with the far end. The numerical model is exercised over a number of cycles to produce a quasi-steady solution.

Figure 11 shows the elastic thickness-averaged fault parallel velocity (normalized by V_p) at 8.7 years into a 160-year earthquake cycle vs perpendicular distance from the fault (normalized by the elastic layer thickness) for the Li and Rice model and the numerical model (the numerical model was run for 5 cycles and the results plotted for corresponding times in the fifth cycle, i.e., 648.7 years = 4 x 160 years + 8.7 years). Agreement between the numerical and approximate analytic models is excellent at three given times during the earthquake cycle.

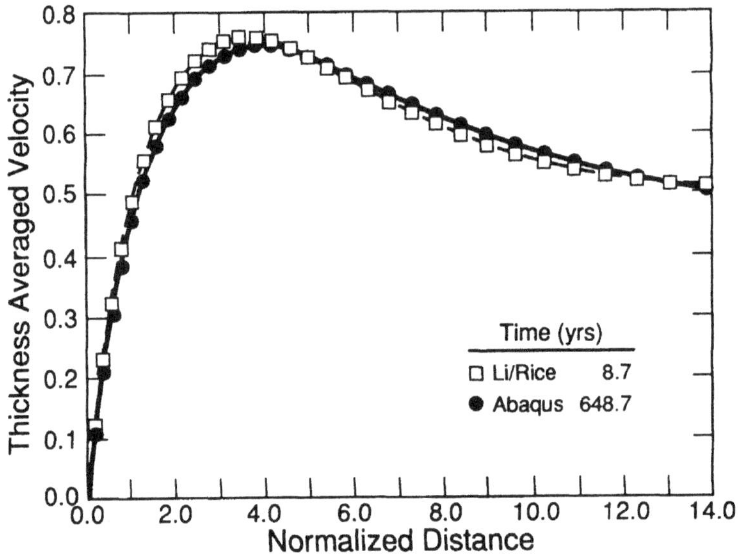

Figure 11. Thickness-averaged velocity vs distance from the fault. A comparison of results computed by ABAQUS using the anti-plane strain model and the analytical results of Li and Rice.

Having benchmarked the finite element code against the Li-Rice solution, finite element calculations were recently completed that use a realistic creep rheology for the behavior of the crust and underlying asthenosphere. The creep law employed in initial calculation is of the form

$$\dot{\gamma} = A \tau \exp(-Q/RT) \quad , \qquad (1)$$

where $\dot{\gamma}$ is the shear strain rate, τ is the shear stress, and T is the absolute temperature, which is taken as a function of depth from the Earth's surface. The quantities A, n, and Q are material constants and R is the universal gas constant.

Figure 12, a plot of the strain rate $\dot{\gamma}$ versus time at a distance of 6 km from a fault that was calculated using the anti-plane strain model discussed previously (three different values of A in Eq. (1) were used), together with the above nonlinear temperature and stress-dependent creep rheology and a temperature field that varied with depth. (The temperature at the surface of the model is 283 K and at the base of the model is 923 K). Also shown in Fig. 12 are experimental data along the San Andreas fault and error bars on the data that have been taken from Ref. 5. Clearly, further adjustment of the material parameters is required to improve the fit of numerical results with experimental data at early times.

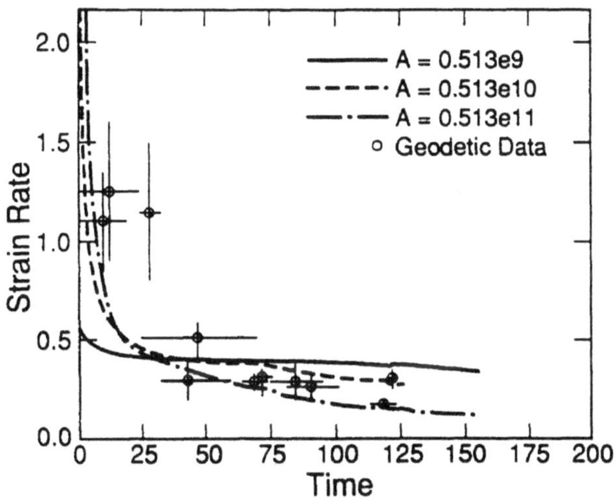

Figure 12. Strain rate versus time at a distance of 6 km from a fault, as calculated by ABAQUS for three different material constants, compared with experimental data from Thatcher [8].

REFERENCES

1. NASTRAN User's Manual, NASA publication 2222(08), June 1986.

2. Hibbitt, H. D., Karlssen, B. I., and Sorensen, E. P., ABAQUS: User's Manual, Version 4.8, Hibbitt, Karlssen, and Sorensen, Inc., Providence, RI, 1989.

3. PATRAN Operations Manual, Rev. 1, PDA Engineering, Costa Mesa, CA, 1990.

4. ASME Boiler and Pressure Vessel Code, Section III, Division 1 Case N-284, "Metal Containment Shell Buckling Design Methods" Approval date: August 25, 1980.

5. Farrar, C. R., Prediction of nonlinear structural response in LMFBR elevated-temperature piping. Los Alamos National Laboratory report LA-10090-MS, 1984.

6. Li, V. C., and Rice, J. R., Crustal deformations in great California earthquake cycles. J. of Geophys. Res., 1987, **92**, B11, 11,533-11,551.

7. Elsasser, W. M., Convection and stress propagation in the upper mantle. In Application of Modern Physics to the Earth and Planetary Interiors, Wiley-Interscience, New York, 1969.

8. Thatcher, W., Nonlinear strain build-up and the earthquake cycle on the San Andreas Fault. J. Geophys. Res., 1983, **88**, 5893-5902.

SOME ASPECTS OF R.C. STRUCTURE CAD SYSTEM

Liu Guang – dong, Shang Shou – ping
(Hunan University, China)

ABSTRACT

In this paper, a brief introduction about developmental trend and characteristic of Civil Engineering Structure CAD System in China is given. Several main problems of developing the Non–interactive CAD System are discussed. Moreover, application of the Artificial Intelligence in above system is described.

1. DATA PRETREATMENT AND DESIGN RESULT TREATMENT IN THE FORM OF GRAPH

In early Engineering Structure CAD System, all original data are input manually, and a great deal of data of the design results are output.

In recent years in China, some interactive data systems in the form of graph have been developed. These systems can draw load diagram directly on screen with digitizer. Then the program collects information automatically from the diagram on screen. Some finite element analysis program can divide element net lattice and number the net nodes automatically, and display or plot out the net lattice diagram.

The design result treatment system is to output the structure analysis results in the form of graph. At present, there had been some successful data treatment systems, such as the FBCAD system[1] developed by Hunan University. It can output automatically the load diagram, the moment, shear force, axial force enveloping diagram, reinforcement area enveloping diagram, deformation diagram and modes of vibration, etc.

2. AUTOMATIC MERGING AND AVOIDING COLLISION

This is a problem of higher level non–interactive CAD system. Automatic Merging means the mergence between structures or structure members which are almost same or completely same in internal force and geometrical dimension; it also means the mergence of the projective overlapping line in the course of forming graph automatically. The Avoiding Collision means to protect graph line or characters from overlapping in the course of forming graph automatically.

In Engineering Structure Design, to merge same structures or members is an important means which can simplify the working drawing and cut down their number. It is natural to merge those completely the same part of structure, merging can also be used when those structure parts are incompletely the same for simplification.

In drawing, the overlapping of projective line is very common, such as the longitudinal bar line in side view of the beam or column. Therefore it is necessary to merge the projective overlapping line in forming graph automatically.

It is obvious that computer have no visual sense. Therefore it is necessary to search, memorize and compare a lot of data before merging. When comparing, the total search and ordinal identification method are often used. So it will take a great of running time and large storage space. The difficulties of merging are not only to judge if merging is needed or not, but also how to represent the graph after merging because we have to use different representation for different merging.

Avoiding Collision is as same as the merging problem. When drafting a drawing, a lot of dimensions and each steel bar number must be marked, which cannot be reduced with graph scale. So it is very easy to come into collision. Fig. 2–1 shows the representative situation. These marks of steel bar number must be plotted into the circles with the diameter of 4 mm. Before drawing these circles, program must search the memory to determine if there was a plotted graph at this position or not and it can draw a new circle or not and whether it is necessary to use a slant line to draw a new circle or not.

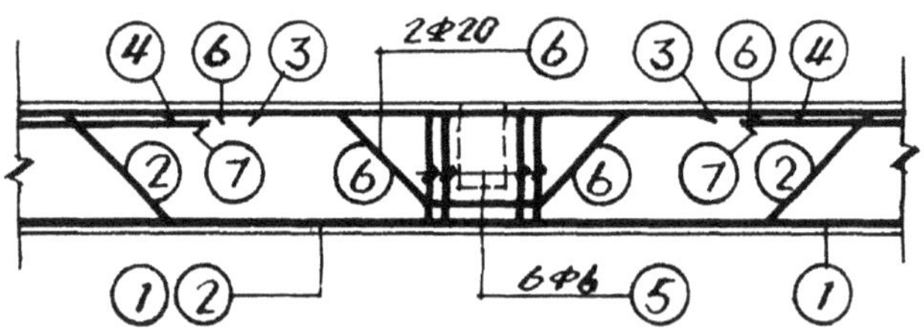

Fig. 2–1. The avoiding collision problem of bar number marks

Of course, we can find some regularity in advance to deal with avoiding collision problem and merging problem, such as leaving some blank place beforehand or drawing some part which is certainly overlapped only once, to solve some problems. But the CAD system is a general program; it must suit various situation. Those regular arrangement cannot solve all problems. Some special conditions will cause and influence the reliability of the system. Up to now, people have not found an effective way to solve the problems of Auto–merging and avoiding collision which is one of the other most difficult in developing Engineering Structure CAD System. Many scholars and experts are engaged in solving these problems. In the FBCAD system, the problems of auto–merging and avoiding collision had been solved better.

3. THE RESEARCH OF REINFORCEMENT DESIGN METHOD

The correctness of the reinforcement design directly influence the reliability and practicability of the CAD system. It is the key link in the Non–interactive Engineering Structure CAD System.

For a long time, the reinforcement design is done manually by engineers and no mature standard computation method can be used. With the computer–aided design technique introduced to structure engineering design, this problem has been urgently presented due to following reasons:

(a) The reinforcement design is not only restrained by design code, but it also deals with engineer's design custom and experience. Every engineer has individual custom and experience, and so the design results are different. It means that the construction design methods are not sole. And it is impossible to work out one program for every engineer to suit the different design custom and experience. So we had better to collect a set of excellent design methods as the basis of the CAD program.

(b) The design code requirements, expert's experience and custom are separated. There is no logical relation between them and have no classical computation method can be used. Therefore how to arrange these design requirements becomes a very difficult problem. A lot of present CAD systems often get wrong at this part, and can not satisfy engineer's design purpose.

Reinforcement construction design is of three dimensional, selecting reinforcement is restrained by a series of technique requirements and control parameters. Some technique requirements are given out not in the form of fixed quantity, but in the form of fixed nature and it is difficult to express them using mathematical function formula.

In China, the initial research work has already begun. Some general methods have been developed such as the dynamic test method[2] of selecting reinforcement on cross section; the basic location method[3] and the multiple location method on cross section when arranging bar.

4. THE DEVELOPMENT OF GRAPH SYSTEM OF NON–INTERACTIVE CAD SYSTEM

Although there are some general graph systems that can be used as the graph supporting software of engineering structure CAD, the civil engineering structure design has yet its own feature and conventional rule. So the engineering structure CAD has its own requirement. It is difficult to fit engineering structure CAD system requirement by the general graph system.

First, it is hoped the graph drawing file is as small as possible. The small graph drawing file holds less storage space and can reduce interchange time between memory and diskette. Therefore it can speed graphic display or graphic plot. Now the popular graph software Auto–CAD has been used in various CAD system, but its graph exchange file (DXF) is very big and complex, and the speed of reading or writing is very slow. Even if the DWG graph file is used, the bigger storage space is still necessary.

Civil engineers raise higher requirements to the graph arrangement on drawing paper. The graph content, location relation between graph blocks should be planned carefully. The range of graph arrangement must fit the requirements of Chinese Drafting Standard.

Engineers hope that all graph blocks can be arranged conveniently according to their own thinking. So graph blocks should be displayed, dragged or rotated rapidly. After arranging the drawing paper, engineers often want to review the details of the graph before plotting for ensuring the correctness. Therefore the graph system should have the window function which can enlarge the part graph.

Of course, the graph modifying function should also be provided. Although it is not necessary for non–interactive CAD system, it is better to provide this function for user. In addition, it is necessary to establish a set of structures for indexing and calling Chinese characters.

According to the characteristics of the automatically forming graph CAD system, to develop a simple and highly effective graph system is a trend in China.

In China, there are three kinds of arranging drawing procedures of the non–interactive CAD system. They are the automatic way, semi–automatic way and the interactive way. According to the position of producing graph module and arranging drawing module in CAD system, the system can be divided into two forms: the pre–arranging drawing and post–arranging drawing (Fig. 4–1 and Fig. 4–2).

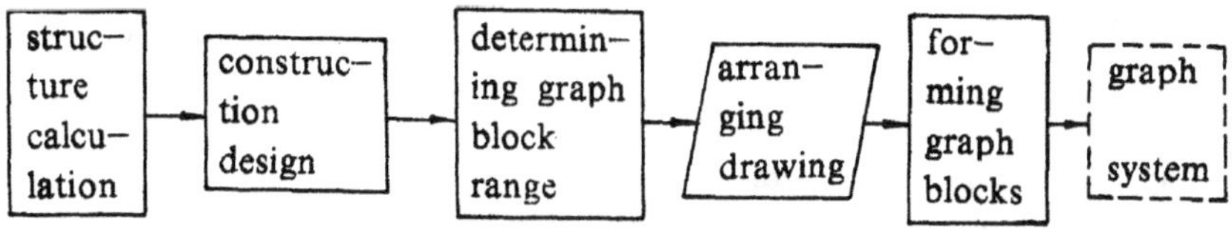

Fig. 4–1. Engineering structure CAD system with pre–arranging drawing procedure

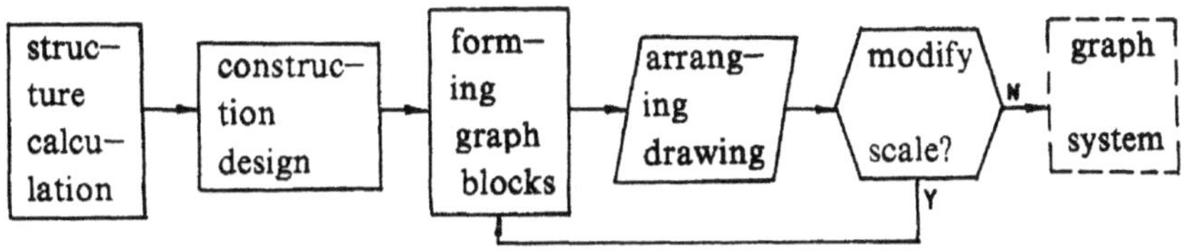

Fig. 4–2. Engineering structure CAD system with post–arranging drawing procedure

At present, a typical non–interactive graph system is GSSE[4] . Its main module structure is shown as Fig. 4–3. The dashdot line frames in Fig. 4–3 represent the modules which are directly related to GSSE but do not belong to GSSE; the curve frames represent disk files, and the rectangular frames represent the main function modules of GSSE.

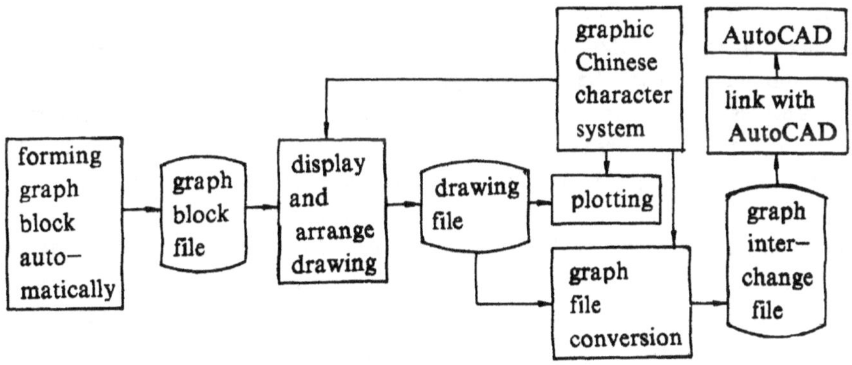

Fig. 4–3. Relation of main modules of GSSE graph system

There are four main modules in GSSE:
a) Display and arrange drawing
b) Plotting
c) Graphic Chinese character system
d) Graph file conversion and the link file with Auto–CAD

Obviously, GSSE relies mainly on the plotting directly; the link file with Auto–CAD is only an auxiliary function. GSSE does not use the Auto–CAD as basic graph supporting system. This is a great feature of GSSE because the graph is formed automatically by CAD system and there is no need to modify usually. So the main task of GSSE is to arrange drawing and graph output. And powerful graph edit function is not necessary in GSSE.

The module of displaying and arranging drawing is the most important part of the GSSE; the graph is displayed rapidly and drawing is arranged agilely by using the GSSE. Its input data is graph block file, and the output information is drawing file. They both are called graph file and are provided to plot module or graph file conversion module as input information.

To record an A1 drawing (851X459 mm), it only needs 100KB disk space when using GSSE graph file, but when using DXF file of Auto–CAD then it needs about 400KB disk space. To display an A1 drawing it needs only about 30 seconds on 286 computer. GSSE has got higher performance on space and time.

5. CHINESE CHARACTER SYSTEM

In comparison with foreign CAD software, Chinese character system in engineering structure CAD system has an important feature. It includes displaying, printing, plotting Chinese characters and prompting Chinese string while taking dialogue between computer and user. Before displaying the dot matrix Chinese character on screen, it is necessary to set the monitor from text mode to graphic mode. At present, there are all types of monitors such as the CGA (with resolution rate of 640 × 200), EGA (with resolution rate of 640 × 350), ATT 400 (with resolution rate of 640 × 400), VGA (with resolution rate of 640 × 480), HERC (with resolution rate of 720 × 348), etc.,with the result that many difficulties are caused for displaying the dot matrix Chinese character. It is necessary to develop different Chinese character system to fit different monitor. Similarly, there is the same problem to print Chinese character on printers. In recent years, various perfect Chinese character systems have been successfully developed by Chinese researchers, such as the Lian–xiang Chinese character operation system, CCBIOS 2.13 and CCDOS 4.0, etc. Good condition has been created for developing CAD software.

Plotting Chinese character on plotter is not similar with printing or displaying

Chinese character. It must draw out strokes of character by pen. Therefore it is necessary to use vector graphic character. The vector graphic character library is larger than the dot matrix character library, but it is easy to control the scale of character.

In China, the vector graphic Chinese character system is successfully used in plotting the working drawing.

6. THE APPLICATION OF ARTIFICIAL INTELLIGENCE IN THE ENGINEERING STRUCTURE CAD SYSTEM

CAD technology is not only a graphic problem. It involves much knowledge of special field, such as mechanics, computational method, structure design theory and design code, design custom and experience, etc.

Specially, the engineering design knowledge is a synthetic knowledge. It is complex and separate, and there are varied selections and design results. So it is a rather complex work to arrange this knowledge into program correctly. The working drawings designed by non–interactive CAD systems often exist some problems about specialistic design. For instance, the construction design has some mistakes, the drawing representation cannot suit the design rule, the custom or the representation depth is not enough, the operation of software is complex or the operation procedure does not suit the design regularity, etc. These defects have reduced the practicability and the reliability of the software, and have made many engineering structure CAD system to have lost their life force. It is obvious that we should put the stress on specialistic design from now on. In recent years, the Artificial Intelligence has been introduced into civil engineering CAD system and has been presented to an important position. People hope that the intellectual Expert System will increase the practicability and reliability of the CAD software.

The core part of Expert System is the Knowledge Base and the Inference Engine. The specialistic field knowledge is stored in the Knowledge Base. This modular structure of Knowledge Base can be modified conveniently to suit the accumulation and the renewal of the knowledge.

The Inference Engine infers according to the knowledge in Knowledge Base, and produces the judgement and design results.

The running course of the Expert System is similar to the judging and thinking course of human brain. It can make multiple selection or best selection.

Since the middle of the '80s, China has begun to research the Civil Engineering Expert System. Each subject of the Intelligence–aided Decision System in Engineering Construction Project which is financed by the National Natural Science Foundation has got some initial achievements[5]. They are as follows:

a) Expert System of High Rise Building Preliminary Design

b) The Expert System of Judging and Defence Decision on Earthquake Damage

c) The Expert System of Judging and Decision on old Factory Building

d) Intelligent Consulting System of Engineering Structure Modularization

e) Intelligent CAD System of Cable–Stayed Bridge Preliminary Design

f) Intelligent CAD System of Reinforcement of R.C. Structure

g) Multiple Target Fuzziness and Intelligent Decision System of Engineering Structures

It is similar to foreign countries that most present Expert System are either consulting or diagnostic or judging or for decision, and the Expert System on engineering structural design is less than the others. The other trend is that the Expert System of plane design or initial design is greatly more than the one of construction design or detail drawing design in design type Expert Systems. What the Engineering Structure CAD System needs is the latter:we know, according to the present information, the Structural Engineering Institute of Hunan University and the Prof Zhong Wan–Xie of Dalian University of technology have done creative work[6] in this respect.

It is thought that the degree of difficulty in developing design type Expert System is higher, because the representation of knowledge and the inference strategy have their own natures.

For R.C. Frame CAD System, the Structural Engineering Institute of Hunan University has established an Expert System Module on reinforcement construction design of beam and column. The study results of the knowledge structure of reinforcement construction design above show that the knowledge of this domain can be divided into three grades: meta–knowledge, the first objective knowledge and the second objective knowledge. According to the knowledge grades, the knowledge acquisition can introduce Domain Expert to input complex knowledge into the Knowledge Base. By problem reduction, it is found that this domain problems can be resolved into multiple layer sub–problem structure taking the shape of tree. Every sub–problem holds itself relative independence. The problem solving may use the depth first search strategy. The Expert System is established by TURBO PROLOG language. Mixed inference including forward chaining and backward chaining is used in this system. It can use rules and facts to represent knowledge and use dynamic data base to record or depict the state of every design stage. Running experiments of the Expert System indicate that this Expert System can imitate the logical thinking process of human brain to inference design. The addition, subtraction or modification of the Knowledge Base is easy and convenient. The design results of the system are satisfactory for engineering design, and the running speed is suitable for practice requirement.

On the basis of this module, it is possible to form an excellent Expert System of reinforcement construction design by expanding the Knowledge Base and completing the interface of knowledge acquisition.

The present Artificial Intelligence is the predicate logical deduction system on the basis of mathematical logic. It is to use a simple structure to make inference formalization and runs on the fourth generation computer in the form of series. Although it embodies intelligence under many conditions, but with many complex problems in practice, it still seems powerless and is not so good as human brain thinking process. The main reason is: human brain realizes intelligence with the help of parallel action of much nerve elements, but the present fourth generation computer has no such structure.

In recent years, the United States and Japan had already begun the study of nerve element[7] . This kind of computer is softly suited to complex and fuzzy condition. We believe optimistically that the further development of Artificial Intelligence will influence the progress of civil engineering CAD more deeply.

The nerve element Expert System has been studied also in China. Prof. Liu xila of Qinhua University together with his doctoral student had studied the Nerve Net Expert System using for High Rise Building Preliminary Design.

7. DEVELOPMENT OF THE CIVIL ENGINEERING STRUCTURE CAD SOFTWARE IN CHINA

The main engineering structure CAD softwares developed in China now are as follows:

(1) Design of steel–pipe structure
(2) Design of R.C. frame, bent–frame and continuous beam
(3) Design of frame shearing–wall structure of high rise building
(4) Design of R.C. floor slab
(5) Design of R.C. independent foundation
(6) Design of structure plan
(7) Design of steel frame with hinges
(8) Design of special structure, such as chimney, water vessel or water tower
(9) Design of prestressed R.C. bridge structure

The characteristics of Chinese engineering structure CAD system are as follows:

(a) The main computers used in China are 80286 and 80386. So the most softwares of civil engineering structure are based on these micro–computers as we develop them.

(b) CAD systems above include construction design and working drawing forming, and are used directly in structural design.

(c) The individual CAD sub–system of specialization for the structure often used is developed.

The typical flow diagram of Engineering Structure CAD Sub–systems is shown in Fig. 7–1. These Sub–systems possess a series of merits such as small storage space, small running space, less original data, simple operation, high speed design and high speed forming of graph.

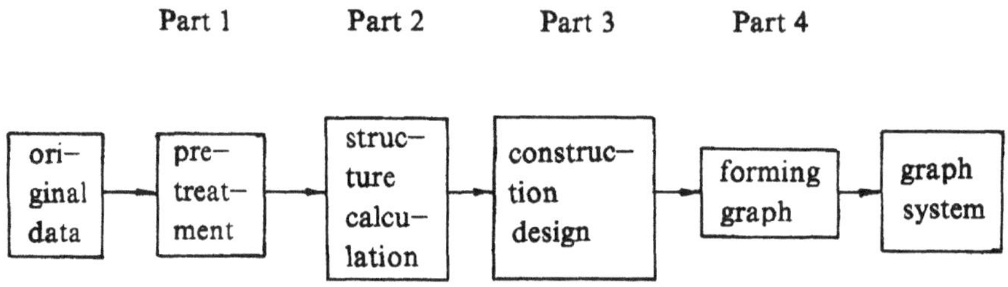

Fig. 7–1. Flow diagram of Engineering Structure CAD Sub–system

(d) The non–interactive or semi–interactive CAD system with higher automatic level is used.

(e) For different structure feature, small finite element analysis program is used individually as internal force analysis module. It is not necessary to fill a number of original data, hold much running space and storage space, and make operation complex when the small finite element analysis program is used individually.

In China there exists already the Space Frame CAD Integrative System (Fig. 7–2) composed by the sub–systems of beam, column, slab and foundation. Through coordinating the data interface between these sub–systems, it can design all structural working drawing of a building according to the plane view, side view and sectional view of architecture design.

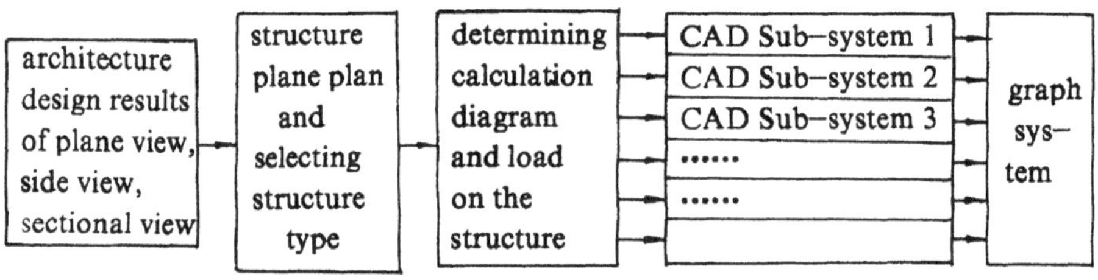

Fig. 7–2. The integrative CAD system of engineering structure

CONCLUSION

Although the research and application of the engineering structure CAD have got a great of progress in recent years, some difficulties and problems have become outstanding more and more. Many subjects remain to be further studied and concluded or summarized into a systematic theory and method, and it is necessary to make the synthetic research of multiple subjects. In China, many scholars are engaged in the study of this respect.

Now in China, the main tendency of developing engineering structure CAD system is still the non–interactive CAD System, and is gradually growing towards the integrative type of CAD system. For instance, the civil engineering CAD will start from architecture design to structure design directly, and it does not need to fill excessive original design data. Moreover, the study of the Expert System has begun, and some results have been achieved. It is believed that Chinese engineering structure CAD will join the advanced team in world.

REFERENCES

1. The proceedings of national conference on building structure computer–aided design, published by building structure committee of Architecture Institute of P.R.C, Beijing, 1987.

2. Shang Shou–ping, The Research of Computation method on Beam and Column Section Reinforcement, Computer, Volume 1, 1989.

3. Shang Shou–ping, Researches of Engineering Structure CAD System, Doctorate Thesis, Supervisors: Prof. Cheng Wen–shan; Prof. Liu Guang–dong; and Prof. Zhou Yin–sheng, Changsha. 1989.

4. Shang Shou–ping, The Graph System of Engineering Structure CAD , published by the Structural Engineering Institute of Hunan University, 1988.

5. Proceedings of Theory and Application of Engineering Intelligence Decision, Computational Structure Mechanics and Application, Volume 2, 1989.

6. Zhong Wuan–xie; Liu Jun–xian, Intelligence CAD System of R.C. Space Frame, the proceedings of the fourth national academic conference on computer application of architectural engineering, held in Guangzhou, 1988.

7. Xu Li–lian, The Nerve Computer with Pretty Future, Computer World, Volume 6, 1989.

FUZZY REASONING AND MACHINE LEARNING OF EXPERT SYSTEM FOR STRUCTURAL DESIGN

Lin Shaopei   Yang Jiaping
Civil & Architectural Engineering Department
Shanghai Jiao Tong University, Shanghai, 200030, China

ABSTRACT

This paper deals with one of the recent advances of AI applications in engineering design issues ——— expert system for design. Analysing the characteristics of design procedures and its logic natures, emphases were made on the architecture of design-type expert system, its inference strategies and knowledge base structure. Based on the conceptual analysis of human being, fuzzy representation and fuzzy manipulation are chosen for simulating the interior laws of human thinking, as well as for deepening the knowledge inference. In the meantime, unified fuzzy reasoning theories of machine learning capability for design type expert system and its implementation are discussed in detail. Numerical examples are presented for determining design parameters of buildings. Finally, some perspectives are indicated.

I. INTRODUCTION

The developments in expert systems have greatly attracted the attention of diverse groups in engineering design. Indeed there is a growing recognition that the artificial intelligence facilitates and becomes essential to a sucessful design. It is reasonable to expect that expert system will have a role to play in a sound engineering design process. Most of the early expert systems appeared in the literature are diagnostic systems. They are characterized by rule-based strategy and shallow knowledge. Generally, it is regarded as the first generation of expert system. However, the second one, which includes deep knowledge, is hierarchy in knowledge structure and reasoning strategy. Nevertheless, civil engineering design is characterized by guidelines in codes and specifications, by data acquisition and retrieval, logical procedures, analytical processes as well as expert knowledge . Clearly, based on the characteristics of the second generation of expert system, the design type expert system is an extension of it. Which includes: mixed type inference machine, hierarchy of knowledge base, incorporating uncertainties and machine learning etc. As a matter of fact, the introduction of uncertainties and machine learning capability to the expert system represents a new era of design type expert system development.

II. CHARACTERISTICS OF DESIGN TYPE EXPERT SYSTEM

2.1. The Nature of Engineering Design

Engineering design, in particular the structural engineering design, is a multi-disciplinary topic, which involves a very diversified range and depth of knowledge. The major portion of this knowledge is either factual or algorithmic. The algorithmic knowledge is based on well-defined scientific theories, such as the principles of mechanics, and is applicable to problems of the same nature. It can range from simple analytical techniques to extremely complicated numerical computations. Factual knowledge includes the information that is needed to describe the physical systems and assumed states addressed in the problem. For a building structure, it may describe the physical attributes such as the geometrical parameters of the structure, the properties and connecting conditions of an individual structural component.

The engineering design strategy is usually characterized by an iterative cycle of analysis and design. Analysis refers to the situation where it is necessary to determine the responses of a physical system generated by a particular excitation. On the other hand, design determines an appropriate system to satisfy a given set of requirements. Experience plays a major role in accelerating the iteration process, without which the huge amount of algorithmic and factual knowledge which related to the design alternatives can lead to a combinatorial explosion.

Engineering design is generally proceeded in stages with certain definite task for specific sequenced stage such that information is cascaded to address the final design solution. Generally, the strategy of design is applied to deduce the overall problem to a set of sub-problems with identified goals. Cybernetic functions are exerted to each sub-problem from one to another, depending upon what goals remain to be pursued and their relative priorities. This problem-deduction strategy is a general principle for most kinds of design issues, but it is particularly appropriate in engineering design because the solution process is usually well-understood in engineering design, so that a realistic goal can be readily defined.

2.2. Architectures

The design process is going through from initial concept to the final solution, which is a process of successful improvement and a series of decision making, where efficient analysis tool as well as comprehensive set of revision strategy are necessary. As the revision of design concept is concerned, an expert system should provide minimum revision time based on domain expert's experiences.

The typical architecture of the system is shown in Fig.1, where two dotted line frames represent the engineering analysis (algorithmic knowledge) and inference machine, these modules will interact with each other by design information or data files, corresponding to two different systems of digital and symbolic representation.

The knowledge base of the system (factual knowledge) is driven by two inference versions: the multiple-pass inference and mixed chaining

inference, which are characteristics of design type expert system and as we know the inference in other type expert system is just one pass.

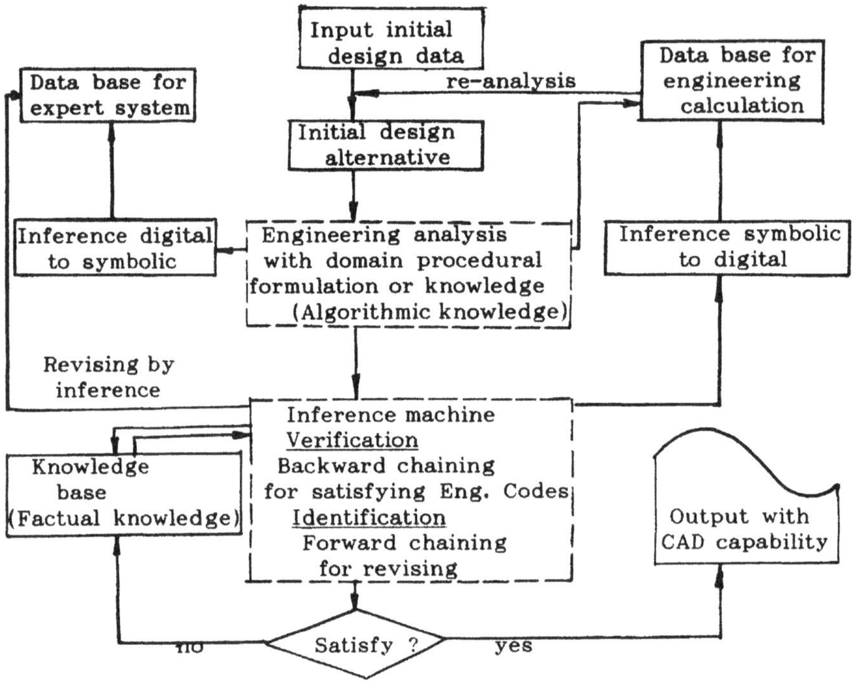

Fig.1

Data base in an expert system and in an integrated system are quite different and should be strictly distinguished. Only by means of interfaces can the data in DB of an engineering system be transformed into the data accessible from DB of an expert system. In the meantime, the DB of expert system is used for storing initial, intermediate and final data for inferring. On the contrary, data in DB for engineering calculation is directly connected to the interfaces. It represents a stack of data in organized way, available for random accessing during execution of engineering calculation.

The contents of a knowledge base can be divided into two categories: the heuristic experience from domain experts and the codes or specifications in engineering community. It is to our understanding that the former one is rather difficult to retrieve and is much time-consuming . Usually, an expert system will have some hundreds of rules and up to 5-6 levels of heirarchy and also some hundreds of nodes in its inference network. The interface between engineering calculation and inference machine is comprised of the interfaces from digital to symbolic and vice versa transformation.

2.3.Mixed Type Knowledge Inference

The inference machine of an expert system plays a role of inferring new knowledge and approaches to problem solving. The forward chaining (driven by data) strategy and backward chaining (driven by goal) strategy are frequently used, particularly for design type expert system, mixed type chaining with both forward and backward chaining is dominant.

According to engineering design processes, one has to submit an initiative design alternative representing the conceptual thinking of designer based on initial data. This procedure is corresponding to forward chaining of inference machine. Yet the verification of a submitted initial alternative by design codes and specifications is a process corresponding to backward chaining of inference machine. The above-mentioned procedures have taken place iteratively in design process; naturally, it emerges the necessity of introducing mixed knowledge inference with both characteristics.

The mixed type inference is compatible to knowledge deepening in design process, since deep knowledge is evidently needed for design selection and decision making rather than some heuristic shallow knowledge in conventional diagnostic and cybernetic expert systems. Hence the design type expert system is characterized in deep knowledge exploration and is essentially ranked as second generation of expert system.

An initial alternative is identified roughly by the system and then verified by the system with improvement. During deep reasoning process, both shallow and deep knowledge are interactively being used for stepwisely improving the design concept. For instance, the profile selection of tall building structure needs the deep knowledge and information of internal forces of member through structural analysis—— an analytic inference of internal force deep knowledge.In the meantime, the profile configuration of member needs code information and the experience of domain experts, which has implemented by data driven forward chaining inference. On the contrary, checking and verifying in strength and stability can be implemented by goal driven backward chaining inference. It is also noticed that the interfaces of each subsystem to the global one should be carefully posed for the reason that the inference is characterized in symbolic operation and whereas the engineering analysis in digital operation. In order to transmit data files with each other, two transmission interfaces of from digital to symbolic and vice versa are provided.

III.FUZZY REASONING IN EXPERT SYSTEM

3.1.Exploration in Thinking Science

The essential of the first generation of expert system is merely a finite series of rule-based "fire up" operation. It is difficult either to represent the comprehension of human being in fuzzy information processing, or to explore the way for knowledge source through fuzzy inference. It is evident that not only pure logic rules are using by experts in problem solving, as a branch of AI, expert system should not be a collection of heuristic knowledge rules; it should play a role by means of software in the computer as a representation of human intelligence, including various abilities of learning, understanding, remembering, inferring, reminding, recurring and analysing.

The interior law of human thinking can be described by specific language in Thinking Science . Except to conceptuation and abstraction, "presentation" will play an active role in human thinking, which is a reappearance of sensational information of the objective facts input to human's brain. It is mentioned that although the information of presentation is far less than what was received, however, it actually represents the essentials of the problem and performs an escalation of "keeping truth and abandoning false". The sensational presentation can be formed by various components, including image, sound, touching sense motive scenario etc. As regard to deepening & expanding of current presentation, the recognizing of objective event can be obtained through multi-level recursion. Then the combination of static and dynamic presentation of the event can be achieved. It is understood that the depth level of human thinking in this period is entering to the stage of "comprehension" and realizing a transformation from intuitive to conscious one. Perceptional presentation, promoting human thinking, can be the sensational presentation combined with static and dynamic presentation escalated; it also can be the intuitive presentation combined with conceptual synthesis and recurred as a new presentation. The infiniteness of presentation deepening process exactly implies the boundless of the evolution of human thinking as well as human understanding.

"Reminding" is an active factor in the evolution of human thinking with the fundamental element "presentation" as its language. It is by remindings that the combination of individual sensational presentation can be possible and a synthetic presentation by arbitrary multi-level combination can be found. Since it is impossible to describe the presentation of human thinking exactly just by natural language, even though it is possible, one may not use it. For instance, the event of "take normal at each node on the plane of a prismatical body" is presented by natural language as: "Visit the node on plane, define a local coordinate X'Y'Z' in the node with a basic plane coincided to the prismatic body, then take the normal line perpendicular to the plane, at last, iterate to every node as assigned". One may convince that both natural language as shown above and simple logic statement

$$(\text{Plane } X'OY') \wedge Z' \vee (\text{Plane } Y'OZ') \wedge X' \vee (\text{Plane } Z'OX') \wedge Y'$$

will not being used in human thinking. What humans really think is just a spatial perspective of prism with normals in each face, hence the human thinking should not be the same with the natural language and it is something similar for the thinking logic of human being to image. The geometrical presentation will always be a direct abstraction of objective world and to be a reflection of interior law of human thinking, the role of image representation is evident.

The introduction of "softening" knowledge treatment as well as fuzzy deep knowledge inference will promote the simplification of knowledge acquisition process and its abstractive formulation, thus making it possible for these fuzzy manipulations to cover broader area of problems and to draw quantitative assessment more reasonably. The implicatives of knowledge in philosophy are described in "quality", "quantity" and "degree" etc. with dynamic natures.Since the emergence of knowledge representation by fuzzy deep knowledge inference, the implicative of knowledge is expanded and the deepening of knowledge is speeding up undoubtedly.

The main objects of treatment of knowledge "softening" are indeterministic information, such as uncertainty or fuzziness caused by indeterminate objective causal law, randomness by imprecise objective definition or criterion without uniqueness etc. Fully utilizing the motivity, creativity and versatility of human being, rating the "quality", "quantity" and "degree" of objective events by fuzzy comprehensive evaluation and interior law of thinking, the fuzzy deepening of knowledge inference is one of the best ways for knowledge acquisition in general sense; it is also an exploration based on the scientific principle of thinking.

3.2. Fuzzy Knowledge Representation

A better knowledge representation would take account of the fact that the essential features of most engineering design involve a high degree of uncertainty. According to above-mentioned analysis, we conclude our consideration on the fuzzy-based design type expert system. To be a potential strategy, it seems much suitable to present an example ——— expert system of earthquake-resistant design of tall building. Since the design environment is full of impreciseness and uncertainties, one can not achieve a reasonable solution without proper consideration of dealing with uncertainties. On the other hand, as we pointed out in paper [1] that in comparison with the probability theory, the fuzzy mathematics can be more suitable for satisfying our goals and show the elasticity in manipulating design issues. Systematically develop the strategy of fuzzy knowledge representation accommodating to design type expert system is needed for exploring this new field, which depicts the framework of this study.

There are two kinds of fuzzy knowledge representation in the knowledge base of design type expert system, both quantitatively and qualitatively. For instance, in the synthetic evaluation of torsional strength of tall building structure [2], the failure membership function of residual carrying capacity of shearwall component can be expressed quantitatively by its probability distribution $Ps(\xi)$ in different levels:

$$\mu^{SW} = \frac{Ps(\xi)_0}{\xi_0} + \frac{Ps(\xi)_1}{\xi_1} + \cdots + \frac{Ps(\xi)_9}{\xi_9} + \frac{Ps(\xi)_{10}}{\xi_{10}} \qquad (3.1)$$

where $\xi = \frac{V_{sw}}{(V_{sw})_0} = \frac{\text{experimental value}}{\text{nominal value}}$, $(\xi_0, \xi_1, \cdots\cdots, \xi_{10})$ is universe of membership function of shearwall component. Fig.2 shows the transformation of membership function of residual carrying capacity of shearwall to the failure membership function itself. The curve truncation represents safety coefficient. Moreover, based on subjective information, one may construct the qualitative membership function empirically by means of table and curve [8].

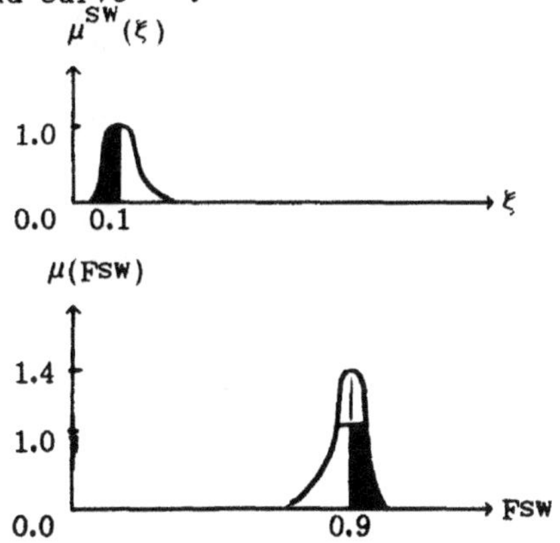

$FSW = 1 - \xi$

Fig.2

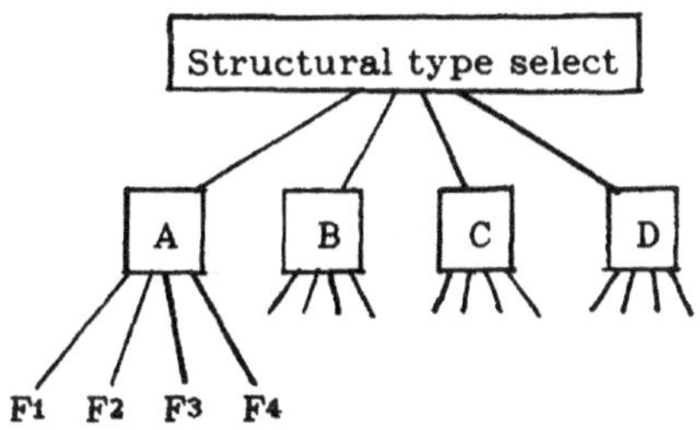

Fig.3

3.3.Fuzzy Knowledge Inference

A problem containing uncertainty or fuzziness is called "improper" problem. The expert system, which is characterized in simulating the experiences of human expert, could be manipulated as improper problem by means of softening of knowledge representation or softening of optimum goal and its constraints, where strict mathematical derivations will not followed. Instead , the design knowledge **was** described by fuzzy factor assessment and quantitation of membership functions.

Fig.3 shows structural type selection of high-rise building with various factors affected. For instance, the type of structure is presented by A, B, C and D with selecting factors numbered by 1, 2, 3 and 4, factor notation F_i (i= 1,2,3,4) and the membership function μ_i (i=1,2,3,4) respectively. For structural type A, the fuzzy subset $\underset{\sim}{A}$ can be expressed by

$$\underset{\sim}{A} = \frac{\mu_{1A}}{F_1} + \frac{\mu_{2A}}{F_2} + \frac{\mu_{3A}}{F_3} + \frac{\mu_{4A}}{F_4} \quad \text{or} \quad \underset{\sim}{A} = [\mu_{1A}, \mu_{2A}, \mu_{3A}, \mu_{4A}] \tag{3.2}$$

where μ_{iA} represents the membership function of factor i to structural type A, given by expert's empirical curve [7] . The membership vector corresponding to different structural type with various influence factors is

$$\underset{\sim}{I} = [\mu_{iA}, \mu_{iB}, \mu_{iC}, \mu_{iD}] = \{\mu_{iI}\} \qquad (i= 1,2,3,4;\ I= A,B,C,D) \tag{3.3}$$

where μ_{iI} is the component of membership vector by factor i to structural type I. Introducing weight vector $\lambda = \{\lambda_1, \lambda_2, \lambda_3, \lambda_4\}$, then (3.3) becomes

$$\underset{\sim}{I}^{\lambda} = \{\lambda_1 \cdot \mu_{1I},\ \lambda_2 \cdot \mu_{2I},\ \lambda_3 \cdot \mu_{3I},\ \lambda_4 \cdot \mu_{4I}\} \tag{3.4}$$

Based on (3.4),the fuzzy selection of structural type can be obtained by means of fuzzy synthetic evaluation [11] .

IV.MACHINE LEARNING IN EXPERT SYSTEM

4.1. Unified Fuzzy Approach for Machine Learning Strategy

The expert system is a program to solve problems based on human knowledge and expertise. For a professional expert system, it depends heavily on domain knowledges. Nevertheless, to acquire knowledge from domain experts is really a tedious and troublesome task, which now forms the main obstacle and bottleneck in building an expert system. Therefore, to incorporate machine learning capability into the system can greatly facilitate and improve the capabilities of the system, especially in the knowledge acquisition aspect.

The essentials of design, as we indicated previously, are a process of sequential decision making in different aspects and at different levels for approaching prescribed design goals. As the design goal can be achieved step by step through the determination of its parameters in different design stage, these parameters serve as the attributes of specific design object (product). The relationship between design parameters can be depicted either by a network or by a tree structure to form a global picture of design state associated with the concept of generalized space. Evidently, each node of the network or tree structure represents a subspace for certain design parameter (or attribute), and the generalized space is implied by topological linkage of those subspaces.

4.2 Fuzzy Mathematical Distance In Design Space

Based on the concept of state in space, the difference between two design solution $\underset{\sim}{A}$ and $\underset{\sim}{B}$ in the same (sub) space can be measured by fuzzy mathematical distance φ ,which may be defined by membership function difference, and is meaningful in view of manipulating design uncertainties. We have

$$\varphi(\underset{\sim}{A},\underset{\sim}{B}) = \sum_{i=1}^{m} W_i(d_i) \cdot d_i(\underset{\sim}{A},\underset{\sim}{B}) \tag{4.1}$$

where m, the dimensions of the (sub) space; W_i, influence coefficient of i-th attribute (factor); $d_i(\underset{\sim}{A},\underset{\sim}{B})$, distance between design solution $\underset{\sim}{A}$ and $\underset{\sim}{B}$ in i-th attribute. Doubtlessly, we could extend the measurement of difference to hierarchy of subspace with uncertain environment of design solution

$$d_i(\underset{\sim}{A},\underset{\sim}{B}) = \mu_{\underset{\sim}{A}}(t_i) - \mu_{\underset{\sim}{B}}(t_i) \tag{4.2}$$

where $\mu_{\underset{\sim}{A}}(t_i)$ and $\mu_{\underset{\sim}{B}}(t_i)$ represent the membership function of fuzzy design solution $\underset{\sim}{A}$ and $\underset{\sim}{B}$ with respect to i-th attribute t_i in its subspace. In (4.1), the influence coefficient W_i might not be constant for different distance; it should be changeable with certain penalty effects. We further define

$$W_i(d_i) = a_i \cdot \zeta_i(d_i) \tag{4.3}$$

where a_i, weight of i-th attribute to the design solution; $\zeta_i(d_i)$, the amplified coefficient of $W_i(d_i)$, suggested to be

$$\zeta_i = 1 + | d_i(\underset{\sim}{A},\underset{\sim}{B}) | \tag{4.4}$$

4.3. Machine Learning Model

Fig.4 shows the function of fuzzy reasoning preprocessor in machine learning of design type expert system for a design solution .

4.3.1. Fuzzy Classification

Classification of design attributes is most important to an expert system. Since the design processes are merely identifying design parameters within ranked samples in the sample base, they must be exactly categoried quantitatively for avoiding abnormality in design solutions.

Providing $\underset{\sim}{B}$ is the learning sample and A_i (i=1,2,....n) are the samples which have been learned and stored in the sample base, the learning sample $\underset{\sim}{B}$ belongs to the same rank of $\underset{\sim}{A}$ in j-th attribute of design if (4.5) holds true:

$$\text{Min}(\ \text{Abs}|\varphi(\underset{\sim}{A}_i^j,\underset{\sim}{B}^j)|,\ A_i \subseteq \underset{\sim}{A}\ , i = 1,2,....n\ ) \geq \Theta_j \tag{4.5}$$

where $\varphi(\underset{\sim}{A}_i^j, \underset{\sim}{B}^j)$ represents the distance of design sample $\underset{\sim}{B}$ and $\underset{\sim}{A}_i$ in j-th attribute and Θ_j represents certain threshold value.

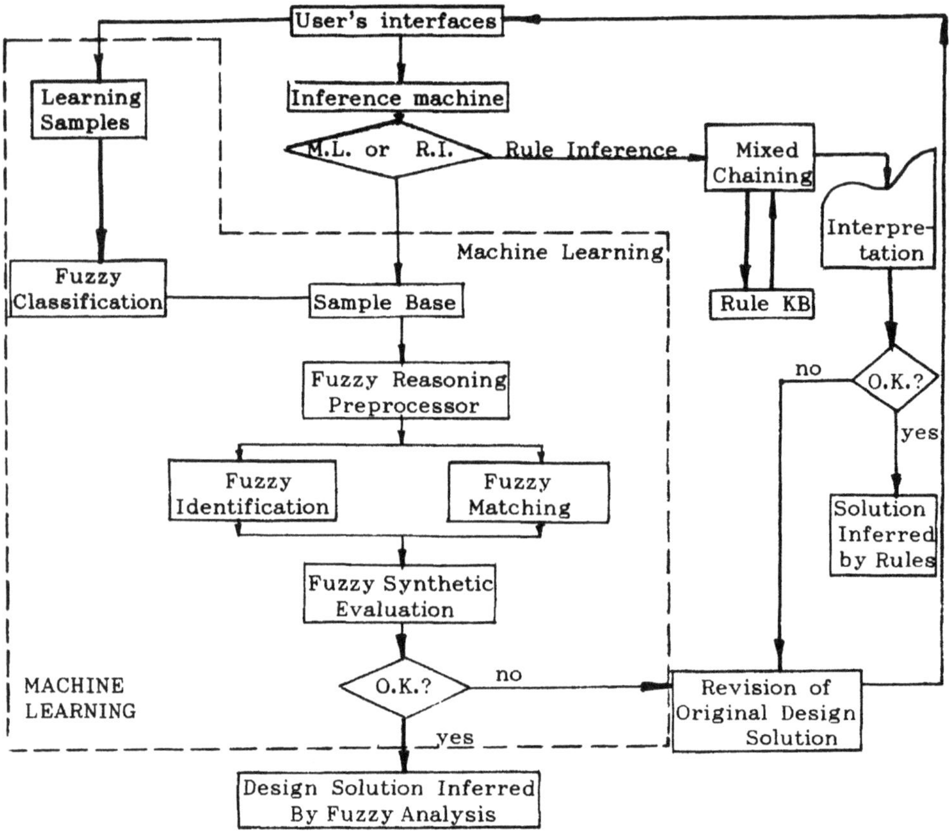

Fig.4

4.3.2. Fuzzy Identification

For the purpose of generating a new design alternative (or parameter), the fuzzy identification model is used through comparison of initial design parameter with those learned samples in the sample base on the basis of similarity, by which the identification process is performed.

Let's denote $\underset{\sim}{A}_i$ $(i = 1,2,\ldots,n)$ the learned samples in sample base, $\underset{\sim}{B}$, the design alternative to be identified; then the distance between $\underset{\sim}{B}$ and $\underset{\sim}{A}_i$ can be determined by (4.1). Taking a new set $\underset{\sim}{A}_j (j=1,2,\ldots,n,n+1)$, the matrix of distance compared term by term is

$$M\varphi\,(\underset{\sim}{A}_p,\underset{\sim}{A}_q) = \sum_{i=1}^{m} W_i(d_i)\cdot d_i(\underset{\sim}{A}_p\,,\underset{\sim}{A}_q\,) \qquad (4.6)$$

where $p,q = 1,2,\ldots,n,n+1$; other signs are the same as presented in (4.1). As we understand m is the dimension of the subspace, e.g. the number of design attributes.

The fuzzy membership function of $\underset{\sim}{B}$ for r-th factor can be expressed by

$$\mu_{br} = (1/n) \sum_{i=1}^{n} (1 - \varphi_{i(1+n)})^{k} \cdot Air \qquad (4.7)$$

where Air, membership function of i-th learned sample with respect to r-th influenced factor, k, amplified factor, suppose k = 2; $\varphi_{i(1+n)}$ represents the fuzzy distance of i-th learned sample to sample B, which is in order of (1+n) and is being classified.

4.3.3. Fuzzy Matching

Although the identification capability can provide reasonable design parameters (solutions) based on the trained samples, nevertheless, in many cases, the designer is aimed to have a reasonable design solution which is not available in the sample base of the system as exactly coincident with the expected design parameters. Thus the matching capability either by extrapolating or interpolating the existing sample parameter to meet the needs of ongoing design project is one of the most interesting features of an advanced expert system for design.

The application of fuzzy approach has naturally provide the possibility of incorporating matching capability into the system, which is just a reasonable extension of fuzzy assessment between some nearest available samples to address the fuzzy matching processing. Therefore, the theoretical basis of fuzzy matching can be performed by membership function approach as discussed in 4.2 based on fuzzy "distance" measurement.

The fuzzy matching model can be expressed by fuzzy "distance" concept which is somewhat different from identification process although the theoretical basis is the same. The matching inference should be based on most closed samples of the design expectation, the difference is measured by fuzzy "positive' distance" and "negative distance" and the matching is just the processing of extrapolation and interpolation respectively. Suppose the design alternative is a fuzzy set B and its k-th attribute B_k is determined by fuzzy matching within trained samples of practical engineering projects A^i (i=1,2,...,n). Using (4.1), calculate the "distance" between k-th attribute of B and A^i, select the top three closest solutions as A_a,A_b and A_c having

$$\varphi_1(B_k,A_{ak}) \le \varphi_2(B_k,A_{bk}) \le \varphi_3(B_k,A_{ck}) < \varphi_j(B_k,A_{jk}) \quad (j=1,2,\ldots,n-3;\ j\ne a, j\ne b, j\ne c) \qquad (4.11)$$

where B^k is the k-th attribute solution of design alternative B, A_{ak},A_{bk},A_{ck} represent the k-th attribute solutions of learned samples A_a,A_b and A_c respectively. φ_1, φ_2 and φ_3 represent the minimum three distances. n is the number of learned samples in the sample base of the system. Then, the k-th attribute of design expectation of B by matching is

$$B_k = \frac{1}{3} \{ (1-\varphi_1) A^j_{ak} + (1-\varphi_2) A^j_{bk} + (1-\varphi_3) A^j_{ck}\} \qquad (4.12)$$

where φ_1,φ_2 and φ_3 are the "distance" presented in (4.11); A^j_{ak}, A^j_{bk}, A^j_{ck} represent the practical value of j-th attribute in sample A_a, A_b, and A_c.
As an example, to determine the column size of a 5-span, 20-storey R.C. framework building, the length of the span is 6m, base shear is

calculated by UBC code as 215 ton. The design requirements for adequate stiffness as the top drift versus height of the building $\delta/H = 1/620$, providing that the parameters of the available samples (A_i, i=1,2,...,10) in the knowledge base are listed in Tab.1.

In order to find out those samples that are considered as the bases for

Tab. 1

No.	Structural type	Number of span	Length of span (m)	Stiffness ratio $\gamma = H/\delta$	Base shear (t)	Column section (cm^2)
B	R.C.frame	4	6.0	620	215	?
A_1	same above	4	6.1	500	255	1600
A_2	same above	4(3)	5.1	480	230	1470
A_3	same above	4	5.5	690	200	1350
A_4	same above	4	5.4	625	180	1280
A_5	same above	4(5)	6.1	610	210	1425
A_6	same above	4	6.15	550	270	1670
A_7	same above	4	5.95	520	266	1650
A_8	same above	4	5.8	570	254	1480
A_9	same above	4	5.75	630	212	1500
A_{10}	same above	4	6.25	560	220	1520

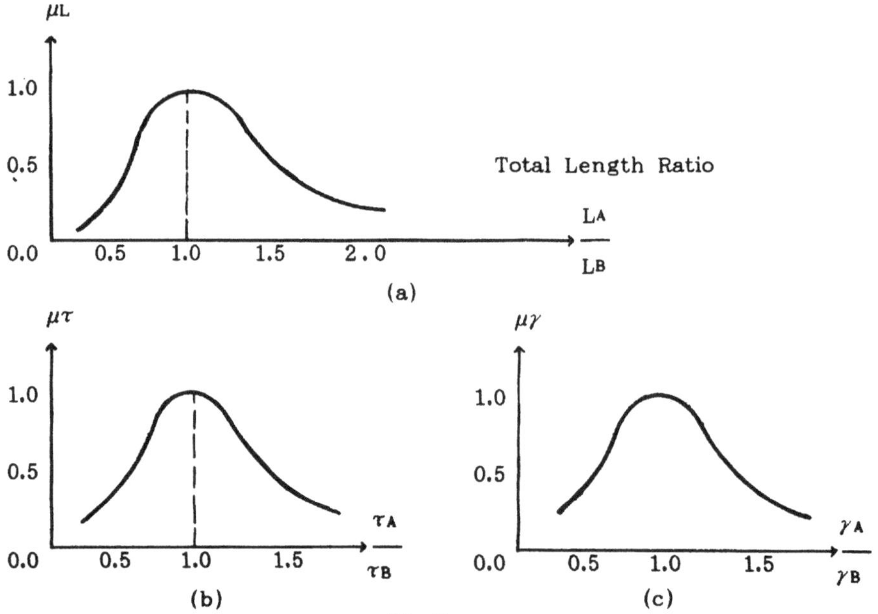

Fig.5

matching, the attributive transformation should be made to accommodate the membership function curve which is prepared in advance by experts' experiences (Fig.5) which are available for specific structural type of building. Fig.5a is the membership function of total length of the span (span number times average single span length), assuming that the design total span L_B corresponds 100% of satisfactory degree μ_L= 1.0 and μ_L

will be changing ($\leq$ 1.0) as the total span of sample L_A varies. Similar cases are Fig.5b, the membership function curve of unit base shear τ_A/τ_B and Fig.5c of the stiffness ratio membership function curve γ_A/γ_B.

Correspondingly, the attributes presented in Tab.1 should be revised in accordance with those presented in Fig.5a,b and c, and the corresponding membership values according to Fig.5. by which the "distance" of design solution with the samples can be evaluated. Tab.2 presents the relative "distance" of each sample A_i (i=1,2,...,10) with respect to the design solution. In which some design attributes are defined and the matching process is mainly dealing with specific design parameters, for instance, matching column section from samples, even though those samples are not necessarily to meet every attribute coincident to design requirements. In Tab.2 d_i^L, d_i^τ and d_i^γ represent fuzzy distances of total span length, unit base shear and stiffness ratio of sample A to design expectation B. φ_i represents the fuzzy distance of sample A_i to design expectation B

Tab.2

No.	d_i^L	a_i^L	φ_i^L	d_i^τ	a_i^τ	φ_i^τ	d_i^γ	a_i^γ	φ^γ	φ_i
A_1	0.012	0.2	0.0024	0.026	0.4	0.0107	0.259	0.4	0.1304	0.1435
A_2	0.212	0.2	0.0514	0.002	0.4	0.0008	0.292	0.4	0.1509	0.2031
A_3	0.088	0.2	0.0191	0.002	0.4	0.0008	0.123	0.4	0.0553	0.0752
A_4	0.108	0.2	0.0239	0.020	0.4	0.0082	0.001	0.4	0.0004	0.0325
A_5	0.012	0.2	0.0024	0.002	0.4	0.0008	0.002	0.4	0.0008	0.0040
A_6	0.022	0.2	0.0045	0.058	0.4	0.0245	0.148	0.4	0.0680	0.0970
A_7	0.001	0.2	0.0002	0.035	0.4	0.0145	0.183	0.4	0.0866	0.1013
A_8	0.026	0.2	0.0053	0.023	0.4	0.0094	0.048	0.4	0.0201	0.0348
A_9	0.037	0.2	0.0077	0.001	0.4	0.0004	0.002	0.4	0.0008	0.0089
A_{10}	0.079	0.2	0.0341	0.001	0.4	0.0004	0.053	0.4	0.0223	0.0568

$$\varphi_i = \varphi_i^L + \varphi_i^\tau + \varphi_i^\gamma = a_i^L d_i^L(1+|d_i^L|) + a_i^\tau d_i^\tau(1+|d_i^\tau|) + a_i^\gamma d_i^\gamma(1+|d_i^\gamma|) \quad (4.15)$$

From Tab.2, the last three minimum distances of φ_i are φ_5, φ_9, and φ_4. Therefore, (4.12) gives the design column section B_j

$$B_j = \frac{1}{3}\{(1-\varphi_5)\cdot A_5^j + (1-\varphi_9)\cdot A_9^j + (1-\varphi_4)\cdot A_4^j\}$$

$$= 0.3333\times(1419.3 + 1486.65 + 1238.4) = 1381.5 \text{ cm}^2 \quad (4.16)$$

(4.16) represents the design column section matched by available samples, which approximately can take a square section with each side 37.2cm width (A= 1383.8 cm^2).

4.3.4. Fuzzy Synthetic Evaluation

After identifying the design alternative B, it is necessary to further evaluate it by fuzzy synthetic evaluation. Assume that the evaluation attribute set is defined by $E = (e_1, e_2, \ldots, e_m)$, the fuzzy evaluation subset of design alternative B with respect to each attribute can be expressed by membership function of satisfaction

$$\mu(\underset{\sim}{B}) = \mu_{\underset{\sim}{B}}(e_1)/e_1 + \mu_{\underset{\sim}{B}}(e_2)/e_2 + \ldots + \mu_{\underset{\sim}{B}}(e_m)/e_m \tag{4.17}$$

the weighted coefficient between e_i (i=1,2,...,m) is b_i (i=1,2,...,m); then we have the fuzzy synthetic evaluation indicator η for the design alternative

$$\eta = \sum_{i=1}^{m} \mu_{\underset{\sim}{B}}(e_i)\cdot b_i \geq I \ (= 0.6) \tag{4.18}$$

where I represents the satisfactory threshold of the design. The design alternative can be accepted if equation (4.18) is satisfied.

V.CONCLUSION REMARKS

The ever-rapid development of AI and its application to design issues imply the perspectives of computer penetration and impact to engineering community in which the expert system for design represents one of the most potential achievements and will play big influence to engineering activities in the future.

It is recognized that the expert system for engineering design is an extension of the second generation of expert system technology, among which deepening and explorating the human intelligence for fuzzily representing knowledge of human being as to simulate human thinking for decision making in design, are key issues. The simulation of human conceptual perception and manipulation are emphasized in the design expert system. Since design is not merely a logic-inferred product, it is a product through perceptional imagination and searching processes by the designers under hierarchically logic control. As a matter of fact, it represents a process of fuzzy nature.

Therefore, the machine learning strategy has to reflect the philosophical considerations of design methodology as to lead the learning capability based on fuzzy method. The "Generalized Space" concept is introduced for accommodating the design procedures. It is much suitable to use this concept in manipulating different design attributes in different design stages.

The hierarchy of design space construction is actually a reflection of real situation of design sequences. The concept of fuzzy distance in the generalized design space, based on fuzzy difference of membership function of two design alternatives, is used for distinguishing the difference of two design solutions which serves as the basis of fuzzy reasoning of machine learning by means of unified fuzzy approaches. It is estimated that in view of the essentials of the problem itself, the human knowledge representation and acquisition can be naturally accommodated with the fuzzy processing method and it implies the potentiality and adaptability of our studies presented in this paper.

VI.ACKNOWLEDGEMENT

Authors would like to express their gratitude to the Chinese National Science Foundation for their partial support. The first author of this paper also appreciates the support from NCEER-88-1006 project grant from U.S.A.. Since the majority of work in this paper was conducted when he was visiting at School of Civil & Environmental Engineering, Cornell University during 1990, gratitude is also due to Professor Peter Gergely and Professor John Abel for their helpful discussions on this topic.

REFERENCES

[1]Lin Shaopei, Fuzzy Reasoning Method — Theoretical Adaptability and Code Compatibility to Design of Earthquake Resistant Structure, Research Report(I) under Contract NCEER-88-1006, Cornell University, U.S.A.,1990.

[2]Lin Shaopei, Yang Jiaping, Fuzzy Evaluation for Torsional Carrying Capacity of Structures, *Proc. of The 3rd East Asia-Pacific Conference on Structural Engineering and Construction*, Shanghai, China, April 1991.

[3]Lin Shaopei, Fuzzy Machine Learning of Design Type Expert System -- An Approach to CU-STRAKE for Earthquake-Resistant Design of Buildings, Report (IV) under Contract NCEER-88-1006 and NSF-ECE-86-07591, School of Civil and Environmental Engineering, Cornell University, U.S.A. 1990.

[4]Lin Shaopei, Yang Jiaping, On Feasibility of Developing CU-STRAKE Expert System on Personal Computer, Research Report (V) under Contract NCEER-88-1006, Cornell University, U.S.A., November 1990.

[5]Subramani, M., Peter Gergely et al., A Knowledge-Based Approach to Structural Design of Earthquake-Resistant Buildings, Technical Report NCEER-89-0006, Cornell University, U.S.A., January 1990.

[6]Wang Guanyuan, Exploration of Structural Soft Design, Press of Harbin Institute of Architectural Engineering, 1987.

[7]Lin Shaopei, Gao Deyin, Deep Knowledge Reasoning for Design Type Expert System, *Proceedings of the International Conference on Expert System in Engineering Applications*, pp.642-646, Wuhan, China, October 1989.

[8]Lin Shaopei, Yang Xiaofeng, Fuzzy Response of Offshore Structure, *Proc. of the 6-th International Symposium of Offshore Mechanics and Arctic Engineering*, Houston, Texas, U.S.A., March 1987.

INDEX OF CONTRIBUTORS

www.ingramcontent.com/pod-product-compliance
Lightning Source LLC
LaVergne TN
LVHW080803011026
849160LV00015B/1517
9780367866129